ATS 0533   3

D0705385

CP 1010 49.5

Introduction to Agribusiness

**Delmar Publishers is proud
to support FFA activities**

Introduction to Agribusiness

Dr. Cliff Ricketts

Dr. Omri Rawlins

Africa • Australia • Canada • Denmark • Japan • Mexico • New Zealand • Philippines
Puerto Rico • Singapore • Spain • United Kingdom • United States

Delmar Staff:

Business Unit Director: Susan L. Simpfenderfer

Executive Editor: Marlene McHugh Pratt

Developmental Editor: Andrea Edwards Myers

Executive Marketing Manager: Donna Lewis

Executive Production Manager: Wendy A. Troeger

Production Editor: Elaine Scull

Cover Design: Charles Cummings Advertising/Art, Inc.

Library of Congress Cataloging-in-Publication Data

Ricketts, Cliff.

    Introduction to agribusiness / Cliff Ricketts, Omri Rawlins.

      p.   cm.

    Includes bibliographical references and index.

    ISBN 0-7668-0024-5

    1. Agricultural industries—United States. 2. Agriculture—Economic Aspects—United States. 3. Food and industry and trade—United States. I. Rawlins, N. Omri (Nolan Omri), 1938– II. Title.

  HD9005.R53   2000

  338.1'0973—dc21   99-40037

                                  CIP

---

*This book is dedicated to my family: my parents, Hall and Louise Ricketts,
who were production agriculturalists during the Great Depression
and overcame great odds to leave a rich heritage for their family
and the agribusiness community; my wife, Nancy; and my
three children, John, Mitzi, and Paul, who are
all employed in the world of agribusiness.*
Dr. Cliff Ricketts

*This book is also dedicated to my family,
who have been a great source of love and inspiration.*
Dr. Omri Rawlins

Hall Ricketts (1919–1998), father of Dr. Cliff Ricketts

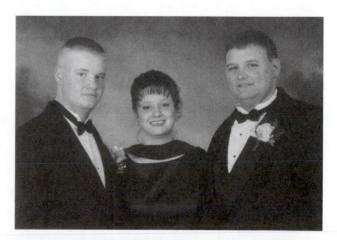

Paul, Mtizi, and John Ricketts, Children of Dr. Cliff Ricketts.

Contents

UNIT 5   AGRICULTURAL ECONOMICS

xiv 🌿 *Contents*

Preface

Agriculture has changed. For years, agriculture was a career and a way of life. Nearly everyone lived on farms, and farmers were self-sufficient. Very few manufactured supplies and materials were available. Today, however, the agricultural industry is a technology-oriented industry that includes production, agriscience, and agribusiness. Agriculture is a big business, and agribusiness itself is massive. This book discusses the kinds, sizes, fundamentals, and applications of this phase of the agricultural industry.

Most people who work in the agricultural industry do not work on farms and ranches but rather are employed in the feed, seed, farm machinery, fertilizer, and chemical supply businesses, which make up the input sector. There are also many agribusiness workers in marketing firms who move food and fiber from production agriculturalists to the consumers, who form the output sector. This book also explains different agribusiness areas, discusses their economic importance to the agricultural industry, and explains basic principles of agricultural economics, marketing, and finance.

Introduction to Agribusiness is divided into five major sections. Unit 1 includes a comparison of agriculture and agribusiness, which sets the stage for the book. The history of agriculture is briefly discussed, including the evolvement of agribusiness and today's technology. Agribusiness is explained in terms of where it fits into the "Big Picture" of the agricultural industry and its relationship with production agriculture and agriscience. The size, importance, and impact of agribusiness on the economy are discussed.

Unit 2 includes four chapters on starting and running an agribusiness. They provide the information necessary for owning and operating an agribusiness. Chapters include planning and organizing an agribusiness, types of agribusinesses, personal financial management, and record keeping and accounting.

Unit 3 explains the agribusiness inputs that go to production agriculturalists. The input sector of agribusiness includes five major areas (a chapter is devoted to each), which include: agribusiness supplies (feed, seed, fertilizer, and chemicals); farm machinery and equipment; private agribusiness services, such as farm and commodity organizations; public agribusiness services such as the U.S. Department of Agriculture (USDA) and state departments of agriculture, and agribusiness credit services.

Unit 4 explains the agribusinesses that are involved with agricultural products once they have left the farm (the agribusiness output sector). Chapters are included on basic principles of marketing, commodity marketing, international agriculture, food and fiber marketing, and agrimarketing channels from the farm to the consumer.

Unit 5 includes a study of the basic economic principles involved in agribusiness. This section, along with other sections in the book, include the twenty-two economic principles that a national association on economic education says should be taught. Two chapters cover, first, agricultural economics and the American economy, and, second, economic activity and analysis.

PEDAGOGICAL FEATURES

Each chapter begins with a list of performance objectives, which include the main topics for each chapter. The objectives are followed by a list of terms to know to enable students to better understand the chapter content. The definitions of all the terms are found in the glossary, and in some cases within the unit.

Each chapter ends with a conclusion and a summary. Also at the end of each chapter are review questions, which encourage students to read the whole chapter, and fill-in-the-blank and matching questions, which are designed to reinforce students' understanding of important ideas. Activities are the last feature of each chapter. These are application exercises that enable students to apply newly learned concepts and practices, and also to draw on their own experiences and share those with others.

ABOUT THE AUTHORS

Dr. Cliff Ricketts came through the ranks of agricultural education and the Future Farmers of America (FFA) and is a product of the program. He has achieved every degree offered by the FFA, including the Greenhand, Chapter, State, and American FFA (Farmer) degrees; Honorary Chapter; Honorary State; and Honorary American FFA; the last of which was received the same year by former President George Bush and Lee Iacocca.

As a high school agricultural education instructor, his students won the State FFA Parliamentary Procedure Contest, State FFA Creed Speaking Contest (four times), and were awarded runner-up in Public Speaking and Extemporaneous Speaking. Ricketts was also a recipient of the National Vocational Agriculture Teachers Association's (NVATA) National Outstanding Teacher Award while a high school teacher.

Ricketts received his B.S. and M.S. degrees from the University of Tennessee and his Ph.D. from Ohio State University. While teaching at Middle Tennessee State University (MTSU), he received both the MTSU Foundation's Outstanding Teacher Award and its Outstanding Public Service Award. He was also a finalist for its Outstanding Research Award.

Ricketts is still involved in production agriculture with a beef cattle operation on his family farm twenty miles east of Nashville, Tennessee. His children, John, Mitzi, and Paul, are the fourth generation on the family farm. Cliff Ricketts is also involved with alternative fuels research and has run engines off ethanol from corn, methane from cow manure, soybean oil, and hydrogen from water. He, along with his students from Middle Tennessee State University, presently hold the world's land speed record for a hydrogen-fueled vehicle.

Dr. Omri Rawlins received his B.S. and M.S. degree from the University of Georgia and his Ph.D. from Texas A&M University. Rawlins has taught courses in agribusiness for over thirty years and was hired as one of the first professors of agribusiness in the country. He is also an authority on agricultural history and has been active in the National Association of Colleges and Teachers of Agriculture (NACTA) for more than twenty years.

ACKNOWLEDGMENTS

The authors are appreciative of several people who had input into this book as a resource for course content or as a consultant.

Beth Cripps, Hendersonville High School

Keith Gill, Lincoln County High School

Becky Thomas, Middle Tennessee State University

Carol Gardner, Middle Tennessee State University

Dawn Mosley, Middle Tennessee State University

Glenn Ross, McEwen High School

Michael Barry, University of Tennessee Extension Service

Kim Webb, Tennessee Farmers Cooperative

Janet Kelly, Middle Tennessee State University

Jamie Mundy, Powell Valley High School

Chaney Mosley, Middle Tennessee State University

Debbie Weston, Lochinvar Corporation

Mitzi Ricketts, Beech High School

Julie Shew, Middle Tennessee State University

The authors and Delmar Publishers also wish to express their appreciation to the following reviewers:

Phillip Johnson
Texas Tech University
Lubbock, TX

Mike McDaniel
North Dakota State University
Miltona, MN

Robert Philpot
Williston High School
Williston, FL

UNIT

1

Introduction to Agribusiness

CHAPTER

1

Agriculture and Agribusiness

OBJECTIVES

After completing this chapter, the student should be able to:

🌿 explain **agribusiness**

🌿 describe the "big picture" of agribusiness

🌿 explain how agribusiness affects us daily

🌿 discuss farming and agriculture before agribusiness

🌿 discuss the beginning of agribusiness in America

🌿 discuss the success of American agribusiness

TERMS TO KNOW

A.D.
agribusiness
agriscience
agronomic
artificial insemination
B.C.
biotechnology
Cooperative Extension Service
cotton gin
crop rotation
cultivate
Dairy Herd Improvement
   Association (DHIA)
domesticate
draft animals

drought
exports
fallow
Federal Land Banks
futures
hog cholera
hybrid
implements
irrigate
indigo
input
organic fertilizer
output
pesticides
production agriculturalists

raw material
reaper
resources
retailers
Soil Conservation Service
   (SCS)
selective breeding
sickle
surveying
terminal
threshing machine
vaccines
wholesalers

INTRODUCTION

When most people think of agriculture, they picture farmers producing animals and crops. Agriculture is often thought of as "cows, sows, and plows" or "weeds, seeds, and feeds." But agriculture has changed. For years agriculture was farming, as previously stated. Nearly everyone lived on farms, and farmers were self-sufficient. Very few manufactured supplies and materials were available. Today, however, agriculture is a technology-oriented industry that includes production, **agriscience**, and agribusiness.

Most people who work in the agricultural industry do not work on farms and ranches but rather are employed in the feed, seed, farm machinery, fertilizer, chemical supply, and food-processing businesses. There are also many agribusiness workers in finance, distribution, and marketing firms who provide service to **production agriculturalists**. Agriculture is big business.

WHAT IS AGRIBUSINESS?

There are many different definitions of agribusiness. Some people interpret the word narrowly to mean only very large businesses within the agricultural industry. However, John Davis and Ray Goldberg, in their early research on agribusiness, defined it as the sum total of all operations involved in the manufacture and distribution of farm supplies; production operations on the farm; and the storage, processing, and distribution of the resulting farm commodities and items.[1] A similar definition of agribusiness describes it as any profit-motivated enterprise that involves providing agricultural supplies and/or the processing, marketing, transporting, and distributing of agricultural materials and consumer products.[2] Ewell Roy defines agribusiness as the coordinating science of supplying agricultural production inputs and subsequently producing, processing, and distributing food and fiber.[3]

Is Farming an Agribusiness?

Some authorities exclude "farming" or the "production" of food and fiber from the definition of agribusiness. If we include production agriculture, agriscience, and agribusiness under the umbrella of the agricultural industry, we must exclude production agriculture (farming) from the agribusiness definition within the context of this book. However, we would be mistaken to think that farming is not a business. A production agriculturalist must make decisions, develop plans, and solve problems, all which require business-related skills. A typical farmer manages interest, taxes, repair and replacement of equipment, fertilizers, wages, fuel, electricity, and many other items.[4] As you can see, production agriculturalists must be financial and business managers or they will fail. Production agriculture is indeed a business, but in the context of this book we will not discuss it as an agribusiness.

THE BIG PICTURE OF AGRIBUSINESS

Agribusiness companies provide **input** supplies to the production agriculturalist (farmer), as shown in Figure 1–1. The production agriculturalist produces food and fiber (cotton, wool, etc.), and the **output** is taken by agribusiness companies that process, market, and distribute the agricultural products, as shown in Figure 1–2.

Many services are needed in agriculture, such as transportation, storage, refrigeration, credit, finance, and insurance. Agribusiness manufacturers furnish production agriculturalists with supplies and equipment needed to produce and protect

Figure 1–1 This Tractor Supply Company (TSC) store is an example of an agribusiness input supply company. Many agribusiness inputs are sold to production agriculturalists. (Courtesy of Tractor Supply Company)

Figure 1–2 Trucks are delivering milk to a milk-producing plant. Both the trucking and the milk-producing plant are examples of agricultural output. (Courtesy of Cliff Ricketts)

their crops. Government agencies inspect and grade agricultural products to ensure quality and safety.[5] Hundreds of agribusiness trade organizations, commodity organizations, committees, and conferences educate, promote, advertise, coordinate, and lobby for their agricultural products. Science, research, engineering, and education help improve agribusiness. Millions of people are employed in agribusiness throughout the world, and people throughout the world also depend on agribusiness for their food, clothing, and shelter. Refer to Figure 1–3 for the "Big Picture of Agribusiness."

AGRIBUSINESS AFFECTS US DAILY

Consider one of our favorite foods, the cheeseburger. To get an idea of how agribusiness is a part of our daily lives, imagine what is involved in assembling all the parts of a juicy cheeseburger with all the trimmings (refer to Figure 1–4).[6]

Many items had to be brought together to obtain the cheeseburger, such as land, labor, seeds, fertilizer, chemicals, machinery, credit, transportation, and marketing, among other **resources**. In addition, agribusiness plants had to be built to handle and process the food products that go into the cheeseburger from their **raw material** state. Many engineering and technological feats were achieved to obtain the quality and standardization required. As the various foods that went into making the cheeseburger left the processing plants, they were transported by either trucks, trains, airplanes, or barges, requiring roads, tracks, runways, waterways, communications, and employees. As the food products arrived in your community, a receiving **terminal**, depot, or warehouse was needed to store these items for later distribution in smaller quantities to retail food stores or restaurants.[7] As you can see, agribusiness is an essential part of our lives and crucial to the economy of the United States.

The Big Picture of Agribusiness

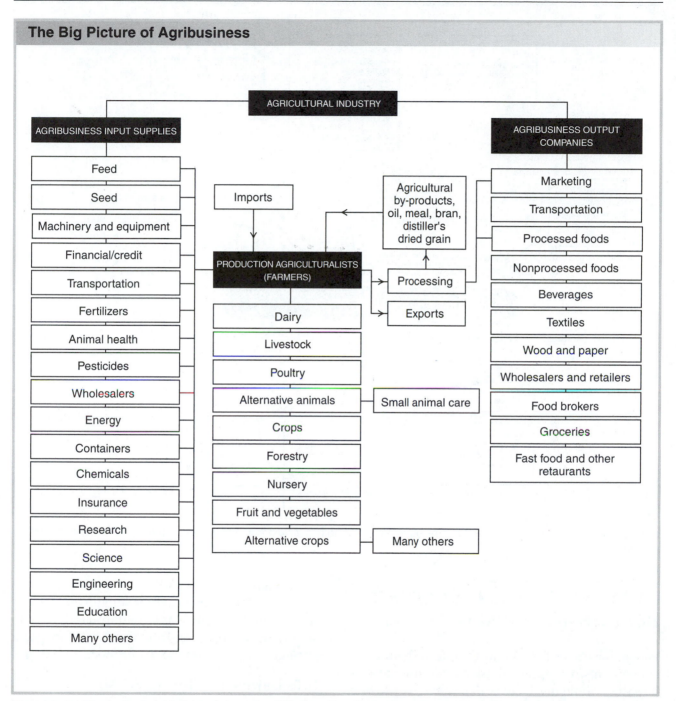

Figure 1–3   The agricultural industry is massive. This flowchart illustrates where agribusiness fits into the "Big Picture" of the agricultural industry.

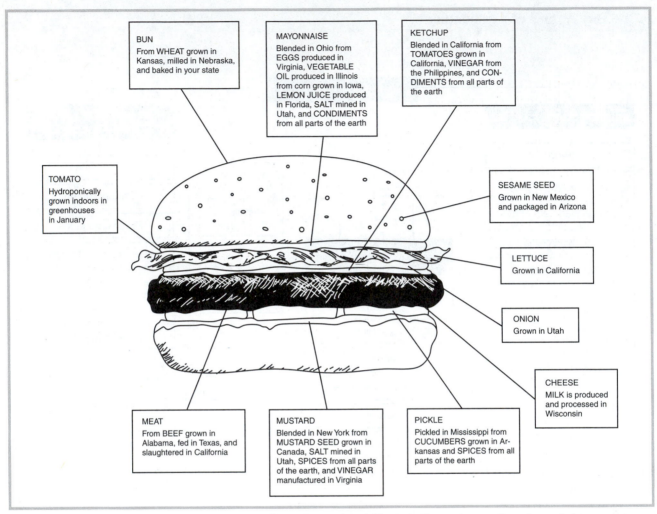

BUN
From WHEAT grown in Kansas, milled in Nebraska, and baked in your state

MAYONNAISE
Blended in Ohio from EGGS produced in Virginia, VEGETABLE OIL produced in Illinois from corn grown in Iowa, LEMON JUICE produced in Florida, SALT mined in Utah, and CONDIMENTS from all parts of the earth

KETCHUP
Blended in California from TOMATOES grown in California, VINEGAR from the Philippines, and CONDIMENTS from all parts of the earth

TOMATO
Hydroponically grown indoors in greenhouses in January

SESAME SEED
Grown in New Mexico and packaged in Arizona

LETTUCE
Grown in California

ONION
Grown in Utah

CHEESE
MILK is produced and processed in Wisconsin

MEAT
From BEEF grown in Alabama, fed in Texas, and slaughtered in California

MUSTARD
Blended in New York from MUSTARD SEED grown in Canada, SALT mined in Utah, SPICES from all parts of the earth, and VINEGAR manufactured in Virginia

PICKLE
Pickled in Mississippi from CUCUMBERS grown in Arkansas and SPICES from all parts of the earth

Figure 1–4   The cheeseburger and all the ingredients that go into it illustrate the vastness, diversity, and closeness to our everyday lives of agriculture.

FARMING AND AGRICULTURE BEFORE AGRIBUSINESS

People have searched for ways to feed themselves since prehistoric times. If people did not eat one day, they would hardly have enough energy to find food the next day. Thus, nearly all their waking time was spent searching for food by hunting or gathering nuts and other naturally grown foods.

As farming developed, it gave people time to develop other things, such as language, art, and religion. Today, if you want to eat, you only need to go to the nearest fast-food restaurant to get a cheeseburger or another favorite food.

Life before Agriculture

Several thousand years ago, early humans roamed great distances to gather plants and hunt animals for food. Nutrition and health were so poor that people seldom lived past twenty-five years of age. They hunted large game, but they were not very successful. They had more luck gathering vegetables and insects. Insects were a favorite. Early humans considered them so tasty that some species were almost exterminated. Common wild vegetables included turnips, onions, radishes, squash, cabbage, and mushrooms. However, two important developments took place before humans discovered agriculture. People learned to fish, and they learned to use fire for cooking. Civilization, however, evolved very slowly until people learned to farm.

Early Agricultural Development

To many people, the idea of farming seems simple. All one does is plant some seed, wait a while, and then harvest the food. But prehistoric humans had a problem with growing food. There was no one around to teach them that seed would grow into plants. Eventually, however, the puzzle to farming was solved and people began to raise crops and **domesticate** animals.

Farming changed the way people lived. They no longer had to roam great distances in search of food. Instead, they could settle in one spot and build a home. They also had time to perfect methods of farming. Simple sticks evolved into wooden hoes and plows. Some clever person invented the **sickle** to make the harvesting of grain and grasses easier.[8]

Bronze Age

During the Bronze Age (around 3,000 **B.C.**), wooden **implements** were made sharper and more durable by using metal. This allowed people to **cultivate** larger areas of land faster. Agriculture spread throughout the world and became a way of life for most people. Some of the developments during the Bronze Age included:

- Bronze tools and plows made for easier and faster farming.

- The Nile River was used by Egyptians to **irrigate** crops.

- The wheel was discovered, making the transport of crops possible.

- World population rose from 3 million before the invention of agriculture to nearly 100 million.[9]

Iron Age

The year 1000 B.C. was the dawn of the Iron Age. The use of iron gave people the ability to produce even more crops. When people could not use all the crops themselves, they had to do something with them, and trade among people developed as a result. Other developments of the Iron Age were:

- Iron hand tools and plows were prepared, some of which are similar to those used today.

- Money was developed because of the need to trade excess crops.

- Leaving land **fallow** gave the soil a chance to rebuild and store moisture.[10]

The Middle Ages (A.D. 400–1500)

The Middle Ages ranged from **A.D.** 400 to A.D. 1500. The fall of the Roman Empire slowed the growth of agriculture. Farming was still a way of life, but only a few developments occurred. Even though there were not many, however, the agricultural developments that were made were important. These included **crop rotation**, a new harness for plowing, and **selective breeding** for livestock.

Farmers in the Middle Ages began to understand the importance of conserving soil moisture and nutrients. They accomplished this by dividing their fields and leaving parts bare during certain years. This innovation led to the development of fences to separate fields.

Before the Middle Ages, oxen were the main **draft animals**. A new harness was developed so that farmers could use horses to work their fields faster. Selective breeding of livestock also began during the late Middle Ages. Livestock producers selected certain animals for breeding based on desirable characteristics. Before the development of this practice, there were few different breeds of domestic animals. Using selective breeding, livestock producers developed many of the breeds we now have.

Near the end of the Middle Ages, Columbus discovered America. Many people began to travel to the "New World." Until this time, most agricultural developments had come from Europe and Asia. Now there was an opportunity to develop American agriculture.[11]

THE EVOLVEMENT OF FARMING AND AGRIBUSINESS IN AMERICA

The first settlers found America sparsely settled by American Indians, who hunted, fished, and farmed. The Native Americans had developed several crops that are still of great importance in agriculture, including corn, kidney and lima beans, squashes, pumpkins, tomatoes, tobacco, and cotton. Potatoes, both sweet and white, were developed by South American Indians but did not reach North America until after the first Pilgrims settled there.[12]

The American colonists found conditions difficult. They faced starvation until they adopted the crops and tillage practices of the American Indians, which differed from practices in Europe. You may be surprised to learn that the first agricultural education teachers in America were the Native Americans; their students were the Pilgrims. For example, American Indians in New England taught the Pilgrims to place fish beneath each hill of corn. While they may not have known that the decaying fish released nitrogen, they knew it helped the corn grow.

In 1613, Virginia colonist John Rolfe shipped tobacco to England, which proved to be a large, untapped market for the crop. Later, export markets for rice and **indigo**, as well as tobacco, provided much of the capital necessary for growth and development in American agriculture.[13] The colonists also shipped furs, timbers, and grain to England and southern Europe, while grain, flour, and dried fish went from the New England colonies to the West Indies. Other markets opened as time passed, marking the beginning of agribusiness in America.[14]

Many developments in agriculture during the seventeenth and eighteenth centuries eventually led to the way we farm today. Some of the other important developments included:

🌿 **Organic fertilizer** was first developed by putting dead fish into the ground along with corn seed.

🌿 Rice, the world's most popular grain, was first grown in the United States.

🌿 George Washington created one of the first experimental farms.

🌿 Thomas Jefferson experimented with seeds and livestock, invented farm implements, and was active in establishing a local agricultural society.[15]

The Era after the American Revolution

After the American Revolution, more people moved to the United States, which meant that more land was needed for the new settlers. People went west and developed new ways to produce food. Some new developments were:

🌿 **Surveying** of land was used to separate property.

🌿 The **cotton gin** was invented by Eli Whitney in 1793.

- Edward Jenner discovered **vaccines** to prevent diseases.

- The first one-piece, cast-iron plow was invented in 1819 by Jethro Wood.

- Interchangeable parts were developed so that people could fix their equipment.[16]

Agricultural and Industrial Revolution

During the 1840s and 1850s another vast change took place that gave a tremendous spur to production agriculture and greatly stimulated the growth of agribusiness. This was the *Industrial Revolution,* a time when many technological inventions came quickly, one after another: these included the steam engine, railroads, the sewing machine, the powered loom for weaving cloth, and many other machines.

A second part of the Industrial Revolution involved the movement of people away from the farms to the cities. They moved to the cities to work in factories, where employers needed many workers to operate all the new machinery. However, factory workers could not grow their own food, as most people had done up until that time. Therefore, in a few years, there was a large new market for farm products.[17]

As a result of the move to the cities, the farmers now had to run their farms with fewer workers. To make their operation more efficient, they began to use some of the newly invented machinery on the farms. They greatly increased production and moved from old-fashioned, simple farming into commercial agriculture. Industry began to produce machinery for use on farms, which increased the farmers' efficiency and output many times over. Mechanical power replaced animal power, and specialized equipment for tilling, planting, and harvesting was developed, which boosted agricultural output even higher.[18]

The Agricultural and Industrial Revolution also brought us the following:

- Henry Ford developed the automobile.

- Crop rotation was promoted by Charles Townsend.

- Advances in livestock breeding were achieved by Robert Bakewell.

- The first workable seed drill was invented by Jethro Tull.

- Cyrus McCormick invented a mechanical **reaper** to reduce hand labor in the harvest of grain.

- A stationary grain **threshing machine** was developed to separate grain from waste.

- John Deere designed a better, one-piece, steel plow (refer to Figure 1–5).

- Barbed wire was invented to keep livestock away from cropland.

- The first gasoline-powered tractor was built in 1892.[19]

- Seed and plant genetics were being developed by Gregor Mendell.

By the end of the 1800s, farming had become the world's most important industry. Many things had been done to make agriculture a better, more rewarding way of life.

Figure 1–5 Agriculture advanced another step with the design of a better one-piece, steel plow by John Deere. (Courtesy of Deere & Company)

First Half of the Twentieth Century

Around the year 1900, things began to improve for the American farmer. New machines made farm work much easier, and better transportation was developed, which meant that farmers were able to market their products to many more people. Farm prices were high, and farmers were making a good living. The extra money made allowed people to pay for the research and development of new ways to farm. Some of these new developments were:

- ❧ The U.S. government established the Bureau of Forestry.
- ❧ A vaccine was developed for **hog cholera**.
- ❧ The Panama Canal opened for shipping.
- ❧ The Smith-Hughes Act established vocational agriculture in high schools.
- ❧ The U.S. government established the **Cooperative Extension Service**.
- ❧ **Federal Land Banks** were established to give credit to farmers.

CAREER OPTION

Agriculturist Scientist—George Washington Carver

There are thousands of agriculturalist scientists working with the U.S. Department of Agriculture (USDA), universities, and agribusiness companies throughout the world. An early agricultural scientist who worked at the beginning of the twentieth century was George Washington Carver. This great African American is known for his work with peanuts, sweet potatoes, and cotton.

Carver was a science professor at Tuskegee Normal and Industrial Institute (now called Tuskegee Institute), started by Booker T. Washington. He was one of the first scientists to teach crop rotation. He found that cotton grew much better after growing sweet potatoes one year in the same field, and growing cowpeas the next. The third year, cotton was grown again. Carver planted peanuts ("goobers") to control the boll weevils. He also extracted the oils, sugars, and starches from peanuts to make products such as oil, rubbing oil, milk, cheese, and margarine.

From cotton, he was able to make paper, rugs, insulating boards, and many other products. He found an amazing 118 uses for the sweet potato.

George Washington Carver at work in his laboratory conducting research on one of his experiments. (Courtesy of George Washington Carver Museum)

- 🌱 **Hybrid** plant seed was developed for better-quality, higher-producing crops.[20]

- 🌱 Many agricultural scientists, such as George Washington Carver, were developing new products (refer to Career Option, page 10).

Bad times occurred after World War I, when farmers could not sell as much food overseas. There were not as many soldiers in foreign countries, and thus there was not as much need for American-grown food overseas. Farm prices began to drop, and many farmers were forced out of business. A terrible **drought** in the midwestern states also caused farmers to go broke. Those days were called the "Dust Bowl" days. Many farmers had bought and used tractors to plow and plant many more acres of land. This added to the problem of the Dust Bowl. Large areas of land were plowed, leaving bare surfaces. When no rain came, the soil blew away.

As farmers made less and less money, the whole country began to lose money. In 1929, the stock market "crashed" and the United States entered the Great Depression. Something had to be done to pull America out of this hole. Many agricultural developments helped get the United States back on track. Some of these developments were:

- 🌱 The U.S. government began to pay farmers for using soil conservation practices.

- 🌱 The **Soil Conservation Service (SCS)** was established in 1935 to keep the Dust Bowl from reoccurring.

- 🌱 The FFA (Future Farmers of America) organization was started in 1928.

- 🌱 Higher crop yields resulted, due to better management practices.

- 🌱 The U.S. government began to pay for more research and education in agriculture.

- 🌱 Antibiotics were first used to treat animals.[21]

- 🌱 The **Dairy Herd Improvement Association (DHIA)** was organized to monitor dairy herds.

- 🌱 Groves of trees were planted along the edge of the farmstead to prevent wind erosion.

Latter Part of the Twentieth Century

The United States came out of the depression, after which World War II (1941–1945) caused an increase in farm prices. This trend got people started on developing more advanced farming methods. Many advances were made in the areas of production, marketing, and agricultural mechanics.

- 🌱 Artificial insemination was more widely used in the livestock industry.

- 🌱 Productivity increased due to new technology.

- 🌱 Farmers began to use electric fences.

- 🌱 Disc plows were widely used.

- 🌱 Chemical fertilizers and **pesticides** were widely used.

- 🌱 **Futures** trading was used to control risk.

- 🌱 Computers became popular as agricultural management tools.[22]

Recently, much effort has been spent on fine-tuning **agronomic** and animal production systems. We have seen the development of plant varieties for specific climatic zones, an increase in per-acre plant populations, and the determination of efficient and specific fertilization rates. As needs were identified and adjustments made, agricultural productivity again began increasing.

Efficiency in farming operations was now observed, especially in the 1980s. Improvements in livestock production were even more substantial than those in crop production. Many technological developments came together to increase livestock productivity, including improved livestock-handling facilities, increased feed efficiency, better and more uniform records to monitor individual animal performance, and improved disease prevention and control.[23] **Biotechnology** is now a part of our lives, with advances in gene splicing, cloning, and gene mapping.

By 1970, a trend started toward the establishment of agribusiness enterprises. These agribusinesses performed functions that farmers had previously been required to handle. Production agriculturalists now specialize in producing the product, while other companies do the hauling, processing, packaging, and marketing. Other supply and service agribusinesses sell and service machinery and other equipment. Agribusinesses now include:

- 🌾 farm machinery dealerships
- 🌾 commodity (futures) brokers
- 🌾 artificial breeding services
- 🌾 research consulting firms
- 🌾 agricultural (AG) chemical companies
- 🌾 veterinary supply companies
- 🌾 livestock supply companies
- 🌾 animal feed companies
- 🌾 biotechnology firms
- 🌾 export companies

Figure 1–6 illustrates the evolvement of farming, agriculture, and agribusiness.

SUCCESS OF AMERICAN AGRIBUSINESS

Because of the development of production agriculture, agriscience, and agribusiness, U.S. agriculture represents one of the greatest achievements in the history of humankind's struggle for food, clothing, and shelter. So efficient has U.S. agriculture become that, due to the effectiveness and efficiency of agribusiness, one American farmer can supply enough food and fiber for over 150 people. In addition, Americans spend less of their income

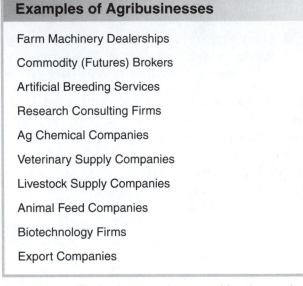

Examples of Agribusinesses

Farm Machinery Dealerships

Commodity (Futures) Brokers

Artificial Breeding Services

Research Consulting Firms

Ag Chemical Companies

Veterinary Supply Companies

Livestock Supply Companies

Animal Feed Companies

Biotechnology Firms

Export Companies

Figure 1–6 The business evolvement of farming, agriculture, and agribusiness.

on food than any other people in the world—just 9 percent of total personal disposable income.[24]

American agribusiness has become the marvel of the world. No nation has better fed itself and still contributed to world food supplies. Besides providing each U.S. citizen with some 1,500 pounds of food annually, U.S. agribusiness also produces **exports** of vast amounts of grains, vegetable oils and fats, cotton, tobacco, and many other products. Because of the success of agribusiness input and output in increasing the efficiency of production agriculture, the federal government had to establish restrictions of various kinds to see that production was controlled. However, many people believe that if American production agriculturalists and agribusinesses were allowed to produce as much as they could, much of the world would have its needs met for food, clothing, and shelter.[25]

The success and power of American agribusiness and its potential to promote world peace and security are enormous. The presence of underfed, ill-housed, and badly clothed people in any part of the world is1 a threat to the peace and security of all.[26]

CONCLUSION

Agribusiness was born in America when John Rolfe shipped the first tobacco to New England and farmers began buying newly invented production machinery and other products from suppliers. With the growth of the market for farm products during the Industrial Revolution, farmers had enough money to buy things they formerly had made on the farm. Horseshoers, barrel makers, harness makers, and other skilled craftspeople contributed the first supplies, thus decentralizing one aspect of agriculture. More decentralization occurred as farmers began to rely on butchers and millers to prepare meat and grain for consumption. As the preparation of products was decentralized, so were distribution and related functions.[27]

The changes in agricultural techniques that came with the introduction of machinery stimulated the growth of agribusiness, which, in turn, brought about more changes in the agricultural industry. Functions that once were performed on farms and ranches now are performed by workers in agribusiness.

SUMMARY

Today, agriculture is a technology-oriented industry that includes production, agriscience, and agribusiness. Most people who work in the agricultural industry do not work on farms and ranches but rather are employed in the feed, seed, farm machinery, fertilizer, chemical supply, food-processing, and related businesses.

There are many different definitions of agribusiness. Some definitions include, and some exclude, production agriculture. Within the context of this text, agribusiness is defined as the manufacture and distribution of farm supplies to the production agriculturalist and the storage, processing, marketing, transporting, and distributing of agricultural materials and consumer products that were produced by production agriculturalists.

Within the big picture of agribusiness are agribusiness companies that provide supplies to production agriculturalists. The production agriculturalists produce food and fiber, and the output is taken by agribusiness companies, which process, market, and distribute the agricultural products. Many other support services, such as research, education, and finance, are also involved. Agribusiness surrounds us in our daily lives. Consider all the agribusiness functions involved in getting a cheeseburger to you, and you will fully appreciate the magnitude of agribusiness.

People have searched for ways to feed themselves since prehistoric times. Some spent nearly all their waking time searching for foods by hunting or gathering nuts and other naturally growing foods. Eventually, the puzzle to farming was solved and people began to raise crops and domesticate animals.

The American colonists found conditions difficult. They faced starvation until they adopted the crops and tillage practices of the American Indians. American agribusiness was started when John Rolfe shipped tobacco to England. Later, many other agricultural exports were shipped, which provided the capital necessary for growth and development in American agriculture.

Many agricultural inventions and practices were developed in America, all of which led up to the agribusiness of today; these include inorganic fertilizer; rice, which was first grown in America; the cottin gin; vaccines; the cast-iron plow; the drill, mechanical reaper, and threshing machine; barbed wire; crop rotation; the gasoline-powered tractor; hybrid seed; artificial insemination; electric fences; disc plows; futures trading; and many more.

Because of the development of production agriculture, agriscience, and agribusiness, U.S. agriculture represents one of the greatest achievements in the history of humankind's struggle for food, clothing, and shelter. Indeed, American agriculture has become the marvel of the world.

END-OF-CHAPTER ACTIVITIES

Review Questions

1. Define the Terms to Know.
2. List twenty businesses or services that are needed to get a cheeseburger ready for consumption.
3. What two important developments took place before humans discovered agriculture?
4. Name five developments during the Bronze Age.
5. Name four developments during the Iron Age.
6. Name four developments during the Middle Ages.
7. What eight crops did American Indians develop that are still of great importance?
8. What were nine crops that were exported by colonists in the New England area?
9. List four agricultural developments during the seventeenth and eighteenth centuries in America.
10. List five agricultural developments during the era following the American Revolution.
11. Name ten developments during the Agricultural and Industrial Revolution.
12. What were nine developments during the first half of the twentieth century?
13. List seven developments during the latter part of the twentieth century.

Fill in the Blank

1. Agriculture, to many, is cows, sows, and _____, or weeds, seeds, and _____ .
2. Several thousand years ago, prehistoric people's nutrition and health were so bad that they seldom lived past _____ years of age.
3. Early humans considered _____ so tasty that some species were almost exterminated.
4. The fall of the_____ _____ slowed the growth of agriculture.
5. The first agricultural education teachers in America were the _____.
6. In 1613, John Rolfe shipped _____ to England, and the Virginia colonists discovered a crop with a large, untapped market.
7. A second part of the Industrial Revolution was the movement of people away from the _____ to the _____.
8. A time of terrible drought in the midwestern states, which caused farmers to go broke, was called the _____ _____ days.

9. One American farmer can supply enough food and fiber for over _____ people.

10. Americans spend just _____ percent of their personal disposable income on food.

Matching

a. A.D. 400–1500   e. 3,000 B.C.   i. John Deere

b. Output   f. Dust Bowl days   j. 1928

c. 1,000 B.C.   g. Great Depression   k. 1892

d. Input   h. 1935   l. Eli Whitney

_____   1. supplies sold to production agriculturalist

_____   2. processing, marketing, and distribution of agricultural products

_____   3. Bronze Age

_____   4. Iron Age

_____   5. Middle Ages

_____   6. result of the stock market crash in 1929

_____   7. SCS was established

_____   8. FFA organization was established

_____   9. first gasoline-powered tractor was built

_____   10. improved on the cast-iron plow built by Jethro Wood

Activities

1. Compile a list of the agribusinesses in your community.

2. Select one of the agribusinesses that you selected and show where it fits into "The Big Picture of Agribusiness." For example, is it an input or output agribusiness company?

3. Determine the agricultural products contained in a pizza. Then, identify and list the types of agribusiness companies involved in getting the pizza ready for your consumption.

4. Select one raw agricultural product and one processed agricultural product and list all the agribusiness and agriservices (example, extension service) involved in getting it ready for your purchase. Present this report to your class.

5. Select an agricultural product (machinery, equipment, fertilizer, medicine, etc.) and complete a two- or three-page report on its development. Present this report to your class.

NOTES

1. John H. Davis and Ray A. Goldberg, *A Concept of Agribusiness* (Boston, Mass.: Harvard Business School, Research Division, 1957).

2. Joseph J. Timka and Robert J. Birkenholz, *Introduction to Agribusiness Unit* (Columbia, Miss.: Instructional Material Laboratory, 1984).

3. Ewell P. Roy, *Exploring Agribusiness* (Danville, Ill.: Interstate Printers and Publishers, 1980).

4. Alfred H. Krebs and Michael E. Newman, *Agriscience in Our Lives* (Danville, Ill.: Interstate Printers and Publishers, 1994).

5. Marcella Smith, Jean M. Underwood, and Mark Baltman, *Careers in Agribusiness and Industry* (Danville, Ill.: Interstate Printers and Publishers, 1991).

6. Jasper S. Lee, *Working in Agricultural Industry* (New York: Gregg Division, McGraw-Hill, 1978).

7. Roy, *Exploring Agribusiness.*

8. *Introduction to World Agricultural Science and Technology* (College Station, Tex.: Instructional Materials Service, 1989), 8379B-2, p. 1.

9. Ibid.

10. Ibid.

11. Ibid.

12. Randall D. Little, *Economics: Applications to Agriculture and Agribusiness* (Danville, Ill.: Interstate Publishers, 1997).

13. Ibid.

14. Ibid.

15. Willard W. Cochrane, *The Development of American Agriculture: A Historical Analysis,* 2nd ed. (Minneapolis: University of Minnesota Press, 1993).

16. Ibid.

17. Lee, *Working in Agricultural Industry.*

18. Ibid.

19. *Introduction to World Agricultural Science and Technology,* p. 17.

20. Ibid.

21. Ibid.

22. Ibid.

23. Little, *Economics: Applications to Agriculture and Agribusiness.*

24. Hiram M. Drache, *History of U.S. Agriculture and Its Relevance to Today* (Danville, Ill.: Interstate Publishers, 1996).

25. Roy, *Exploring Agribusiness.*

26. Ibid.

27. Lee, *Working in Agricultural Industry.*

CHAPTER 2

The Size and Importance of Agribusiness

OBJECTIVES

After completing this chapter, the student should be able to:

- discuss the size and importance of production agriculture
- analyze the efficiency of production agriculture
- discuss the size and importance of agribusiness
- describe the agriservice sector of the agricultural industry
- explain the importance of agribusiness and foreign trade
- describe the relationship between agribusiness and energy
- describe the relationship between agribusiness and the environment

TERMS TO KNOW

aerobic
agribusiness input sector
agribusiness output sector
agriservices sector
distillation
ecologists
economists
enterprises

environmentalists
ethanol
exports
fermentation
gasohol
gross domestic product (GDP)
humus
imports

industrialization
outstanding loans
private agriservices
production efficiency
public agriservices
value-added

INTRODUCTION

In Chapter 1, we discussed the fact that the agricultural industry is changing. We learned that nearly everyone was a farmer in early America. Today, however, the effectiveness and efficiency of the production agriculturalist is so great that less than 2.5 percent of the American population is actively involved in the production of food or fiber.

Because farming (production agriculture) and agribusinesses are often presented as the same thing, distorted pictures of the agricultural industry are reinforced. For example, because the number of farms has been decreasing for a number of years, many people have concluded that the agricultural industry is declining. Mistakenly, high school counselors often advise students that opportunities in agricultural occupations are declining because, they note, the number of farmers is decreasing. However, they do not take into consideration the fact that agricultural exports and imports and agricultural inputs and outputs are at all-time highs.

Today's modern agricultural industry of production agriculturalists, agrisciences, and agribusinesses encompasses an extremely massive and complex group of organizations and the people who run them. In this chapter, we will discuss the size and importance of various areas of agribusiness that make up the agricultural industry. However, first let us look at production agriculture, which makes all agribusiness possible.

SIZE AND IMPORTANCE OF PRODUCTION AGRICULTURE

There could be no agribusiness without production agriculture. It all starts with the land. Goods from the earth come from the land, from farms and the hard work of production agriculturalists.[1] Day by day, the role of the American production agriculturalists becomes more important as more and more people are becoming dependent upon them.

Land

Because land is the major resource in modern farming, we are especially interested in how much is available and how it is being used. The United States has about 2.3 billion acres. About 21 percent of the land is used for crops, 25 percent for livestock, and 30 percent for producing forest products. The remaining 26 percent is used for nonagricultural purposes.

Some people contend that because of increasing nonagricultural uses of land, we are running out of good farmland. The facts do not support this conclusion. Even though the number of acres of farmland is decreasing, the economic land base is increasing. The productivity of land is increasing faster than the acres are decreasing.

Farm numbers have declined since the 1920s, but at different rates. Currently, about 2.1 million farms are producing the total food supply for Americans and numerous foreign countries. Although farm numbers are decreasing, the average size of farms is increasing. Currently, the average American farm is 469 acres.[2] This means that most farms that go out of business are bought by other farmers. **Economists** predict that the trend toward fewer farms will continue, but the decrease will occur in small- and medium-sized farms. Larger farms are expected to increase in number and volume of sales. Presently, about 169,000 farms contain 1,000 acres or more; 200,000 farms contain 500 to 1,000 acres; over 700,000 farms have 10 to 100 acres; and approximately 183,000 farms have less than 10 acres.

Products

To see the impact of production agriculture on the U.S. economy, we need to reference the yardstick that measures the value of goods and services that

America produces in a year, the **gross domestic product (GDP)**. The agricultural industry accounts for 17 percent of the GDP and provides more than 20 percent of all the jobs in the country. Two percent of the GDP comes from firms or people that sell goods and services to production agriculturalists. However, 13 percent of the GDP comes from related industries. These related industries include ice cream makers, textile mills, flour mills, tanneries, breakfast food makers, and a host of others (mentioned in Chapter 1). These related industries purchase food and fiber from production agriculturalists and then process and package it so they will have a **value-added** product to sell to consumers.

In 1998 America's GDP was $8.5 trillion, and the agricultural industry accounted for 17 percent, which comes to an impressive $936 billion.[3] To place the $936 billion in perspective, it would be enough money to feed the entire U.S. population for almost five years.

The products produced by production agriculturalists are very diverse. The relative percentages of the various farm products, from production agriculturalist estimates, are shown in Figure 2–1. There are many kinds of production agriculturalists in America, depending on conditions such as the type of soil, the topography, the climate, rainfall, and markets.

The animal **enterprises** with the greatest value of production for the United States are beef cattle and calves, dairy cattle, hogs, and poultry. The leading plant enterprises, in value and production, in America are corn for grain, soybeans, wheat, and cotton. Among the fruits, the leaders are grapes, oranges, and apples. Almonds are the leader in the nut category. Refer to Figure 2–2 for a list of enterprises produced by production agriculturalists.

Percent of Food Dollar to the Production Agriculturalist

The production agriculturalists' share of each dollar spent for food by the consumer is about 30

Percentage of Farm Products Sold

Item	Percentage of total output
Meat animals	39.6
Feed grains	17.9
Dairy products	11.8
Poultry and eggs	6.9
Vegetables	5.3
Food grains	4.7
Other crops	4
Fruits and tree nuts	4
Cotton	3
Tobacco	2.4
Other livestock	0.4
Total	100

Figure 2–1 These percentages represent the value of farm products sold by production agriculturalists. (Courtesy of USDA)

cents. This means that approximately 70 cents of every dollar the consumer spends for food products go to pay the cost of transportation, processing, other marketing services, and profits after the products leave the hands of the production agriculturalist. Currently, farmers take in about $150 billion a year from the sale of farm products, plus another $45 billion from off-farm sources.[4]

EFFICIENCY OF PRODUCTION AGRICULTURE

One of the most outstanding characteristics of production agriculturalists in America has been the tremendous increase in **production efficiency**.

Production Agricultural Enterprises

Crops

Barley	Oats	Sunflowers	Nursery Crops
Beans	Peanuts	Tobacco	Others
Corn	Popcorn	Wheat	
Cotton	Rice	Timber (lumber)	
Hay	Sorghum	Timber (paper)	
Maple Syrup	Soybeans	Christmas Trees	
Mint Oil	Sugarcane	Floriculture	

Vegetables

Asparagus	Cabbage
Beans, green/lima	Carrots
Beans, snap	White Potatoes
Beets	Sweet Potatoes
Broccoli	Pumpkins
	Lettuce
	Others

Animals

Broilers (meat)	Sheep and Lambs	Ostriches
Cattle and Calves	Wool	Emus
Chicken (eggs)	Turkeys	Other Fowl
Hogs and Pigs	Bees for Honey	Others
Dairy (milk)	Horses	

Fruits and Nuts

Apples	Grapefruit	Peaches	Strawberries
Apricots	Grapes	Pears	Tangerines
Blueberries	Lemons	Pecans	Tangelos
Cantaloupes	Limes	Persimmons	Tomatoes
Cherries	Oranges	Pomegranates	Watermelons
Cranberries	Papayas	Plums and Prunes	Other

Figure 2–2 There are many agriculture products produced by production agriculturalists. Those listed here represent a majority of them.

The efficiency of American agriculture is second to none. With less than .3 percent of the world's production agriculturalists, the United States produces a major percentage of the world's total food supply. Refer to Figure 2–3 to see America's share of the world's food production.

There are many different ways to measure farm efficiency, and the degree of efficiency depends on the type of measurement used. One of the most common indicators used is the number of persons supplied by the average farmworker. Presently, the average farmworker supplies over 150 persons with food and fiber, compared with only 20 in 1955 and 15 in 1945. Annual consumption of food in the United States is 1,365 pounds per person. Production agriculturalists produce an average of about 108,000 pounds, which is almost 54 tons of food! This productivity frees others from food production. It allows them to pursue careers in science, government, medicine, computers, and many other fields that help make our lives better.[5]

American farmers are recognized throughout the world for their ability to continually increase output while inputs remain relatively constant. U.S. farm output has increased more than 60 percent since 1950, while total inputs into farming have remained the same. In the last twenty years, agricultural productivity has increased more than three times faster than industrial productivity per hour worked. Today, an hour of farm labor produces sixteen times as much food and fiber as it did sixty years ago. One production agriculturalist creates six agribusiness jobs for people who produce the things production agriculturalists need and who process, transport, and merchandise the things farmers produce.[6]

In 1946, bread cost ten cents a loaf and steak was 50 cents a pound. These prices may sound like a bargain, but the pay for an hour's work was a little over a dollar before taxes or other deductions. While retail food prices are more than four times higher than a half century ago, the average worker's paycheck is 11 times higher. When comparing costs over time, effort required is a better way to visualize the cost to consumers. An hour of work today is the same as it was at any time in history, while a dollar today is not the same as it was in 1930. Refer to Figure 2–4.

Today, Americans spend less than 9 percent of family income on food. In some countries, as much as 60 to 70 percent of income goes for food. Americans spent 16.2 percent of income on food in 1973 and 18.7 percent in 1963.[7]

America's Share of World Food Production

Product	Share	Product	Share
Soybeans	64%	Grapefruit	56%
Corn	46%	Sorghum	31%
Oilseeds	42%	Eggs	17%
Poultry	24%	Oranges	25%
Beef	23%	Green Peas	23%
Pork	13%	Wheat	17%
Cotton	17%	Milk	15%

Figure 2–3 American agriculture is a showcase for the world. These percentages represent the impact of American agriculture on the world food supply. (Courtesy of USDA)

Why Is Efficiency So Important?

Increased efficiency of production agriculture releases manpower for other work and for increasing **industrialization**. Greater industrialization leads to a healthier economy, the result of which has been the gradual elevation of the standard of living of all people.

Minutes of Work Equal to the Price of Selected Food Items[1]

Item	Amount	1930	1950	1970	1980
		Minutes			
Round Steak	1 lb.	48.4	43.8	28.8	29.4
Potatoes	10 lb.	40.9	21.6	19.9	20.2
Bacon	1 lb.	48.3	29.8	21.0	15.5
Eggs	1 doz.	50.6	28.3	13.6	8.9
Bread (2 loaves)	3 lb.	29.3	20.1	16.1	16.2
Butter	1 lb.	52.7	34.1	19.2	20.0
Milk	1 gal.	64.1	38.6	29.2	22.3
Coffee	1 lb.	44.9	37.2	20.2	33.3
Sugar	5 lb.	34.7	22.7	14.4	22.8
Rice	5 lb.	54.0	39.3	21.2	27.2
All the above	1 ea.	467.9	315.5	203.6	215.8

Source: USDA.

[1]Price of food item relative to manufacturing wage rate after taxes and employee social security contributions. Through the Internet or other sources determine what the current prices would be.

Figure 2–4   This figure represents the efficiency and effectiveness of American agriculture. Notice the eggs: it took 50.6 minutes to buy a dozen eggs in 1930, but only 8.9 minutes in 1980. What do you suspect the figure is today? (Courtesy of USDA)

SIZE AND IMPORTANCE OF AGRIBUSINESS

Agriculture is the largest industry in the United States.[8] As mentioned earlier, agribusiness accounts for approximately 17 percent of the nation's total economic output. It is also the country's largest single employer, providing more than one out of every five jobs. Production agriculturalists sell approximately $123 billion in farm products, but by the time they are purchased by the consumer, the value added brings the amount to over $546 billion.[9] This means that the dollar increase from farmer to consumer is nearly four times as much. The importance of agribusiness in America is reflected throughout the world because agribusiness is the single most important contributor to the world's economy.[10] Agribusiness worldwide represents approximately one-fourth of the total world economic production and provides employment for nearly half the population on earth.[11]

The Agribusiness Input Sector

One of the most neglected areas of study in agriculture is that of the supplier of inputs for farm products, the **agribusiness input sector.** Agriculturists have assumed that this group constituted a section of general business and was not a part of agriculture. However, it plays a major role in the production of food and fiber and is currently recognized as a major phase of agribusiness. In 1996, over $180 billion was spent on agribusiness inputs by American production agriculturalists.[12]

Agricultural input provides production agriculturalists with the feed, seed, fertilizer, credit, machinery, fuel, chemicals, and similar things that they need to operate. Improvements in the quality of agricultural inputs have been the major reason for the outstanding efficiency of production agriculturalists. Examples of some larger firms within the agribusiness input sector include John Deere (farm equipment), Ralston Purina (feed), Pioneer Seed, International Mineral and Chemical (fertilizer), and Monsanto (chemicals).

Farm equipment, including tractors, other motor vehicles, and machinery, cost production agriculturalists approximately $8.8 billion each year and requires 140,000 employees to produce. Industry requires 40,000 workers to produce the 7 million tons of steel needed for farm equipment, trucks, cars, fencing and building materials.[13]

Fuel, lubricants, and maintenance for machinery and motor vehicles used by production agriculturalists cost $10.1 billion each year. Production agriculture requires more petroleum than any other single industry.

Other yearly expenditures for agricultural inputs include:

- ❧ $3.9 billion for seed
- ❧ $18.4 billion for feed
- ❧ $10.5 billion for livestock
- ❧ $8.8 billion for fertilizer and lime
- ❧ $10.3 billion for hired labor
- ❧ $21.7 billion for depreciation and other consumption of farm capital[14]

In view of the trend toward specialized farming, the input sector is expected to become an even more important factor in food production. Production agriculturalists spent over $22 billion more in 1992 than in 1987, and this trend is expected to continue due to the demand for more inputs by farmers as they attempt to supply the world's need for more food. Figure 2–5 shows the major agribusiness expenditures as a percentage of the total.

The Agribusiness Output Sector

The **agribusiness output sector** includes all agribusiness and individuals that handle agricultural products from the farm to the final consumer. This includes agribusinesses involved in buying, transporting, storing, warehousing, grading, sorting, processing, assembling, packing, selling, merchandising, insuring, regulating, inspecting, communicating, advertising, and financing. Restaurants, fast foods chains, and grocery stores are also a part of the agribusiness output sector. Figure 2–6 illustrates a few of the various agribusinesses involved in the output sector.

The agribusiness output sector needs approximately 20 million workers to handle the output of the nation's farms. The following are examples of where these workers are employed:

Meat-processing industry	350,000
Dairy-processing industry	175,000
Baking industry	225,000
Canned, cured, and frozen foods industry	240,000
Cotton/textile industry	175,000[15]

The USDA estimates that over 600,000 businesses are involved in marketing food products.[16] Nearly $425 billon was spent by consumers for food products alone within the agribusiness output sector. This represents about 9 percent of the consumers' disposable personal income. Some of the better known agribusiness firms include Kellogg's,

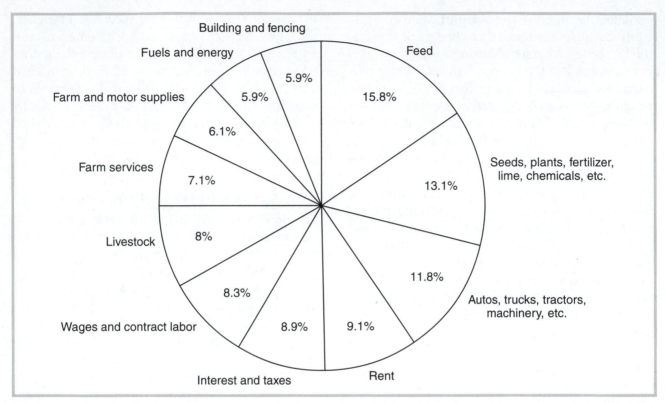

Figure 2–5 Many agribusiness inputs are needed by productive agriculturalists. This graph shows the percentage of total expenses of each of the agribusiness inputs. (Courtesy of USDA)

Pillsbury, General Mills, Campbell's Soup, Beatrice Foods, Carnation, H. J. Heinz, Hershey Foods, Archer Daniels Midland, IBP Meat Packers, Mid-America Dairymen, and Sunkist. Figure 2–7 shows sales and rank of the top agribusiness (both input and output) corporations in America.

AGRISERVICES

The **agriservices sector** of the agriculture industry is concerned with researching new and better ways to produce and market food and to protect food producers and consumers, and with providing special, custom-type services to all the other phases of agriculture. Public agencies have played

a dominant role in the agriservices area, but private agencies are increasing their offerings of farm services at a rapid pace.

Public Agriservices

The **public agriservices** group provides special services at the federal, state, and local levels. The major areas of emphasis include research, education, communication, and regulation. There are over 100,000 employees in the USDA, 12,000 of whom work in Washington, D.C. Often their work is closely allied to that of state and county agricultural workers. Some 17,500 employees work in the state and federal extension services, and 24,000 more are attached to the state agricultural experiment stations. About 25,000 county com-

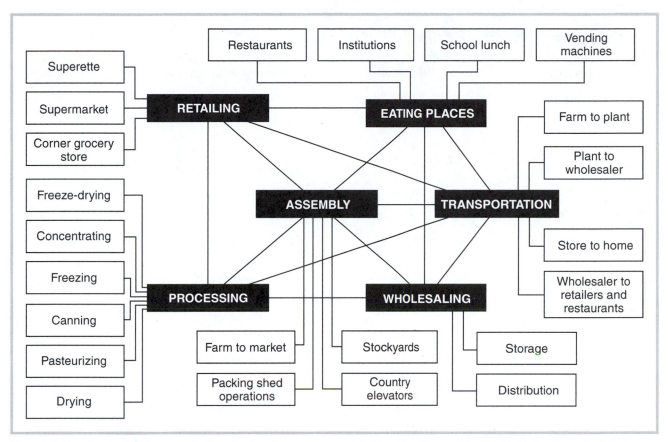

Figure 2–6 The agribusiness output sector involves many agencies and functions after the agricultural products leave the farm. (Courtesy of USDA)

mittee persons administer the agricultural programs established by Congress.[17] There are approximately 12,000 teachers of agricultural education. Other public agriservices include the Food and Drug Administration; Department of the Interior, which includes the National Park Service; and a small portion of the Department of Commerce. More will be discussed about public agriservices in Chapter 11.

Private Agriservices

Recently, according to the USDA, production agriculturalists paid more than $1 billion yearly for **private agriservices** such as veterinary care, feed grinding and mixing, machine harvesting, contract labor, and spraying. Over 30,000 firms provide agricultural services, some of which are nonfarm. This group employs over 100,000 workers and has a payroll of about $600 million. This does not include about 30,000 nonpaid owners and family members and other part-time workers. Specifically, there are three major areas of private agriservices available to the agricultural industry: financial services, trade associations, and agricultural cooperatives.

Financial Services. Financial services are a vital part of the agricultural industry. Lending

Largest Agribusiness Corporations in America

Agribusiness Company	Rank	Sales ($ Millions)	Agribusiness Company	Rank	Sales ($ Millions)
Wal-Mart Stores[1]	4	93,627.00	Weyerhaeuser	104	11,787.70
Philip Morris	10	53,139.00	Deere	124	10,290.50
State Farm Group	12	40,809.90	Pfizer	127	10,021.40
E. I. Du Pont De Nemours	13	37,607.00	McDonald's	132	9,794.50
Procter & Gamble	17	33,434.00	Publix Super Markets	136	9,470.70
Pepsico	21	30,421.00	Monsanto	146	8,962.00
Conagra	26	24,108.90	CPC International	154	8,431.50
Kroger	27	23,937.80	General Mills	156	8,393.60
Dow Chemical	36	20,957.00	H. J. Heinz	162	8,086.80
International Paper	60	19,797.00	Eli Lilly	171	7,535.40
American Stores	45	18,308.90	Campbell Soup	177	7,278.00
Coca-Cola	48	18,018.00	Farmland Industries	178	7,256.90
Sara Lee	50	17,719.00	Ralston Purina	180	7,210.30
Fleming	52	17,501.60	Kellogg	187	7,003.70
Supervalu	57	16,563.80	Champion International	188	6,972.00
Safeway	58	16,397.50	James River Corp. of Virginia	194	6,799.50
Caterpillar	63	16,071.00	Coca-Cola Enterprises	196	6,773.00
RJR Nabisco Holdings	64	16,008.00	Quaker Oats	206	6,365.20
Georgia-Pacific	75	14,292.00	American Brands	221	5,904.90
Kimberly-Clark	78	13,788.60	W. R. Grace	228	5,784.20
Archer Daniels Midland	92	12,671.90	Aramark	235	5,600.60
IBP	94	12,667.60	Tyson Foods	239	5,511.20
Albertson's	96	12,585.00	Mead	256	5,179.40
Anheuser-Busch	97	12,325.50	Case	261	5,105.00
Winn-Dixie Stores	103	11,787.80	Vons	263	5,070.70

(continued on the following page)

Figure 2–7 Few people outside the agricultural industry are aware of the magnitude of the agribusiness input and output sectors. (Courtesy of *Fortune* magazine)

Largest Agribusiness Corporations in America *(concluded)*

Agribusiness Company	Rank	Sales ($ Millions)	Agribusiness Company	Rank	Sales ($ Millions)
Boise Cascade	265	5,057.70	Circle K	348	3,565.60
Union Camp	300	4,211.70	Food 4 Less Holdings	353	3,494.00
Supermarkets Genl. Holdings	303	4,182.10	Smith's Food & Drug Centers	399	3,083.71
Dole Food	304	4,152.80	Hormel Foods	409	3,046.20
Stop & Shop	308	4,116.10	Louisiana-Pacific	449	2,843.20
Chiquita Brands International	312	4,026.60	Sonoco Products	462	2,706.20
Giant Food	318	3,695.60	Dean Foods	472	2,630.20
Hershey Foods	336	3,690.70			

Source: *Fortune,* April 29, 1996.

[1]Note: Wal-Mart is included because of the grocery division. However, these revenues represent all Wal-Mart sales.

money to all three sectors of the agricultural industry (input, production, and output) is big business. **Outstanding loans** for farm real estate alone amount to nearly $80 billion, in addition to other outstanding loans and nonreal estate debt, which is nearly another $65 billion.[18] Commercial banks, the Farm Credit System, the Farm Service Agency, individual businesses and cooperatives, and insurance companies all provide funds that production agriculturalists need to buy land, equipment, livestock, machinery, seed, fertilizer, and the other products they use in their daily farming operations. Many of these same agencies are also involved with providing financial services to the agribusiness input suppliers and those that buy, transport, process, and market agricultural products. These agribusinesses need credit and capital for their day-to-day operations, buildings, and equipment, just as production agriculturalists do.

Trade Associations. Trade associations, as well as dairy and livestock associations, are vital to the agricultural industry. Every agribusiness and pro-

duction agriculture enterprise has a trade association, society, or institute, which are supported by members who are active in a particular enterprise. Thousands of trade associations operate in America. They serve agribusiness and promote certain agricultural products, and they have become an essential part of agribusiness relationships.

Associations perform a wide variety of services for their members in areas such as public relations, promotion, legislative lobbying, communications, sales training, auditing and record keeping, publicity, transportation, research, and legislative and marketing information. These and similar services can be provided better by organized groups rather than by individuals. Members may pay dues to an association to receive its benefits, or producers may pay a check-off from sales of their products.[19]

Agricultural Cooperatives. Agricultural cooperatives serve many needs and engage in a variety of essential services to the agricultural industry. Over 4,200 cooperatives market agricultural products and furnish the agricultural industry with production supplies and services. Their annual business volume is over $83 billion. Cooperatives market over one-quarter of all agricultural products and provide one-quarter of the production supplies for farmers.[20] They also furnish electricity to the agricultural industry. Cooperatives enable livestock producers to better market their animals and improve dairy products through dairy herd improvement associations. (Cooperatives will be discussed further in Chapter 10.)

AGRIBUSINESS AND FOREIGN TRADE

The United States has long played a major role in world agricultural trade and is rapidly increasing its role relative to most other countries. Currently, the United States is the major participant in international trade of agricultural products. The five major farm commodities sold in world markets in 1994 were feed grains and feed grain products, soybeans and soybean products, wheat and wheat products, live animals, meat and meat products, and vegetables, for a total value of $46.7 billion.[21] In comparison, the total value of all agricultural **exports** was only $6 billion in 1966. Almost every country of the world purchases some agricultural products from the United States.

In addition to being a major world exporter of agricultural products, the United States is also a major importer of agricultural products. Over a ten-year period, annual farm imports increased from $4.45 billion to $10.50 billion, a 136 percent increase. Although the major quantity is imported from Latin America, Asia, and Europe, U.S. **imports** come from every continent except Antarctica. Although imports are greater than ever, the U.S. trade surplus (dollars exported minus dollars imported) was $18.9 billion in the mid-1990s.[22]

In addition to providing markets for American farmers, foreign exports provide additional advan-

tages. Between 25,000 and 30,000 jobs are created in the United States for each billion dollars of agricultural exports. In 1994, between 1.2 and 1.4 million full time jobs could be attributed to our agricultural exports. Exports account for approximately one-third of U.S. agricultural production. In many states, as much as one-half of farm income comes from agricultural exports.[23]

AGRIBUSINESS AND ENERGY

Agriculture as a User of Energy

The agricultural industry is both a producer and user of energy. Recent reports indicate that the total agricultural industry consumes 10 to 20 percent of the nation's energy. About one-third of the energy is used by production agriculturalists, and the remaining two-thirds is used by agribusiness.

Energy Produced from Renewable, Agricultural Products

Agriculture is becoming a larger producer of energy. Four major areas include direct burning, ethanol production, biodiesel, and methane gas production.

Direct burning. A rebirth in the use of wood-burning stoves developed because of rising costs of home-heating fuels. An estimated 5 million American homes rely on wood-burning stoves for all their heating needs, and wood-burning stoves or fireplaces partially heat another 20 million homes. Direct burning of wood and wood wastes provides a sizable energy source. Estimates predict this resource can meet 20 percent of the energy needs in the United States in the future.[24]

Ethanol Production. **Fermentation** and **distillation** of grains into drinking alcohol existed for thousands of years. In **ethanol** production, sugars and/or high starch grains feed yeast cells in an oxygen-free environment. Corn liquor, or "moonshine" as it is sometimes called, is highly combustible when it is 80 percent pure (160 proof) or higher. When blended with gasoline, **gasohol** is the result. Brazil pioneered the development of gasohol. Large-scale fermentation plants used local sugarcane to produce this form of alcohol. The use of gasohol increased rapidly in the early 1980s. Ethanol production consumed 535 million bushels of corn in 1994 (5.3% of the record 10 billion bushel corn crop). The ethanol industry is responsible for more than 40,000 direct and indirect jobs, creating more than $1.3 billion in increased household income annually, and more than $12.6 billion over the next five years.[25] Refer to Figure 2–8.

Biodiesel. Worldwide, biodiesel is the most used renewable agricultural alternative energy source. Sometimes referred to as soy diesel, this fuel is made from soybeans and is used either as a pure soybean oil or mixed with diesel fuel. Diesel engines can use this fuel without changes. However, the fuel storage system on the vehicle has to be equipped for heating and mixing the fuel. Refer to Figure 2–9.

Methane Gas Production. Methane gas production is the product of bacteria feeding on organic matter under anaerobic conditions. Nearly two-thirds of biogas is methane, a natural gas. Methane production also occurs as a product of sewage sludge and landfill decomposition. Production of pure methane depends on the availability of concentrated quantities of manure, as in feedlots or dairy lots. There is one strong advantage of methane production. Use of the manure eliminates the need for its immediate disposal. The residue material is nutrient-rich **humus** which serves as a valuable soil conditioner.[26]

Agribusiness Energy as a Viable Alternative. Dr. Cliff Ricketts and his students at Middle Tennessee State University have converted engines to run off ethanol, biodiesel oil (soybean oil), and hydrogen. These agriculture students have entered and won races in their alternative fuel divisions in Utah, California, North Carolina, and Florida. Certainly agribusiness energy is a viable alternative.

Figure 2–8 This vehicle has been driven over 26,000 miles on pure ethanol. It was converted to ethanol by agribusiness and agriscience students at Middle Tennessee State University. After 25,000 miles, it raced and won in its division at the Bonneville Salt Flats. (Courtesy of Cliff Ricketts)

Figure 2–9 This Corvette, "The Senator," is powered by an Allis-Chalmers 4-cylinder turbo diesel engine that is fueled by soybean oil. Built by agriculture students of Dr. Cliff Ricketts, Professor of Agricultural Education at Middle Tennessee State University, "The Senator" holds several speed records for a soybean oil–fueled vehicle. (Courtesy of Cliff Ricketts)

AGRIBUSINESS AND THE ENVIRONMENT

The environment is becoming more important to the business of agriculture. Unfortunately, our resources and environment took a back seat to production and profit in American industries for most of this century. Concerned citizens and scientists began to speak out about the various types of pollution when they realized that the quality of air and water around them was deteriorating. In the early days of the environmental movement, these people were known as **ecologists**. At first they were considered radicals and alarmists, and very few people paid any attention to their cries of doom. However, in the late 1960s, an increasing number of Americans joined the effort to reduce pollution of all types.

Rachel Carson's *Silent Spring*, published in 1962, was the first major publication calling attention to environmental issues. The Sierra Club is a major pioneer that called attention to environmental problems. The club's membership tripled between 1965 and 1970 and since that time has increased yearly by about 10,000 members, mostly young people. Other major environmental groups include the Audubon Society, the National Wildlife Federation, Friends of the Earth, the Environmental Defense Fund, and the League of Conservation Voters.

Pollution control became a major priority for the nation. Federal, state, and local legislators began to develop and pass laws to prevent and control pollution. The National Environmental Policy Act of 1969 created the Council on Environmental Quality (CEQ), which initiated the requirement of environmental impact studies for controversial environmental activities. The Office of Environmental Quality was established in 1970, under the Environmental Quality Improvement Act, to support the work of the CEQ.

The Environmental Protection Agency (EPA) and the National Oceanic and Atmospheric Administration (NOAA) were established in 1970. The Federal Water Pollution Control Act Amendments of 1972 set up the National Commission on Water Quality and established water discharge requirements. Numerous other legislation has been approved to develop and enforce air, pesticide, noise, drinking water, occupational safety, and health requirements.

Many job openings have developed that have environmental agriculture implications. High schools have courses and colleges have environmental agriculture majors and produce graduates ready to enter the work force. Research and development efforts are creating jobs with ecology as the focal point. Agricultural careers within environmental areas include soil, water, air, and energy sources. Refer to Figure 2–10 for a list of job opportunities for **environmentalists** in each of these four areas.

Environmentalists in Agribusiness

Those involved in the use of the soil for growing crops should be environmentalists. Most fruit, vegetable, and grain producers have been environmentalists for hundreds, or perhaps thousands, of years. To control soil erosion and protect the soil is to show concern for our environment. Soil conservation has been advocated for centuries. Ancient peoples of Rome, India, Peru, and elsewhere valued soil highly enough to build anti-erosion terraces.[27]

Natural Resources Conservation Service

The first field research on soil and water conservation in the United States was conducted by M. F. Miller and F. L. Daley at the University of Missouri.[28] This study was started in 1917 and the first results were published in 1923. This provided the foundation for the soil conservation movement. After

Environmental Agricultural Job Opportunities

Job Opportunities Related to Soil Resources

Agricultural Economist	Environmental Control Scientist	Mineralogist
Agronomist	Fertility Expert	Nursery Gardener
Biologist	Field Auditor	Nursery Manager
Botanist	Forester	Plant Breeder
Chemical Sales Representative	Geologist	Plant Propagator
Chemist	Horticultural Garden Curator	Range Conservationist
Conservation Officer	Horticulture Extension Agent	Seed Specialist
Conservation Scientist	Land Planner	Soil Conservation Field Inspector
County Agricultural Extension Agent	Landscape Architect	Soil Engineer
Crop Research Scientist	Landscape Contractor	Soil Scientist
Disease and Insect Control Technician	Landscape Drafter	Soil Science Technician
Drainage Design Coordinator	Lobbyist	Soil Tester
Ecologist	Management Specialist	Urban Planner
Education Consultant	Mediator	Wastes Engineer
Entomologist	Miner	Wastes Technician

Job Opportunities Related to Water Resources

Agriscience Teacher	Fish Culturist	Lawn Sprinkler Installer
Aquaculturist	Fish Hatchery Manager	Lobbyist
Aquatic Facility Manager	Food Processing Manager	Marine Biologist
Attorney	Game Warden	Marine Life Hatchery Manager
County Agricultural Extension Agent	Gravity Flow Irrigator	Park Naturalist
Dairy Operator	Hydroelectric Facility Supervisor	Park Ranger
Drainage Design Coordinator	Hydroponics Nursery Manager	Plumber
Ecologist	Ichthyologist	Public Health Microbiologist
Environmental Control Technician	Information Scientist	Public Works Engineer
Financial Institution Officer	Irrigation Engineer	Rice Producer

(continued on the following page)

Figure 2–10 It is evident from this list that many jobs are available in the area of environmental agriculture. (Courtesy of USDA)

Environmental Agricultural Job Opportunities *(concluded)*

SCUBA Diver	Sprinkling System Irrigator	Water Quality Control Engineer
Shellfish Hatchery Superintendent	Technical Publications Writer	Water Quality Tester
Slaughtering and Processing Manager	Water Purification Chemist	Water Treatment Plant Engineer
Soil and Water Conservation Agent	Water Quality Board Member	Waterworks Supervisor

Job Opportunities Related to Air Resources

Agricultural Engineer	Furniture Manufacturer	Plant Pathologist
Allergist	Government Agency Employee	Public Health Technician
Automobile Designer	Industrial Chemist	Research Technician
Automobile Mechanic	Market Research Analyst	Tobacco Market Analyst
Diesel Engine Mechanic	Meteorologist	Tractor Mechanic
Environmental Research Technician	Planning Engineer	Vehicle Inspection Station Worker
Environmental Scientist	Plant Ecologist	Weather Observer

Job Opportunities Related to Energy Resources

Agricultural Economist	Line Repairer	Safety Engineer
Building Inspector	Loan Officer	Safety Inspector
Computer Specialist	Materials Scientist	Sales Representative
Construction Worker	Mechanical Engineer Technician	Service Technician
Crew Boss	Methods Study Analyst	Surveyor
Equipment Repairer	Research Director	Utilities Supervisor
Geologist	Research Engineer	Welder

several name changes, the soil conservation agency of the USDA is now called the National Resource Conservation Service. The objective of soil conservation has been stated as "the use of each acre of agricultural land within its capabilities and the treatment of each acre of agricultural land in accordance with its need for protection and improvement."[29] Some traditional soil and water conservation practices that protect the environment include:

- ❧ miminum tillage
- ❧ no-till practices
- ❧ strip cropping
- ❧ contour farming
- ❧ surface drainage

❦ subsurface drainage

❦ farm ponds

❦ grassed waterways

❦ irrigation systems

❦ field windbreaks

❦ brush management pastures and ranges

❦ rotation grazing

❦ pasture and hay land seedings

❦ range reseeding

❦ improved tree harvesting

❦ tree planting

❦ windbreak renovation[30]

Many careers are needed in order to design, construct and manage these conservation practices. Refer to the career option about environmental managers and conservationists.

CAREER OPTION

Career Areas: Environmental Manager, Conservationist

*E*nvironmental managers and conservationists are involved in the management, protection, and preservation of the environment and natural resources. Natural resources include all the things that help support life, such as sunlight, water, soil, and minerals. Plants and animals are also natural resources. Without the work of environmental managers and conservationists, most of the earth's resources would be wasted, degraded, or destroyed.

Environmental managers may specialize in wildlife, air and water quality, fire control, automotive emissions, and forestry emissions. Conservationists may specialize in soil, forest, grazing lands, minerals, energy, and water conservation. Many people with an interest in the outdoors pursue a career in forestry, wildlife ecology, or water, soil, air, or energy conservation. Refer to Figure 2–10 for a list of several environmental agriculture and conservationist job opportunities.

Soil scientist Marife Corre prepares to analyze soil samples from a riparian buffer. The carbon and nitrogen status of riparian zone soils indicates their potential to remove nitrate from shallow groundwater and to improve water quality. (Courtesy of USDA)

SUMMARY

Today's modern agricultural industry of production agriculturalists, agriscience, and agribusinesses encompasses an extremely massive and complex group of organizations and the people who run them.

There could be no agribusiness without production agriculture. It all starts with the land. The United States has about 2.3 billion acres. About 21 percent of the land is used for crops, 25 percent for livestock, and 30 percent for producing forest products. The remaining 26 percent is used for nonagricultural purposes. The agricultural industry accounts for 17 percent of the GDP and provides more than 20 percent of all the jobs in the country. The production agriculturalists' share of each dollar spent for food by the consumer is about 25 cents.

One of the most outstanding characteristics of production agriculturalists in America has been the tremendous increase in their production efficiency. The average farmwork supplies over 150 persons with food and fiber. Annual consumption of food in the United States is 1,365 pounds per person. One production agriculturalist produces an average of 108,000 pounds, which is about 54 tons of food per year. Americans spend about 9 percent of their personal disposable income on food.

Agribusiness is the largest industry in the United States. Products from production agriculturalists valued at $123 billion increase in value to $546 billion once they are purchased by the consumer, a four-fold increase. In 1996, over $180 billion was spent on agribusiness inputs by production agriculturalists. Nearly $425 billion was spent by consumers for food products alone within the agribusiness output sector. The USDA estimates that over 600,000 businesses are involved in the agribusiness output sector.

The agriservice sector of the agricultural industry includes both public and private services, which research new and better ways to produce and market food and fiber, disseminate new technology, develop and enforce laws to protect food producers and consumers, and provide special, custom-type services to all other phases of agriculture within the public agriservice sector. The USDA alone employs over 100,000 employees. Within the private agriservice sector, there are over 30,000 firms, which employ over 100,000 workers with a payroll of about $600 million. The three major areas of private agriservices available to the agricultural industry are financial services, trade associations, and agricultural cooperatives.

Foreign trade is a big part of agribusiness. Almost every country of the world purchases some agricultural products from the United States. Exports account for approximately one-third of U.S. agricultural production. The top five exported agricultural commodities sold for $46.7 billion. In 1994, between 1.2 and 1.4 million full-time jobs could be attributed to our agricultural exports.

The agricultural industry is both a producer and user of energy. The agricultural industry consumes 10 to 20 percent of the nation's energy. Agriculture is becoming a greater producer of energy, including direct burning, ethanol production, biodiesel, and methane gas production.

The environment is becoming more important to the business of agriculture. For much of this century, our resources and environment took a back seat to production and profit in American industries. Today, however, many job openings have developed that have environmental agriculture implications. High schools have courses and colleges have environmental agriculture majors and produce graduates ready to enter the workforce.

END-OF-CHAPTER ACTIVITIES

Review Questions

1. What percent of the gross domestic product is attributed to the agricultural industry?
2. List five crop, five vegetable, five animal, and five fruit and nut enterprises.
3. Why is production agriculture efficiency so important?
4. What is the largest industry in the United States?
5. What type agribusinesses are included in the agricultural input sector?
6. Name five agribusiness input firms.
7. List the yearly national expenditures for each of the following agricultural inputs: (a) seed, (b) feed, (c) livestock, (d) fertilizer and lime, and (e) hired labor.
8. List 18 types of agribusiness outputs.
9. Name ten agribusiness output firms.
10. How many workers are employed in each of the following? (a) meat-processing industry, (b) dairy-processing industry, (c) baking industry, (d) canned, cured, and frozen food industry, (e) cotton/textile industry, and (f) agricultural exports.
11. Explain the role of agriservice sector.
12. What are the three major areas of private agriservices?
13. List ten things that trade associations do.
14. What are the four major areas of energy produced from renewable agricultural products?
15. List the names of six environmental groups.
16. What are four areas of concern to environmentalists?

Fill in the Blank

1. Only _____ percent of the American population is actively involved in the production of food and fiber.
2. There could be no agribusiness without _____ _____.
3. The average American farm consists of _____ acres.
4. The production agriculturalists' share of each dollar spend for food by the consumer is about _____ .
5. America produces _____ percent of the world's soybeans, _____ percent of the world's corn, _____ percent of the world's poultry, and _____ percent of the world's beef.
6. Annual consumption of food in the United States is _____ pounds per person.
7. One production agriculturalist produces an average of about _____ pounds.
8. One production agriculturalist creates _____ agribusiness jobs.

9. Agribusiness worldwide represents approximately _____ of the total world economic production and provides employment for nearly _____ of the population on earth.

10. Cooperatives market over _____ of all agricultural products.

11. Between 25,000 and 30,000 jobs are created in the United States for $_____ in agricultural exports.

12. Exports account for approximately _____ of U.S. agricultural production.

Matching

a. 30%
b. 5,000
c. 26%
d. $300 billion

e. 2.3 billion acres
f. 600,000
g. 3 million acres
h. 21%

i. 100,000
j. 80 billion
k. 25%
l. 21 million

_____ 1. land in United States
_____ 2. percent land in United States used for crops
_____ 3. percent land in United States used for livestock
_____ 4. percent land in United States used for forests
_____ 5. percent land in United States used for nonagricultural purposes
_____ 6. number of farms producing the total food supply for Americans and numerous foreign countries.
_____ 7. amount spent by consumers on food products
_____ 8. number of businesses involved in the agribusiness output sector
_____ 9. approximate number of USDA employees
_____ 10. outstanding loans from farm real estates alone
_____ 11. number of cooperatives marketing agricultural products

Activities

1. The author uses the terms *agricultural industry, production agriculture, agriscience,* and *agribusiness.* Draw a picture illustrating the difference and relationship of these terms. Use whatever words are needed to explain your picture.

2. Prepare a brochure titled, "Reasons for Taking Agricultural Education Courses." Include material from this chapter and other places illustrating the size and importance of agriculture. Ask the school guidance counselor to select the best brochure from your class.

3. Identify five agribusiness input suppliers and five agribusiness output firms in your community.

4. Identify three public agriservices and three private agriservices in your community or county.

5. Identify three trade associations in your community. Briefly describe the purpose of each.

6. Name and describe two cooperatives in your community.

7. Prepare a short two-to-three-page essay on an agricultural alternative energy resource.

8. Copy an article from a magazine or newspaper on an environmental agriculture issue. Share the article with your class by making a presentation about it.

9. With the aid of the Internet or a USDA source, determine the minutes of work equal to the prices of selected food items (as shown in Figure 2–4) from the latest available data.

NOTES

1. *Bureau of Economic Analysis,* 1999, United States Department of Commerce (Washington, D.C.: U.S. Government Printing Office, 1999).
2. United States Department of Agriculture, *Agricultural Statistics 1995–96* (Washington, D.C.: U.S. Government Printing Office, 1996).
3. Bob R. Stewart, *Introduction to Agricultural Business, Unit for Animal Science Core Curriculum* (Columbia, Miss.: Instructional Materials Laboratory, 1984).
4. Alfred H. Krebs and Michael E. Neuman, *Agriscience in Our Lives* (Danville, Ill.: Interstate Publishers, 1994).
5. *Agribusiness Management and Marketing* (College Station, Tex.: Instructional Materials Service, 1988), 8720B, p. 4.
6. Ibid.
7. Ibid.
8. James G. Beierlein, Kenneth C. Schneeberger, and Donald D. Osburn, *Principles of Agribusiness Management* (Prospect Heights, Ill.: Waveland Press, 1993).
9. Agricultural Statistics, 1998. United States Department of Agriculture (Washington, D.C.: U.S. Government Printing Office, 1998).
10. N. Omri Rawlins, *Introduction to Agribusiness,* 3rd ed. (Murfreesboro, Tenn.: Middle Tennessee State University, 1998), p. 16.
11. Ibid.
12. *USDA Agricultural Statistics,* 1998.
13. Rawlins, *Introduction to Agribusiness,* 1998, p. 80.
14. Krebs and Neuman, *Agriscience in Our Lives,* p. 3.
15. Ibid., p. 2.
16. Rawlins, *Introduction to Agribusiness,* p. 29.
17. Ibid., pp. 28, 115.
18. USDA, *Agricultural Statistics, 1995–96.*
19. Smith, Underwood, and Bultman, *Careers in Agribusiness and Industry,* p. 287.
20. Rawlins, *Introduction to Agribusiness,* p. 150.
21. USDA, *Agricultural Statistics,* 1995–96.
22. Ibid.
23. United States Department of Agriculture, *Agricultural Fact Book, 1997* (Washington, D.C.: U.S. Government Printing Office, 1997).
24. Ibid.
25. "Ethanol Information: Economic Benefits." Available online: http:///www.ethanol.org
26. United States Department of Agriculture, *Agricultural Fact Book, 1997* (Washington, D.C.: U.S. Government Printing Office, 1997).
27. Frederick R. Troch, J. Arthur Hobbs, and Roy L. Donahue, *Soil and Water Conservation for Productivity and Environmental Protection* (Englewood Cliffs, N.J.: Prentice-Hall, 1980), p. 11.
28. Ibid., p. 629.
29. Ibid.
30. Ibid., p. 638.

CHAPTER

3

Emerging Agribusiness Technologies

OBJECTIVES

After completing this chapter, the student should be able to:

- ❧ discuss the value of global positioning
- ❧ explain the importance and use of genetic engineering
- ❧ explain the importance and use of cloning
- ❧ describe the procedure of determining gender selection
- ❧ discuss four types of production hormones
- ❧ give examples of animal research that results in human medicine
- ❧ describe domestic animals of the future
- ❧ describe computer and electronic animal management technologies
- ❧ explain emerging mechanical technologies in the plant industry
- ❧ explain the importance and use of tissue culture
- ❧ discuss selected emerging plant technologies
- ❧ discuss the importance and use of integrated pest management
- ❧ explain the use of hydroponics and aquaculture as a solution for extended space travel

TERMS TO KNOW

augmentation	gene splicing	pessary
bovine somatotropin (BST)	genetic engineering	pheromone
callus	hormones	physiologically
cloning	host	porcine somatotropin (PST)
deoxyribonucleic acid (DNA)	hybrids	precision farming
embryo splitting	hydroponics	restriction enzyme
embryo transfer	implant	solarization
flaming	Integrated Pest Management	tissue culture
gender selection	(IPM)	transgenic
gene mapping	pathogens	xenotransplantation

38

INTRODUCTION

Agricultural technologies are applied sciences. However, once they have been developed, it is the agribusiness input or agribusiness output sector that sells or markets the emerging technologies. These technologies are used to improve agricultural production and to improve methods of processing, transporting, and distributing agricultural goods.

Agricultural technology has expanded at a rapid rate since the beginning of recorded history, but technology development has a "snowball effect." Many of America's older production agriculturalists started their careers behind teams of horses or mules. Today, production agriculturalists operate air-conditioned tractors, which permit them to do more work in a few hours than the previous generation used to complete in a week.

The agricultural industry is moving at such a rapid rate that it is difficult, if not impossible, to label any process or procedure as "new" or "futuristic." However, it is the goal of this chapter to give a brief overview of outstanding technological steps in the agricultural industry, as viewed by any generation. For example, we cannot forget the value of hybrids, artificial insemination, and the round haybaler. Although these technologies are not new, the agricultural industry made significant advances as a result of their use.

Max Lennon, former president of Clemson University, made a most appropriate evaluation when he stated: "In the area of technology alone, there will be more new developments in the next fifteen years than in all of history to this point. Simply put, the world is changing, and whether the agricultural industry will benefit from these changes, or suffer because of them, depends on the industry itself."

GLOBAL POSITIONING

The Global Positioning System (GPS) is an experimental technology that uses satellites to view specific fields and crops. The GPS uses its view from space to provide a computerized picture that will aid field crop management and increase yield per acre.[1]

Purpose of Global Positioning and Field Management

In days past, before agriculture was industrialized, production agriculturalists were familiar with their fields as a whole, and successful production agriculturalists adapted farming techniques to the variations of their fields. Now, crop producers are concerned, not only with managing large fields, but with field conditions, which can vary from one square yard to the next. A single crop producer's field can contain three or four different soil types, with varying fertility, drainage, organic content, and nitrogen levels. Traditionally, agricultural production methods required that fields be treated uniformly despite these within field variances in soil type, weed and disease incidence, and topography. Often this led to over- or underapplications of inputs such as fertilizer, seed, water, and pesticides. Overuse often results in wasted resources and pollution. By mapping fields using the Global Positioning System (GPS), production agriculturalists can vary their seeding, irrigation, fertilization and cultivation practices to accommodate each section of their field and thus increase yield.[2] Refer to Figure 3–1.

Origin of the Global Positioning System

The GPS was originally designed for the military defense systems of the U.S. Department of Defense. The system consists of twenty-four satellites that circle the earth every twelve hours, orbiting 10,900 miles above the earth. Infrared vision found in satellites can distinguish healthy crops from sick ones,

Figure 3–1 In Missouri, corn is harvested by a combine that is linked to the satellite-based Global Positioning System. Precise yield and location data will be correlated with soil samples taken earlier at sites throughout the field. This information will help growers plan optimal fertilizer rates for the next crop. (Courtesy of USDA)

locate problem spots, and distinguish high weed and other, less productive areas.[3] The GPS system consists of four components: a GPS receiver, a crop yield monitor, a digital soil fertility map, and variable rate application technology (VRT). In many cases, the crop producers may use an inexpensive hand-held GPS receiver or attach the GPS system to equipment that will automatically vary application rates. The application rate of seed, fertilizer, pesticides, and water is varied by sensors mounted on a truck or tractor that allow computer-controlled nozzles to know the location in the field; here, soils and crop needs are amended by specific applications of inputs.[4]

Global Positioning and the Environment

By mapping the field using GPS, crop producers will obtain information that will help them make decisions that will increase productivity by better management of all areas of the field. Not only will this reduce the cost of production, but also, many argue, this type of **precision farming** will improve farm productivity while reducing the

application of fertilizer, herbicides, and pesticides. Site-specific information could thus help decrease the impact of adverse environmental agricultural practices. Additionally important for the environment, water use and management are enhanced by the GPS system.[5]

Precision Farming

Precision farming may become another of the crop producers' wide range of tools to help increase productivity while maintaining the environmental integrity of the land. Precision farming combines crop planning, including tillage, planting, chemical applications, harvesting, and post-harvest processing of the crop.[6]

The five major objectives of comprehensive precision farming are:

- 🌿 increased production efficiency
- 🌿 improved product quality
- 🌿 more efficient chemical use
- 🌿 energy conservation
- 🌿 soil and ground water protection

Precision farming requires the use of digitally geographically referenced data in cropping operations. Precision farming is principally associated with soil fertility and soil management.[7] Simply put, the farm field and crop are referenced as a computer model of each specific section of a crop producer's field. Then, each section is individually assessed for increased production, decreased cost, and environmentally sound techniques.

GENETIC ENGINEERING

When selective breeding stock depends on certain traits being present, this process can be speeded up. **Genetic engineering** bypasses the natural selection process by inserting the desired gene into the chromosomes of a living organism. Refer to Figure

3–2. The target cells receive the new genes, which were obtained from another organism. Refer to Figure 3–3.[8]

Advantages of Using Genetic Engineering

Transforming a Single Gene. With traditional and classical breeding methods, a plant scientist has to cross and backcross several generations of a plant to eliminate undesirable traits or characteristics. With genetic engineering, the plant scientist can transfer a single gene for the desired trait into the plant's genetic material without changing the rest of its genetic makeup.

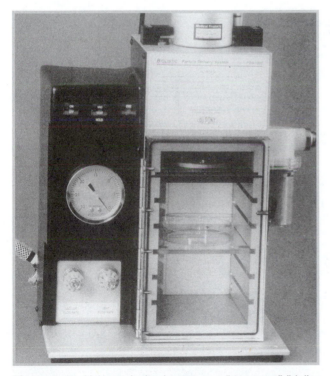

Figure 3–2 Using a device known as a "gene gun" (biolistic particle delivery system), a gene can be propelled into growing tissue by putting material containing the gene on a projectile and shooting it at the target cells. The projectile, which is similar to a BB, is stopped, but the genetic material continues to move and penetrates the target cells. (Courtesy of DuPont Co. Biotechnology Systems)

Thus, a breeder working with genetically engineered plants can easily predict the characteristics of the modified plant, need not hunt for a rare desirable plant, and need not spend many years backcrossing, since no undesirable genes have been added. The result will be the faster production of new varieties.

Transfer to a Totally Unrelated Plant. Classical breeding is also limited to plants that are cross-fertile. With genetic engineering it is possible to transfer genes containing desirable traits from one plant to another, totally unrelated plant. The genetic engineer can take a gene from any organism and add that gene to the chromosome of another organism. The recipient of the gene does not have to be related to the donor. Scientists are now adding bacterial genes to plants, plant genes to bacteria, and animal genes to plants.

Making Hybrids from Plants That Cannot Be Cross-Pollinated. Another well-known breeding technique is hybridizing. One can create **hybrids** by pollinating plants back to themselves and planting the seed. This inbreeding concentrates the undesirable characteristics. There is more chance that recessive ("bad") genes will pair up and express themselves, which will then affect the plant.

You save seed from the better quality plants. You inbreed still further and discard still more plants with undesirable characteristics. Usually, inbred plants lose vigor. But when you cross these inbred lines, you get a boost in vigor and better quality than you started with. You have created a hybrid.

Through the use of biotechnology, we can now make hybrids from plants that cannot be crossed using pollen. For example, tomatoes can be crossed with potatoes. Further, in a year or less, we can produce drought resistance or other qualities that would take fifteen years in classical breeding.

Producing Disease-Resistant Plants. Through genetic engineering it is now possible to produce plants that are resistant to certain diseases. Refer to

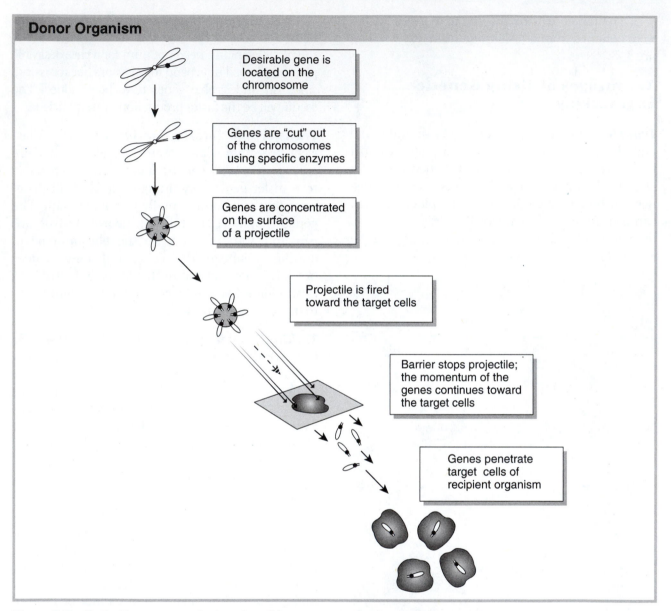

Donor Organism

Desirable gene is located on the chromosome

Genes are "cut" out of the chromosomes using specific enzymes

Genes are concentrated on the surface of a projectile

Projectile is fired toward the target cells

Barrier stops projectile; the momentum of the genes continues toward the target cells

Genes penetrate target cells of recipient organism

Figure 3–3 Desirable genes may be transferred from one organism to another.

Figure 3–4. An example would be a tobacco plant that is resistant to crown gall disease.

Producing Plants That Are Toxic to Insects But Not Humans. It is also possible to concentrate gene messages that tell a leaf to produce materials that are toxic to insects, but are not produced in the edible fruit of the plant.

Producing Crops That Are Tolerant of Herbicides. Plants are being engineered to be tolerant of certain herbicides so that environmentally safer

Engineering Virus-Resistant Plants

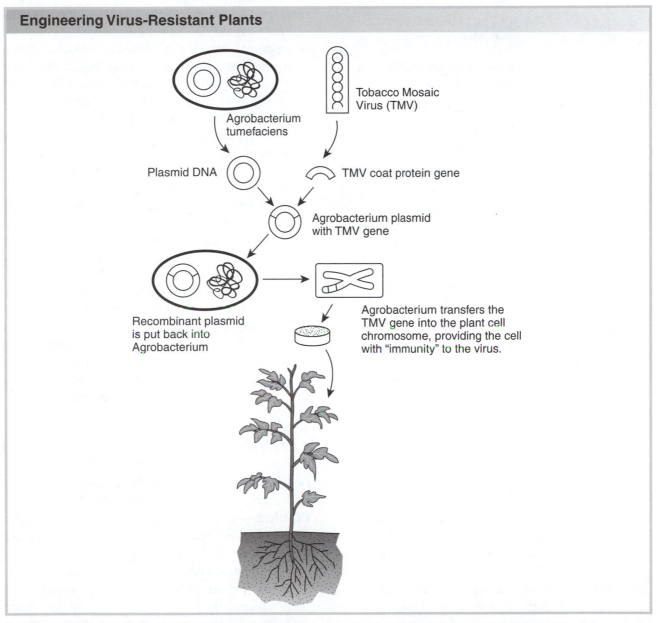

Figure 3–4   The genetically engineered cells are cultured and grown into whole plants; each cell contains the tobacco mosaic virus (TMV) gene. When the plants are reinfected with the virus, they do not contract the disease.

chemicals can be used to kill weeds and not the cultivated crop plants. This should enhance competition among agribusiness chemical companies because the winners will be those that use the her-bicides that are most desirable to crop producers, other consumers, and the regulatory agencies.[9] Refer to Figure 3–5 for examples of plants that have been genetically engineered.

Examples of Plants That Have Been Genetically Engineered

- Corn plants with a higher photosynthesis rate, in order to manufacture food more efficiently.

- Salt-tolerant barley and tomato plants.

- Strawberry plants that produce fruit all summer long, sometimes well into October.

- Black cherry trees that produce higher-quality fruit.

- Flower varieties that produce colors that were previously nonexistent for those particular varieties. A brick-red petunia flower is an example.

- Walnut trees that resist the codling moth and blackline disease. (These pests can kill walnut trees or severely affect production.)

- More perfect potatoes that chip, french fry, or bake better, resist local or regional potato diseases, and mature at the right time for states of different latitudes.

Figure 3–5 These are just a few examples of plants that have been genetically engineered.

Gene Splicing

Some of the most important and intense biotechnology research is focused on **gene mapping**. This is the process of finding and recording the location of a particular gene on a chromosome. Refer to Figure 3–6. Plant and animal cells contain millions of genes located on many different chromosomes, but only a few gene locations are known. The exact location of a gene on a chromosome must be known before the gene code can be modified.[10]

The scientist must isolate the chromosome upon which the desired gene is located and identify the locus on the chromosome where the gene is positioned. Once the location of the gene is known, research is conducted to find a **restriction enzyme** capable of cutting a particular gene out of a chromosome.[11] A scientist can cut and splice a chromosome using enzymes in much the same way that a person uses a pair of scissors to remove a blemish from an object. Enzyme preparations are very specific and can be targeted to specific genes. Using the proper enzyme, a scientist can remove a gene from its position on a chromosome and replace it with another. This process is called **gene splicing**. Refer to Figure 3–7.[12]

Gene-splicing techniques allow scientists to transfer genes within and between species to obtain desired characteristics without waiting for useful gene mutations to occur naturally. Refer to Figure 3–8 on how the process works.

Although selective breeding continues to be a valuable tool, it can now be used to propagate individuals that carry modified genetic traits. Gene splicing allows science to control and direct genetic changes in organisms in a much more orderly manner and at a much more rapid rate than was possible using selective breeding alone.[13]

CLONING

Dolly, the recently cloned sheep, caused scientific and ethical tidal waves throughout the world. **Cloning** is a process through which genetically identical individuals are produced. By replacing the nucleus of an unfertilized ovum with the nucleus from a cell of the organism being duplicated, individuals of identical genetic makeup can be produced. Each cell composing an organism contains a full component of genes. When different tissues are produced, only a portion of the available genetic information is used, and genes performing other functions are turned off. Although the cloning of Dolly was apparently successful, cloning of domestic farm animals for purposes of production is still being researched.

CHAPTER 3 – Emerging Agribusiness Technologies 🌾 45

Gene Mapping of DNA

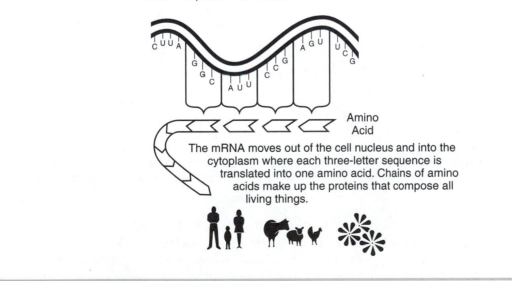

DNA is a double-stranded helix. The two strands are connected by the chemical bases A, C, G and T. A pairs with T; G pairs with C. A gene is a segment of DNA that has a specific sequence of these chemical base pairs.

Duplication—The DNA strand separates; new chemical bases attach to each single strand and two new DNA strands, identical to the original, are formed.

Protein synthesis—Special enzymes copy a single DNA strand to make a single messenger RNA (mRNA) strand in which U replaces T as a base.

Amino Acid

The mRNA moves out of the cell nucleus and into the cytoplasm where each three-letter sequence is translated into one amino acid. Chains of amino acids make up the proteins that compose all living things.

Figure 3–6 Before gene splicing, a gene must be mapped.

Gene Splicing

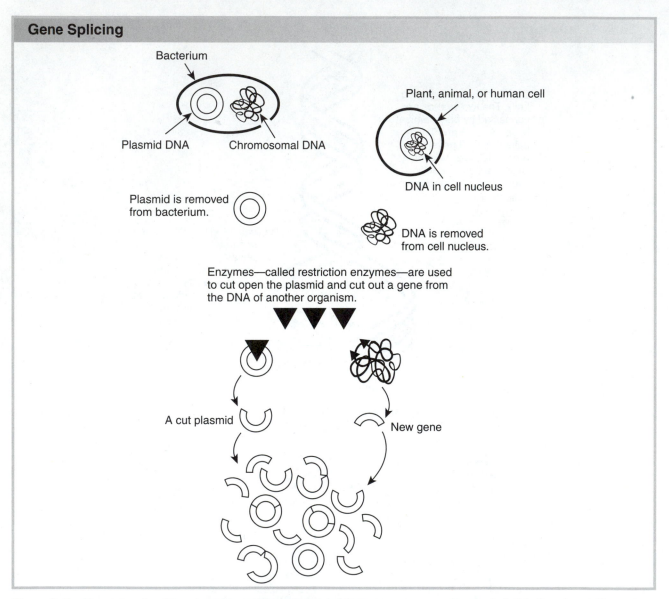

Bacterium

Plant, animal, or human cell

Plasmid DNA   Chromosomal DNA

DNA in cell nucleus

Plasmid is removed from bacterium.

DNA is removed from cell nucleus.

Enzymes—called restriction enzymes—are used to cut open the plasmid and cut out a gene from the DNA of another organism.

A cut plasmid

New gene

Figure 3–7   The cut ends of the plasmids and the cut ends of the new genes are chemically "sticky," so they will attach to each other—recombine—to form a new loop containing the inserted gene. This technique is called *gene splicing* of recombinant DNA.

Cloning Plants

Cloning may be accomplished in plants by stimulating cell division in a single plant cell to produce a complete plant. To do this, the cell must be returned to a state in which it is undifferentiated into special tissues. If the cells in the culture medium came from leaves, they must be changed by adding hormones to cause all the genes to become functional again. The cells resulting from this treatment are called **callus**. Once this is accomplished,

Plant Genetic Engineering

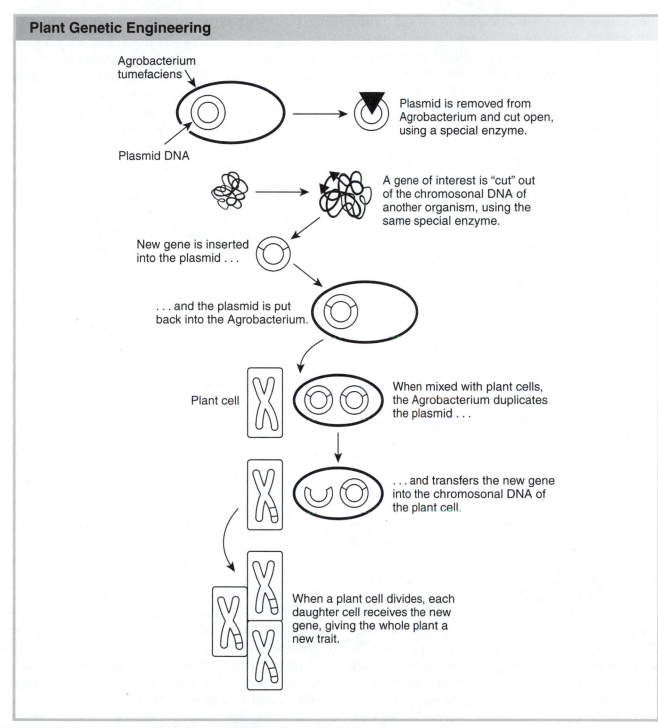

Figure 3–8   Examples of how plants are genetically engineered.

the culture medium is changed to include nutrients and hormones that cause the genes to switch on and off in a proper order to stimulate growth of roots, stems, leaves, and other plant organs.[14] This procedure is frequently referred to as **tissue culture**. (Tissue culture will be discussed later in the chapter.)

Plant breeders have used the cloning process to speed up the development and propagation of new plant varieties. When cloning is used in combination with gene splicing, it is possible to stimulate the rate of development of useful plant varieties.

Embryo Splitting

One approach to cloning of mammals has proved to be very effective, however. **Embryo splitting** is a form of cloning in that individuals produced through the use of this technology are genetically identical. It is accomplished by dividing a growing embryo using a form of microscopic surgery. An embryo in the 16- to 32-cell stage can be successfully split into two or more viable embryos. The procedure is expensive, but identical individual animals can be produced when the embryos are implanted in receptor females. The practice has been most widespread in the dairy and beef cattle industries.[15]

Embryo Transfer

Thousands of calves have been produced from split embryos since the practice of **embryo transfer** has gained acceptance. If the embryo is transferred without first splitting it, the pregnancy rate is about 65 percent. By splitting embryos prior to transferring them, technicians have been able to obtain success rates approaching one calf produced for each healthy embryo recovered. Half the split embryos produce pregnancies, but since twice as many embryos are available, the net result is about 35 percent more calves. Genetic progress is enhanced greatly when this procedure is used with superior females.[16] Refer to Figure 3–9.

Advantages of Embryo Transfers

Cloning will make it possible to make multiple, identical copies of a champion bull, prize steer, or any domestic animal. Also, USDA Agricultural Research Service scientists at Beltsville, Maryland, are a step closer to breeding cattle with natural immunity against diseases that have plagued ranchers for decades and are estimated to cost them $9 billion a year. Using split embryos implanted in surrogate mother cows, the scientists have produced five animals with identical immunity genes—possibly the foundation for a whole herd of genetically matched cattle, which could offer an unprecedented opportunity to study the immune system.

GENDER SELECTION

Gender selection is the ability to control the sex of offspring at the time of mating. Animal scientists have attempted for many years to learn to control the sex of newborn animals.

Sperm-Sexing Procedure

The only measurable difference between X (female) and Y (male) sperm is their **deoxyribonucleic acid (DNA)** content. The X chromosome is larger and contains slightly more DNA than the Y chromosome. An instrument called a *flow cytometer* can identify X and Y sperm cells after they have been treated with fluorescent dye, which stains the DNA after the sperm is passed through a laser beam. The amount of fluorescent light given off is measured by a computer. Because the X chromosome is larger and contains more DNA, the female sperm gives off more light. From this analysis, the ratio of X to Y sperm in a semen sample can be determined. Also, the X and Y sperm can be separated after identification. Refer to Figure 3–10.

The sperm cells can be encased in a droplet of liquid, given a positive or negative charge, and

How Embryo Transfer Works

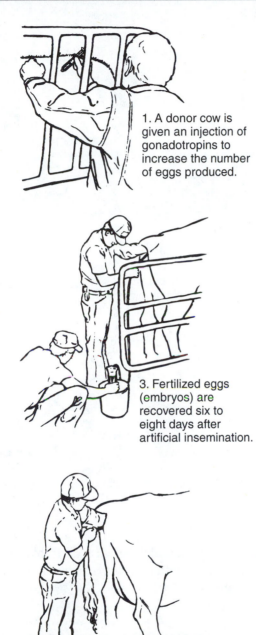

1. A donor cow is given an injection of gonadotropins to increase the number of eggs produced.

2. The donor cow is artificially inseminated five days after receiving the injection.

3. Fertilized eggs (embryos) are recovered six to eight days after artificial insemination.

4. Fertile embryos are isolated for storage in liquid nitrogen or for placement in the recipient cow.

5. Embryos are transferred to the recipient cow.

6. The recipient cow gives birth nine months after the embryo transfer. The calf has the genetic makeup of the donor cow.

Figure 3–9   Explanation of the embryo transfer procedure.

passed between two high-voltage steel deflection plates. Each plate will attract the sperm of the opposite charge, separating them into collection tables. At present, the separated sperm cannot fertilize eggs because the tails have been removed, but researchers are seeking to separate intact sperm.

Advantages of Sperm Sexing

Dairy farmers could produce female calves for herd replacement, thus reducing the time and expenses of raising bull calves for slaughter. Beef producers want mainly male cattle because they gain weight faster than females and usually command a higher price at slaughter.

Sorting Embryos According to Sex

One form of gender selection that exists in the livestock industry today is accomplished by collecting embryos from donor females, sorting them on the basis of sex, and implanting those of the desired sex. A high-quality embryo can be divided into as many as four identical embryos. This is done before they are implanted.

Modifying Gender in Fish by Ultraviolet Light

The fish industry has overcome many of the problems of gender selection by modifying the process of meiosis. Scientists have discovered that they can modify the sperm obtained from male fish using ultraviolet light to make it incapable of combining with the chromosomes found in the eggs from the female. Eggs fertilized with the treated sperm are capable of developing, however, and the result is a hatch of female fish. These females are then treated with male hormones to modify their sex characteristics to those of males. The chromosome makeup of the sperm produced by these modified males consists entirely of X chromosomes. Since the eggs of all fish contain only X chromosomes, all of the fish eggs that are fertilized will produce female fish. Refer to Figure 3–11. In the trout

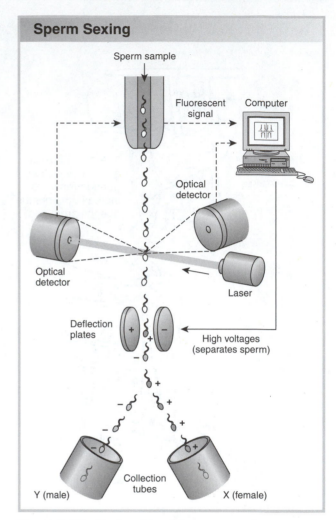

Figure 3–10 Explanation of the sperm-sexing procedure.

industry, female fish are desirable because they produce higher quality meat at market weights than male fish.[17]

PRODUCTION HORMONES

Hormones are substances that are formed in the organs of the body. They are carried by body fluids

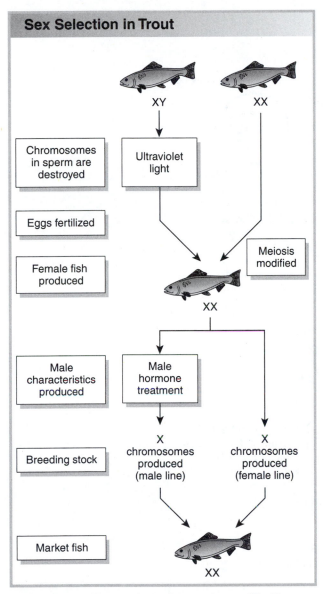

Figure 3–11 The gender in fish can be modified by ultraviolet light.

laboratory. Hormones obtained in this manner are often called synthetic hormones.

Animal Growth Hormones

Animal research scientists have learned that some kinds of hormone treatments increase the rate and efficiency of growth in meat animals. Sometimes the hormones are provided in very small amounts in the feed of animals. Other kinds of hormones are injected into the animals on a regular basis. Hormones that are injected into the tissues of an animal are absorbed rapidly. Some hormones are formed into tiny pellets and injected under the skin in the animal's ear.[18]

Bovine Somatotropin (BST). Production of milk can be increased as much as 20 to 40 percent by treating dairy cows with growth hormones. This material is also known as **bovine somatotropin (BST)**. The BST hormone treatment has been cleared for use by the U.S. Food and Drug Administration (FDA) and the USDA. BST is a protein that occurs naturally in milk. It is a hormone substance that stimulates milk production. BST can be produced in the laboratory using a fermentation process, and when it is injected into dairy cows, they become more efficient at converting their feed to milk. Milk from treated cows cannot be distinguished from the milk of untreated cows.[19] Refer to Figure 3–12.

Porcine Somatotropin (PST). The pork industry may soon have available the experimental **porcine somatotropin** (PST) to enhance muscle growth on young pigs. This growth-enhancing hormone has been shown, in recent experimentation at Auburn University, to increase slaughter weight and litter size by an average of one pig per litter.[20]

At the U.S. Experiment Station in Beltsville, Maryland, a strain of hogs has been developed

to other tissues, where they cause specific body functions to occur. Scientists have learned how to make or refine some important hormones in the

Bovine Somatotropin Production

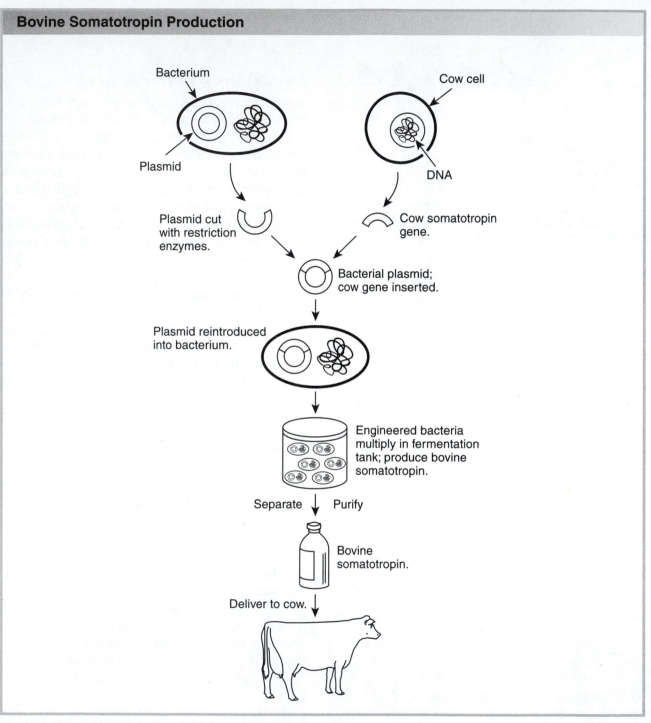

Figure 3–12 Explanation of how bovine somatotrophin is produced.

that produces much larger amounts of growth hormones than are usually present. The result of this experimental work is that the genetically engineered hogs grow much more rapidly than would be expected. They also produce less fat than normal pigs.[21]

Fish Growth Hormones. Scientists in Canada have proven that fish injected with growth hormones grow 50 percent faster than other fish. Growth hormones were developed using genetic engineering procedures. Scientists in the United States have injected a growth hormone gene obtained from trout into carp eggs, resulting in **transgenic** fish that grow 20 percent faster than other carp.

Heat Synchronization

The estrus cycles of all species of farm animals can be modified by changing the hormone balance. This is usually accomplished by injecting carefully measured amounts of hormones into the animal's body or by placing a hormone **implant** under the skin of the animal. Some hormone treatments are applied by placing a sponge **pessary** containing the hormone material in the female reproductive tract. Hormones are absorbed slowly from implants and pessaries over a period of several days before breeding occurs. The pessaries are removed prior to breeding.

One advantage that may be gained from controlling estrus in farm animals is that large numbers of animals may be bred in a short period of time. This is important, for example, when artificial insemination is practiced with cattle in the range areas of the western United States. Cows become scattered over great distances when they are turned out to the open range. By controlling estrus, the majority of the herd may be bred in a period lasting only a few days instead of the three weeks that would be required for all of the cows to exhibit estrus naturally.[22]

ANIMAL RESEARCH AND HUMAN MEDICINE

Animals have been used for scientific investigation in veterinary and human medicine for generations. There is a close relationship between the health of humans and the health of animals. Illnesses that are devastating to humans and animals are studied using animal research to determine the cause, transmission, mechanism of infection, prevention, treatment, and control of the organisms which cause these diseases. The result is a longer, healthier, and more productive life for both animals and humans.

Swine Research for Human Health

The anatomy of the pig most closely resembles the human than almost any other species. **Physiologically**, swine have organs similar to the human heart, circulatory system, and digestive tract, as well as similar enzyme production and teeth. Swine and humans often suffer from similar diseases, such as tuberculosis, influenza, peptic ulcers, brucellosis, and cardiovascular disease. Pigs have similar nutritional requirements and digest and process their food in much the same manner as humans. For these reasons, swine have become subjects for medical research that may aid in reducing human death and disease. For decades the swine have produced insulin for enzyme replacement use by diabetics.[23] Refer to Figure 3–13 to see how swine serve as a wide source of medical products.

Transplantation Technology

Animal organs may soon be routinely placed for transplantation in critically ill humans. According to statistics, 300 patients die each year awaiting an organ suitable for transplantation, while another 100,000 more die without qualifying for the transplantation list. Specialists agree that

Medical Products from Swine

Adrenal glands—produce steroids, cortisone, and epinephrine

Intestines—produces heparin, an anticoagulant

Ovaries—produces estrogen and progesterone

Stomach—produces pepsin, an intrinsic factor that aids in vitamin assimilation

Liver—produces desiccated liver to aid in iron metabolism

Pancreas—produces insulin, glucagon, and a host of digestive enzymes

Heart and Heart Valves—Hog heart valves, specially preserved and treated, are surgically implanted in humans to replace heart valves weakened by disease, injury, or congenital malformation. Since the first operation in 1971, tens of thousands of hog heart valves have been successfully implanted in human recipients of all ages.

Pituitary Gland—antidiuretic hormone (ADH), which helps regulate fluid balance; thyroid-stimulating hormone (TSH), which regulates metabolism.

Skin—for skin grafts in severe burn cases

Figure 3–13 Examples of several medical products from swine.

thousands of lives might be saved if organs available for transplantation were readily available.[24] It is believed that pigs raised without known diseases would make excellent sources of transplantable organs. The technology now exists for controlling the molecular basis of the immune system rejection that previously prevented the use of animal organs for transplantation to critically ill humans.

Xenotransplantation. Xenotransplantation is the transplantation of organs and tissues between species. It is now possible for drugs to be designed that will counteract the rejection process. Swine may also be genetically altered to produce proteins that are sufficiently alike to human proteins that they will withstand the assault of the rejection process of the human immune system. In cases where patients must have a human donor organ, the animal organ may serve as a stopgap measure, allowing the patient to buy time while waiting for an acceptable organ from a human donor. This has been studied especially with pig livers, which may be genetically altered to enhance their already close anatomical and physiological similarities.[25]

Pharmaceutical and Other Medical Products Made from Animal Products

Help for Heart Patients. Scientists are working on ways to perfect the use of genetically engineered cows that will be living factories for producing such products in their milk as TPA (a new biotech product given to heart attack victims). The blood-clotting factor used to treat hemophilia is an example of a rare protein that may someday be extracted from milk.

Help for Burn Victims. Doctors are cultivating cells from burn victims' healthy areas of skin in layers of cow protein. This new technique is showing great promise and is allowing grafts to anchor to the patients' bodies within nine days, compared to the usual four weeks to six months or a year associated with conventional techniques.[26]

Help for AIDS Patients. Cows may hold the key to a measure of relief for some AIDS patients. Because of their limited immunity, certain AIDS patients become infected with an internal parasite that also infects animals. The parasite—found in contaminated food or water—causes severe diarrhea in humans, but it does not cause AIDS. USDA Agricultural Research Service scientists have obtained an immunoregulatory chemical from the lymph nodes of cows infected with the parasite. When medical researchers from New York University gave the chemical to AIDS patients, 75 percent stopped having diarrhea without experiencing unwanted side effects. (The treatment has no effect on the AIDS itself.)

Production of Hepatitis B Vaccine. In Japan, scientists have developed a method that uses silkworms to produce a superior hepatitis B vaccine.

Help for Leprosy Patients. Approximately 4,000 people in the United States are affected every year by the disease known as leprosy. Leprosy is rarely fatal but does produce skin and nerve lesions, and when left untreated, it can cripple or severely disfigure. Scientists had trouble testing new drugs that may rid humans of leprosy because there was no other animal known to get the disease. However, a biologist in Florida doing reproductive and other studies on armadillos noticed a similarity between these animals and humans. Leprosy attacks the cooler areas of the body, such as the nose and ears. Armadillos have a normal body temperature (between 92° and 95°F), which is lower than a human's normal body temperature. Armadillos are therefore susceptible to leprosy and can be used as laboratory test animals in finding a cure for the dreaded disease.

Using Leeches to Increase Blood Circulation. Leeches are making a comeback in medical science. But they have become harder to find in the wild, which has spurred ingenious animal producers to start leech farms to answer the need for quantities of leeches. When burn patients start to reject skin transplants, doctors apply leeches to the affected area to increase the blood flow, and, in turn speed healing. Leeches also produce an anticlotting agent called hirudin. This substance is used to treat hemorrhoids, rheumatism, thrombosis, and contusions. Another enzyme produced by leeches, hyaluronidase, is used to speed the spread of injected drugs or anesthetics.

DOMESTIC ANIMALS OF THE FUTURE

Dairy Cattle

The big news in dairy cattle for the twenty-first century will be milk output—herds yielding as much as 40,000 pounds of milk per animal per year, or approximately 15 gallons of milk per day. Through nationwide genetic improvement efforts, cows will be capable of producing 70 pounds of milk more per year than the cows of today because they will be descendants of animals that produce more milk. This is compared with an annual gain of 10 to 20 pounds prior to innovations in genetic evaluation and improvement starting in the early 1960s.

Beef Cattle

In the future the speed to market size of beef cattle will change drastically. Cattle will grow 50 percent faster because producers will know more about nutritive requirements and will feed their cattle precisely to those requirements to more fully use their genetic ability for meat production.

Cloning of cattle, as discussed earlier, is certain to be part of the twenty-first-century scene, as will shipping whole "herds" of cattle into and out of the country in the form of frozen fertilized embryos. Animal geneticists predict that in the United States, producers will likely see new breeds, selected genetically to match the climate. They are conducting experiments now to see how vigorous these crosses are, and their preliminary results are very encouraging.

Agricultural scientists also expect an increase in twinning of calves. A project is underway where scientists are selecting intensely for twinning, using daughters of cows with outstanding twinning capabilities and crossing them with bulls with unusual twinning capabilities in their daughters. Presently, they have had 13 percent twins in their fall calves and 8 to 9 percent in their spring calves, whereas the norm is only 1 to 2 percent.[27]

Swine

The pigs of tomorrow will be bigger, but not fatter. With somatotropin, producers can increase the yield of muscle tissue 20 percent and simultaneously reduce the amount of body fat up to 70 percent. Currently producers slaughter pigs at 220 to 240 pounds. The carcass is about 70 percent of the pig's live weight. Of that, the fat-to-lean ratio is now two-to-one. A ratio of one-to-one is within the producer's grasp.

The ability to control fat levels will change the market weight of hogs. Producers slaughter them when they do mainly because beyond that weight, they become too fat. But if they can control the amount of fat, there is no reason to stop at 240 pounds. They could go as high as 300 pounds. The key is body composition.

Poultry

In 1950, it took 84 days to produce a 4-pound broiler with a feed conversion rate of 3.25 pounds of feed for every pound of meat produced. Today it takes 42 days, with a 1.9 feed conversion rate. In the twenty-first century, farmers should be able to produce a 4-pound broiler in 25 days. They will make these gains the same way they reduced production time from 84 to 42 days: by making the system better via genetics, disease control, nutritional control, and environmental control. But these processes will have to be synchronized.

Also, it has been predicted that breeders will probably make a dwarf breeder hen; there are some of these already. That will have no effect on the size of the chick, but it will save on feed for the hens.

Sheep

Some scientists are predicting that average lamb slaughter weight at 5 months of age will rise to 170 pounds by 2040, compared with 120 pounds

today and 80 pounds 50 years ago. Producers have gone up almost a pound a year, and there is no reason, some scientists believe, that the trend will not continue.

Catfish

Fish farmers are hoping a hormone called dihydrotestosterone (DHT) will do the trick in producing all male catfish. Their aim is to make the most of the finding that male catfish grow faster than females, because even a small advantage could mean big profits to producers in one of the nation's fastest growing agricultural endeavors. (Earlier in the chapter, we learned that female trout were preferred.)

Bees

In the future, honey producers will use specific bee species to pollinate particular plants. An example is that when using *Habropoda laboriosa*, known commonly as the southeastern blueberry bee, as opposed to the common European honey bee, the blueberry honey yields increase. This is due to the fact that blueberry bees have longer tongues and can pollinate rabbiteye blueberries. Honey bees' tongues are not long enough to do the job on this type of blueberry plant. Also, pollination by blueberry bees on rabbiteye blueberries produces more and better-quality berries earlier in the season.

Transgenic Animals

As discussed earlier, gene transfer is a method whereby scientists transfer a novel gene from one species to another. Some scientists are predicting that transgenic animals will find a place in the human food supply within the next decade. Consider the following examples:

🌿 Scientists have introduced cattle genes into pigs in an attempt to produce pork with more meat and less fat.

🌿 Biologists at the University of California at Davis have produced a "geep" by fusing a sheep embryo and a goat embryo in a test tube, and then implanting the fused embryo into a goat's womb for development of the new animal.

🌿 Some scientists are replacing traditional or classical breeding methods with gene transfer to develop new breeds of swine. Nonnative swine species are being imported into the United States to act as donors of genetic material in crossings with U.S. breeds of swine.

🌿 A rare, wild, ox-like animal called the Gaur was crossed with dairy cattle, and the resulting animals proved to be superior in meat production. Scientists in the future may use genes from exotic species of animals to improve domestic species, whether they are related genetically or not.[28]

COMPUTERIZED AND ELECTRONIC ANIMAL MANAGEMENT TECHNOLOGIES

Agricultural Computer Use

As in most areas of modern life, the computer has become an important tool for efficient and time-saving progress in agriculture. Computer software applications record milk productivity in dairy herds, and compare milk output to feed consumption. Other agricultural packages offer farm tax information, electronic tax filing, and specific bookkeeping programs. Other software applications help gardeners and landscaping professionals prepare low-maintenance and energy-saving beautification projects for rural, urban, and suburban home sites.

Computers aid the livestock industry by calculating balanced animal diets consisting of effective combinations of feeds that will provide all necessary nutrients at the least cost to the producer. Many computer programs evaluate production

records in order to aid in the selection of superior animals for breeding. Refer to Figure 3–14. The agriculture student and scientist use the computer as a storage vessel for the many records and scientific reports, which can be recalled instantly for review of the scientific literature in any given research area.

Electronic Animal Management Systems

Robotic Milking System. Electric devices have done much to eliminate the manual labor associated with livestock chores. Very few cows are milked by hand these days due to the development of electronic milking systems. Routine tasks, performed the same way time after time, are being performed in some research facilities by specially designed robots. Milking cows is a task capable of being solved by robotic devices.

One new experimental facility is developing a modern milking system that can wash, milk, and feed a cow. The computer records milk production and feed consumption for each cow. The cow enters the milking area voluntarily several times each day. The milking machine is attached to the

Figure 3–14 These agricultural biologists are analyzing data to confirm the DNA sequence for receptor genes. (Courtesy of USDA)

cow by a robotic arm, and it is automatically removed when the cow has been milked.[29]

Electronic Sensors. Electronic sensors control livestock ventilation in confinement livestock production. They may also be used for applying insecticides or delivering a cool mist of water to animals on a hot day.[30]

Robotic Sheep Shearing. The task of shearing may soon be performed by a robotic arm that first measures the sheep, calculates the movements of the shearing tool, and then removes the wool. Human labor will still be required to put the sheep in a restraining device, but the tedious labor of removing the wool will be performed by the robotic arm.[31]

Birthing Sensors. Electronic sensors implanted in an animal's body cavity can help producers detect when she is ready to give birth. This helps the producer to prevent losses due to difficulties during parturition. The same sensors have proven to be more than 90 percent accurate in detecting when cows are in heat.

Live Animal and Carcass Evaluation Devices

X-ray Scanners. X-ray scanners can provide an excellent way for evaluating meat quality while the animal is still alive. The technique used is called computerized tomography. It produces images on a television screen of segments right through the animal's body. This technique offers a complete picture of conformation without having to slaughter the animal.

Measuring Percentages of Red Tissues. A commercial company has developed a machine that can measure percentages of red tissues as the animal walks through a scanning chamber. The way the machine does this is to sense the conductivity of electricity. Red tissues conduct electricity twenty times better than white tissues. The higher the conductivity reading that is recorded, the greater the lean meat percentage. The machine can measure 240 hogs per hour at better than 98 percent accuracy.

Ultrasound Imaging. Animal scientists are borrowing from human medicine in their search for harmless ways to see inside livestock. Ultrasound imaging (sonogram) permits researchers to measure body fat and lean tissue in growing animals.

EMERGING MECHANICAL TECHNOLOGIES IN THE PLANT INDUSTRY

Many technological developments are occurring in the plant industry. Space does not permit an explanation of each, but short introductory comments are made here on several mechanical technological developments:[32]

❧ A device has been developed that helps farmers in the early detection of insect pests. When early detection is used, the farmer can use various methods to prevent costly and damaging insect infestations. The system is called "sodar" (for sound detection and ranging) and has a second system with infrared light detectors. The device looks like a rocket-shaped weather vane and is mounted on a pole. A rudder orients the device into the wind and releases pheromone downwind. Male insects pick up the scent and follow it to the echo receiver. Each echo causes one radio signal to be sent to a receiver and computer located away from the field.

A second pest detection system is a basic cone trap baited with a **pheromone** that has been modified for counting. The pheromone attracts a male, and as he realizes that this is a false alarm, he flies up into a wire mesh cone. As the insect flies up through the top of the cone, he passes an infrared light beam and again passes a radio signal to the computer receiver. The devices can detect and count the moths of several major pests.

🌿 A computerized apple has been developed that helps shippers pinpoint problem areas where major fruit damage is occurring. Apple losses after harvest, caused by damage from handling, shipping, and storing, total millions of dollars annually. Because fresh apples get higher prices than processed apples, the growers want to minimize damage. The artificial apple automatically records the bumps that real apples receive as they are moved from orchards to retail stores.

🌿 Someday in the future, crop dusters may be flying planes by remote control. The planes will only have a wingspan of 8 feet and a 4-horsepower chainsaw engine, but they will be able to fly at much lower speeds than conventional crop dusting aircraft. The lower airspeeds enable crop dusters to apply whatever chemical or natural biological control at more efficient rates. This method of crop dusting will, in many cases, also be less expensive to operate.

🌿 A high-tech microphone can tell grain operators when insects are most active inside stored grain. The microphone is part of a durable acoustic system that can detect the feeding sounds—amplified up to 75,000 times—of the lesser grain borer, rice weevil, and Angournois grain moth. This information can cut costs for crop producers and grain operators by telling them, with no need for grain samples, when insecticidal fumigants should be used to do the most good.

🌿 An underground plow originally designed to pulverize hard, thin planes of soil in dryland wheat-growing regions may become a boon to conservation tillage farmers. The inventor

is patenting a sweep plow he modified by attaching four steel shanks underneath to break up the soil. A sweep is a flat, V-shaped blade with a "wingspan" of up to 5½ feet. In operation, the blade is pulled horizontally through the soil, point first, about 3 to 4 inches below the surface.

🌿 A robot that does monotonous, backbreaking work could one day help to automate the labor-intensive raising, shipping, and transplanting of vegetable and tree seedlings into fields, nurseries, and greenhouses. Researchers have designed and filed for a patent on the first component of a robotic transplanting system that could work at least four times faster than human-dependent systems, with each row in the machine processing and planting 180 to 240 seedlings per minute, compared to a typical 40 by hand.

🌿 Melons that look and smell ripe at the supermarket are often picked too soon to be sweet. To solve this problem, Agricultural Research Service (ARS) engineers have developed a device that uses light rays to measure the sweetness of melons like honeydew, watermelon, and cantaloupe. The breadbox-size device can monitor sweetness in melons by measuring the amount of near-infrared light the fruit absorbs. The more infrared light absorbed, the sweeter the fruit.

🌿 Laser is an acronym for light amplification by stimulated emission of radiation. Lasers contain a medium that, when stimulated by energy such as electricity, produces a light that is amplified by a reflection process between their mirror surfaces. A portion of the amplified light is allowed to come out of the laser to produce a thin, straight, extremely bright beam. It is this characteristic that makes lasers suitable for many measurement applications.

In agriculture, lasers are used to alter the land to improve irrigation and in building levees and terraces. Fields can also be leveled with consistently high accuracy. Irrigation systems can be built that improve crop yield and reduce water usage at the same time. Drainage materials can be installed at a more economical rate. With proper use, lasers can be a valuable tool in soil, water, and energy conservation. They can also be a tool for increasing food production.

- Scientists are researching weather modification and analysis. The United States and other countries have been working on perfecting techniques for seeding clouds. Some progress has been made in tests in Florida, where a 50 percent increase in rainfall was obtained.

- To get more accurate information on long-term weather patterns, scientists use remote sensing technology. By analyzing atmospheric variations, they are able to predict general weather patterns.

TISSUE CULTURE

Plant propagation using tissue culture methods is a relatively new tool for plant scientists. Plants are grown from tissue obtained from the parent plant, such as buds, leaf parts, or terminal shoots. It allows many new plants, which are genetically identical to the parent stock, to be propagated rapidly. The tissue is sterilized and placed on a sterile nutrient agar jell or similar growing material along with growth-regulating hormones, and then sealed to prevent contamination. Tiny plant sprouts soon begin to grow from the plant tissue. The sprouts are carefully removed with tweezers and placed in a new medium to stimulate root growth.

Great care must be taken at every step of this procedure to maintain a sterile environment and avoid contamination of the materials. This new procedure makes it possible to produce exact duplicates of valuable plants in large numbers. Plant tissue culture is one of the most valuable plant technologies to be discovered in recent times.[33]

Special Skills Required

An understanding of general laboratory procedures is needed, as is an understanding of general cleanliness. Anyone can be trained to perform the cutting, prepare the media, and clean the lab. However, someone in the operation must be familiar with horticulture and plant physiology for the laboratory to be successful and profitable.

Advantages and Disadvantages of Tissue Culture

Advantages. The major advantage of tissue culture is that one leaf-tip cutting can produce 4,096 plants in a year. In the nursery industry, many hard-to-propagate, woody ornamental plants are produced economically through tissue culture. Another advantage is that each plant produced is a clone, an exact replica of the parent plant. This assures that the plants that are produced and sold will be true to species and culture. During the early phases of tissue culture, the only facilities required are enough room for jars of media and growing plants. It is only after the plants have been transplanted that a greenhouse for storage is needed. The plants can be transplanted into a container with media or into a hydroponic system.

Disadvantages. The largest disadvantage of tissue culture is the chance of introducing a contaminant into the laboratory, which could spread and cause the death of the plants in the lab. This would result in great monetary loss as well as the loss of time and labor.

CAREER OPTION

Career Areas: Aquaculture Research, Genetic Engineering, Hydroponics, Plant Breeding, Plant Propagation, and Tissue Culture

Aquaculture enterprises are big business in many nations of the world. Catfish farming is one of the fastest-growing food production enterprises in the United States. The rapid growth in aquaculture has spurred research and development activities. These, in turn, have stimulated career opportunities in animal science, nutrition, genetics, physiology, aquaculture construction, facility maintenance, pollution control, fish management, harvesting, marketing, and other areas.

Genetic engineering cuts across many fields of endeavor. Procedures for the genetic modification of organisms have been developing for over a decade. Biologists, microbiologists, plant breeders, and animal physiologists are some examples of specialists who might use genetic engineering in their work. Work settings include the field, laboratory, classroom, and commercial operations.

The practice of hydroponics is not new, but hydroponics for commercial production has recently captured the world's imagination. The recent popularity of hydroponics operations provides new career opportunities, especially in urban areas.

Plant breeders' objectives might include making plants faster growing, disease resistant, frost tolerant, more beautiful, or better flavored, depending on the particular uses.

Tissue culture, a procedure developed in biotechnology, permits the production of thousands of new plants identical to one, superior plant. The procedure is relatively cheap and easy and is used extensively to reproduce ornamental plants. Many jobs are available in the area of plant reproduction.

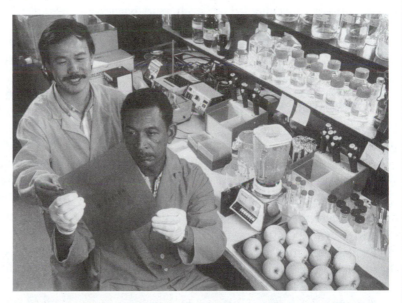

These agricultural scientists (a plant pathologist and a biological technician) are checking apples that have been genetically engineered to produce larger amounts of rot-inhibiting protein. (Courtesy of USDA)

There are many careers available in emerging agriculture technologies. Refer to the Career Option on page 61 for examples.

EMERGING PLANT TECHNOLOGIES

Frost-Protected Plants

Many fruit and vegetable plants are highly sensitive to freezing temperatures. Some plants sustain severe damage anytime freezing occurs during the growing season. Other plants are damaged only when they are flowering or when fruit is immature. The length of time the temperature is at or below freezing and the number of degrees below freezing make a major difference in the amount of damage that occurs.[34]

Biotechnology has enabled researchers to develop a special frost-free spray, that, when applied to plants, prevents them from freezing. Some bacteria help ice to form on plants at temperatures just above the freezing point.

Plant Growth Regulators

Plant growth regulators are chemical compounds that are found naturally in plants. They act as plant hormones by causing the plants to express certain genes. They play a role in regulating the growth of roots, stems, and leaves, and in controlling the development of fruits and seeds.[35] This is needed by nursery owners and flower growers to help them better market their plants for the appropriate season. For example, there is no demand for Easter lilies after Easter. Moreover, regulating the size of plants also has marketing advantages.

Salt-Tolerant Plants

Genetic engineers have been successful in transferring salt-tolerant genes into some species of useful forage crops. Using ocean water as the only source of water available to these new plants, good yields have been produced. As additional plants are developed that are tolerant to salt, crop production using ocean water for irrigation may become a reality on some desert lands.[36]

INTEGRATED PEST MANAGEMENT (IPM)

The modern concept of **Integrated Pest Management (IPM)** uses many methods to control pests, rather than depending on a single approach. Integrated Pest Management uses an array of cultural, mechanical, biological, and chemical methods to keep pest damage below economic levels. All this is done while keeping the balance of the ecosystem in mind. It involves a single, unified program whose goal is to reduce a pest population to an acceptable level and then maintain it at that level.[37] Furthermore, Integrated Pest Management is an ecosystem approach to controlling insect problems. It takes into account the effects that a particular form of insect control might have on the other living things that are found in the ecosystem.

Rationale for IPM

Two major factors have accelerated the adoption of IPM: Resistance by some consumers to the use of chemical insecticides on food, animals, and crops; and the ability of insects to develop genetic tolerances for insecticides. An effective alternative to insect control other than complete reliance on insecticides is needed.[38] A brief discussion follows illustrating the cultural, mechanical, biological, and chemical control of pests.[39]

Cultural Control of Pests

Cultural practices make the environment less favorable for the survival, growth, and reproduction of the pest. Some cultural methods that are used include frequent cultivation, adjusting dates of planting and harvest, crop rotation, appropriate choice of good seed and proven varieties, water management, and **solarization**.

Mechanical Control of Pests

Mechanical methods use machines and equipment to remove or destroy pests outright. To be effective, action must be immediate. The advantage of using mechanical methods is that there is no chemical residue. Mechanical methods (depending on the use of a field or greenhouse) include:

- 🌱 use of barriers such as screen and netting
- 🌱 artificially raising or lowering temperatures
- 🌱 direct mechanical destruction, such as the use of shredders, rollers, plows, and soil pulverizers
- 🌱 sterilizing of soil
- 🌱 **flaming**

Biological Control of Pests

In biological control, we depend on the action of parasites, predators, and **pathogens** on a **host**. This results in reducing the pest population. Although biological control does not effect an immediate reduction in pest numbers, over a long period of time it is more effective and economical than chemicals on certain plants. Traditional biological control includes:

- 🌱 introduction of species
- 🌱 conservation of parasites and predators
- 🌱 **augmentation** of parasites, predators, and pathogens

Example of Biological Control. *Interference with the natural life cycle of insects* is one means that has been shown to be effective. Sterile males of certain insect populations are introduced, which alter the reproductive success of the entire population.

For example, the screwworm fly, a livestock pest, has been controlled by the introduction of large numbers of sterile male insects. These males, which have been made sterile through irradiation, are released into the natural environment to mate with fertile females, which then lay infertile eggs.

Pheromones (naturally occurring organic compounds that act as hormones) may be introduced into the environment to alter insect behavior and reproduction.

Genetic insect resistance: Some plants are naturally resistant to insects. They may give off an odor that insects avoid. Other plants contain natural insecticides in their plant juices.

Production of natural insecticides through use of genetic engineering techniques: It is now possible to transfer the genes responsible for producing natural insecticides to plants that have no insect-resistant traits. The advantage of this technology is that only those insects attempting to eat the plant are killed. Chemical insecticides are not needed for such a crop. Pollinating insects and natural insect enemies are not subjected to insecticides, so they survive in the field to aid in controlling the damaging insect species.[40]

Pesticide Control of Pests

Pesticides (chemicals) often have a primary role in pest management. The potential dangers of pesticide chemicals to humans, food products, animals, and the environment make them the least desirable method of controlling agricultural pests. However, pesticides are often the only method of control. Pesticides provide a barrier between the plant and the attacking pest. The major benefits derived from the use of chemical pesticides include effectiveness, ease and speed of controlling pests, and reasonable costs compared to other alternatives. Some valuable crops are so susceptible to devastation by insects or pathogens that without chemical control they would simply disappear.

HYDROPONICS AND AQUACULTURE AS A SOLUTION FOR EXTENDED SPACE TRAVEL

The return to space by Senator John Glenn has aroused new interest in space travel. For extended space travel involving several people, a way to produce food in space is being researched. Scientists at the Kennedy Space Center are conducting a project called Closed Ecological Life Support System (CELSS). This project is a system that will allow plants to be grown in space. Plants are grown hydroponically in an environmentally controlled chamber called a biomass production center.

Aquaculture in Space

Scientists are researching the possibility of raising fish in space as a source of protein. Presently, the National Aeronautics and Space Administration (NASA) is experimenting with Tilapia. Research is being conducted to see if the fish can be raised using **hydroponics,** in a solution along with nutrients produced from residue. Refer to Figure 3–15.

Utilization of Plant Residue (Stalks) in Space.
By using microbiological digestion, plant residue (stalks) is broken down into a "soupy" liquid. At present, 75 percent of the cellulose can be broken down into sugar. From this stage there exist other possibilities. First, the digested sugar solution can be fermented and ethanol can be produced. Second, the solution can be consumed by fish. Grass carp are a possibility.[41]

Hydroponic Plant Growth

The types of hydroponic systems are classified according to the support medium available for the plants. A brief discussion of each follows:

Tank Culture.
The first type is called tank (or water) culture. In this type of system, the root hangs in the nutrient solution while the plant is supported on wire, line, wood, or other materials. Greenhouse tomatoes and "floatbed" tobacco plants are raised this way. Refer to Figure 3–16.

Modified Drip System.
The modified drip system is an open system in which the nutrients are not recycled. Plants are grown in a fine-textured substance. Refer to Figure 3–17. The roots of the plant support it by anchoring themselves in inert media like sand, vermiculite, perlite, or peat moss. The nutrient solution is held within the particles of the media. Today, this system is widely used in the nursery industry for propagation beds. This type of system generally has an intermittent-mist watering system, which works in conjunction with the nutrients to produce a healthy, rooted cutting.

Flood System.
The flood system often has an irrigation system installed beneath the gravel, which pumps the nutrients and water into a reservoir beneath the plant roots. The gravel holds the plants in place and has so far been the easiest system to sterilize between crops. Refer to Figure 3–18, on page 68.

Airoponics.
The airoponics system simply sprays a hydroponic mist on the plants and root system suspended in air. Disney's EPCOT Center demonstrates such a system on "The Land Pavilion." Refer to Figure 3–19 on page 69.

Differences When Using Hydroponic Systems in Space and on Earth

Unlike hydroponic systems on earth, for space farming, plants will be unable to absorb water or nutrients in the form of water droplets. Capillary action in micro-gravity will cover the surface with a film of water (nutrients). Refer to Figure 3–20. Liquids will have to be absorbed as a thin film to keep them from floating away.

Agricultural scientists, Steven Britz and Todd

Closed Hydroponics and Aquatic System

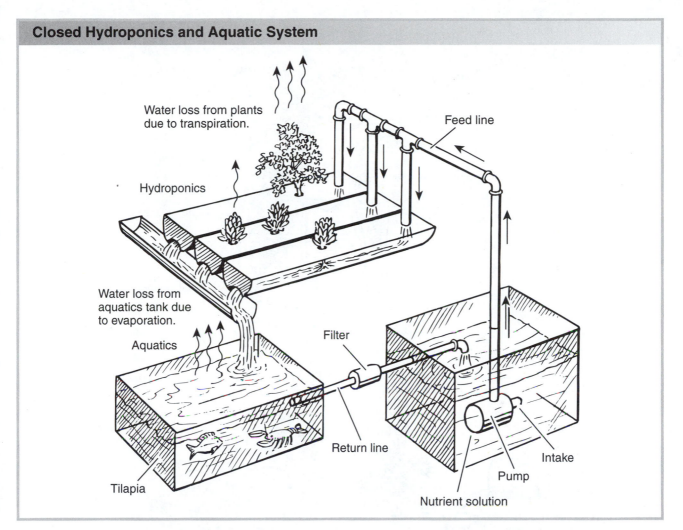

Figure 3–15 Hydroponics and aquatics in a closed system where the nutrient solution is used to feed plants, and the runoff is used to feed aquatic animals. All water and excretions are recycled to keep the system going and basically self-sustaining. The temperature and light are synthetic and controlled. (Note: The original supply of water will come from the engine fuel cell, where hydrogen and oxygen produce electricity, with the by-product being water.)

Peterson at the USDA Research Station in Beltsville, Maryland, are using two plastic pipes, fitted one inside the other. The inner tube is wettable and porous. It serves as the source from which the roots draw nutrients and as a support surface for the roots.[42]

By increasing the suction in the pipes against which the plant must extract moisture or nutrients through a porous membrane (the inner tube), the researchers will be able to subject plants to a measurable range of water stress while delivering sample nutrients.[43]

Simple Hydroponic Units

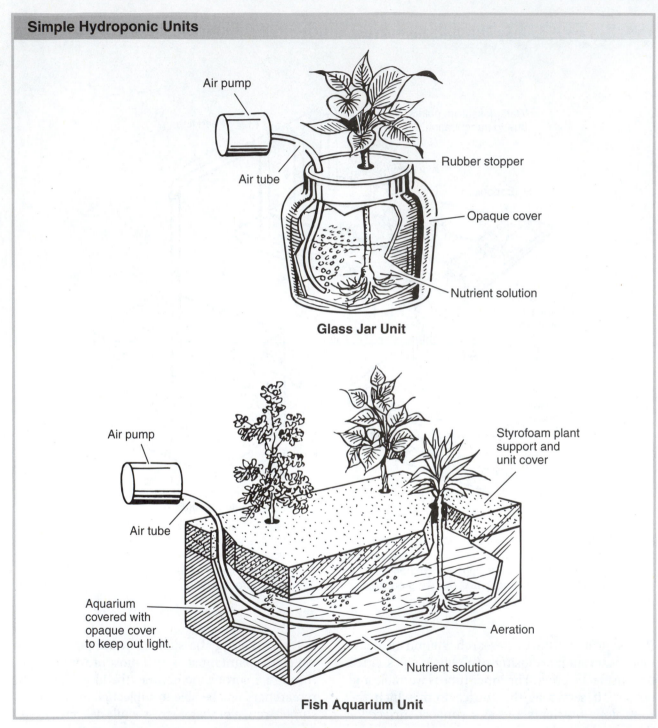

Glass Jar Unit

Fish Aquarium Unit

Figure 3–16   Two simple examples of hydroponics tank culture systems.

Modified Drip System

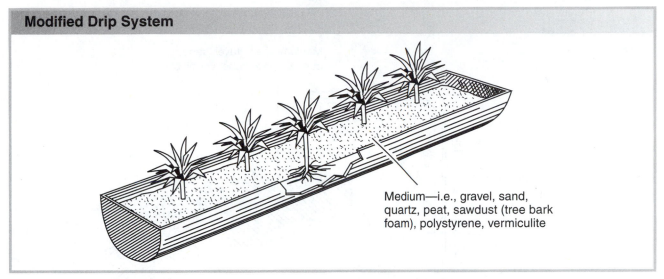

Medium—i.e., gravel, sand, quartz, peat, sawdust (tree bark foam), polystyrene, vermiculite

Figure 3–17   With the hydroponics modified drip system, an intermittent-mist watering system provides the essential nutrients for plant growth.

Thirteen Essential Elements for Growing Plants Hydroponically

Macroelements.   Essential elements are needed by all plants in predetermined quantities and must be provided by the grower. Often these required elements are called macroelements because they are needed in larger quantities than other trace elements. The plants cannot produce essential elements by itself. The essential nutritional elements required by a plant are:

✿ Nitrogen (N)—the element demanded most by a plant. Nitrogen accumulates in the tissues of young plants. Nitrogen can readily move from one area of a plant to another, from the area of older growth to younger leaves.

✿ Phosphorus (P)—Most often found in great concentration in the fruiting tissue of a plant rather than the vegetative tissue. Phosphorus can move throughout a plant, but it generally gathers in the actively growing tissue.

✿ Potassium (K)—Cell plasma and leaf tips will be the site of potassium storage.

✿ Magnesium (Mg)—The element moves from one plant tissue to another with ease.

✿ Sulfur (S)—This element is not mobile in a plant and is required in small amounts.

✿ Calcium (Ca)—Calcium is not a transferable element. It does not move from an area of older tissue to younger tissue. Saline water usually contains sufficient calcium for hydroponic growing.

Microelements (Trace Elements).   Seven micronutrients are required to produce hydroponically grown plants. Since plants can absorb some nutrients through the leaf tissue, a topical spray is often used to apply microelements to a crop. The micronutrients needed by hydroponic crops are:

✿ Iron—Iron is used to prevent chlorosis in the tissue of the plant.

✿ Boron—Boron is supplied to a plant as boric acid or borax.

✿ Manganese—This microelement helps keep tissue healthy.

Flood System

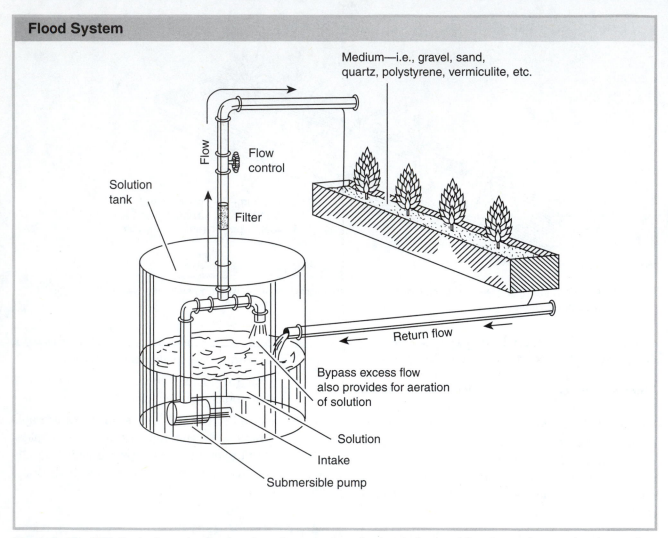

Figure 3–18 With the hydroponics flood system, the water is recirculated after providing the nutrients to the plant and returning to a holding tank.

❧ Zinc—This trace element works with light quantity and helps leaf tissue grow.

❧ Copper—Copper helps head off chlorosis.

❧ Chlorine—About 10–20 ppm is needed by plants for optimum growth.

❧ Molybdenum—This microelement is needed only in minute quantities and is sometimes identified as an impurity in the nutrient solution.

Figure 3–21 shows the ingredients that would provide enough nutrients from 1,000 gallons of water. Several agribusiness companies sell a premixed solution that can be mixed with water.

Figure 3–19 Besides growing plants hydroponically, in Disney's Epcot Center, genetic engineering is conducted in a biotechnology laboratory. (Courtesy of USDA)

Keeping the Hydroponic Solution Fully Active

Once the solution is in the tank, 1 gallon (3.8 liters) of water containing 1 ounce (28 grams) of manganous sulfate and 3 to 5 drops of concentrated sulfuric acid should be added to each 1,000 gallons (3,800 liters) of solution once a month. Four ounces (113 grams) of ferrous sulfate in 1 gallon (3.8 liters) of water should be added once a week. Oxygen in the water solution must be replaced as fast as it is taken up by the roots. Therefore, air must be continuously pumped or mixed into the solution.

Technologies Used in Hydroponically Grown Plants

Scientists have developed a computer program that can automatically control the hydroponic

Nutrient Film System (Bare Root System)

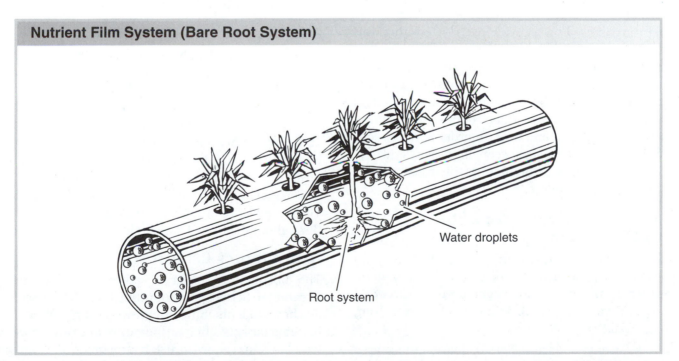

Water droplets

Root system

Figure 3–20 In space, due to a lack of substantial gravitational pull, plants will have to be grown in suction pipes to keep the hydroponics solution from floating away.

Hydroponic Mixture for 1,000 Gallons of Water

Potassium Nitrate, 5 pounds, 13 ounces
(2.6 kilograms)

Ammonium Sulfate, 1 pound (.5 kilograms)

Magnesium Sulfate, 4.5 pounds (2 kilograms)

Monocalcium Phosphate, 2.5 pounds
(1.1 kilograms)

Calcium Phosphate, 5 pounds (2.3 kilograms)

Figure 3–21 These are the ingredients necessary to provide enough nutrients for 1,000 gallons of water. Several agribusiness companies sell such a mixture.

solution. Through the use of probes and sensors, the essential elements are added as needed to get the optimum mixture. A Zybetron sterilizer is used to kill algae as the water circulates. Metal halide lighting is used on a revolving track. In this way, fewer lamps can be used and yet the same growth rate can be achieved. A simple kerosene lamp is utilized to substantially increase growth rate due to the additional CO_2 produced. Also, research is continuing in the use of fiber optics as a means of transporting light and heat. In order to create conditions similar to space (except for gravity and oxygen), scientists are conducting this research in a cave-type atmosphere.[44]

NASA Discoveries from the CELSS Project

The crops being tested in the CELSS Project are wheat, rice, soybeans, white potatoes, tomatoes, green beans, cow peas, sugar beets, lettuce, and sweet potatoes. The following are discoveries from the NASA research:

- ❧ Twenty square meters of plant surface area will produce all the requirements to keep one person alive in space indefinitely. The plants will use enough carbon dioxide (CO_2) and generate enough oxygen (O_2).

- ❧ High-pressure sodium lamps are more efficient than fluorescent lamps for growing plants in space, but they do not provide a full spectrum for reproduction.

- ❧ To better utilize space, there is only one meter between the light and plants. Therefore, dwarf varieties are selected that will not get as tall.

- ❧ Wheat is grown in plastic trays with one-quarter square meter surface area for every 400 plants. This will provide enough food for one person per day.

- ❧ By using a ⅛-inch-thin film hydroponic solution, 1 liter used continuously will grow 25 to 30 plants.

- ❧ It takes 65 to 85 days to grow a crop of wheat in space.[45]

Adoption of NASA Research

Due to the research at NASA and creativity of high school agricultural education instructors, demonstration projects combining hydroponic and aquaculture are being conducted at several high school and agricultural education programs. A program at Mountain City, Tennessee, under the direction of Harvey Burniston, Kenneth McQueen, and Thomas Boyd is combining a greenhouse, hydroponics, and aquaculture heated by geothermal energy. Their innovative work has attracted visitors from throughout the United States and foreign countries.

Hydroponics Projects

The horticulture program at Johnson County Vocational School in Mountain City, Tennessee, teaches students how to grow bedding plants, hanging baskets, chrysanthemums, and other potted plants, and how to use hydroponics. Concentration is placed on three economical crops: tomatoes, Bibb lettuce, and European cucumbers.

Tomatoes are grown using bag culture. Both rockwool and perlite bags are used to show potential vegetable producers two different types of medium. Students grow tomatoes in the aquacenter and in the traditional hydroponics greenhouse. Tomatoes are sold to local restaurants. Students grow a variety called "Trust," a tomato plant that will produce for a year or more.

Salina Bibb lettuce is grown using the hydroponic nutrient film technique (NFT) system, in which nutrient water is continually recycled. Seeds are started in rockwool cubes and transferred to NFT channels when the plants are two to three weeks old. Lettuce is grown at the aquacenter using exclusively recycled fish water. The system is designed as a symbiotic relationship between the plants and the fish. A major portion of the lettuce is used by the school system's food service program each year.

European seedless cucumbers are also grown hydroponically at the aquacenter using perlite bag culture. Three to four ounces of nutrient solution is injected into each plant once an hour during daylight hours, very similar to the tomato production techniques. Cucumber crops last 90 to 120 days, resulting in three to four crops a year.

All traditional crops of bedding plants and hanging baskets at the aquacenter are watered with water from the fish tanks using drip irrigation and sprinklers.

Geothermal Heating Systems. The Alternative Farming Center (Aquacenter) utilizes an innovative "geothermal" heat pump system to both heat and cool the structure and water. Refer to Figure 3–22. This system is designed on geothermal or ground source heating and cooling principles, based on the fact that, at a depth of 5 feet or more, the earth maintains a constant temperature of 53–54°F in this area. This principle is used to equalize the temperature of the system by pumping water through a continuous pipe in the ground

called the ground loop. In the system used at the aquacenter, a total of 2 miles of continuous pipe is run through fifteen vertical wells at a depth of 250 to 300 feet. The water is pumped through the pipe, where it equalizes to 54 degrees, then returned to the heat pumps.

Aquaculture Projects

Aquaculture is a growing agricultural industry in the United States. The Johnson County agricultural education program in Mountain City, Tennessee, is studying the feasibility of aquaculture as an alternative crop for the region and has designed a system to raise fish as well as hydroponic plants, traditional bedding plants, and hanging baskets.

The aquaculture system, which is designed for fish, consists of four tanks 72 by 8 by 5 feet and one tank 50 by 6 by 5 feet. The system holds 90,000 gallons of water, which is recirculated through bead filters for solid removal and some bacterial exchange nine times per day.

Bio Towers are used to help degasify the water.

Figure 3–22 The agricultural education program at Mountain City, Tennessee, combines hydroponics and aquaculture in a geothermally heated and cooled greenhouse. (Courtesy of Harvey Burniston)

The water is pulled across the tanks from the bottom and sent through filters by pumps; gravity then causes it to flow back to the raceways at six intervals along the sides. The system can handle anywhere from 200 to 400 pounds of feed per day.

Air is added to the water through air stones placed 24 to 32 inches deep in the water at 1-foot intervals throughout the tanks. Liquid oxygen is added through saturators after the water has circulated through the heat exchanger. The water, which is heated by geothermal heat pumps, is sent through a heat exchanger and returned to the raceways by 1-inch pipes located in the bottom of the tanks.

Currently, Tilapia and Koi are being studied. Presently, they seem to work well together, as one is a food fish and the other, an ornamental fish. Tilapia are also top feeders whereas Koi are bottom feeders. Tilapia have been cultivated for years all over the world. The fish were first believed to have been grown and eaten by the Egyptian pharaohs over 4,500 years ago. Presently, they have a strong market. Koi has also been recently studied as an alternative type of fish. With a growing market for indoor and outdoor water gardens, a market for the colorful fish is also growing.

CONCLUSION

Although most of the emerging technologies discussed in this chapter are applied agriscience technologies, there must be a way to get these technologies to the consumer. Research completed that is not then applied is research lost. This is where the agribusiness input or agribusiness output sectors, which sells or markets these technologies, enter the picture. Many of the companies that market these emerging technologies will be discussed in later chapters.

SUMMARY

Agricultural technologies are applied sciences. However, once they have been developed, it is the agribusiness input or agribusiness output sector that markets these emerging technologies. The agricultural industry is moving at such a rapid rate that it is difficult, if not impossible, to label any process or procedure as "new" or "futuristic." In the area of technology alone, there will be more new developments in the next fifteen years than in all of history.

The Global Positioning System (GPS) is an experimental technology that uses satellites to view specific fields and crops. The view from space can provide a computerized picture that will aid field crop management and increase yield per acre. Because of GPS, precision farming is now a reality. In brief, the farm field and crop are referenced as a computer model of each specific section of a crop producer's field. Then, each section is individually assessed for increased production, decreased cost, and environmentally sound techniques.

Genetic engineering bypasses selective breeding and natural selection by inserting the desired gene in the chromosomes of a living organism. The target cells receive the new genes that were obtained from another organism. With gene splicing, a scientist can cut and splice a chromosome using enzymes in much the same way that a person uses a pair of scissors to remove a blemish from an object.

Cloning is a process through which genetically identical individuals are produced. By replacing the nucleus of an unfertilized ovum with the nucleus from a cell of the organism being duplicated, individuals of identical genetic makeup can be produced. Embryo splitting and embryo transfer are parts of the cloning process. Cloning may be accomplished in plants by stimulating cell division of a single plant cell to produce a complete plant.

Gender selection involves the ability to control the sex at the time of mating. The only measurable difference between X (female) and Y (male) sperm is their DNA content. Instruments called *flow cytometers* can identify X and Y sperm cells after they have been treated with fluorescent dye. After identification, the X and Y sperm can be separated after being given a positive or negative charge and passed between two high-voltage, steel deflection plates. Each plate will attract the sperm of the opposite charge, separating them into collection tubes. Also, with cattle, embryos can be collected from donor females, selected by sex, and implanted into the recipient cow. With fish, ultraviolet light can determine the sex.

Hormones are substances that are formed in the organs of the body. Some hormones are formed into tiny pellets and injected under the skin in the ear of an animal. This increases the rate and efficiency of growth in meat animals. Production of milk can be increased by treating dairy cows with a growth hormone, bovine somatotropin (BST). The pork industry is experimenting with porcine somatotropin (PST) to enhance the muscle growth of young pigs. Heat synchronization is the result of hormone injection.

Animals have been used for scientific investigation in veterinary and human medicine for generations. There is a close relationship between the health of humans and the health of animals. Illnesses that are devastating to humans and animals are studied via animal research to determine the cause, transmission, mechanism of infection, prevention, treatment, and control of the organisms that cause these diseases. Many pharmaceutical and other medical products are made from animal products.

Domestic animal husbandry of the future will be much more efficient. Dairy cows will produce as much as 40,000 pounds of milk per year. Beef cattle will grow 50 percent faster. Swine will be bigger, but not fatter. Poultry broilers will reach 4 pounds in twenty-five days. Transgenic animals will become more common.

As in most areas of modern life, the computer has become an important tool for efficient and time-saving operations in agriculture. Computer software applications record milk productivity in dairy herds; offer farm tax information, electronic filing, and specific bookkeeping programs; help gardeners and landscaping professionals; and aid the livestock industry by calculating balanced animal diets. Electronic animal management systems include robotic milking systems, electronic sensors, robotic sheep shearing, birthing sensors, and ultrasound imaging.

Many mechanical technologies are occurring in the plant industry. Some examples include a device for early detection of insect pests, a computerized apple to pinpoint shipping damage, remote-control crop dusters, development of an underground plow, lasers to alter the land, and many others.

Plant propagation using tissue culture is a relatively new tool for plant scientists. Plants are grown from tissue obtained from the parent plant such as buds, leaf parts, or terminal shoots. It allows many new plants that are genetically identical to the parent stock to be propagated rapidly.

Many technologies are emerging with plants. Biotechnology has enabled researchers to develop a special frost-free spray that can be applied to plants to prevent them from freezing. Plant growth regulators regulate the growth of roots, stems, and leaves, which helps nursery owners and flower growers market their crops at the appropriate season. Also, scientists are researching plants that are tolerant to ocean water.

Integrated Pest Management uses an array of cultural, mechanical, biological and chemical methods to keep pest damage to acceptable levels. All of this is being done while keeping the balance of our ecosystem in mind. It involves a single, unified program whose goal is to reduce a pest population to an acceptable level and then maintain it at that level.

For extended space travel, scientists are researching hydroponics and aquaculture as solutions. The hydroponic medium, consisting of water and the essential nutrients for plant growth, will be used to grow wheat, rice, soybeans, and several vegetables. Also, scientists are researching the possibility of raising fish in space to see if the fish can be raised in the hydroponic solution. Further, scientists want to know if the fish can consume crop residue, such as stalks or stems from wheat plants.

Although most of the emerging technologies discussed in this chapter are applied agriscience technologies, someone also must get these technologies to the consumer. This is where agribusiness enters the picture.

END-OF-CHAPTER ACTIVITIES

Review Questions

1. Define the Terms to Know.
2. Explain the purpose of global positioning and field management.
3. What are four components of the Global Positioning System?
4. What are the five major objectives of comprehensive precision farming?
5. List five advantages of genetic engineering.
6. Briefly explain the gene-splicing process.
7. Briefly explain the cloning process.
8. Briefly explain embryo splitting.
9. Briefly explain embryo transfer.
10. List two advantages of embryo transfers.
11. Briefly explain the sperm-sexing procedure.
12. List two advantages of sperm sexing.
13. List six reasons why illnesses that are devastating to both humans and animals are studied using animal research.
14. Give twelve examples of how swine are physiologically similar to humans.
15. List six pharmaceutical and other medical products made from animal products.
16. Name one characteristic or technological development in the future of each of the following domestic animals: (a) dairy cattle, (b) beef cattle, (c) swine, (d) poultry, (e) sheep, (f) catfish, and (g) bees.
17. List four examples of research resulting in transgenic animals.
18. List eight examples of computer use in the agricultural industry.

19. Briefly explain four electronic animal management systems.
20. Briefly explain three technological live animal and carcass evaluation devices.
21. Briefly explain nine emerging mechanical technologies in the plant industry.
22. Briefly explain the basic procedure for tissue culture.
23. List two advantages of tissue culture.
24. Briefly explain three emerging plant technologies.
25. What two major factors have accelerated the adoption of Integrated Pest Management (IPM)?
26. List five examples of mechanical control of plants.
27. List three traditional biological controls.
28. Briefly discuss two examples of biological controls.
29. List three major benefits derived from the use of chemical pesticides.
30. Explain how fish could possibly be raised in space.
31. Name and briefly discuss four types of hydroponic systems.
32. What is the main difference between using hydroponic systems in space and on earth?
33. List six essential elements (macro-elements) in a hydroponic solution and give one purpose of each.
34. List seven trace elements (micro-elements) required to produce hydroponically grown plants.
35. List five technologies used in hydroponically grown plants.
36. List six discoveries made by NASA from the CELSS project.

Fill in the Blank

1. Although agricultural technologies are applied science, it is _____ that will sell or market these technologies.
2. _____ _____ _____ is an experimental technology that uses satellites to view specific fields and crops.
3. Scientists have discovered that they can modify the gender in fish by _____.
4. The anatomy of the _____ most closely resembles that of the human than almost any other species.
5. _____ _____ may soon be routinely placed for transplantation in critically ill humans.
6. The largest disadvantage of tissue culture is the chance of introducing a _____ into the laboratory.
7. Integrated Pest Management uses an array of _____, _____, _____, and _____ methods to keep pest damage below economic levels.
8. _____ is a NASA project researching a system that will allow plants to be grown in space.
9. Due to the research at NASA and elsewhere, demonstration projects combining hydroponics and aquaculture are being conducted in several high school _____ _____ programs.

10. _____ and _____ are two agricultural practices being researched to be used in a combined effort in space.

Matching

a. porcine somatotropin

b. ultrasound imaging

c. micronutrients

d. heat synchronization

e. pesticide control of pests

f. animal growth hormones

g. mechanical control of pests

h. biological control of pest

i. cultural control of pest

j. bovine somatotropin

k. macroelements

_____ 1. treatments that increase the rate and efficiency of growth in meat animals

_____ 2. increases the production of milk by as much as 20 to 40 percent

_____ 3. enhances muscle growth on young pigs

_____ 4. modifying the estrus cycle of farm animals by changing the hormone balance

_____ 5. permits researchers to measure body fat and lean tissue in growing animals

_____ 6. practices that make the environment less favorable for the survival, growth, and reproduction of the pest

_____ 7. uses machines and equipment to remove or destroy pest outright

_____ 8. depends on the action of parasites, predators, or pathogens on a host

_____ 9. chemical used on plants

_____ 10. essential nutrients

_____ 11. trace elements

Activities

1. Research current periodicals (magazines, newspapers, Internet, professional journals, etc.) to determine current economical trends concerning the effects of biotechnology. This could include concentration on the use of BST and/or any other product. State your opinion as to whether you think the biotechnology product should be used, taking into consideration the current status of the economy. Support your opinion with facts, and cite all sources used. Present your findings to the class.

2. Biotechnology has given production agriculturalists new crops that are herbicide resistant. The crop producer can then apply herbicides that normally kill a broad range of plants without killing the herbicide-resistant crop. The benefit is that production agriculturalists can now grow these crops in areas where it was previously impossible due to high weed infestations. However, some people are predicting that herbicide-resistant crops will backfire on the crop producers. It has been

proposed that the natural evolutionary process will produce weeds that will be resistant to the herbicides. This could then cause the crop producer to need even more herbicide.

New data is being published constantly that could support either side of this debate. Research biotechnology using articles in periodical publications or the Internet. Also use library sources such as magazines and newspapers. After analyzing the data you collected, decide how you think the possible problems can be solved. Give evidence and bibliographical information for supporting your opinion.

3. Genetically engineered organisms are being released into the environment to fight agricultural problems such as frost damage on strawberries, insect and weed pests, and plant diseases. Organisms can mutate and exhibit new traits, which may be beneficial or detrimental. Some people believe that engineered organisms could mutate after being released into the environment and thereby become a new agricultural or environmental pest. Research the theory of the causes of mutations. Biology textbooks covering genetics are a good source. If you were a lawyer protesting the release of engineered organisms into the environment, could you cite cases where problems were caused because of this practice? Research this information using library sources such as magazines and newspapers. List bibliographical information to support your case.

4. Select a prominent biotechnology scientist and study the individual's education, work, and contribution to society. Examples include: Barbara McClintock and Paul Mangeledorf (corn), Norman E. Borlaug (wheat), and Steven F. Lindow (frost-free bacteria). Use library sources, and list all bibliographical information for the sources used. Write a report on the biotechnology scientist of your choice and present it to the class.

5. Write a two-page report on a current article concerning a hydroponic system. Present your report to the class.

6. Design a plan of an ideal tissue culture laboratory. Consider using "off-the-shelf items" or salvage parts. Present your findings to the class.

7. Design a hydroponic demonstration operation. Along with the help of three to five classmates, construct the hydroponic demonstration unit. Consider using the following materials: tank, nutrients, water pump and/or aerator, sheet of Styrofoam, litmus paper, plants.

8. Space within this chapter did not permit the inclusion of all the emerging technologies. Research and present a report to the class on a new or emerging technology not discussed in this chapter.

9. Approximately fourteen different categories of technologies were presented in this chapter. Identify an agribusiness in your community that exists because of one of these technologies. Write a brief report and present it to the class.

10. Select and write a brief report on new technology in either the agribusiness input or agribusiness output sector. Present your report to the class.

NOTES

1. "Satellite System Can Control Tractors," *USA Today*, December 8, 1997, vol. 126, no. 2631, p. 11.
2. "Applications and Products of GPS in Agriculture"; available from http://www/age/uluc.edu/age221/project/1996/team3/index.html); Internet.
3. "Hitting the Spot," *Economist*, January 27, 1996, vol. 338, no. 7950, p. 74.
4. Stephen A. Wolf and Spencer D. Wood, "Precision Farming: Environmental Legitimization, Commodification of Information, and Industrial Coordination," *Rural Sociology*, 62, no. 2 (summer 1997) pp. 180–206.
5. Michael D. Weiss, "Precision Farming and Spatial Economic Analysis: Research Challenges and Opportunities" (Paper presented at the annual meeting of the American Agricultural Economics Association, 28–31 July 1996, San Antonio, Texas).
6. Jess Lowenberg-DeBoer, "Precision Farming and the New Information Technology: Implications for Farm Management, Policy, and Research: Discussion" (Paper presented at the annual meeting of the American Agricultural Economics Association, 28–31 July 1996, San Antonio, Texas).
7. Gary T. Roberson, "Precision Agriculture: A Comprehensive Approach" (North Carolina State University, College of Agriculture and Life Science, Department of Biological and Agricultural Engineering); available from http://www.bae.ncsu.edu/programs/extension/agmachine/precision/; Internet.
8. L. DeVere Burton, *Agriscience and Technology* (Albany, N.Y.: Delmar Publishers, 1998), p. 30.
9. Cliff Ricketts, *Science IA (Agriscience)* (Nashville, Tenn.: Division of Vocational and Technical Education, 1994).
10. Burton, *Agriscience and Technology*, p. 33.
11. Ibid.
12. Ibid., p. 34.
13. Ibid., p. 37.
14. Ibid.
15. Ibid., p. 38.
16. Ibid., p. 75.
17. Ibid., p. 76.
18. Ibid., pp. 52–53.
19. Ibid., p. 53.
20. "Injections Increase Muscle," *Hogs Today*, June 1997, p. 40.
21. Burton, *Agriscience and Technology*, p. 74.
22. Ibid., p. 73.
23. "In Praise of Research Animals," chap. 22 in *The Science of Animals That Serve Humanity: Animal Research in Retrospect and Prospect*, New York: McGraw-Hill, 1985, pp. 730–741.
24. Fisher, Lawrence M., "Down on the Farm, a Donor: Genetically Altered Pigs Bred for Organ Transplants," *New York Times*, January 5, 1996, p. C1.
25. Insulin-Free World Foundation, "Immunology: The Essential Ingredient"; available from http://www/insulin-free.org/research/platt.htm; Internet.
26. Ricketts, *Science IA (Agriscience)*, pp. 20–14.
27. Ibid., AgS, 20–17.
28. Ibid., AgS, 20–65.
29. Burton, *Agriscience and Technology*, p. 55.
30. Ibid.
31. Ibid.
32. Ricketts, *Science IA (Agriscience)*, AgS 26 (44–48).
33. Burton, *Agriscience and Technology*, p. 114.
34. Ibid., pp. 84–85.
35. Ibid., p. 92.
36. Ibid., p. 94.
37. Ricketts, *Science IA (Agriscience)*, AgS 25–19.
38. Burton, *Agriscience and Technology*, p. 89.
39. Ricketts, *Science IA (Agriscience)*, AgS 25 (20–23).
40. Burton, *Agriscience and Technology*, p. 90.
41. Ricketts, *Science IA (Agriscience)*, AgS 27–16.
42. Ibid., AgS 27–6.
43. Ibid., AgS 27–13.
44. Ibid., AgS 27–17.
45. Ibid., AgS 27–16.

UNIT

2

Starting and Running an Agribusiness

Planning and Organizing an Agribusiness

OBJECTIVES

After completing this chapter, the student should be able to:

- explain the importance of small businesses
- examine whether entrepreneurship is for you
- describe the challenges of entrepreneurship
- describe why agribusinesses fail
- analyze your potential agribusiness venture
- prepare a business plan
- establish goals for an agribusiness

TERMS TO KNOW

business cycles
business plan
business survey
capital
capital intensive
capitalism
collateral

diligence
entrepreneur
financial institutions
financial resources
free enterprise
investors
niches

rapport
rationale
small business
Small Business Administration
  (SBA)
undercapitalization
venture capital

INTRODUCTION

Part of the American dream is that everyone who wants to should own his own business and be his own boss. It is still possible for a person to start with very little **capital** and build a successful business, but it is a very difficult process and becoming more difficult each year. Even though most agribusiness students will never own their own businesses, it is important to understand the basic processes involved in business planning and organizing.

As we enter the world of agribusiness, we must decide to either work for ourselves or work for someone else. We are usually motivated by our needs, preferences, and desires. The style of life we want to live; our abilities, training, experience; and our financial circumstances lead us to set goals in order to satisfy our internal drives.[1] The importance that we place on our needs for security, self-expression, recognition from others, and **rapport** with others all affect our decisions relating to self-employment or working for others.[2] The student shown in Figure 4–1 is assessing whether to start an agribusiness or work for someone else.

Figure 4–1 Starting an agribusiness involves much planning and decision making. This student is accessing information for deciding whether to start a business or work for someone else. (Courtesy of Cliff Ricketts)

IMPORTANCE OF SMALL BUSINESSES

Most all agribusinesses are included in business and economic terminology as small businesses. There is no single definition of **small business**, but the U.S. **Small Business Administration (SBA)** defines it as a business that is independently operated, is not dominant in its field, and meets certain size standards in terms of number of employees and annual receipts.

Small Businesses Are Not Really Small

When it comes to jobs, small businesses are not small in the number of people they employ. In the 1980s, large companies lost 4.1 million jobs; small businesses created 1.5 million jobs in 1992 and 1993 alone. Ninety percent of the nation's new jobs in the private sector are in small businesses. Two-thirds of the new jobs are in companies with fewer than 25 employees.[3] That means that there is a very good chance that you will either work in a small business someday or start one. The following are some interesting statistics illustrating the importance of small businesses:

- There are about 25.4 million full- and part-time home-based businesses in the United States.[4]

- Of all nonfarm businesses in the United States, almost 97 percent are considered small by SBA standards.

- Small businesses account for over 40 percent of gross domestic product (GDP).

- The total number of employees who work in small business is greater than the populations of Australia and Canada combined.

- The first jobs of about 80 percent of all Americans are in small business.

- The number of women owning small businesses has increased eleven-fold since 1960, and studies predict that women will own half of the small businesses by 2000.

- The number of minority-owned businesses increased more than 64 percent between 1962 and 1987 (the latest date for government figures).[5]

- Small companies produced 90 percent of the new jobs in 1994.[6]

As you can see, small businesses are really a big part of the U.S. economy. The reason why small businesses, including small agribusinesses, will always have a market can be explained by looking at the differences between rocks (big businesses) and sand (small businesses). If you fill a hole with sand, there are no spaces between the grains of sand. However, if you fill it with big rocks, there are many empty spaces between them. That is how it is in business. Big business cannot serve all the needs of the consumer. Therefore, there is plenty of room for small businesses to make a profit filling those **niches**.[7]

ENTREPRENEUR IN AGRIBUSINESS—IS IT FOR ME?

The American economy is based on the **free enterprise** system, or **capitalism**. This simply means that we have the right to own our own business and to make a profit. It also means that we can lose money, perhaps even all the money that was invested. There is a big risk factor for those who start their own business. Therefore, entrepreneurship may not be for everyone.

General Characteristics of Entrepreneurs

Entrepreneurs are people who have the initial vision, **diligence**, and persistence to follow through. An **entrepreneur** is a person who accepts all the risks pertaining to forming and operating a small business. This also entails performing all business functions associated with a product or service and includes social responsibility and legal requirements.[8] There are many reasons why people are willing to take the risks of starting a business:

- Entrepreneurs work for themselves, are independent, and make their own business decisions.

- Whatever income they earn above their financial obligations is theirs to keep.

- They can test their own theories and ideas on how to run their business.

- They set their own working hours.

- They themselves set prices, determine production levels, and control inventory according to the market.

- They determine the product or service offered and control its quality as well as the overall reputation of the business.

🌾 They solve the problems.

🌾 They perform all the human resource functions such as hiring, training, and firing.

🌾 They set the company policy.[9]

Not everyone satisfies the qualifications to be an entrepreneur. It takes a lot of hard work, and there may be little job security. Many small businesses fail each year for a number of reasons, as will be discussed later. You should thoroughly investigate all the aspects of starting your own business. Make sure it is what you really want and can actually handle. Refer to other characteristics of the entrepreneur in the Career Option section.

Personal Characteristics of Entrepreneurs

Every person is different and unique. This includes entrepreneurs. However, entrepreneurs generally have common qualities that set them apart. The skills needed by entrepreneurs vary widely, depending on the type and nature of the business. One person may possess technical knowledge about an agribusiness but lack leadership and management skills. A good guideline is that an entrepreneur should have the following qualities, as well as knowledge of or expertise in the particular agribusiness that he or she is pursuing. The characteristics listed here are found most often in successful entrepreneurs.

Independent. Entrepreneurs believe that they can do the job better than anyone. They prefer to control their own destiny.

Self-confident. People who believe in themselves and what they can do have a definite edge in self-employment.

Energetic. Entrepreneurs usually are sick less often than other people. They enjoy good health and use it to improve their business.

Organized. These people are able to organize their work according to their own, unique system.

Vision. Entrepreneurs can keep tabs on the whole business and see how the different parts of the business fit together. They are in command of the entire business operations.

Persistent. They are able to keep their business moving forward even when tough times are prevailing. If they do encounter a roadblock, they will find another way.

Optimism. No matter what the situation, the outlook is always good for the entrepreneur. Such people are natural optimists.

Committed. They accomplish what is necessary to carry out their ideas.

Problem Solvers. Entrepreneurs willingly take on new challenges. They capitalize on opportunities to make use of their time, talents, and ideas.

Self-nurturing. They have little concern for what others think about them. Entrepreneurs have a conviction that they are on the right track regardless of what others may think.

Risk Takers. They are willing to give up things like a steady job in order to achieve their personal and business goals. They are not gamblers; rather, they take risks based on the confidence in their own abilities.[10]

Action Oriented. Great business ideas are not enough. The most important thing is a burning desire to turn dreams into reality.

A Sense of Urgency. Entrepreneurs are restless unless they are working, and they may often seem to want challenges if everything is going too smoothly.

Flexibility. If one option falls through, they will find another way to do the job or finance the change.

CAREER OPTION

Career Area: Entrepreneur

The entrepreneur is the person who organizes a business or trade or improves an idea. Entrepreneurship is the process of planning and organizing a small business venture. It is the entrepreneur who visualizes the business strategy and is willing to take the risk of getting the business started.

Agribusiness entrepreneur opportunities cut across the food and nonfood spectrum of agribusiness. This includes the agribusiness input sector and the agribusiness output sector. Entrepreneur opportunities exist in every sector where goods are bought, sold, or produced.

Many high school and college students become well established as entrepreneurs before they finish their educations. Some popular agribusiness entrepreneurships include lawn services, lumber businesses, greenhouse or nursery operations, machinery repair, agricultural supplies, home and garden centers, florist shops, and retail flower sales.

Entrepreneurs may sell services such as animal care, crop spraying, or recreational fishing privileges. Careful planning is very important before starting an agribusiness because many new businesses end in failure. Careful planning increases the chances of success.

Preparation for careers in agribusiness management includes both formal education and on-the-job training. High school agribusiness and agriscience programs provide excellent training for business. Such programs should include classroom and laboratory training, supervised agriscience experience, and leadership development. Advanced agribusiness programs at technical schools, colleges, and universities offer appropriate training in economics, finance, and management.

The entrepreneur is constantly seeking ways to provide better products, to improve profits, and to meet the needs of consumers. (Courtesy of USDA)

Emotionally Stable Entrepreneurs are not given to emotional highs and lows as their moods change.[11]

CHALLENGES OF ENTREPRENEURSHIP

Being successful is not easy. You do not simply start a business and enjoy the profits. Three major challenges will have to be addressed by entrepreneurs.

Total Responsibility

A beginning entrepreneur is in charge of everything. The agribusiness's success or failure depends on just one person—the owner. Entrepreneurs must manage workers, manufacturing, and shipping. They have to find customers, sell the product, and be certain that orders are met. No matter what size the business may be, the owner always has total responsibility. Refer to Figure 4–2.

Long, Irregular Hours

Being your own boss requires much work. People who start their own business work more hours than those who work for someone else. It is not uncommon for an entrepreneur to work more than sixty hours per week. Weekends are often spent working in the business.

Financial Risks

The most serious disadvantage of a small business is the need for money. Obviously, it takes money to get a business going. While the business is getting started, the entrepreneur has to pay bills and wages, which will likely be more than income. The owner will probably try to borrow money, but lending institutions are often reluctant to lend to a new business because of the high risk.

The chances for a business reaching its fourth birthday are only about 50-50. The chances are

Figure 4–2 Although this student is attending a national FFA Convention, during a break she is attending to her agribusiness back home, for which, as the owner, she has total responsibility.

that about 40 percent will fail in the wholesale business or in manufacturing and over 50 percent will fail in retail trade during the first four-year period.[12]

WHY AGRIBUSINESSES FAIL

There are many potential reasons why agribusinesses fail. Usually failure will fall under the umbrella of either management, labor, or **financial resources**. Many people go into business or expand without adequate planning, and without analyzing the added cost associated with the additional returns or the potential risk. Too many agribusinesses have poor record-keeping and inadequate management information systems, which results in their inability to control cost, spot and correct problems, and recognize profit opportunities. Agribusiness failures usually display one, several, or many of the characteristics that follow within the categories of management, labor, or financial resources.

Management

People do not plan to fail, they only fail to plan. Once income and expenses have been projected in the planning process and a profit is not foreseen, there will probably be no profit. Therefore, stop the plans for your business. However, even if the projections show a profit, many factors can still cause a business to fail. The point is that: if the business will not work "on paper," it will not work in the real world; even if the business works on paper, there is still no guarantee it will succeed. Avoid the following pitfalls as you plan and manage your business:

- 🌾 plunging in without first testing the waters on a small scale
- 🌾 buying too much on credit
- 🌾 underestimating how much time it will take to build a market
- 🌾 going into business with little or no experience and without first learning something about it
- 🌾 attempting to do too much business with too little capital
- 🌾 not allowing for setbacks and unexpected expenses
- 🌾 not understanding **business cycles**[13]
- 🌾 lack of an effective marketing program
- 🌾 spending too much time and/or money on nonproductive and nonprofitable activities
- 🌾 unnecessary capital investments made to minimize income taxes

- 🌾 underpricing or overpricing goods or services
- 🌾 extending credit too freely
- 🌾 extending credit too rapidly[14]

Labor

The strength of a business is in its people. Hiring undependable and unqualified employees can quickly destroy your business. Failure can also come if you attempt to support too many people from the business. For example, bringing your children into the business or helping them get started can result in business failures. Too much debt compounded by too much pride or a sense of obligation presents many parents from making the necessary adjustments until it is too late.[15] As mentioned previously, the entrepreneur has to work hard. Some owners mistake the freedom of having their own business for the liberty to work only as they wish.

Financial Resources

Management of financial resources is not merely an exercise to be carried out at the beginning and ending of every year. It is an ongoing activity, which involves comparing plans to actual performance and taking appropriate action. Avoid the following as you manage your financial resources:

- 🌾 starting with too little capital
- 🌾 starting with too much capital and being careless in its use
- 🌾 borrowing money without planning just how and when to pay it back
- 🌾 failing to keep complete, accurate records, which causes you to drift into trouble without realizing it
- 🌾 carrying out habits of personal extravagance into the business
- 🌾 forgetting about taxes, insurance, and other costs of doing business[16]

🌿 overdependence on **collateral**

🌿 improper loan structuring due to an improper match of loan repayment period and loan repayment ability

🌿 failure to control living expenses because of withdrawing more from the business than the business actually earns[17]

Failure is always a possibility. However, if we become afraid to fail, we will not even try. Plan, study, and research every possible angle—then give it your best. Refer to Figure 4–3.

Don't Be Afraid to Fail

You've failed many times, although you may not remember. You fell down the first time you tried to walk. You almost drowned the first time you tried to swim, didn't you? Did you hit the ball the first time you swung a bat? Heavy hitters, the ones who hit the most home runs, also strike out a lot. R. H. Macy failed seven times before his store in New York caught on. English novelist John Creasey got 753 rejection slips before he published 564 books. Babe Ruth struck out 1,330 times, but he also hit 714 home runs. Don't worry about failure. Worry about the chances you miss when you don't even try.

Figure 4–3 When it comes to selling, or, generally, trying to achieve productive goals in life, it is better to have tried and failed rather than never to have tried at all.

Undercapitalization

Undercapitalization is an important factor in small business failure. Many small businesses do not have adequate start-up capital to survive the initial one- to two-year period of business establishment. This is analogous to building a car without the wheels. It may run, but it will not move forward.

ANALYZING YOUR AGRIBUSINESS VENTURE

Most people would like to become wealthy by starting a business that proves to be successful. However, it is one thing to dream; it is another to make that dream become reality. Careful planning from the beginning is one way to ensure the success of your agribusiness. Planning includes recognizing a need, considering start-up factors, applying business fundamentals, and conducting a business survey.

Recognizing a Need

A business succeeds only when it fills an economic need, and it is the responsibility of the prospective business organizer to determine what type of businesses are needed. A free enterprise system does not guarantee anyone a successful business just because it is of a desirable type. It is up to the individual to select the right type of business and to make it work. Thousands of firms go bankrupt each year because they provided the wrong type of product or service or were unable to make a potentially good business work. Many new businesses succeed because they provide a product or service that others need and for which they are willing to pay.

Agribusiness Start-up Factors to Consider

Anyone trying to analyze whether to start an agribusiness should consider the following factors.[18]

What Financial Resources Are Needed for Getting Started? For each kind of agribusiness being considered, the entrepreneur should calculate all the capital requirements. This analysis is especially critical if the entrepreneur must borrow money to get started.

What Are the Labor Needs for the Agribusiness? Careful analysis of available capital and the potential cash flow are needed so that the entrepreneur can determine the extent to which money would be available to pay for hired labor.

What Management Requirements Must Be Met? A major qualification is a knowledge of the technical aspects of the business: in this case, both the agricultural and business operations. In addition to the technical knowledge required, an entrepreneur needs to know about the records to be kept, labor laws, sales taxes, local business ordinances, sources of materials, markets, pricing, inventory management, credit, human relations, and general business procedures.

Is There a Market for the Product or Service?
A product or service must be marketable if a profit is to be made. Thus, a study of the market potential for an agribusiness is a very important step in helping an entrepreneur decide on which type of agribusiness to start.

Where Should the Agribusiness Be Located?
The location of the agribusiness may determine its success. The location is especially important if the agribusiness will depend on people passing by as customers.

Should You Buy an Agribusiness or Start a New One? Sometimes it may be possible to buy a small agribusiness rather than start one. Buying an agribusiness has advantages and disadvantages. Consider the following:

Advantages
- allows a quicker start
- provides ready-made customers
- eliminates some competition
- reduces the cost of getting established
- yields a base of financial information for estimating costs and profits[19]

Disadvantages
- greater capital resources needed at the very beginning
- no time to learn while the business is developing
- possibility of misjudging and buying a loser and
- problem of having to either accept the location or move the business
- loss of the safety that comes from expanding and growing into a business as resources permit
- risk of missing some critical and costly aspect of the business in closing the deal
- cost of legal assistance needed for making the purchase
- cost of going out of business if the experience proves disappointing[20]

Applying Business Fundamentals

There are many fundamentals of an agribusiness that entrepreneurs need to learn as they enter the business world as owners. These fundamentals apply regardless of the size of the agribusiness. A few of the important fundamentals for getting started in an agribusiness are as follows:

- Keep the size of the business consistent with the capital resources available and with progress in developing management capability.
- Select a business that is labor intensive rather than **capital intensive**.
- Devote adequate time to the development of management capability.
- Keep all excess family labor fully employed in the business or elsewhere.
- Prepare a sound plan of operations and follow it.

- Study and improve the business continually.
- Maintain an adequate cash operating fund so that borrowing at high interest rates will not be necessary.
- Maintain an inventory level that provides for responding to customer requests promptly.
- Plan to use as much of the profits as possible to expand the business.
- Use the physical facilities of your home to the fullest extent possible.
- Establish the prices charged for products and services at a level that will result in a reasonable profit while still being lower than or equal to the prices charged by competing businesses.
- Buy good-quality materials and necessary supplies at the lowest price possible.
- Treat customers and potential customers courteously and fairly.
- Treat employees fairly.[21]

Conducting a Business Survey

Some individuals go into business based on a "feeling" or "intuition" that the business will be successful. A business founded on this basis may, with good luck, be successful. However, the risk is extremely high, and most businesspeople prefer a more solid foundation on which to start. The procedures to follow in investigating the economic potential for a new business vary with the specific type of business under consideration. However, certain basic information on income, expenses, and market potential should be collected and analyzed regardless of the type of business. Regardless of the type of **business survey**, some basic questions should be answered:

- What resources are needed?
- Are these resources available?
- What are the costs of these resources?
- What level of management is required?
- Does the prospective owner have the experience necessary to operate the business?

Furthermore, the economic feasibility of marketing the products or services should be analyzed. Some of the questions that should be answered include the following:

- Are there one or more markets for the products or services?
- At what prices can the products or services be sold?
- Are the prices sufficient to cover the costs and give the seller a satisfactory profit?
- How reliable are the potential markets?
- What is the potential for future growth?

The answers to these and other relevant questions should be collected and analyzed in determining the market potential for the products or services.

PREPARING A BUSINESS PLAN

It's amazing how many people are eager to start an agribusiness yet have only a vague idea of what they want to do. Eventually, they come up with an idea for their agribusiness and begin discussing the idea with friends. It is at this stage that the entrepreneur needs a business plan.

A **business plan** is a written description of a new business venture that describes all aspects of the proposed agribusiness. It helps you focus on exactly what you want to do, how you will do it, and what you expect to accomplish. The business plan is essential for receiving help from potential **investors** and **financial institutions** for starting the business. Although there is no standard business plan format, there are many similarities. Refer to Figure 4–4 for one example.

The business plan for each venture is unique. Writing a business plan helps you further analyze the agribusiness that you want to start. It helps you be realistic, honest, detailed, and objective about your plans. A business plan forces you to set your goals and objectives. Furthermore, a business plan will help convince financial institutions that they are investing in a business that has a good chance of succeeding. Only a clear and logical business plan can persuade them.[22] Remember, business plans vary, but the following represents the components of most.

Introduction

Begin with a two- to three-page management overview of the proposed venture. Include a short

---

Business Plan Outline

- ◆ Introduction:
- ◆ Company Description:
- ◆ Product and/or Service:
- ◆ Management Plan:
- ◆ Marketing Plan:
- ◆ Legal Plan:
- ◆ Location Analysis:
- ◆ Business Regulations:
- ◆ Capital Required:
- ◆ Financial Plan:
- ◆ Financing Arrangements:
- ◆ Competition:
- ◆ Operating Plan:
- ◆ Appendix:

Figure 4–4 People do not plan to fail, they only fail to plan. Proper planning, by developing a business plan outline, is essential for the success of an agribusiness.

---

description of the business, and discuss your major goals and objectives.

Company Description

Explain the type of company, and give its history if it already exists. Tell whether it is a manufacturing, retail, service, or other type of business.

Product and/or Service

Describe the product and/or service. Detail any unique features. Explain why people will buy the product or service.

Management Plan

Identify key people who will direct and manage the company. Cite their experience and special skills. Include an organizational chart and job descriptions of the employees.

Marketing Plan

Show who the potential customers are and what kind of competition the business will face. Outline the marketing strategy and specify what makes the company unique. Do not underestimate the competition. Review industry size, trends, and the target market segment. Discuss the strengths and weaknesses of the product or service.

Legal Plan

Show the proposed type of legal organization the ownership will take: sole proprietorship, partnership, corporation, cooperative, or franchise (these will be discussed in Chapter 5). Point out any special legal concerns.

Location Analysis

Provide a comprehensive study of consumers in the area of the proposed agribusiness as well as a traffic-pattern analysis and automobile and pedestrian counts. Your plan should reveal a location that is easily accessible and highly visible. In many businesses, location is one of the most important factors.

Business Regulations

Include any federal, state and local regulations that will have to be addressed by your agribusiness. For example, your local planning commission may specify waste disposal. Zoning restrictions regulate the location of the business. Be aware of insurance requirements. Federal law regulates the working environment you provide your employees and the tax structure of your business.

Capital Required

Indicate the amount of capital needed to begin your agribusiness and describe how these funds are to be used. Make sure that the totals are consistent with the financial plan. This area will receive a great deal of review from lending institutions, so make sure it is clear and concise.

Financial Plan

Provide a five-year projection for income, expenses, and profits. Do not assume that the business will grow in a straight line. Adjust your planning to allow for funding at various stages of the business's growth. Explain the **rationale** and assumptions used to determine the estimates. Assumptions should be reasonable and based on similar agribusiness trends. Make sure all totals add up and are consistent throughout the plan.

Financing Arrangements

A bank or another lending institution may be willing to lend you a portion of the amount you need. However, it may be unable to lend the total amount you desire. Therefore, seek other sources of funds for your business, such as family, friends, or investors. The money they invest is called **venture capital**. Other potential sources for financing are personal savings, the government (through the Small Business Administration), and suppliers. Include these financial commitments in your business plan.

Competition

Identify your competition, and explain how you can offer a better-quality or less-expensive product in order to capture a profitable share of the market. Show how you can target your product or service to a particular area in which you are able to specialize.

Operating Plan

Explain how you are going to run the day-by-day operations of your agribusiness. Explain the type of manufacturing or operating system you will use. Describe the facilities, personnel, materials, and processing requirements of your business.

Appendix

Include all marketing research on the product or service (reports, articles, etc.) and other information about the product concept or market size. Provide a bibliography of all the reference materials you consulted. This section should demonstrate that the proposed company will not be entering a declining industry or market segment.[23]

GOAL SETTING

People do not plan to fail—they simply fail to plan. Prior planning prevents poor performance. Setting goals is a necessity for your agribusiness. You must have a goal, because it is just as difficult to reach a destination you have not located as it is to come back from a place to which you have never been. You must have definite, precisely written, clearly set goals if you are going to realize the full potential of your agribusiness.

Setting Your Agribusiness Goals

There are some general rules that can help you set goals for yourself.

Write Down Your Goals. The best place to start as you begin to think about the goals of your agribusiness is by writing them down.

Organize Your Goals. You will have both specific goals that you want to reach in a few days, weeks, or months and goals toward which you will work for many years. In between, you have goals that will take a year or two. Arrange your goals according to these three groups: immediate, short term and long term.

❦ *Immediate goals* are the goals that you would like to accomplish within a day, a week, or a month or two. As an entrepreneur, these immediate goals will probably include the first steps needed to get your business started.

❦ *Short-term goals* include the things you want to accomplish in a year or two. These goals often include the steps you need to build toward your long-term goals. For an entrepreneur, business expansion or marketing perfection would fit between immediate and long-term goals.

❦ *Long-term goals* are the ones toward which you intend to work for many years. They give you an idea of what you want to do with your business several years from now.

Reaching Your Goals

Goals are worthless if you do not attempt to reach them. Follow these steps as you start working to reach your goals.

Manage Your Time. The first step toward reaching your goals is to learn to manage your time. In order to do this, you must understand how you spend your time. Everyone has 24 hours to spend in each day and 365 days to spend in each year. Some people accomplish a great deal using the amount of time that they have; others do not. Consider the 24 hours that you have each day.

How do you use them? Three techniques that lead to successful time management are as follows:

❦ *Avoid procrastinating*: Procrastinators are especially good at putting off anything that does not have to be done immediately. Good time managers use all the time they have to progress toward their goals.

❦ *Judge your time*: The second technique you will want to develop is the ability to judge the time needed to accomplish a task. With experience, you will become able to have a very accurate idea of how much time different activities require. This allows you to use the third time management technique.

❦ *Schedule your time*: Be realistic. You will not perform a task well if you do not allow enough time to complete it properly. On the other hand, do not budget so much time for a project that you find yourself wasting time between appointments or projects.[24]

Establish Priorities. Some of your goals will be more important to you than others. Using a scale of 1 to 10, rate each goal on your list according to its importance, giving those with the highest priority a 10. Rate the others, comparing them to your most important objectives. When you begin to work toward your goals, begin with the ones having the highest priority. When you have accomplished all that you can toward reaching those goals, focus on the goals with the second highest priority. By establishing your priorities, you will be using your time, not to work harder, but to work smarter.[25]

Breaking Goals into Manageable Units. The best goals are the big ones. These are the ones that you set your sights on—the exciting ones. Your long-term goals can help you overcome the daily frustrations that you will encounter as you work on your immediate goals. However, big goals cannot be accomplished in a day. If you break them into manageable units, which become immediate goals, you can see your progress. While continuing to work for the larger goal, you will earn a sense of accomplishment that is encouraging. Think about how you can work toward your goals on a daily basis, and then strive to accomplish that much each day.[26]

Example of Business Goals

Nancy Baker is about to start on a new business venture. The small town where she lives does not have a flower shop, and people in her community order flowers from a larger town twelve miles away. Nancy has always liked working with flowers, and her friends often compliment the floral arrangements in her home. For two years she has made a hobby of growing flowers in a small greenhouse that she built in her backyard. Recently, several friends have asked Nancy to make up floral arrangements for gifts, and she believes she can build a profitable business with her talent. Nancy prepared a list of goals for her venture. Refer to Figure 4–5 for a list of Nancy's goals.

Immediate Goals

◆ Draw up a business plan in order to know exactly how much capital will be needed and how and when it will be repaid.

◆ Obtain capital and enlarge greenhouse, buy a refrigerating unit, and open a small shop.

◆ Produce enough income to ensure the business will have an effective cash flow.

◆ Stay within a specified debt limit.

◆ Sell fresh-cut local floral arrangements and potted plants to people in the community.

◆ Locate suppliers for vases and other needed items.

Short-Term Goals

◆ Repay original loan.

◆ Capitalize expansion; perhaps include silk floral arrangements and garden plants.

◆ Provide flowers for weddings and add other floral-catering services.

◆ Have an income that provides a satisfactory profit over expenses.

Long-Term Goals

◆ Hire and train employees in order to expand the capacity of the company (and take vacations).

◆ Broaden the market and price products reasonably in order to entice new clients from the larger town.

◆ Establish financial security.

◆ Sell the company and retire with a substantial profit.

Figure 4–5 If you do not know where you are going, you will not know it when you get there. Nancy prepared these goals in order to measure the success of her floral shop.

PROBLEM SOLVING AND DECISION MAKING

People planning and organizing an agribusiness will have many problems and decisions to make. How the owner or manager handles these will determine his or her success. In the process of problem solving, a new or different course of action may be required to correct a problem. The process by which the new or different action is selected is called decision making. At times, after careful consideration, you may find that your best decision is to do nothing because the problem simply cannot be solved. At other times, the decision may be to change your goals to eliminate the problem.[27]

Skills Needed in Problem Solving and Decision Making

Lloyd J. Phipps suggests that individuals need to develop certain problem-solving and/or decision-making skills. Among these are the ability to:

- recognize problem situations
- clearly distinguish the problem from the problem situation
- clearly define goals and/or objectives
- develop creative, imaginative solutions to problems
- gather information related to the possible solutions
- be open-minded toward possible solutions offered by others
- carefully evaluate information in accepting or rejecting solutions
- work with others to solve problems
- avoid jumping to unwarranted conclusions (be flexible), accept the fact that you may make mistakes, and put aside opinions, feelings, emotions, and self-interests that may interfere with objective thinking

- understand different types of problems and techniques for solving them
- understand, and use, a systematic approach to problem solving and decision making[28]

Various Styles of Decision Making

Not everyone makes decisions the same way. What may be of importance to one person may be of little concern to another. The criteria used in making decisions may also differ. Criteria are used to make up a standard or test by which something can be judged or compared. Consider the following three distinct decision-making styles:

- *Reflexive style:* People with this style tend to make quick decisions. This style does not often allow the individual an appropriate amount of time to consider all options and consequences. However, people who use this style are not likely to put matters off to the last minute.

- *Reflective style:* Individuals who possess this style of decision making weigh all the facts before jumping to conclusions. All the needed information is considered before the decision is made. This eliminates a lot of possible problems associated with snap decisions. Because of the time used in considering a problem, however, the decision may be made too late to be of any use.

- *Consistent style:* The consistent style involves the best attributes of the previous two styles. The individual considers all the facts and then acts in a timely manner. This makes for the most reasonable decision.[29]

Steps in Problem Solving and Decision Making

A systematic approach to problem solving can be of great benefit when making important decisions. These steps are similar to those of the scientific method, which scientists use in their experiments. There are seven steps to this method.

🌿 *Step 1, Recognize the Problem.* Remember, problems are inevitable and must be solved, not ignored, if you are going to be successful. Obviously, the first step toward solving a problem and making a decision is realizing that you have a problem that requires a decision.

🌿 *Step 2, Determine Your Alternatives.* Once you have identified your problem, you need to determine your alternatives. Alternatives are the different courses of action you can take to solve your situation. There are likely to be many alternatives to any problem, and you need to consider each one carefully before making a decision. It may be helpful to list your alternatives on paper to keep a clear focus.

🌿 *Step 3, Gather Information.* Once you have listed your alternatives, you need to gather information relative to each one. Look at factual information. Relying only on opinions, emotions, and feelings may lead to hastily made decisions that you will later regret. In gathering factual and objective information, ask yourself the following questions: What do I need to know about each alternative? What materials or information may be needed to implement each alternative? What costs will be involved? Is the alternative feasible in the first place?

🌿 *Step 4, Evaluate the Alternatives.* Once you have gathered information for each of your alternatives, you will need to evaluate (and perhaps list) the advantages and disadvantages of each in relation to solving the problem and in relations to each other.

🌿 *Step 5, Select a Workable Solution.* After you have evaluated each alternative and its possible results, you must choose the one that is the most practical, reasonable, and effective to solve your problem.

🌿 *Step 6, Carry Out Your Solution.* Once you have determined your course of action, follow through. If you fail to carry out your solution, you have wasted the time and effort you have used to this point.

🌿 *Step 7, Evaluate Your Results.* The problem-solving/decision-making process does not end when the proposed solution is carried out. It ends in the future when you decide whether your problem has been solved, the same problem persists, or new problems have been created. Evaluation may lead you to accept your solution as a good decision, make further adjustments to improve your solution, or discard your decision and start the process over again.[30]

CONCLUSION

Remember that you will encounter obstacles as you pursue your goals. Some of these obstacles can be foreseen. If you are prepared for the foreseen problems, you will have more time and energy to deal with the unexpected ones. If your business goals require a great amount of your time, you will want to consider how this will affect your commitments, such as your family. Your goals may conflict with the goals of others. Communication is the key to human relations. Be sure that those who will be affected understand what your goals are.

Circumstances will change with your age, your health, and your family obligations, among other things. Be prepared to reevaluate your goals from time to time. You may find that they need to be restructured because your priorities have changed along with your values. Do not be afraid to make decisions. Do not feel defeated if a decision turns out to be wrong—sometimes even the best decisions can have bad results. The most successful entrepreneurs are the ones who can profit from their mistakes and move ahead. Develop a methodological approach to breaking your goals into manageable units and then pursuing those accomplishments one step at a time. It is up to you to make daily goals a habit as you pursue your dreams. You must think carefully about what kind of business you want. You are not likely to find everything you want in one business—easy entry, security, *and* reward. Choose those characteristics that matter the most to you; accept the absence of the others; plan, set your goals, and then go for it!

SUMMARY

Even though most agribusiness students will never own their own business, it is important to understand the basic processes involved in planning and organizing a business. As we enter the world of agribusiness, we must decide whether to work for ourselves or work for someone else.

Small businesses, which include most agribusinesses, are not small in terms of the number of people they employ. Small businesses are really a big part of the U.S. economy. In fact, ninety percent of the nation's new jobs in the private sector are in small business.

Entrepreneurship is not for everyone. Entrepreneurs are people who have the initial vision, diligence, and persistence to follow through. An entrepreneur is a person who accepts all the risks pertaining to forming and operating a small business. Entrepreneurs tend to be independent, self-confident, energetic, organized, persistent, optimistic, committed, self-nurturing, action-oriented, flexible, and emotionally stable; they are generally visionaries, problem solvers, and risk takers.

Being successful in a new agribusiness is a challenge. Three major challenges of entrepreneurship are total responsibility; long, irregular hours; and financial risks. The chances for a business reaching its fourth birthday are only about 50-50.

There are many potential reasons why agribusinesses fail, but most failures are attributed to either bad management, improper use of labor, or misuse of financial resources. Many people go into business or expand without adequate planning, without analyzing the added cost of the added returns, and without analyzing potential risk. They also fail to realize that the strength of a business is in its people. Hiring undependable and unqualified employees can quickly destroy your business.

Careful planning from the beginning is one way to ensure the success of your agribusiness. Every potential angle of your agribusiness venture has to be analyzed. Planning includes recognizing a need, considering start-up factors, applying business fundamentals, and conducting a business survey.

A business plan is a written description of a new business venture that describes all aspects of the proposed business. It helps you focus on exactly what you want to do, how you will do it, and what you want to accomplish. The business plan is essential for the potential investors and financial institutions on whom you will depend for starting the business.

You must have definite, precise, clearly set, written goals if you are to realize the full potential of your agribusiness. You should write down your goals and organize them into immediate, short-term, and long-term goals. As you strive to achieve your goals, you must manage your time, establish priorities, and break your goals into manageable units. It is up to you to make daily goals a habit as you pursue your dreams.

The ability to solve problems and make decisions can mean the difference between excellence and mediocrity. Whether you are the chief operating officer (CEO) of a large corporation or an individual agribusiness owner, your success may be determined by the ability to solve an individual problem or make a decision. Those who succeed are most often those who plan, and those who plan are often accomplished goal setters, problem solvers, and decision makers.

END-OF-CHAPTER ACTIVITIES

Review Questions

1. Define the Terms to Know.
2. What are nine reasons why people are willing to take the risks of starting a business?
3. What are fifteen characteristics found most often in successful entrepreneurs?
4. What are three major challenges of entrepreneurship?
5. What are five responsibilities of the entrepreneur?
6. What are thirteen management decisions that could cause your agribusiness to fail?
7. What are nine things to avoid as you manage your financial resources?
8. What are six questions to analyze when considering whether to start an agribusiness?
9. What are five advantages of buying an agribusiness rather than starting one?
10. What are eight disadvantages of buying an agribusiness rather than starting one?
11. List fourteen fundamentals of business that entrepreneurs need to learn as they enter the agribusiness world as owners.
12. What are five basic questions that should be answered when giving a business survey?
13. What five questions should be answered when determining the economic feasibility of marketing a particular product or service?
14. What are three crucial steps to follow as you attempt to reach your goals?
15. List eleven skills needed in problem solving and decision making.
16. Briefly explain the three distinct decision-making styles.
17. Briefly explain the seven steps in the problem-solving and decision-making process.

Fill in the Blank

1. When it comes to the number of jobs, small businesses are _____ in the number of people they employ.
2. _____ percent of the nation's new jobs in the private sector are in small business.
3. There are about _____ full- and part-time home-based businesses in the United States.
4. Of all nonfarm businesses in the United States, almost _____ percent are considered small by SBA standards.
5. Small businesses account for over _____ percent of gross domestic product (GDP).

6. It is not uncommon for an entrepreneur to work more than _____ hours per week.

7. The chance of a business reaching its fourth birthday is _____ percent.

8. Agribusiness failures usually fall under the umbrella of either _____, _____, or _____.

9. The strength of a business is in its _____.

10. A business succeeds only when it fills a/an _____ need.

11. People do not plan to fail, they simply _____ to _____.

12. Three techniques that lead to successful time management are _____, _____, and _____.

Matching

a. type of legal plan

b. company description

c. location analysis

d. financing arrangements

e. financial plan

f. capital required

g. competition

h. management plan

i. business regulations

j. marketing plan

_____ 1. adhering to working environment, local planning commission, state, and federal guidelines

_____ 2. explains how you can offer a better-quality or less-expensive product in order to capture a profitable share of the market

_____ 3. shows who the potential customers are and what kind of competition the business will face

_____ 4. identifies key people, citing their experience, special skills, and job descriptions

_____ 5. explains the business and tells whether it is a manufacturing, retail, service, or other type of business

_____ 6. source of funds needed to start a business

_____ 7. sole proprietorship, partnership, corporation, cooperative

_____ 8. five-year projections from income, expenses, and profit

_____ 9. amount of money needed to start the agribusiness and explanation of how the funds will be used

_____ 10. study of traffic-pattern analysis and automobile and pedestrian counts.

Note: All the matching relates to the parts of a business plan.

Activities

1. Look through the yellow pages or any other source and list all the agribusinesses in your area. Describe the product or service each sells. Identify whether the business is an agribusiness input sector or part of the agribusiness output sector.

2. Listed below are categories of skills needed by most small business owners.[31] Read each one carefully and decide how adept you are at each one. Do this by placing a number from 1 to 5 in the

space provided beside each item, with 1 meaning that you have no knowledge or expertise in the area. You will use these rankings in your personal inventory later. When you have completed each category, total the ratings within each section. Divide the total number by the number of items in each section. This will give the average rating in each section. Place the average score in the space provided for the category.

A. Managing Money   Average Score____

1. Borrowing Money

2. Keeping Business Records

3. Avoiding Losses

4. Handling Credit

5. Figuring Taxes

B. Managing People   Average Score____

1. Hiring Employees

2. Supervising Employees

3. Educating Employees

4. Motivating Employees

C. Directing Business Operations Average Score____

1. Purchasing Supplies

2. Purchasing Equipment

3. Purchasing Merchandise

4. Managing Inventory

D. Directing Sales Operations   Average Score____

1. Identifying Various Customer Needs

2. Developing Products for Different Customer Needs

3. Learning to Answer Customer Objections

4. Closing the Sale

5. Instructing Others in Selling Techniques

E. Marketing   Average Score____

1. Developing New Ideas for Product or Services

2. Recognizing Community Needs

3. Recognizing Potential Customers

4. Creating Promotional Strategies

5. Designing Promotional Materials

F. Setting Up a Business   Average Score____

1. Choosing a Location

2. Obtaining Licenses or Permits

3. Determining Initial Inventory

4. Obtaining Financing

5. Planning Long- and Short-Term Cash Flow

6. Choosing the Form of Ownership (e.g., Partnership)

When rating your personal skills, you can use any type of experience relating to the particular skill. For example, you might have organized a fund-raiser for your club or organization. This would boost your ratings in the "Setting Up the Business" section. Do this with the other sections. Try to think of any personal experience that would relate to each item. If you do have such experience, then enter a number from 2 to 5 beside the item, depending on the amount. If not, place a 1 beside the item. Use your score as you consider starting an agribusiness. Compare your score with scores of your classmates.

3. Some entrepreneurs "recognize" a need in society and start a business to fill that need. Name three businesses that follow that pattern.

4. Some entrepreneurs "create a need." They supply goods or services with the hope that the product is attractive enough that people will decide they "need" it. List three companies that have offered new and unusual products.

5. Select an agribusiness that you believe would be successful. Answer the following questions.

    a. What product or service did you select?

    b. Is there a market for the product?

    c. What are the financial resources needed for getting started?

    d. What are the labor needs of the business?

    e. What skills or tasks would you have to learn how to do in order to succeed?

6. Using the agribusiness that you selected above, complete a business plan. Use the format in Figure 4–4.

7. Using the same agribusiness again, write three to five immediate, short-term, and long-term goals for each. Refer to Figure 4–5 for an example.

NOTES

1. William H. Hamilton, Donald F. Connelly, and D. Howard Doster, *Agribusiness: An Entrepreneurial Approach* (Albany, N.Y.: Delmar Publishers, 1992).

2. Ibid.

3. "Where the New Jobs Are," *Business Week,* March 20, 1995, p. 24.

4. Cynthia E. Griffin, "Going Home," *Entrepreneur,* March 1995, pp. 120–125.

5. Wilma Randle, "Sowing Seeds," *Chicago Tribune,* October 27, 1993, sec. 7, pp. 3, 14; Cynthia Todd, "Black Businesswomen Make Their Own Success," *St. Louis Post-Dispatch,* April 4, 1994, pp. 1A, 6A.

6. Gene Koetz, "Small Business Is Putting Some Snap in the Job Market," *Business Week,* April 25, 1994, p. 26.

7. William G. Nickels, James M. McHugh, and Susan M. McHugh, *Understanding Business* (Chicago, Ill.: Richard D. Irwin, 1996).

8. *Entrepreneurship in Agriculture* (College Station, Tex.: Instructional Materials Service, 1988), 8747-A, p. 1.

9. Ibid., pp. 1–2.

10. Ibid.

11. Hamilton, Connelly, and Doster, *Agribusiness: An Entrepreneurial Approach*.

12. Ibid, p. 8.

13. The Service Corps of Retired Executives (SCORE), a part of the Small Business Administration; and Judith Gross, "Autopsy of a Business," *Home Office Computing,* October 1993, pp. 52–60.

14. *Advanced Agribusiness Management and Marketing* (College Station, Tex.: Instructional Materials Service, 1990), 8735B, pp. 2–3.

15. Ibid.

16. SCORE and Gross, "Autopsy of a Business."

17. *Advanced Agribusiness Management,* pp. 2–3.

18. Alfred H. Krebs and Michael E. Newman, *Agriscience in Our Lives* (Danville, Ill.: Interstate Publishers, 1994), pp. 656–659.

19. Ibid., p. 659.

20. Ibid.

21. Ibid., pp. 659–661.

22. Betty J. Brown and John E. Clow, *Introduction to Business* (New York: Glencoe/McGraw-Hill, 1997), pp. 215–216.

23. Brown and Clow, *Introduction to Business;* Hamilton, Connelly and Doster, *Agribusiness: An Entrepreneurial Approach;* Nickels, McHugh, and McHugh, *Understanding Business.*

24. Hamilton, Connelly, and Doster, *Agribusiness: An Entrepreneurial Approach.*

25. Ibid.

26. Ibid.

27. Robert N. Lussier, *Human Relations in Organizations: A Skill Building Approach* (Homewood, Ill.: Richard D. Irwin, 1990), p. 276.

28. Lloyd J. Phipps, *Handbook on Agricultural Education in Public Schools* (Danville, Ill.: Interstate Printers and Publishers, 1965), pp. 118–119.

29. Lussier, *Human Relations in Organizations: A Skill Building Approach*, p. 278.

30. Cliff Rickets, *Leadership, Personal Development, and Career Success* (Albany, N.Y.: Delmar Publishers, 1997), pp. 313–316.

31. *Entrepreneurship in Agriculture*, pp. 2–3.

CHAPTER 5

Types of Agribusiness

OBJECTIVES

After completing this chapter, the student should be able to:

- ❧ compare proprietorships, partnerships, and corporations
- ❧ explain the characteristics of the single (sole) proprietorship
- ❧ explain the characteristics of partnerships
- ❧ discuss the different types of corporations
- ❧ explain the characteristics and value of cooperatives
- ❧ describe the characteristics of franchises

TERMS TO KNOW

board of directors
common stock
cooperative
corporate charter
corporation
dividends
double taxation
franchise
franchisee

franchisor
general partnership
legal classification
legal entity
legal structure
limited partnership
marketing cooperatives
parent company
partnership

preferred stock
prepackages
single (sole) proprietorship
stock
stockholders
Subchapter C
Subchapter S
Subchapter T
unlimited liability

INTRODUCTION

As our society becomes increasingly complex, the process of selecting the most desirable type of agricultural business, or **legal structure**, for an agribusiness becomes very important. The amount of restriction by government agencies, the tax structure, and other legal requirements are determined, to a large degree, by the legal classification, or type, of agribusiness.

The type of agribusiness, or **legal classification**, you choose may make the difference between success or failure. Although there are many specific legal structures available, most legal specialists agree that the three major types of business organizations are (1) the **single (sole) proprietorship**, (2) the **partnership**, and (3) the **corporation**. The three types of corporations are regular corporations (**Subchapter C**), family farms and small businesses (**Subchapter S**), and **cooperatives** (**Subchapter T**). Cooperative corporations are increasing in importance, and many authors list the **cooperative** as a fourth major type of business. Another corporation that is a special form of business organization is a **franchise**, which is the fifth type of agricultural business.

If you are going to operate a tractor repair shop, it is more likely that you will be the sole owner. If you plan to have a manufacturing facility, you will probably have partners or form a corporation. If you want to have a fast-food chain, you will have a franchise business. You need to assess all forms of legal classifications to decide which is best for you. You can also change your initial decision about your type of agricultural business. As your business grows, changing conditions may require you to change your form of ownership.

COMPARISON OF PROPRIETORSHIPS, PARTNERSHIPS, AND CORPORATIONS

Some agribusinesses are relatively easy to start, such as a lawn service or landscaping business. An organization that is owned, and usually managed, by one person is called a single (sole) proprietorship. It is the most common form of business ownership, with over 12 million businesses.[1]

Many people lack the money, time, or desire to run a business on their own. They prefer to have someone else or some group of people get together to help them form an agribusiness. When two or more people legally agree to become co-owners of a business, the organization is called a partnership. In the United States, there are approximately 1.4 million partnership businesses.[2]

It is sometimes best to create a business that is separate and distinct from the owners. A legal classification with authority to act and have liability separate from its owners is called a corporation. Although there are only 2.8 million corporations in the United States, comprising only 17 percent of all businesses, they do 87 percent of the sales volume.[3] Refer to Figure 5–1 for a comparison of proprietorships, partnerships, and corporations.

No single type of legal structure is best for all businesses. Therefore, when developing a new agribusiness, all structures should be considered relative to their advantages and disadvantages. Although single proprietorships are more numerous in all types of industries except manufacturing, they are especially dominant in production agriculture, including forestry and fisheries. Corporations are the least numerous for these groups, but they exceed partnerships in all other types of industries. We will now take a closer look at each

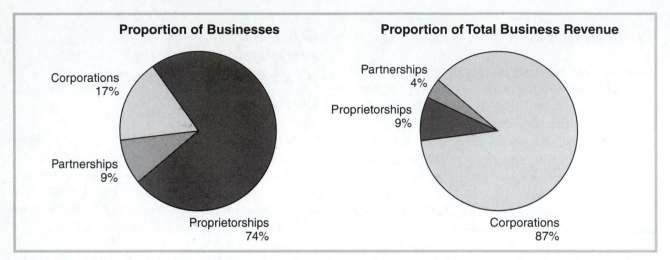

Figure 5–1   Although 74 percent of businesses are sole proprietorships, corporations account for 87 percent of total business revenue. Source: U.S. Internal Revenue Service, *Statistics of Income Bulletin,* Summer 1995.

of these legal classifications, or types, of agricultural businesses.

SINGLE, OR SOLE, PROPRIETORSHIP

The single proprietorship is the major type of legal structure in the agricultural industry as well as in most other industries. It is the simplest type of business and is the easiest to organize. It is the type closest to the American dream, in which the owner is in complete control. On average, single proprietorships are the smallest types of businesses. As a result, sole proprietorships are especially suited to areas that require personalized service and are flexible enough to meet the specific needs of individual customers.

Advantages of Sole Proprietorship

Sole proprietorships have several advantages. As mentioned, they are simple to start. There are few government regulations or restrictions. Depending on local laws, the only requirement may be a license, which is a legal permit for doing business. Management and control are solely in the owner's hands. Therefore, no voting is necessary. This allows owners to make quick decisions, without waiting to talk with others. Sole proprietors can choose their own products and set their own hours. Another advantage is that the owners receive all the profits from the business. Finally, sole proprietors pay taxes only once on the income from the company.[4] Refer to Figure 5–2 for a list of the advantages of a single or sole proprietorship.

Disadvantages of Sole Proprietorship

Even though we tend to stress the advantages of single proprietorships, they also have certain disadvantages. The major one is the claim that creditors can make on the owner's personal assets, as well as the business assets, in the payment of business debts. This relationship is known as **unlimited liability**.

Another major disadvantage of one-owner firms is that the accumulation of the large amounts of capital required to begin and operate many businesses today is limited.

In some cases, an owner may be good in the "product" of the business (repairing tractors, for instance) but he or she may not have business skills in other areas, such as leadership, human and public relations, or tax preparation. Owners may have to seek outside help to keep the business going. Another disadvantage of the sole proprietorship is its limited life. The business legally ends when the owner sells the business or dies. Someone may buy the business from the owner, but the business's success may depend on the former owner's special skills. Refer to Figure 5–2 for a list of the disadvantages of the sole proprietorship.

PARTNERSHIPS

If starting a sole proprietorship sounds too risky for you, you may decide to share the responsibility and benefit from someone else's skills and get a partner to join you. A partnership is usually defined as a business association of two or more persons. Someone who wants to start a business but does not have the capital or management skills required usually seeks one or more individuals who can meet these requirements. If these individuals can agree that a business is needed and a profitable arrangement can be made, the business is organized as a partnership. When formal agreements are made and recorded, it is easy to identify the partnership.

Business partnerships make it possible for persons to use their specialized skills to organize a company in which they can use their abilities to the fullest extent. For example, three people might form a partnership to start an agricultural machinery dealership because one is a skilled mechanic, another is excellent in sales, and the third is an excellent manager. Partnerships are found in many other areas of the agricultural industry. A few examples include agricultural supply stores, landscaping businesses, and hunting preserves. Many times the names of the business will give you a clue about how many partners are

---

Single (Sole) Proprietorship

Advantages	Disadvantages
◆ Ease of starting and ending the business	◆ Unlimited liability—the risk of losses
◆ Being your own boss	◆ Limited financial resources
◆ Pride of ownership	◆ Difficulty in management
◆ Retention of profit	◆ Overwhelming time commitment
◆ No special taxes	◆ Few fringe benefits
◆ Simple and flexible	◆ Limited growth
◆ Independent decision making	◆ Limited life span
◆ Change organization or operation quickly and easily	◆ Business and owner are a single entity
◆ Relatively low initial capital requirement	◆ Restriction of expansion potential
◆ Fewer government regulations	◆ Owner assumes all risks
◆ One owner in control	◆ Owner is often tied to the business and may be unable to spend time away without closing it
◆ Offers opportunity for personal advancement	

Figure 5–2   There are many advantages and disadvantages of a single (sole) proprietorship.

involved: a two-person partnership might be called "Walker and Jones Feed Store," and a three-person partnership might be named something like "Rodriguez, Rochelle and Jordan Meat Processors."

Types of Partnerships

General Partnership. A general partnership is an association of two or more people who, as owners, manage a business together. General partnerships contain certain recognizable factors. An agreement should be formalized explaining the terms of the partnership and outlining who is to contribute what, how decisions are to be made, and how profits are to be split. A general partnership is dissolved by death, agreement, or bankruptcy.

Generally, voting and profit sharing are based on the amount each partner contributes to the partnership. Each partner is fully liable for the partnership's activities. The partnership must file tax returns, but it pays no taxes. The partnership's tax return provides information so that each partner can file an individual return. Any profit earned by the partnership is divided between the partners based on previous agreements.[5]

Limited Partnership. There is a special type of partnership in which some partners are not completely liable for their partners' debts. This type is usually called a **limited partnership**, and most states recognize its legality. To be a limited partner, one must invest in the business but may not participate in the management phase. The limited partner's name cannot even appear in the partnership name. Limited partnerships are for investment purposes only. "Limited partners are liable for partnership obligations only up to the amount of their investment in the partnership and may not be held personally liable."[6] This provides incentive for investing in a business without the fear of losing personal or other business assets. However, limited partners can no longer deduct a partnership loss

from other personal income (since the Tax Reform Act of 1986). To be safe, limited partners should have legal documentation of their limited relationship in the business to protect them should creditors question their responsibility for business debts.

Advantages of Partnerships

Partnerships offer an immediate advantage in obtaining capital as two or more persons are the founders, and therefore are sources of start-up capital as well as operational funds. Decision making is shared, as are all other aspects of the business. Start-up costs are lower than those in corporations. Like the income of the sole proprietorship, the income is taxed only once.

Often, it is much easier to own and manage a business with one or more partners. Your partner can cover for you when you are sick or go on vacation. Your partner may be skilled at inventory keeping and accounting, while you do the selling or servicing. A partner can also provide additional money, support, and expertise. Refer to Figure 5–3 for a list of the advantages of partnerships.

Disadvantages of Partnerships

The major disadvantage of a partnership, its major distinguishing characteristic, is unlimited liability. This means that each partner is completely liable for all obligations of the partnership. The personal assets of either partner may be taken by a creditor in payment of a business debt; thus, a business partner should be selected very carefully. In fact, some businesspeople contend that this one disadvantage outweighs all the advantages of a partnership, and many refuse to take the responsibility of this type business.

Another major disadvantage of partnerships is its instability and the lack of continuity in the event one partner dies. Generally, if one partner dies, the remaining partners close the business and settle with the deceased partner's estate. However, the courts have recognized the rights of the remaining partners to continue the business after the death. In some cases a new partnership of the surviving family members family is formed and the business continues. "For federal tax purposes, a partnership does not terminate on death of a partner unless the partnership ceases to operate or there is a change of 50 percent or more in partnership capital and profit within a 12–month period."[7] Additional disadvantages are summarized in Figure 5–3.

Written Partnership Agreements

It is not hard to form a partnership, but it is wise for each partner to get the advice of a lawyer who is experienced with such agreements. Lawyers' services are usually expensive, so study and be prepared before calling a lawyer. For your protection, be sure to put your partnership agreement in writing. The Model Business Corporation Act recommends including the following in a written partnership agreement:

🌿 The name of the business: many states require the firm's name to be registered with state and/or county officials if its name is different from the name of any of the partners.

🌿 the names and addresses of all partners

🌿 the purpose and nature of the business, the location of the principal offices, and any other locations where the business will be conducted

🌿 The date the partnership will start and how long it will last: Will it exist for a specific length of time or will it end when one of the partners dies or when the partners agree to discontinue?

🌿 The contributions made by each partner: Will some partners contribute money while others provide real estate, personal property, expertise, or labor? When are the contributions due?

🌿 The management responsibilities: Will all partners have equal voices in management or will there be senior and junior partners?

🌿 the duties of each partner

Partnerships

Advantages
- More financial resources
- Shared management and pooled knowledge
- Longer survival
- Special skills of the partner
- Legal aspects of forming a partnership are relatively simple
- Pay no taxes as a business
- May be ended anytime partners agree
- Limited government regulation
- Greater management base than with one owner
- Partners, like sole proprietors, often feel pride in owning and operating their own company

Disadvantages
- Unlimited liability
- Division of profits
- Disagreements among partners
- Difficult to terminate
- Dissolved when a partner dies or leaves the partnership
- Size is limited by resources
- Difficult to manage if there are too many partners
- Lack of continuity
- Divided management authority

Figure 5–3 Business partnerships make it possible for persons to use their specialized skills to organize a company in which they can use their abilities to the fullest extent.

🌿 the salaries and drawing accounts of each partner

🌿 provision for the sharing of profits or losses

🌿 provision for accounting procedures: Who will keep the accounts? What bookkeeping and accounting methods will be used? Where will the books be kept?

🌿 the requirements for taking in new partners

🌿 any special restrictions, rights, or duties of any partner

🌿 provision for a retiring partner

🌿 provision for the purchase of a deceased or retiring partner's share of the business

🌿 provision for how grievances will be handled

🌿 provision for how to dissolve the partnership and distribute the assets to the partners[8]

CORPORATIONS

Corporations are another way of doing business in the United States, especially where large investments of money are needed. Although the word *corporation* makes people think of big businesses like General Motors, Ford, IBM, FedEx, Exxon, and John Deere, it is not necessary to be big in order to incorporate. Obviously, many corporations are big. However, incorporating may be beneficial for small businesses, too.

Although agribusiness corporations are limited in number relative to single proprietorships, they are increasing in number and importance. This increase is due primarily to the increasing requirements for large amounts of capital in modern agriculture and the limited liability characteristic, which allows individuals to invest in a business without personal liability beyond the amount of investment.

Characteristics of Corporations

A corporation is an organization owned by many people but treated by law as though it were a person. A corporation is a **legal entity**, separate from the people who own it or work for it. It can own property, pay taxes, make contracts, sue and be sued in court, and do other things a person can do. If you want to form a corporation, you issue **stock**, or shares in the ownership of your corporation. The new owners, called **stockholders**, pay a set price for their shares. Each stockholder has one vote in the major decisions made by the corporation for each share of purchased stock. Some large corporations have over a million stockholders.

Because a corporation often has numerous owners who are not involved in the management of the business, strict regulations are designed to protect each owner. Therefore, corporations have more legal regulations to meet than any other type of business. For example, each corporation must obtain a legal charter from a state government before it can operate; no other type of business is required to do this. Corporations can be chartered in any state. They must then elect a **board of directors**. Most state laws governing the formation of corporations are similar; it generally begins with filing an articles of incorporation application. If the articles are in agreement with state law, the state will grant you a **corporate charter**, which is a license to operate from that state.[9]

The stockholders of a corporation elect a board of directors, which is a group of individuals chosen to make the decisions for the company. They appoint officers to make day-to-day decisions for the corporation. The officers, such as the president, vice-presidents, and treasurer, make most of the day-to-day decisions for the corporation.

Types of Corporations

There are three types of corporations. The Subchapter C is a regular corporation, which sells stocks to investors. The Subchapter S is basically for small businesses or a family, and Subchapter T is for cooperatives. Both Subchapter C and Subchapter S are profit-making corporations. Subchapter T (for cooperatives) is a nonprofit corporation.

Subchapter C (Regular Corporations). In this type of business, the original owners sell stock to investors to raise capital. The investors, who also become owners, buy the stock in hopes that the company will do well. If the company does do well, it may pay **dividends** on their shares of stock. Money paid to a shareholder on a share of stock is called a dividend. The amount of the dividend is set by the board of directors and is based on the amount of profit made by the corporation. If no profit was made, there may be no dividend. However, if the company is reasonably profitable, the value of the stock will increase because more investors will want to own it. Of course, shareholders can loose money if the share price falls below the price they paid for it.

In regular corporations, voting is done by those owning common shares of stock. Each share of stock is worth one vote. Policy decisions are made by the stockholders and the board of directors. The corporation is financially liable.[10] Agribusiness corporations of this type include Ralston Purina, Case International, Ciba-Geigy, and many others. Refer to Figure 5–4 for a list of additional agribusiness corporations.

Subchapter S (Small Business or Family Corporation). Subchapter S is a unique government creation that has the characteristics of a corporation but is taxed like sole proprietorships and partnerships. These corporations have the benefit of limited liability. The paperwork and details of the Subchapter S corporations are similar to those of regular corporations. They have shareholders, directors, and employees, but the profits are taxed as the personal income for the shareholders—thus avoiding the **double taxation** encountered by regular corporations.[11]

Agribusiness Corporations

Phillip Morris	Safeway	Monsanto
Procter & Gamble	RJR Nabisco	General Mills
Pepsico	Georgia-Pacific	H. J. Heinz
Conagra	Archer Daniels Midland	Eli Lilly
Kroger	IBP Packers	Campbell Soup
Dow Chemical	Winn-Dixie Stores	Farmland Industries
International Paper	Deere	Ralston Purina
Sara Lee	McDonald's	Kellogg
Quaker Oats	Tyson Foods	Hershey Food
Dole Food	Hormel Foods	Dean Foods

Figure 5–4 Agribusiness is big business. The agribusiness corporations listed here represent only a few examples of agribusinesses in the United States.

Regular corporations (Subchapter C) must first pay taxes on the profits of the firm, and then the income is taxed again as individual income when stockholders receive their dividends. This is commonly referred to as double taxation. Although regular corporations are usually required to pay the corporate tax, Subchapter S corporations are not taxed as regular corporations and do not pay the double tax. Avoiding the double corporate tax rate was enough reason to persuade 1.6 million small U.S. businesses to operate as Subchapter S corporations.[12] Not all businesses can become Subchapter S corporations. For a corporation to qualify for taxation under Subchapter S, it must meet the following requirements:

- The owners must be individuals or estates.
- All stockholders must be U.S. citizens or resident aliens.
- Only one type of stock is allowed.
- The owners must not be members of an affiliated group of corporations.
- The owners may not own 80 percent or more of the stock of another corporation.
- Complete agreement of all stockholders must be achieved to pass profits to the owners.
- Not more than 25 percent of the gross income can come from rents, royalties, dividends, and interests.
- No more than 80 percent of gross income can come from outside the United States.

This type of corporation also allows small groups such as production agriculturalists and small agribusinesses to gain the advantages of incorporating without the disadvantage of the double taxation. This type of corporation is very popular in the agriculture industry and its numbers are increasing rapidly.

Subchapter T (Cooperatives). This is the third type of corporation. Since cooperatives are nonprofit and are very popular in the agricultural industry, they will be discussed in detail later in the chapter.

Advantages of Corporations

There are several advantages to the Subchapter T type of business. A large corporation will generally find it easy to raise capital, or money, for expansion by issuing stocks. However, this is not true for small agribusiness corporations. In fact, it is difficult to sell stock in a small corporation since it represents a minority interest in a company that is not publicly traded. Large corporations are often formed when large amounts of money are needed to start or expand a business. Each stockholder has limited liability. This means that the stockholder is responsible for the losses of the corporation only to the extent of his or her investment. Because a corporation is a legal entity apart from the owners, the corporation does not dissolve if the owners sell their shares. Ownership can be transferred to new stockholders and the corporation goes on.[13]

There are added benefits in estate planning, such as giving gifts of corporate stock to children to transfer ownership of the corporation. In time, these gifts can pass the ownership of an estate to the children and avoid huge estate taxes at the death of the owner.[14] Refer to the advantages of profit corporations in Figure 5–5.

Disadvantages of Corporations

Corporations are complicated and costly to organize. More procedures are involved in forming corporations than for other types of business organizations. Expenses include filing fees, articles of incorporation, and initial legal and accounting expenses. Corporations are often saddled with complicated record-keeping systems. Double taxation is also a problem (as discussed earlier). Also, the limited financial liability afforded by the corporate structure is often restricted in small, family corporations due to requirements by lenders for personal guarantees.

Corporations

Advantages	Disadvantages
◆ More money for investment	◆ Initial cost
◆ Limited liability—personal assets are protected	◆ Paperwork
◆ The right size to do needed things	◆ Two tax returns
◆ Perpetual life	◆ Size causes slow response time to market changes
◆ Ease of ownership change	◆ Difficulty of termination
◆ Ease of drawing talented employees	◆ Double taxation
◆ Separation of ownership from management	◆ Possible conflict with the board of directors
◆ Corporation is a legal entity	◆ Complicated to establish
◆ Combined resources of shareholders	◆ Must follow state laws
◆ Stable level of production	◆ Complex organization
◆ Owners (stockholders) do not have to devote time to the company to make money on their investment	◆ Owners have limited control of business

Figure 5–5 An outstanding advantage of corporations is the ease of raising capital or money for expansion by issuing stocks.

Ending a corporation may be complicated and expensive. For example, for tax purposes, land is valued at market value instead of its original value when a corporation is forced to dissolve. Because the corporation pays taxes on the difference between market value and original value, if the value of the land increases, a corporation will probably end up paying extra taxes. The other types of business structures do not face this problem. Refer to Figure 5–5 for other disadvantages of such profit-making corporations.[15]

Establishing a Corporation

The process of forming a corporation varies somewhat from state to state. The articles of incorporation are usually filed with the secretary of state's office in the state in which the company incorporates. The articles contain:

❦ the corporation's name

❦ the names of the people who incorporated it

❦ its purposes

❦ its duration (usually perpetual)

❦ the number of shares that can be issued, their voting rights, and any other rights of the shareholders

❦ the corporation's minimum capital

❦ the address of the corporation's office

❦ the name and address of the person responsible for the corporation's legal service

❦ the names and addresses of the first directors

❦ any other public information the incorporators wish to include[16]

Before a potential corporation can so much as open a bank account or hire employees, it needs a federal tax identification number. To apply for one, you must get an SS-4 form from the IRS.

In addition to the articles of incorporation listed, a corporation also has bylaws. These describe how the firm is to be operated from both legal and managerial points of view. The bylaws include:

❦ how, when, and where shareholders' and directors' meetings are held, and how long directors are to serve

❦ director's authority

❦ duties and responsibilities of officers and the length of their service

❦ how stock is issued

❦ other matters, including employment contracts[17]

A typical corporation will have a structure similar to that in Figure 5–6.

COOPERATIVES

A cooperative is a corporation formed to provide goods and services to members either at cost or as near to cost as possible. Cooperatives are not formed to make profits, but to serve the people who own shares in the organization. Agribusiness cooperatives are very popular.

Kinds of Cooperatives

In the agricultural industry there are three kinds of cooperatives: supply (purchasing) cooperatives, marketing cooperatives, and service cooperatives.

Supply (Purchasing) Cooperatives. These act mostly as purchasing associations for such things as feed, seed, fertilizer, and fuel. Supplies for production agriculturalists are bought in quantity for resale to members. The big advantage is that by buying in large quantities, cooperative members are usually able to save money over what they would have paid individually.[18] In some cases, the supply cooperative can manufacture its own supplies instead of buying from a private company.

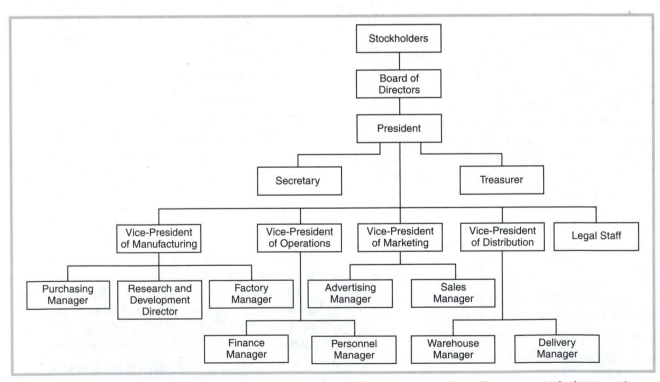

Figure 5–6 Many different structures may exist for a corporation. The schematic above illustrates a typical corporation structure.

Market Cooperatives. These mostly assist production agriculturalists in marketing agricultural products by finding buyers who will pay the highest price. Some marketing cooperatives process agricultural products such as milk and vegetables and sell them directly to consumers and retailers. Examples of marketing cooperatives are fruit growers' cooperatives, such as Sunkist, and dairy marketing, such as Land O' Lakes and Mid-America Dairy.[19]

Service Cooperatives. These cooperatives provide their members with a specific service, rather than a product, that members cannot afford to provide individually. Service cooperatives are not as numerous as marketing and supply cooperatives, but they do provide valuable services to many farmers. Specific examples of service-type cooperatives include farm credit services, banks for cooperatives, rural credit unions, mutual irrigation cooperatives, dairy herd improvement associations, and artificial breeding cooperatives. Rural electric and telephone cooperatives were once primarily made up of farmers, but now they serve all rural residents, which includes more nonfarmers than farmers in many areas. Therefore, these are now considered consumer cooperatives rather than agricultural cooperatives.

Cooperative Statistics

According to the National Cooperative Bank, 2 million businesses belong to 20,000 cooperatives (co-ops) nationwide.[20] A good example is the agricultural cooperative. Agricultural cooperatives have become a multibillion-dollar industry.

To give you an idea of the size of typical agricultural cooperatives, consider Farmland Industries, Inc. It is the country's largest agricultural cooperative, with an annual revenue of $4.5 billion. Its 1,800 member cooperatives include 250,000 farmers in nineteen states.[21] Farmland Industries owns manufacturing facilities, oil wells

and refineries, fertilizer plants, feed mills, and plants that produce everything from grease and paint to steel buildings. It also offers insurance and financial and technical services and owns a network of warehouses.[22]

Cooperatives and Membership

Some co-ops serve the general public; others serve members only. For those that serve the general public, the major emphasis is on the members. Voting stock and investment stock are separate in a cooperative, and only the former, also called **common stock**, gives a person the right to vote on business matters of the business. Investment stock, also called **preferred stock**, only gives a person the opportunity to invest in the business and, it is hoped, to receive a reasonable return on investment. In some cases, preferred stock is available only to holders of common stock, but some co-ops sell preferred stock to anyone who wishes to invest. In any event, members of the cooperative are encouraged to purchase preferred stock, whereas nonmembers are discouraged from doing so. The law currently limits the return on cooperative investment to an established percentage above the discount rate (the interest rate that the federal government charges for loans to member banks), regardless of how successful the business is in a given year.

Cooperatives and Control

Most cooperatives use a democratic system of control whereby each member has one vote only, regardless of how much business the member does with the co-op or how much investment stock he or she holds. Critics often contend that co-ops are run by a few members only, and in some cases this is true, but only because many members fail to exercise their right to vote during the annual stockholders' meetings.

Distinguishing Characteristics of Cooperatives

Three major distinguishing characteristics of cooperatives are:

❧ service at cost; any excess earnings are returned to patrons in the form of patronage dividends

❧ democratic control, or one member, one vote

❧ limited returns on investment

Cooperatives and Taxes

Many cooperatives are also exempt from paying a corporate tax provided they meet minimum cooperative requirements. At least 50 percent of their business must be with members, and all profits must be returned to the patrons in the form of patronage dividends. However, some cooperatives choose to pay the corporate tax and not be restricted by tax-exempt requirements.

Advantages of Cooperatives

Members risk only the amount they have invested. Cooperatives are owned and controlled by their members; therefore, the cooperative should be responsive to the members' needs. Cooperatives are designed to operate on a low level of profit in order to hold down costs for members. As mentioned earlier, when profits reach an established percentage above the discount rate, a portion is returned to the members. Because a portion of the profits go to the customer, the income tax paid by the cooperative is limited.[23] Refer to Figure 5–7 for other advantages of cooperatives.

Disadvantages of Cooperatives

The customers will almost never receive the full amount of their shares of the profit because some of the patronage dividend is generally withheld by the cooperative. However, the customer pays tax on the full amount of the patronage dividend received. Another disadvantage is that current customers will suffer from past business losses because the cooperative must either raise prices or use patronage refund money to cover those losses.[24] Refer to Figure 5–7 for other disadvantages of such cooperatives.

Cooperatives

Advantages

◆ Owners have limited liability
◆ Continues at death of shareholders
◆ Benefits go to members
◆ Economic benefits for members
◆ Members-shareholders share in direction of business
◆ Broad capital base
◆ Special tax advantages
◆ Legal entity
◆ Special antitrust exemptions

Disadvantages

◆ More legal formalities than sole proprietorship or partnership
◆ Members-shareholders have limited control over business
◆ Expensive to form, maintain, and dissolve
◆ A few persons may gain excessive power
◆ Restrictive charter requirements
◆ Lack of member understanding about cooperative (co-op) structure
◆ Lack of member participation
◆ Business community resentment against cooperatives
◆ Divided management authority

Figure 5–7 Agribusiness cooperatives are very popular. They are formed to provide goods and services to members at a reduced cost.

FRANCHISES

Many of us eat often at fast-food restaurants. Therefore, we are already familiar with some of the most popular franchises such as McDonald's, Wendy's, and Domino's. A franchise is a contract in which a **franchisor** sells to another business the right to use its name and sell its products. The **franchisee** (person purchasing the franchise) buys a system of operation that has proven successful.

Recently, there were over 540,000 franchised outlets in the United States.[25] Outside the United States, McDonald's alone had over 4,500 restaurants.[26] Government projections show that by the year 2000, franchising sales could account for nearly 50 percent of all retail sales in the United States.[27]

Characteristics of Franchises

The **parent company** (franchisor) **prepackages** all the business planning. Prepacking generally includes management training and assistance with advertising, selling, and day-to-day operations. In return, the franchisee agrees to run the business in a certain way.[28] Some people are more comfortable not starting their own business from scratch. They would rather join a company with a proven track record by entering into a franchise agreement. A franchise can be formed as a sole proprietorship, partnership, or corporation.[29]

Advantages of Franchises

The biggest advantage of a franchise is that the parent company helps the owners start their business. Often, the parent company will have a training program to teach the franchise about the business and set the standards of operations. It will help in choosing a location for the building and in arranging credit. The owner of the franchise benefits from advertising campaigns, and travelers will identify with the national franchise chain. By sharing its experience and proven techniques, the parent company cuts down the risk of failure. Franchising has offered products and services that are reliable, convenient, and cost

Franchises

Advantages
- Nationally recognized name and reputation
- Help with finding a good location
- Management system with successful track record
- Successful methods for inventory and operations
- Financial advice and assistance
- Training for owners and staff
- National advertising and promotional assistance
- Periodic management counseling
- Proven record of success
- Personal ownership

Disadvantages
- Large start-up and franchise costs
- Additional cost may be charged for marketing
- A monthly percent of gross sales may go to the parent company
- Possible competition from other, nearby franchises
- Little to no freedom to select decor or other design features
- Management regulations
- Many rules and regulations to follow
- *Coattail* effect if other franchises fail nationwide
- Restrictions on selling

Figure 5–8 Franchising sales will soon account for nearly 50 percent of all retail sales in the United States.

Career Areas: Agribusiness Owner/Manager, Agribusiness Partnership, Agribusiness Corporation, Agribusiness Cooperative, Agribusiness Franchise

The manager of an agribusiness, whether it is a sole proprietorship, partnership, corporation, cooperative, or franchise, provides expert advice, supplies, or services to the agricultural industry. The agribusiness manager needs a good educational background in business and accounting procedures. A college degree in an agricultural discipline is often required.

Education for agribusiness managers should begin at the high school level or earlier. The agribusiness is expected to have first-hand experience and detailed knowledge of procedures and techniques. Most agribusiness managers work their way up in the management system of the company or cooperative with which they are employed. As their management skills improve, they are given added responsibilities.

In order to be successful, an agribusiness manager needs to pay attention to detail and stay aware of the financial condition of the business. (Courtesy of USDA)

effective. Refer to Figure 5–8 for other advantages of franchises.

Disadvantages of Franchises

If franchises are so great, why aren't all businesses franchises? Unfortunately, franchises do have drawbacks. First, it takes a large amount of money to purchase most franchises. Second, the owner must share sales or pay a predetermined yearly fee to the parent company. Also, the parent company will limit the owner's management of the franchise. Refer to Figure 5–8 for other disadvantages of franchises. Refer to the Career Option for a variety of careers in different types of agribusinesses.

CONCLUSION

There are several different type legal structures for an agribusiness. You can start your own sole proprietorship, partnership, corporation, or cooperative, or you can buy into a franchise. There are many characteristics, advantages, and disadvantages of each. Before deciding on the appropriate legal structure, carefully evaluate each type of alternative and its risks. Remember, everyone has to start somewhere. Names like J. C. Penney, John Deere, Levi Strauss, and (Henry) Ford are businesses named after the real people who founded them. They started small, accumulated capital, grew, and became successful. Remember, you have to go out on a limb to get to the fruit.

SUMMARY

The major types of agricultural businesses are single (sole) proprietorships, partnerships, and corporations. Although cooperatives are really a form of corporation, they are sometimes listed as a fourth type of business. Another special form of business organization is the franchise, which is the fifth type of agricultural business.

No single type of legal structure is best for all agricultural businesses. Therefore, when developing a new agribusiness, all structures should be considered relative to their advantages and disadvantages. Although corporations are only 17 percent of the business establishment, they do 87 percent of the sales volume.

The single proprietorship is the major type of legal structure in the agricultural industry as well as in other major industries. It is the simplest type of business and the easiest to organize. The major disadvantage is that creditors can take the owner's personal assets as well as the business assets in pay of business debts.

Another type of agricultural business is a partnership, which is defined as a business association of two or more persons. There are two types of partnerships, general and limited. General partners own and manage a business together, whereas a limited partner invests in the business but may not participate in the management phase. The major advantages of a partnership include greater resources and specialization of talent. The greatest disadvantage is its unlimited liability.

Corporations are another way of doing business. There are three types of corporations: Subchapter C is a regular corporation, which sells stocks to investors; Subchapter S is basically for small businesses or a family; and Subchapter T is for cooperatives. The major advantage of corporations is the ability to raise capital or money for expansion by issuing stock. The major disadvantage is that they are complicated and costly to organize.

Cooperatives are corporations formed to provide goods and services to members at cost or as near to cost as possible. There are three types of cooperatives in the agricultural industry: supply (purchasing) groups, which act as purchasing associations for such things as feed, seed, fertilizer, and fuel; service cooperatives, which provide production agriculturalist members with a specific service rather than a product; and marketing cooperatives, which assist production agriculturalists in marketing agricultural products by finding the buyers who will pay the highest price. The major advantage of cooperatives is in being able to purchase lower-cost supplies. The major disadvantage is that members or shareholders will not receive the full amount of their shares of the profit (dividends) because a portion is withheld by the cooperative.

A franchise is a special type business with the right to use its name and sell its products. The parent company prepackages all the business planning, management training, and assistance with advertising, selling, and day-to-day operations. The biggest advantage of a franchise is that the parent company helps the owners start their business. The biggest disadvantage is that it takes a large amount of money to purchase most franchises.

END-OF-CHAPTER ACTIVITIES

Review Questions

1. Define the Terms to Know.
2. What are the three major types of agricultural business (business organizations)?
3. What are the three major types of corporations?
4. What are ten advantages of single (sole) proprietorships?
5. What are ten disadvantages of single (sole) proprietorships?
6. What are five recognizable factors of general partnerships?
7. What are four characteristics of limited partnerships?
8. What are nine advantages of partnerships?
9. What are nine disadvantages of partnerships?
10. Name five corporations.
11. Name five characteristics of corporations.
12. Name five agribusiness corporations.
13. What are eight requirements for a corporation to qualify for taxation under Subchapter S (small business or family corporations)?
14. What are ten advantages of profit corporations?
15. What are ten disadvantages of nonprofit corporations?
16. What are three major distinguishing characteristics of cooperatives?
17. What are nine advantages of cooperatives?
18. What are nine disadvantages of cooperatives?
19. What are five advantages of franchises?
20. What are five disadvantages of franchises?

Fill-in-the-Blank

1. Corporations comprise 17 percent of all business, but they do _____ percent of the sales volume.
2. A _____ is an organization owned by many people but treated by law as though it were one person.
3. There are _____ small businesses in the United States operating as Subchapter S corporations.
4. The articles of incorporation are usually filed with the _____ office in the state in which the company incorporates.
5. A _____ is a corporation formed to provide goods and services to members at cost or as near to cost as possible.

6. In the agricultural industry there are three kinds of cooperatives: _____ cooperatives, _____ cooperatives, and _____ cooperatives.

7. An example of a market cooperative for fruit growers is _____.

8. An example of a market cooperative for dairies is _____.

9. There are 2 million businesses that belong to _____ cooperatives nationwide.

10. Farmland Industries has 1,800 member cooperatives, which include _____ farmers in nineteen states.

11. Most cooperatives use a _____ system of control whereby each member has one vote only.

12. There are over _____ franchised outlets in the United States.

Matching

a. corporation

b. Subchapter C

c. marketing cooperative

d. Farmland Industries, Inc.

e. single (sole) proprietorship

f. Subchapter T

g. agricultural cooperatives

h. franchise

i. Subchapter S

j. partnership

_____ 1. 2.8 million of this type of business

_____ 2. 1.4 million of this type of business

_____ 3. 12 million of this type of business

_____ 4. John Deere

_____ 5. regular corporation that sells stocks to investors

_____ 6. small business or family corporations

_____ 7. cooperatives

_____ 8. act mostly as purchasing associations for such things as feed, seed, and fertilizer

_____ 9. country's largest agricultural cooperative

_____ 10. assist the production agriculturalist in selling agricultural products by finding the buyers who are willing to pay the highest price

_____ 11. buys the right from a parent company to use its name and sell its products

Activities

1. Jake has his own business, and Maria is interested in becoming a part of it. They need your help in avoiding problems when forming their partnership. In the following table, list any problems they need to avoid as they form their partnership and list any steps you think you could take to avoid problems.

Problems to Avoid	Steps to Avoid Problems

2. Draw the following table on a separate sheet of paper and fill in the blanks to describe the characteristics of each type of business.

Type	Ownership	Start-up Costs	Taxes
Single (sole) Proprietorship			
Partnership			
Corporation			
Cooperative			
Franchise			

Type	Liability	Responsibility for Decisions	Major Advantages and Disadvantages
Single (sole) Proprietorship			
Partnership			
Corporation			
Cooperative			
Franchise			

3. Answer each of the following:
    a. Have you ever considered starting your own business?
    b. What opportunities seem attractive?
    c. Write down the name of a friend (or friends) whom you might want for a partner.
    d. List all the financial resources and personal skills you need to start the business.

e. Make a separate list of the capital and personal skills that you and your friend(s) might bring to your new venture.

f. What capital and personal skills do you need but neither of you have?

4. Prepare a sample written partnership agreement. Use the partnerships section in this chapter, which lists everything that you need to include.

5. Prepare a sample articles of incorporation. Use the corporations section in this chapter, which lists everything that you need to include.

6. Prepare a sample set of by-laws. Use the corporations section in this chapter, which lists everything that you need to include.

7. Secure either a sample written partnership, sample articles of incorporation, or sample set of by-laws from the library, a local business, or any other source. Share the document with the class.

8. Identify the franchises in your community. If the owner of one of the franchises is available, ask him or her to identify three advantages and three disadvantages of franchises. Share your results with the class.

9. Identify the agribusinesses in your community. List whether it appears to be a single (sole) proprietorship, partnership, corporation, cooperative, or franchise. Also, identify whether it appears to use more than one legal structure.

NOTES

1. William G. Nickels, James M. McHugh, and Susan M. McHugh, *Understanding Business* (Chicago, Ill.: Richard D. Irwin, Inc., 1996), p. 148.
2. Ibid.
3. Ibid.
4. Betty J. Brown and John E. Clow, *Introduction to Business* (New York: Glencoe/McGraw-Hill, 1997).
5. Robert J. Birkenholz and Ronald L. Plain, *Agricultural Management and Economics* (Columbia, Mo.: University of Missouri, Instructional Materials Laboratory, 1988), p. II-2.
6. Neil E. Hard, *Farm Estate Business Planning*, quoted in N. Omri Rawlins, *Introduction to Agribusiness* (Englewood Cliffs, N.J.: Prentice-Hall, 1980), p. 31.
7. Ibid., p. 209.
8. Nickels, McHugh, and McHugh, *Understanding Business*, p. 153.
9. Brown and Clow, *Introduction to Business*, p. 230.
10. Birkenholz and Plain, *Agricultural Management and Economics*.
11. Peter Weaver, "Heightened Scrutiny of S Corporations," *Nation's Business*, February 1995, p. 37.
12. Nickels, McHugh, and McHugh, *Understanding Business*, p. 159.
13. Brown and Clow, *Introduction to Business*, p. 231.
14. Birkenholz and Plain, *Agricultural Management and Economics*, p. II-5.
15. Ibid.
16. Stephen L. Nelson, "How Not to Incorporate," *Home Office Computing*, February 1994, pp. 28–30.
17. Ibid.
18. Jasper S. Lee and Max L. Amberson, *Working in Agricultural Industry* (New York: Gregg Division/McGraw-Hill Book Company, 1979), pp. 95–97.
19. Jorean Huber, "All Together Now," *Entrepreneur*, April 1994, p. 14.
20. Nickels, McHugh, and McHugh, *Understanding Business*, p. 169.
21. Ibid.

22. Ibid.
23. Birkenholz and Plain, *Agricultural Management and Economics*, p. II-5.
24. Ibid.
25. Jeffrey A. Tannenbaum, "FTC Says Franchisers Fed Clients a Line and Failed to Deliver," *Wall Street Journal*, May 16, 1994, pp. A1, A10.
26. Carol Steinburg, "What's behind the Global Franchise Boom," *World Trade*, April 1993, p. 148.
27. Greg Matusky and John P. Hayes, "Franchising's Great Transformations," *Inc.*, April 1990, pp. 87–99.
28. Brown and Clow, *Introduction to Business*, p. 213.
29. Nickels, McHugh, and McHugh, *Understanding Business*, p. 163.

CHAPTER

6

Personal Financial Management

OBJECTIVES

After completing this chapter, the student should be able to:

- ❦ discuss earning money
- ❦ select a financial institution
- ❦ manage a checking account
- ❦ identify where your money goes
- ❦ plan and prepare a budget
- ❦ describe financial management and financial security tips and hints
- ❦ explain four ways to potentially retire as a millionaire
- ❦ explain key factors that make the millionaire plans work
- ❦ discuss how to achieve financial security

TERMS TO KNOW

balancing a checkbook
brokerage
compatible
compound interest
enterprises
entrepreneurship
extrapolate
fidelity shares
financial security
fixed expenses
gross pay

Individual Retirement
  Account (IRA)
investment portfolio
liquid income
locked into
mutual funds
net pay
part-time (avocational)
  enterprises
perseverance
"plastic prosperity disease"

prospectus
reconcile
risk management
tax-deferred
tax-deferred savings
term life insurance
therapeutic activity
Tax Shelter Account (TSA)
universal life insurance
utilization of resources
variable expenses

INTRODUCTION

In the opening ceremony of an FFA meeting it says, "George Washington was better able to serve his country because he was financially independent." The first step toward managing the finances of a business in the agricultural industry is becoming better able to manage your personal finances. People can better lead and receive greater respect when they have control of their money. Washington could better lead the country because he had no monetary problems. He could not be influenced financially by outside forces and would not be distracted from his presidential duties. Washington also used the same sound management techniques that made him successful in his personal life to further the status and well-being of his country.

You have many decisions to make in life. The decisions you make about how to handle money, however, will have a profound impact on your life. This chapter will show you how to manage your money in the most effective and responsible way. You should seek the help and advice of a trustworthy financial advisor as the need arises. Also, there is a High School Financial Planning Program published by the College for Financial Planning. Some banks also have high school financial planning curriculums. In the meantime, this chapter can serve to open your eyes to financial opportunities that you may not have been aware of previously.

Everyone needs to be a student of money management. J. Paul Getty, billionaire oil magnate, once said, "No man's opinions are better than his information." Hopefully, the information in this chapter, which includes how to budget, tips on money management, how to potentially retire as a millionaire, and achieving financial security through agricultural **entrepreneurship**, will help you form healthy opinions of money management and accumulating wealth.

EARNING MONEY

The career or type of work you choose and whether or not you choose to work once you get your job can be paramount to your financial well-being. You must plan your work and then work your plan. Select a job for which you have a natural aptitude and then do that job extremely well. The ideal job is one in which you have a vacation as a vocation.

Working hard at something that you are good at will sooner or later put you where you want to be. In most careers, you can simply outwork 80 percent of your cohorts and outsmart (knowledge) 15 percent of the rest, putting you in the top 5 percent, which always pays very well.[1] If you are good at something, it normally makes you more intense.

Gross versus Net Pay

This leads us to the point of receiving the paycheck. There are hidden costs to earning money. These include taxes and optional employee benefits, which are deducted from an individual's paycheck. The total earned before deductions is called **gross pay**. The amount of money left after deductions is called **net pay**, or "take-home" pay. Figure 6–1 illustrates a typical pay statement showing gross versus take-home pay.

Mandatory Deductions

Money is withheld from an employee's pay to cover the cost of government-mandated taxes and programs. These vary from state to state, and include:[2]

🌿 *Income taxes (federal, state, and local):* These taxes are based on the amount of money earned on a job or through investment income. Income taxes are used by federal, state, and local governments to provide public goods and services for the benefit of the community as a

PAY STATEMENT

AGRIBUSINESS SUPPLY
Main Street and First Avenue
Our Town, USA 54321

Employee: ARTIE F. DAVIS   **Pay period: 9/22 to 9/28**

HOURS		EARNINGS				
Regular	Overtime	Regular	Overtime	Bonus	Other	Gross Pay
20	-	100.00	-	-	-	100.00
DEDUCTIONS						
FICA	Federal With. Tax	State With. Tax	Health Ins.	Retirement	Other	Net Pay
7.20	10.80	3.60	-	-	-	$78.40
Year to Date	Regular	Overtime	Gross Pay	FICA	Federal With. Tax	State With. Tax
1988	1,920.00	-	1,920.00	144.00	216.00	72.00

Figure 6–1   Pay statement illustrating gross versus take-home pay.

whole. This includes taxes for education, military, roads, the legal system, and police and fire departments. State and city taxes vary, based on where you live and work. For example, seven states impose no income tax.

❦ *Social Security tax (FICA):* The Federal Insurance Compensation Act (FICA) is used to provide retirement or disability income for individuals or their survivors who have contributed to the fund.

❦ *Unemployment insurance:* This insurance provides income for qualified individuals whose employment has been terminated.

❦ *Workers compensation insurance:* This insurance protects individuals in the event of injury or illness that happens on, or because of, a job.

Employee Benefits

Most employers will offer special benefits, which help them attract and keep good employees. These benefits may be paid for by the employer, by the employee, or by a combination of the two. Sometimes the benefits are optional. Benefits may include:[3]

❦ *Life insurance:* pays a beneficiary (usually a family member) in the event of an employee's death.

❦ *Long-term disability insurance:* provides income for employees who become disabled and are unable to work.

❦ *Medical insurance:* covers some portion of employee medical and/or family dental expenses, often including preventive, diagnostic, basic, and major orthodontic services.

❦ *Retirement savings plan:* provides employees with an opportunity to save money on a **tax-deferred** basis. More will be discussed about this later in the chapter.

❦ *Profit sharing:* allows employees to share a portion of the profits the company makes.

❦ *Other:* some employee benefits may not require pay deductions. These include paid holidays, vacation time, sick pay, performance bonuses, and the use of a company vehicle.

Taxes and Benefits Summary

Federal, state, and local taxes are deducted on a withholding basis. This is simply a means of spreading taxes out over the year, rather than waiting until the April 15 filing deadline to pay all the money owed. Benefits cost an employer between 25 and 50 percent of an employee's pay. While the employee cannot spend them, benefits are worth money because they reduce out-of-pocket expenses. Refer to Figure 6–2 for a summary of taxes and benefits.

SELECTING A FINANCIAL INSTITUTION

Financial institutions help consumers manage, protect, and increase their money. People have a variety of financial needs, which vary at different stages of life.

Types of Institutions

There are several different types of institutions. These include:

❦ banks

❦ credit unions

❦ savings & loans (S&Ls)

❦ brokerage firms

In the past, each type of institution had a specific, limited range of services. Banks took deposits in the form of checking accounts, savings accounts, and certificates of deposits (CDs), and they granted credit to qualified individuals. S&Ls offered savings accounts and home mortgages. Credit unions made low-interest loans available to their members. **Brokerage** firms bought and sold stocks and bonds on behalf of their customers. Due to deregulation of the financial services industry, the lines among the different types of financial institutions have begun to blur. For instance, many banks now sell stock and bond mutual funds, while many credit unions, brokerages, and S&Ls now offer accounts that are very similar to checking accounts.[4]

Factors to Consider in Choosing a Financial Institution

Note the following suggestions when choosing a financial institution.

❦ *Types of service:* Do they offer the financial services I need?

❦ *Convenience:* What are their hours? Is there a location convenient to me? How many locations do they have? Can I do business with them from a remote location (by mail, over the phone, or by computer)?

❦ *Costs:* How do the costs compare with other institutions? What are the charges for the product or service I need? Examples include monthly service charges, annual fees, transaction charges, and interest rates.

❦ *Relationships:* Do I already have a relationship with this institution? Do I like the people? Do they know my financial history? Will it be to my advantage to do more business with them, since we already have a relationship?

❦ *Security:* Is my money insured? Will I get my money back if something happens to the institution?

Taxes and Benefits Summary

Mandatory taxes and benefits

Deduction	What is it?	Who pays?
Federal income tax	Funds those services provided by the federal government, such as defense, human services, and the monitoring and regulation of trade.	Employee
State income tax	Funds those services provided by state government, such as roads, safety, and health. (Not all states levy an income tax.)	Employee
Local income tax	Funds those services provided by the city or other local government, such as schools, police, and fire protection. (Not all areas levy an income tax.)	Employee
Social Security tax (FICA)	Provides income for retired or disabled employees or their survivors.	Employee and employer
State unemployment insurance	Provides income and other benefits to qualified people who have lost their jobs.	Employer
Worker's compensation insurance	Provides protection in the event that an employee sustains an injury or illness in the course of a job, or insurance because of it.	Employer

Employee benefits*

Benefit	What is it?	Who pays?
Life insurance	Pays a beneficiary in the event that an employee dies.	Employer or employee, or shared
Long-term disability insurance	Provides benefits in the event that an employee is completely disabled.	Employer or employee, or shared
Medical insurance	Employee and family insurance coverage for medical care expenses, including hospitalization, physician service, surgery, and major medical expenses.	Employer or employee, or shared
Dental insurance	Employee and family insurance coverage for dental care expenses, including preventative, diagnostic, basic, major, and orthodontic services.	Employer or employee, or shared
Retirement savings plan	A tax-deferred savings plan for retirement.	Employee (employer may contribute)
Profit sharing	A distribution of company profits to employees.	Employer

*Note: Whether or not these benefits are offered, and who will fund them, varies by company.

Figure 6–2   Taxes and benefits summary. (Courtesy of NationsBank)

❦ *Customer qualifications:* Am I qualified to do business with this institution? Do they require membership or have other criteria in order to do business?[5] For example, to buy a mutual fund, some companies require a $2,000 minimum investment to open the account.

Figure 6–3 offers an overview of financial services. Study this chart as you select the appropriate financial institution for you.

CHECKING ACCOUNTS

Use Caution

Keep your checkbook properly recorded and balanced. It is simple, but many people fail to do it. Bankers tell horror stories of people who bring in checking accounts so far out of balance that the only thing that can be done is close them out and start over with a new account. The math involved in keeping and balancing a checkbook is basic addition and subtraction.[6] Some students may wonder why checking account information is included here when you may have had a checking account for several years. The authors' experience, however, is that many high school and college students do better in algebra and trigonometry than they do in balancing a checkbook.

The biggest caution in keeping a checkbook is being rushed. People become hurried and forget to record the proper amount, especially when you are in a grocery store or some other busy checkout line (Figure 6–4). When they get home, they cannot remember the right amount. Sometimes the check is recorded, but the balance is never brought forward.

If you have trouble recording your checks, try using duplicate checks. Most banks sell an NCR paper or carbon check, which automatically records your check as you write it. However, you still have to carry your balance forward and **reconcile** your checkbook to the statement each month.[7]

Reasons for Writing Checks

Perhaps writing a check is the simplest way of withdrawing money—maybe even too simple. We tend to spend more money when we write a check compared to when using cash. However, checks are needed for many purchases. They are a convenient way to pay bills and protect money from loss or theft, and they are usually accepted as legal proof of payment. For many, checks are used as the accounting procedure to determine expenditures, especially in production agriculture. A properly written check will have the type of expense written on the check. As expenses are being calculated for income tax preparation, the checks are used as official records for the year, along with the cash receipts.

Figure 6–4   Even though you feel rushed in a checkout line, be patient and take the necessary time to record your purchase amount in your checkbook register.

Overview of Financial Services

DEPOSIT SERVICES

Checking accounts	The convenience and safety of paying by check instead of cash
Savings accounts and certificates of deposit (CDs)	Safe places to let your money grow
Automated teller machines (ATMs)	Easy access to your money from multiple locations, twenty-four hours a day
Direct deposits and automatic withdrawals	The ability to deposit money or pay bills automatically
Deposit insurance (such as FDIC)	The knowledge that your deposits are insured by the federal government for up to $100,000 per depositor. Agencies that provide this insurance are FDIC (banks) SAIF (savings & loans), or NCUA (credit unions)

CREDIT SERVICES

Credit cards	The ability to access credit conveniently up to the amount of your approved credit limit
Installment loans and credit lines	The opportunity to borrow for major items such as a new or used automobile, education, home improvement, and other personal or household items
Mortgages	The opportunity to borrow for a home purchase
Home equity loans	The ability to borrow against the equity in your home
Student loans	The ability to borrow at below-market rates to pay for a college education
Small business loans	The ability to borrow for financing the growth of a small business

INVESTMENT SERVICES

Retirement accounts (IRAs, SEPs, KEOGHs)	The ability to save money on a tax-deferred basis toward retirement
Stocks, bonds, and mutual funds	The ability to invest in corporations and governments in order to meet your financial needs for the future

TRUST SERVICES

Living trusts, testamentary trusts, estate planning and administration	The ability to ensure that your property will pass smoothly to your heirs with a minimum amount of tax

OTHER MONEY MANAGEMENT SERVICES

Safe deposit boxes	Safekeeping for important documents and valuables
Wire service	The ability to transfer money quickly and safely over long distances
Traveler's and cashier's checks	The safety of checks and the convenience of their accepted by parties that usually require cash
Remote banking	The ability to conduct banking transactions by mail, over the phone, or by Internet hookup

Figure 6–3   This overview of financial services will help you select the appropriate financial institution. (Courtesy of NationsBank)

Points to Remember When Writing Checks

Checks must be written accurately because the bank needs to understand exactly how much you want to give to whom. Also, a poorly written check can be tampered with. Some points to remember when writing checks include:

❦ Write in ink.

❦ Write clearly.

❦ Fill in all blanks.

❦ Use correct date.

❦ Place the decimal correctly.

❦ Write the amount close to the dollar sign so it cannot be changed.

❦ Write the amount in words as far to the left as possible on the space given; fill in the unused portion with a wavy line.

❦ Do not cross out the name or erase on the check. If you make a mistake, tear up the check and write another.

❦ Never sign a blank check.

❦ Keep your checkbook in a safe, protected place.[8]

Steps in Writing a Check

❦ Write the correct date.

❦ Write the name of the person or company you would like to pay.

❦ Enter the amount of the check in numbers, including a decimal point to show cents. Start the numbers as close to the dollar sign as possible.

❦ Enter the amount of the check in words. Start writing from the far left side of the line. Follow the dollar amount by the word and then write the cents amount as a fraction, over 100. (If there are no cents, use 00.) As stated previously, draw a wavy line from the

end of your writing to the end of the line, so there is no additional room to insert words or numbers.

❦ Sign your check the same way you signed the signature card when you opened your account.

❦ Write down the purpose of the check. You may also use this space to write the account or invoice number, if you are paying a bill.

Figure 6–5 shows a check completed using the preceding steps.

Keeping a Checkbook Register

A checkbook register is the small booklet that you keep with your checkbook to keep a record of deposits made and checks written. It allows you to tell at a glance the amount of money in the bank. There is nothing hard about keeping a checkbook register. You simply must take the time to do it. Remember being rushed in the checkout lane? Reasons for keeping a check register are that it

❦ shows balance at all times

❦ helps avoid overdrafts

❦ keeps a record of checks written

❦ keeps a record of deposits made

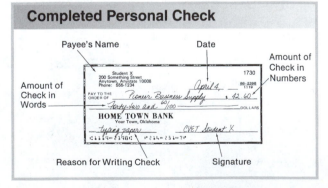

Figure 6–5 Completed personal check (Courtesy of CIMC, Oklahoma Department of Vocational and Technical Education)

🌿 shows dates of checks and deposits

🌿 provides a record to compare with bank statements.[9]

Figure 6–6 shows a checkbook register with an example of an entered check, deposit, and service charge.

Entering Checks Written. By referring to Figure 6–6 you can see the following steps:

🌿 Write the check number (173).

🌿 Write the date (1/22).

🌿 Write the name of the person or business to whom the check will be written (Tractor Supply Company).

🌿 Write the amount of the check in the correct column ($34.00).

🌿 Subtract the check from the balance on scratch paper or use a calculator.

🌿 Write in the balance ($140.00).

🌿 Repeat each of the steps each time a check is written.

Entering Deposits. By referring to Figure 6–6 you can see the following steps:

🌿 Write the date (1/28).

🌿 Write the word *deposit* on the line.

🌿 Write the amount of the deposit in the proper column ($100.00).

🌿 Add the balance on scratch paper or use a calculator.

🌿 Write in the balance ($240.00).

🌿 Repeat each of the steps each time a deposit is made.

Entering Service Charges. Service charges can be located on the monthly bank statement. These may be monthly bank services, checks or (hopefully not) overdraft charges. By referring to Figure 6–6 you can see the following steps:

🌿 After locating any services charges on the bank statement, write service charge on the line in the check register.

🌿 Write the amount of the service charge in the check column ($5.00).

🌿 On scratch paper or with a calculator, subtract the service charge from the balance.

🌿 Write the new balance ($235.00).[10]

Balancing (Reconciling) Your Checkbook

Your checking account statement is a complete record of all the activity in your account (deposits and withdrawals) for the previous thirty-day cycle, or period. You will want to take time each month to compare, or reconcile, this information with your checkbook register, to make sure your records agree with those of the bank. This process is called **balancing a checkbook**.

Importance of Balancing Your Checking Account Monthly. Although banks do make mistakes, the monthly account statement is an accurate picture of how much money is really in the account and available to be spent. If your checkbook is not balanced, you are in danger of spending more money than you have available. Other dangers include:

Checkbook Register

NUMBER	DATE	DESCRIPTION OF TRANSACTION	PAYMENT/DEBIT (-)	√ T	FEE (IF ANY) (-)	DEPOSIT/CREDIT (+)	BALANCE 174 00
173	1/22	Tractor Supply Company	$ 34.00	√	$	$	140 00
	1/28	DEPOSIT				100.00	240 00
		SERVICE CHARGE	5.00	√			235 00

Figure 6–6 Checkbook register with an example of an entered check, deposit, and service charge.

❦ There may be a steep financial penalty for "overdrawing" your account.

❦ If you overdraw your account often, it could affect your ability to get credit; the bank may even ask you to close your account.

❦ Unauthorized use of your automated teller machine (ATM) card will go unnoticed and unreported.

❦ Bank errors can go unnoticed; it is your responsibility to notify the bank of these errors in a timely manner.

❦ You may actually have money available to you of which you were not aware.[11]

Common Reasons Why Checkbook Registers and Account Statements Differ

❦ Checks you have written have not yet been cashed.

❦ Checks you have deposited have not yet cleared and been added to your balance.

❦ Interest has been automatically deposited to your account (on interest-bearing accounts).

❦ Fees have been charged to your account, such as monthly service fees and check printing fees.

❦ You have neglected to record checks or deposits in your checkbook register, especially ATM transactions, direct deposits, and automatic withdrawals.

❦ You have made an error in addition or subtraction when calculating the running balance, or you have transposed numbers (for example, $89.30 instead of $98.30).

❦ There has been unauthorized use of your ATM card or checkbook.

❦ The bank has made an error.[12]

Steps in Balancing (Reconciling) a Bank Statement. Compare the bank statement and your checkbook register. When your checkbook

total and the bank statement total agrees (after subtractions and additions have been made), you have reconciled your checkbook with your bank statement. Here are the steps to balancing your checking account:

❦ Subtract service charges from the checkbook register that are shown on the bank statement. Refer to Figure 6–7.

❦ Arrange canceled checks numerically (170, 171, 172, etc.).

❦ Compare the amounts on canceled checks and deposits in the bank statement with amounts written in the checkbook register. (Figure 6–8 shows a simplified bank statement).

❦ Check off all canceled checks and deposits in the checkbook register. (Refer to Figure 6–7.)

❦ List and add total outstanding checks (those not returned to bank or checked in your register) on the back of your bank statement (Figure 6–9). Be sure to list the check number and the amount of the check.

❦ List and add total deposits.

❦ Add total outstanding deposits to balance ($100.00).

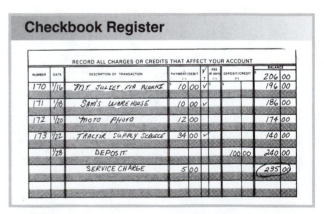

Figure 6–7 Checkbook register showing subtracted service charges.

Checkbook Register

Date of Statement: January 31, _____

DATE	#	CHECK	DEPOSITS	BALANCE
1-17	170	$10.00		$196.00
1-19	171	$10.00		$186.00
1-23	173	$34.00		$152.00
1-29		$5.00 SC		$147.00
				LAST AMOUNT IN THIS COLUMN IS YOUR BALANCE

Figure 6–8   Bank statement showing checks, check number, and date that each check cleared the bank. This bank statement also shows a $5.00 service charge.

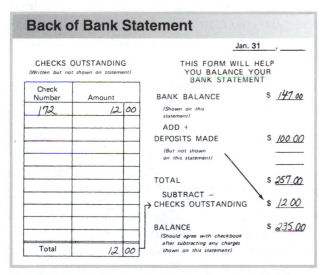

Figure 6–9   Back of bank statement used in balancing (reconciling) your checking account.

🌿 Subtract total outstanding checks from total of bank balance ($147.00) and outstanding deposits ($100.00), or ($257.00 – $12.00 = $235.00).

🌿 Make sure that the adjusted bank balance and adjusted checkbook balance are equal.

🌿 Circle the checkbook total to show that you agree with the bank.[13] If you forget how to reconcile your statement, check the back of the statement for instructions, or your bank can help you.

WHERE YOUR MONEY GOES

You must gain control over your money or your money will gain control over you. Money is active. Time, interest rates, amounts, cash flows, inflation, and risk all intermingle to create a current that is ever-flowing. Dave Ramsey in his 1993 book, *Financial Peace*, said that money is similar to a beautiful horse. It is very powerful and always moving and in action, but if not trained when young, it will become an out-of-control dangerous animal when it grows to maturity.

Handling Money Takes Discipline

Handle your money correctly, or it will handle you. Money and finance are not passive. They require you to take the initiative. This takes discipline. You will have conflict, worry, shortages, and a general lack of fun and funds unless you achieve some discipline in the handling of your money.[14] You discipline your spending with both fixed and variable expenses.

Fixed Expenses

A fixed expense is a set amount of money due on a set date. As a rule, **fixed expenses** must be paid when due. Although we do not have control over the types of fixed expenses, we do have some control on their amounts. Although you are still a student and many of the examples will not apply to you, they will certainly apply in the next one to five years. Several examples of fixed expenses follow.

Mortgage or Rent Payments.   Obviously, this is an unavoidable expense once you "get out on your own." This could also include property tax and insurance if these are a part of your house payment. Your rent or house payments should not exceed 25 to 35 percent of your income.

Other Real Estate Payments. These could include a second mortgage, home improvement loan, storage rental, pasture for horses, or some other item.

Income and Property Taxes. This refers to taxes not taken directly out of the paycheck. Examples are federal and state taxes, past taxes, and property taxes not included with the house payments.

Installment Contract Payments. Many financial counselors advise against incurring debt. (This will be discussed fully later.) Examples include car payments, furniture, appliances, and many other items.

Insurance. Some insurance is not taken directly out of the paycheck or not included in the house payment. These payments can be real "budget busters" since they are often due every three months. An envelope technique (discussed later in the chapter) can help manage these "budget busters." Examples include real property (fire, theft, liability), personal property (homeowner's, renter's, auto), life, and health insurance.

Regular Contributions. Examples are contributions to the United Way or other charities. Many people believe this is the humanitarian thing to do. Steel magnate Andrew Carnegie spent the first half of his life attaining wealth and the second half giving it away. St. Ambrose said, "Just as riches are an impediment to virtue in the wicked, so in the good they are an aid of virtue."

Dues. The type of job that you have determines whether dues will be a fixed expense. Teachers, doctors, lawyers, and real estate agents pay dues to professional organizations. Many hourly workers pay dues to labor unions.

Savings More will be discussed about savings later. You must develop the conviction that savings constitute a fixed expense even though you could argue that they do not. If you do not develop an emergency fund, you are destined for pain, embarrassment, and potential disaster.

Variable Expenses

Variable expenses vary in amount and frequency and offer the best opportunities for adjustments in a money plan. Be conservative with your variable expenses, or they will control you. Carefully consider your buying motives. Why do you want or think you need this item? Could you live without it? Do you want it for selfish reasons, like showing up the neighbors, or does the item have utility to you? You must develop power over purchase rather than letting the purchases continue having power over you.

Utilities. You probably have the most control over gas and electricity bills. Moderating thermostat settings, conserving electricity, and making maximum use of insulation can make a big difference in monthly expenses. Telephone and water bills can also be monitored. Garbage, sewage, and cable TV are other possible utility charges.

Charge Accounts. Many an expert has written on the evils of credit cards. These can include department stores, oil and gas companies, major credit cards, and mail order companies. Credit cards are dangerous and horrible financial tools. The convenience of plastic makes you buy things you would never buy otherwise. You can easily get **"plastic prosperity disease."** You see, no cash passes from your hand, so you register very little emotional realization that you just spent money.[15]

Medical/Dental Bills Not Covered by Insurance. Even though you have insurance, most plans only cover 80 to 90 percent. My son recently had surgery on his elbows. He checked in the hospital at 6:00 A.M. and got out the same day at 3:00 P.M. The total bill for the hospital, doctor, and related services was $14,200 for nine hours. Even with insurance, unless you have an emergency fund set aside, medical and dental bills can cause financial hardship. Prescriptions, office calls, and eyeglasses are also included in this area.

Transportation. Car payments are a fixed expense; however, gas, oil, repairs, license plates, your driver's license, and servicing must also be considered when making a budget.

Household Maintenance and Repair. Some examples in this area include gardening, cleaning supplies, appliance and TV repairs, electrical repairs, and the dreaded unexpected expenses.

Child Care. Besides food and clothing, there will be some babysitting expenses. Depending on your lifestyle, there could also be day care expenses. Make sure that the second income is not an economic myth. What may appear on the surface to be an economically correct thing to do may prove, with some serious budgeting, to be infeasible (Figure 6–10).

Food. This also includes all the nonfood items that are part of your supermarket bill. Eating out can go into this category, or it can go with recreation/entertainment. A little hint to consider is this: The power of cash is an emotional negotiating tool. We buy less when we use cash. It is much easier to sign a check or a credit card receipt than it is to lay down cold, hard cash to pay for something. When you spend cash, it hurts, so you will end up spending less. Switching to paying cash for groceries and eating out, instead of using checks and credit cards, has saved many people over $100 per month.[16]

Personal Maintenance. A part of self-concept includes making ourselves look and feel good. Examples of expenses are clothing purchases; laundry, barber and beauty salon visits; and grooming and health products.

Second Income ($18,000 Annual Salary)		
Monthly	**Yearly**	
$1,500	$18,000	Income
		Minus:
–550	–6,600	Taxes and payroll deduction
=$950	=$11,400	**Take-home pay**
		Minus:
–100	–1,200	More extensive wardrobe
–45	–540	Extra dry cleaning
–50	–600	Extra mileage depreciates car
–50	–600	Extra maintenance on car
–560	–6,720	Day care, two kids, $65 each per week
–50	–600	Extra meals out due to fatigue
=$95	$1,140	**Real Net Income**

Figure 6–10 In some cases, earning a second income may not be the economically correct thing to do, as shown by this real-life example. (Courtesy of Dave Ramsey, *Financial Peace.* Lampo Group, Nashville, TN, 1993)

Self-Improvement/Education. We all want to stay up-to-date and know what is happening. Leaders continually try to improve themselves and learn. Examples of expenses are books, magazines, newspapers, seminars, music lessons, tuition, and travel expenses.

Recreation/Entertainment. Recreation and entertainment are very important and should be included in a budget. Examples of expenses in this area are movies, sports, restaurants, vacations, parties, cassettes, CDs, and photographs.

Other Expenses. There are always unexpected expenses so budget for them. Expect the unexpected. It can come in the form of breaking down, children, injury, loss of work, and so on. There are gifts (graduation, birthday, or wedding), traffic tickets, Christmas decorations, and presents, among many other expense items.

Figure 6–11 shows a pie-shaped graph giving the portion of your income for fixed and variable

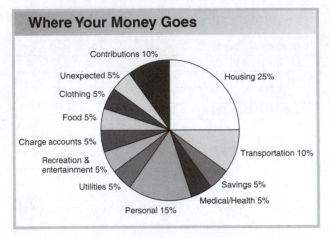

Where Your Money Goes

Contributions 10%
Unexpected 5%
Clothing 5%
Food 5%
Charge accounts 5%
Recreation & entertainment 5%
Utilities 5%
Personal 15%
Housing 25%
Transportation 10%
Savings 5%
Medical/Health 5%

Figure 6–11   A compilation from several sources showing recommended percentages of fixed and variable expenses. Obviously, these percentages will change somewhat according to your income.

expenses. It is a compilation of several sources. These are only recommended percentages, and they will change dramatically accordingly to your income.

I hope this section on fixed and variable expenses did not depress you, but I hope it did motivate you to give special attention to the next section, which covers budgets and budgeting.

PLANNING AND PREPARING A BUDGET

A budget is a plan for spending and saving money. A good budget is an essential tool in every household. This fact comes as no surprise to most of us, yet 90 to 95 percent of American households operate without a detailed, accurate written outline of income and expenses (a budget). By failing to plan, we are planning to fail, and there will always be too much *month* left at the end of the *money*.[17]

Benefits of a Budget

Everyone needs a written budget. Creating a budget guards you from being blindsided by an unexpected expense or a forgotten bill. It also serves as a tangible reminder that you cannot afford to take out another loan or you do need to eat leftover spaghetti tonight rather than eating out at the local restaurant.

Budgets are constricting at first, but after a time they actually provide greater freedom. A correctly prepared budget is not a form of torture. On the contrary, a proper, simple, written plan will actually give you more free time, and the money with which to enjoy it. Budgets can help build your confidence.

Other benefits of a budget include:

❧ A budget puts you in control of your financial future.

❧ A budget ensures that you do not spend more money than you earn.

❧ A budget helps you prepare for major periodic expenses, such as car insurance, medical insurance, and vacations.

❧ A budget helps you save money to prepare for unpredictable expenses, such as medical bills or major car repairs, as well as for special opportunities, such as a sale on needed furniture or a special family event in another city.[18]

Still Living at Home Budget

You obviously do not have as many expenses while still living at home. However, you should start immediately preparing a personal budget. Figures 6–12 and 6–13 show two simple examples. Figure 6–12 shows a weekly budget, and Figure 6–13 shows a budget that is more appropriate as a monthly or even yearly budget.[19]

Personal Budget
One Week

Income (after taxes)	
Part-time job	$53.00
Allowance	12.00
Total	$65.00
Savings & Spending	
Savings	$15.00
Food	8.00
Clothing	13.00
Entertainment	10.00
Gifts & contributions	6.50
Transportation (bus)	7.50
Miscellaneous	5.00
Total	$65.00

Figure 6–12 Example of a personal weekly budget while still living at home

Personal Budget

NAME: ______________________

TIME PERIOD: ______________________

INCOME (after taxes)	
1	$ ________
2	________
Allowance	________
Other ____________	________
TOTAL	$ ________
EXPENDITURES	
Savings (emergency/opportunity)	$ ________
Food	________
Clothing/shoes	________
Entertainment	________
School supplies	________
Gifts and contributions	________
Transportation	________
Insurance	________
Loans/credit payments	________
Other ____________	________
Miscellaneous	________
TOTAL	$ ________

Figure 6–13 Example of personal monthly or yearly budget while still living at home.

Living on Your Own Budget

Once you move out of your family's home to attend college or simply start living on your own, the household and monthly expenses "step up another notch." Many things that you took for granted now are major expense items. Examples are rent, electricity, telephone, laundry and dry cleaning, groceries, and even "medicine cabinet" supplies. If you have a roommate, some of the expenses can be shared. Figure 6–14 gives an example of items to consider in a budget for people living on their own. It also includes columns in case there are any roommates.

Family Budgets

For those who choose to start a family, the budget is somewhat different due to added responsibility. A good budget changes as your life changes. You will need to make periodic adjustments. You may have budgeted too little for some areas and be strained; you will need to adjust. In other areas you will have budgeted too much. It will take you a few minor adjustments to adjust the plan to a realistic level.

A budget is not intended to complicate your life; on the contrary, when you begin to know where your cash is flowing, it will make life easier. You cannot possibly realize you are spending too much on a category if you do not track your spending.

Figure 6–15 offers an example of a family budget. A few items require some explanation. Charitable gifts are technically optional, but they may

Living on Your Own Household Expenses

START-UP EXPENSES

Expense	Cost	Cost per roommate			
		#1	#2	#3	#4
First/last month's rent					
Security deposit					
Telephone deposit					
Moving expenses					
Refrigerator, if needed					
Small appliances, if needed					
Total					

MONTHLY EXPENSES

Expense	Monthly cost	Annual cost	Fixed, flexible, discretionary?	Shared? (yes/no)	Cost per roommate			
					#1	#2	#3	#4
Housing								
Rent								
Gas, oil, and water								
Telephone								
Renter's insurance								
Food								
Groceries								
Dining out								
Transportation								
Car payments								
Gasoline, oil								
Car insurance								
Car maintenance and repair								
Public transportation								
Personal								
Clothes								
Dry cleaning, laundry								
Education (tuition and books)								
Personal allowances								
Health								
Doctors and dentists								
Prescriptions								
Health insurance								
Savings and investment								
Savings								
Recreation								
Vacation								
Movies, theater, cable TV								
Subscriptions								
Other								
Gifts and contributions								
Credit card payments								
Bank charges								
Total expenses								

Courtesy of Nations Bank

Figure 6–14 Examples of a budget of household expenses for those living on their own or with roommates. (Courtesy of NationsBank)

not be an option to many people psychologically. Some people value helping humankind in some fashion. Emergency and retirement funds will be discussed later. What is the discretionary income item? Plan to blow, waste, or not account for some portion of your money. Even if you do not plan for this, you will do it anyway.[20] The problem is not that we do not recognize this category, the problem is that most people's discretionary income involves their entire budget. A regimented budget that is too tight is not realistic. If you do not allow some cushion in your plan, you will fail and say that the budget did not work for you.[21]

FINANCIAL MANAGEMENT AND SECURITY TIPS

Sixteen tips are given here to help you to better manage your money. These "sweet sixteen" money tips can change your financial life if you adhere to them.

Limit Debt

Do not believe that debt is a way of life. Except for a home mortgage, there are ways to stay out of debt if you develop the proper mindset. Unfortunately, we want our dessert first. We borrow as if we had forgotten that we have to pay it back.

If you had no debt, how much money could you save every month? We will learn later in this chapter about the magic of compound interest. With a totally freed-up budget, you could be wealthy— wildly wealthy—within just a few years. However, we are strapped with debt. An old Mideast proverb says, "The rich rule over the poor and the borrower is servant to the lender."

Do Not Use Credit Cards

The typical card holder carries five to seven cards. Many financial counselors warn about the evils of credit cards. They are dangerous and horrible financial tools. Their convenience makes you buy "stuff" you would never buy otherwise. You get the "plastic prosperity disease," where you appear prosperous but are really digging your own grave. You see no cash pass from your hand, so you register little emotional realization that you just spent money.

Most people say they can control their credit card use. However, therapists who specialize in addictions say that the first level of treatment always involves denial, which they consider to be a strong indicator of an addiction problem.[22]

Save Big Money on Home Mortgages

If you must borrow, borrow on short terms and only on items that go up in value. That means you never borrow on any consumer items except homes, land, or livestock. The terms are very important. If you can, buy less so you can pay the loan off faster. Make sure you get the lowest possible interest rate. You may pay more than three times the amount borrowed on a thirty-year mortgage if you pay normal payments for the full term. Less than 10 percent of your payments are applied to the principal for the first fifteen years at an interest rate of approximately 10 percent. Furthermore, it takes approximately twenty-four years to reduce your mortgage to one-half the amount financed. Consider the following examples:[23]

Fifteen-Year Rather than 30-Year Loans. If you were to finance $80,000 on a home at 10 percent, here are two ways:

    30 years: 360 payments at $702 per month = $252,720 total

    15 years: 180 payments at $860 per month = $154,800 total

The total difference is $158 more in payments monthly, but savings of $97,920.

Pay an Additional Payment per Year. The effect of paying one additional payment per year

Family Budget

	Amount Paid Monthly	Total Annual Amount
Charitable Gifts		
SAVINGS		
Emergency fund		
Retirement fund		
College fund		
HOUSING		
Mortgage/Rent		
Furnishings		
Property taxes		
Electricity		
Heating		
Water		
Garbage collection		
Telephone		
Yard work		
Repairs/Maintenance		
Remodeling		
Cable		
TOTAL		
FOOD (Toiletries, Cosmetics, etc.)		
CLOTHING		
TRANSPORTATION		
Car payments		
License plates		
Gas and oil		
Maintenance/Repairs		
Parking		
New tires		
TOTAL		
ENTERTAINMENT RECREATION		
Eating out		
Babysitters		
Periodical subscriptions		
Vacations/Holidays		
Clubs		
Hobbies		
TOTAL		

	Amount Paid Monthly	Total Annual Amount
MEDICAL EXPENSES		
Hospitalization Ins.		
Physicians		
Dentists		
Optometrists		
Pharmacists		
TOTAL		
INSURANCE		
Home		
Life		
Disability		
Automobile		
TOTAL		
CHILDREN		
School lunches		
Allowances		
Tuition		
Lessons		
Club dues		
Child care		
TOTAL		
GIFTS		
Graduations		
Holidays		
Birthdays		
Anniversaries		
Weddings		
TOTAL		
MISCELLANEOUS		
Organizational dues		
Cleaning/Laundry		
Pet care		
Beauty/Barber		
Credit Card 1 (Hopefully 0)		
Credit Card 2		
Credit Card 3		
Discretionary $$		
TOTAL		
TOTAL LIVING EXPENSES		

Figure 6–15 Example of a budget for those who choose to start a family. Special features of this budget include savings for an emergency fund, retirement fund, and college fund. There is also a discretionary item.

on your home mortgage substantially reduces the principal.

> 30 years: $75,000 at 10.75 percent; monthly payment = $700.11
>
> (360 payments = $252,740 principal and interest)

Assume from each monthly payment of $700.11 that $534.00 is interest per month, and $157.11 is paid on the principal. By paying one additional payment per year of $700.11, you will pay four and one-half months ($700.11 divided by $157.11) of principal, thus reducing your mortgage payoff in direct proportion. In other words, you can pay off your loan in approximately one-third less time by making one extra payment per year. The savings is approximately $84,000.

Divide the Total into Two Monthly Payments. By dividing the monthly amount into two monthly payments rather than one lump sum, you will have savings similar to the additional payment per year just discussed. However, all loans are not set up this way. If your principal can be reduced by extra payments, then it can be reduced quicker by the divided payments.

Increasing Retirement Benefits. To retire with comfort, use one of the three techniques just discussed. This will lead to a thirty-year mortgage being paid in full nine years, eight months, early. Since your budget is adjusted to this payment, using self-discipline, continue until the remainder of the proposed mortgage length is paid off, but place this monthly payment in an investment. Depending on the rate of interest earned at the end of thirty years, not only will your home be paid for, but over $100,000 will now be available for your retirement. If your original payments were $700.11 per month and you put that amount in investments for the remainder of the original mortgage term, you will see what happens by examining Figure 6–16.

Reduce Debt by Using the Debt Snowball Technique

Rationale. If you accumulate debt, what is the best way to get out of it? The best way is the "debt snowball" technique. The first step in this technique is to put your debts in ascending order, with the smallest remaining balance first and the largest last. Do this regardless of the interest rate or payment. These will be paid from top to bottom. This works because you get to see some success quickly and are not trying to pay off the largest balance just because it has a high rate of interest. Consider the example in Figure 6–17.

Interest Rates	Total Cash
At 0%	$ 81,213
At 5.5%	$106,880
At 8%	$121,969
At 10%	$135,988
At 12%	$152,036

Figure 6–16 By paying your home off early and then continuing to put the remainder of the payments in savings, a substantial amount of money can be accumulated.

Debt Snowball Payoff Technique

ITEM	BALANCE	PAYMENT	RATE
Sears	400	60	18%
Castner Knott	700	70	18%
Den Charge	1200	200	18%
Medicount	6500	250	12%
Credit Union	7000	123	9%
House	60,000	540	9%

Figure 6–17 The debt snowball payoff technique works by arranging debts in ascending order, with the smallest remaining balance first and the largest last, regardless of interest rate. Once the first bill is paid off, apply that amount of money to the next bill. Continue this procedure until all bills are paid, except, perhaps, your mortgage.

Procedure. Follow this strategy as you refer to Figure 6–17. Suppose you get an extra job or have a garage sale and can pay off the Sears card in the first month. Next, do not spend the $60 per month you used to spend on the Sears card; instead, add $60 to the next payment on the list. You are then paying Castner Knott $130 per month until paid. Castner Knott will be paid off in the seventh month. Next, add $130 to the $200 Den Charge payment to make your payments $330. Because you have already been paying on the Den Charge for seven months, it will be paid off in the eighth month (the next month).

Continue the "snowball" by adding $330 to your Medicount payment of $250, making your Medicount payment $580. This will cause Medicount to be paid off in seventeen more months, only twenty-five months into the program. Continue this process for the Credit Union to pay it off in four additional months.

What About the House? Since you are having so much fun, you may be wondering how to pay off the house. Now you have $1,243 per month with which to pay on the house, and it will be paid for in five more years.

I Need a Car. As an alternative, you could save $275 per month of the $1,243 to purchase a car with cash. When you pay the remaining $968, the house will be paid off in just seven years, while the $275 per month saved will grow to $11,490 in just three years (at 10 percent) for your cash car purchase.[24]

Attitude on Debt. As these bills are paid off, you will change your attitude about financial matters. Debt will be used only cautiously. Depression and misery do not have to control you. Do not get into a debt mess again. If you do, remember you will instantly become the servant of the lender.

Develop Power over Purchases

You must gain control over your finances or they will gain control over you. Money has no morals. Money is not the root of all evil. The love of money is the root of all evil. Extreme amounts of money or extreme lack of it magnifies character.[25] Develop power over purchase and be aware of "buyer's fever." You can lose all sense of reason. You want it, and you justify it by saying that you desire it. Quite frankly, you go crazy. Do not get the fever. You will lose all patience and negotiating power. If you feel you have to have it, sleep on it and wait forty-eight hours before making your buying decisions. There will be another "one-time" opportunity. Develop good spending habits as you develop power over purchase.

Live within and below Your Income

Avoid the lifestyles of the rich when you are not rich. Henry David Thoreau said, "Almost any man knows how to earn money, but not one in a million knows how to spend it."[26] You must limit your style of living, figure out what your income is, and proceed to live far below that mark.

Paying Cash for a Car versus Going into Debt

Suppose you were to purchase a new car for $16,000 with a debt is $300 per month. What if instead you bought a preowned car for $5,400 at $100 per month and saved the other $200 per month at 10 percent for seven years, giving you $24,190 at the end. At the end of seven years, either car will be nearly worn out, but let's say you saved that $24,190 and went out and bought a $16,000 car for cash from that savings, leaving $8,190.26.

You now have a new car with no car payments and $8,190 in the bank, but let us go for yet another seven years. You have no car payment, so instead of the $100 payment and the $200 savings you were putting aside every month, now you decide just to save $100 per month, freeing up $200 per month. At the end of seven more years, that leftover $8,190, plus $100 per month and the 10 percent interest, will grow to $28,539. Again, you have a seven-year-old car that is worthless, so you buy another $16,000 car for cash, leaving $12,539 in savings. For the next seven years you will have no car payment, and though you contribute no more to savings, the $12,539 at 10 percent will grow to $24,436.[27]

Negotiate and Buy Only Big Bargains

Some people spend more time deciding which brand of cat food to buy than in selecting a major household appliance. If you do not spend time in comparison shopping, you may spend several hundred dollars needlessly and miss out on features that you could have received for the same price. There are three essential things to remember when buying: you must learn to negotiate, you must learn to find bargains, and you must have patience.

Everything you buy is negotiable at some time and at some place, and you must find the bargain. Do not be an impulse buyer. Most people are very anticonfrontational by nature. If you understand that people do not want to meet head-on, it will aid you in your negotiations. Make sure your purchases involve a "win-win" situation, and you will buy things very cheaply. If someone needs to sell very badly, you help them by making a purchase. You must win, too, and your winning can occur in the area of price or terms. Here are seven basic principles of negotiating.

1. *Always tell the truth.* There is never enough money to be made, or lost, on a deal to warrant a lack of integrity.

2. *Use the power of cash.* People get silly when they see cash. There is something highly emotional about the effect of flashing cash when making a purchase.

3. *Understand and use "walk-away power."* You must be prepared to walk away and not make the purchase. If the seller senses (and he or she usually can) that you are married to that purchase you will receive no discounts. Sometimes this is not a bluff, which means you must simply walk away to buy another day.

4. *Talk sparingly.* When faced with a purchase, we may sense the inherent confrontation, get nervous, and talk too much. Just make small comments and let the opposing negotiator rattle. For example, say, "But it seems your price may be too high." Then, talk sparingly and see just how far the seller will come down.

5. *"That is not good enough."* Everyone has more room to expand or reduce anything, and the same principle holds true when negotiating. When the price is given, reply with, "That is not good enough—what can you really do?" You will see the price drop, sometimes lower than you would have offered. Instead of giving a lower offer, try using that statement.

6. *Good guy–bad guy.* For example, your Mom, Dad, or another person who is not with you should always seem mean to the other side. "My Mom would kill me if I took that price," or "You know, my Dad would not like it if I came home with that if I didn't get a good buy." Note: When you are using this technique, you must be telling the truth.

7. *If I give, I also take.* When you reach a point where you must give up something (money, for example), be sure you take something at the same time. You should say, for example, "If I give you $2,500 for that entertainment center, then you have to throw in the cabinet

(or something else) and the sales tax at no extra charge."[28]

The second principle you must understand to get great bargains is you can sometimes find buried treasures. Consider buying from individuals, public auctions, garage sales, flea markets, classified ads, and coupons, and also take advantage of real estate foreclosures. There are bargains everywhere if you take time to look. Do not get the idea that only junk can be bought at a bargain.

Once you have learned the basic negotiating techniques and where to find bargains, you must learn patience. This requires discipline. Even if you have the negotiation skills and have looked in the right places, without patience, you will still miss the bargains.

We must begin to think like vultures. Vultures are not very pretty, but the patience they display in the wild is something we should emulate.

Build an Emergency Fund

A good financial planner will tell you that to start you should have three to six months of income in savings that are liquid for emergencies. An example of **liquid income** is a simple bank savings account or a money market fund that has check-writing capability. If you make $24,000 annually, you should have $8,000 to $12,000 located where you can easily get it before you do any other investing. As mentioned previously, you have to expect the unexpected.

Manage Your Budget Bursters

Payments that you make on a nonmonthly basis can destroy your budget unless they are planned. Converting them to a monthly basis and using the "envelope system" (discussed next) to set money aside monthly can be used to avoid strain or borrowing when these payments become due. Refer to Figure 6–18 for examples of nonmonthly payments. When you prepare your own sheet of nonmonthly payments, adjust it to your budget. If you

Nonmonthly Payments For Which to Plan

ITEM NEEDED	ANNUAL AMOUNT		MONTHLY AMOUNT
Real Estate Taxes	_____	/ 12 =	_____
Howeowner's Ins.	_____	/ 12 =	_____
Home Repairs	_____	/ 12 =	_____
Replace Furniture	_____	/ 12 =	_____
Medical Bills	_____	/ 12 =	_____
Health Ins.	_____	/ 12 =	_____
Life Ins.	_____	/ 12 =	_____
Disability Ins.	_____	/ 12 =	_____
Car Ins.	_____	/ 12 =	_____
Car Repair/Tags	_____	/ 12 =	_____
Replace Car	_____	/ 12 =	_____
Clothing	_____	/ 12 =	_____
Tuition	_____	/ 12 =	_____
Bank Note	_____	/ 12 =	_____
IRS (Self-Empl.)	_____	/ 12 =	_____
Vacation	_____	/ 12 =	_____
Gifts	_____	/ 12 =	_____
Other	_____	/ 12 =	_____
Other	_____	/ 12 =	_____

Figure 6–18 This example will hopefully help you manage your budget. Without proper planning, bills that are not on a monthly cycle can destroy your budget.

make a payment quarterly, then adjust the payments monthly for this sheet.

Envelope System

The time-honored envelope system of cash management is a budgeting system recommended by most good financial counselors. Items for which you use a credit card or a check, out of supposed convenience, will likely destroy your budget if you continue the practice.

If you get paid once a month, put cash in an envelope for each budgeted category. If your grocery and food budget is $300 monthly, put $300 cash in an envelope for food each payday. As you buy food, take the money from that envelope, and from nowhere else. This provides you with instantaneous cash management in that you will almost never spend more than you allotted. If you have had a hard week and you "deserve" to go out to eat on Friday night, simply pull out the "Food" envelope and look into the envelope to see if you can afford to do what you want to do.

Again, refer to Figure 6–18 for the nonmonthly budget bursters. Prepare an envelope for each of the items. Put either cash in the envelope, or, for security, put play money in the envelope and put the real money in a savings account. However, make sure there is enough real money in the savings to cover all the play money in the envelopes. Then, as the payments become due, the money will be in place and you will be able to manage the payments.

Have Enough Insurance

Any person should take needed insurance to financially secure the well-being of themselves and their family. You may not be able to afford it, but, more important, you cannot afford to be without it. Just do it, if at all possible. Make sure you have life insurance, health insurance, disability insurance, auto insurance, and homeowner's insurance. Without these, you have great potential for financial disaster, including bankruptcy. By having enough insurance, you will have greater control over your destiny and show others that you are competent and capable.

Master the Magic of Compound Interest

You do not have to have a fortune to invest to enjoy the magic of **compound interest**. You can accumulate a substantial amount of money by "investing" in a small way.

If you invest the money used to purchase your afternoon soft drink, see Figure 6–19. If you quit smoking cigarettes, see Figure 6–20. If you save $2.40 a day, see Figure 6–21.

You see, it does not take a lot of money to begin "cashing in" on the magic of compound interest. It truly is amazing.

Use an Automatic Savings Plan

If you have the opportunity, use a payroll deduction savings plan. If you do not receive the money, you will hardly miss it. The accumulation in your savings will motivate you to do even more. If you get a raise, put a percentage of your raise in savings. This is an excellent way to build your emergency fund.

Take Advantage of Matching Stock Purchases

Some larger corporations offer the opportunity for employees to buy stock in their companies. Some companies match dollar for dollar. Some match 50 cents per dollar, and there are many other options. In reality, you will get 100 percent return on your investment if every dollar is matched dollar for dollar. Do not miss this opportunity if it is provided to you.

Use Your Hobby to Make Money

You can have a hobby that costs money, or you can have a hobby that makes money and have just as much fun. We will spend more time on this as we discuss "The $2,000 Work-Unit Approach" later in this chapter. Examples are coins, antiques, books, stamps, phonograph records, autographs, baseball cards, old car restoration, undeveloped plots of land, metals, bonds, art, stock, and rental property.

The first step is to become knowledgeable about the type of investments in which you want to put your money. The next step is to commit a certain amount of your time and energy (perspiration equity) toward making your investments prosper. The third step is getting started. Explore a variety of investments, find mentors to guide you, constantly upgrade your investments—and then enjoy them.

$.50/day @ 5 days/week = $10.00 month							
	10 yrs.	20 yrs.	30 yrs.	40 yrs.	50 yrs.	60 yrs.	70 yrs.
5%	$1,559	$4,127	$8,357	$15,323	$26,797	$45,695	$76,820
10%	$2,065	$7,656	$22,793	$63,767	$174,687	$474,952	$1,287,781

Figure 6–19   Example showing the magic of compound interest through saving fifty cents per day on a soft drink.

$2.40/day for 30 days = $72.00 month		
Years	5%	10%
10	$11,178.92	$14,745.79
20	$29,593.08	$54,664.39
30	$59,922.59	$162,721.41
40	$109,877.56	$455,224.63
50	$192,157.16	$1,247,011.58
60	$327,677.84	$3,390,326.69
70	$550,890.58	$9,192,139.44

Figure 6–20   Example of the magic of compound interest for a person who quits smoking.

$1/day @ 5 days/week = $20.00 month		
Years	5%	10%
10	$3,118	$4,131
20	$8,254	$15,313
30	$16,714	$45,586
40	$30,647	$127,535
50	$53,595	$349,375
60	$91,390	$949,904
70	$153,640	$2,575,561

Figure 6–21   Example of the magic of compound interest for a person who saves $1.00 a day, five days a week, or who saves $20.00 per month.

You can use the profits from your hobby money to accumulate great wealth, as will be discussed in the next section.

RETIRING AS A MILLIONAIRE

Is it possible for an agribusiness manager or worker to retire as a millionaire? It is very possible. Although nothing in life is guaranteed, things certainly will not happen unless you make them happen. If you think you can, you can. If you think you cannot, you cannot. Either way, you are correct. Your attitude determines your altitude. The concept of the four following plans hinges on a 12 percent interest rate. This will be discussed later.

Being a millionaire may not be a goal of yours. However, I believe most people would like to know how it could be done.

Plan One—$2,000 a Year for Six Years

Starting at age 22, put $2,000 a year into a tax-free retirement account, **Individual Retirement Account (IRA),** or **Tax Shelter Account (TSA)** 403(b), for six years, and then stop. At age 65, $12,000 will compound to $1,348,000 if a 12 percent interest rate is attained. Figure 6–22 shows an illustration of this. Notice the importance of time and consistency. If you wait until age 28 to start, you will have to put in $2,000 a year until age 65 ($76,000) to get the same amount of money. You can see the importance of starting early: it's $12,000 if you start at age 22, but $76,000 if you start at age 28, to have $1 million by age 65.

Because of the impact of compounding interest over a long period of time, it pays big dividends to start saving early in life. The following example again proves the point:

Rose scrapes and saves every penny she can, and at age 19, she's able to stash away $2,000 in a long-term investment that earns 10 percent compounded annually. She does that for eight years. At age 27, Rose simply cannot afford to save any more because of family responsibilities and other expenses. But the $16,000 investment made early in life enables her to retire as a millionaire. Refer to Figure 6–23.

Nancy has a good time while she's young and spends her money on a fancy car. At age 27, she figures it is time to settle down, so she starts saving $2,000 per year until she is 65. It, too, is invested at 10 percent compounded annually. Because Nancy got a later start, she only retires on $805,185. As you can see from Figure 6–23, that is $213,963 less than Rose's retirement. Nancy had to work a lot harder at it—saving $2,000 per year for 39 years for a total of $78,000.

Obviously, these are hypothetical examples. There are not many Roses who will save $2,000 per year. The principle of compounding interest remains the same, regardless of the amount of money involved. There is nothing wrong with enjoying life by spending some of your hard-earned money while you are young. But you need to weigh those expenditures against their impact on your future. In the long run, Nancy's new car cost her $213,963!

You can give your children financial security. Give your child $2,000 when he or she is born, with the understanding that it remains untouched until age 65. Invested at a rate of 10 percent compounded annually, that $2,000 birthday gift will be worth $980,740 at retirement!

Plan Two—Buy a Used Car Instead of a New One and Invest the Difference

Give up on owning a new car before age 23 and buy a used car instead. Invest the difference between a new car and a used car in an IRA or TSA. Consider the following illustration: Jeff, a 17-year-old senior in high school, bought a nice pickup with payments of $250.00 per month for five years. If $166.66 monthly were put into an IRA or TSA, $83.34 would still be left to pay monthly for a used vehicle. By analyzing Figure 6–22, one can **extrapolate** and see that the new truck cost Jeff $2.6 million. The Rule of 72 states that by dividing the interest rate into 72 (72 ÷ 12 = 6), one can determine how long it will take for one's money to double. The additional $1,300,000 in six more years would double to $2.6 million. The Rule of 72 will be discussed later in the chapter.

Don't be lured by the new car pride syndrome. Millions of dollars are spent every day with the express purpose of enticing you, the American public, to believe that you will look better, feel better, and be more successful if you drive a brand new car. Driving a new car is not a measure of success, and the repercussions of buying for that purpose can be devastating to a family! Buying a new car—especially if you go into debt for it—can look harmless on the surface, but in reality it can be a very dangerous trap. The debt of the American family is a crushing weight that threatens to destroy the very people it supposedly is helping to "get ahead." The reality is that you become a slave to a piece of metal! You have to fuel it, clean it, service it, and make the payments! The average car payment in America today is somewhere between $350 to $450. Most people do not realize that

Plan One — $2,000 A Year for Six Years

	EXAMPLE A		EXAMPLE B	
Age	Payment	Accumulation End of Year	Payment	Accumulation End of Year
22	$2,000	$2,240	0	0
23	2,000	4,749	0	0
24	2,000	7,559	0	0
25	2,000	10,706	0	0
26	2,000	14,230	0	0
27	2,000	18,178	0	0
28	0	20,359	$2,000	$2,240
29	0	22,803	2,000	4,749
30	0	25,539	2,000	7,559
31	0	28,603	2,000	10,706
32	0	32,036	2,000	14,230
33	0	35,880	2,000	18,178
34	0	40,186	2,000	22,599
35	0	45,008	2,000	27,551
36	0	50,409	2,000	33,097
37	0	56,458	2,000	39,309
38	0	63,233	2,000	46,266
39	0	70,821	2,000	54,058
40	0	79,320	2,000	62,785
41	0	88,838	2,000	72,559
42	0	99,499	2,000	83,507
43	0	111,438	2,000	95,767
44	0	124,811	2,000	109,499
45	0	139,788	2,000	124,879
46	0	156,563	2,000	142,105
47	0	175,351	2,000	161,397
48	0	196,393	2,000	183,005
49	0	219,960	2,000	207,206
50	0	246,355	2,000	234,310
51	0	275,917	2,000	264,668
52	0	309,028	2,000	298,668
53	0	346,111	2,000	336,748
54	0	387,644	2,000	379,398
55	0	434,161	2,000	427,166
56	0	486,261	2,000	480,665
57	0	544,612	2,000	540,585
58	0	609,966	2,000	607,695
59	0	683,162	2,000	682,859
60	0	765,141	2,000	767,042
61	0	856,958	2,000	861,327
62	0	959,793	2,000	966,926
63	0	1,074,968	2,000	1,085,197
64	0	1,203,964	2,000	1,217,661
65	0	$1,348,440	$2,000	$1,363,780

Figure 6–22 "Plan One" for potentially retiring a millionaire involves saving $2,000 per year for six years, starting at age 22, and getting a 12 percent interest rate on your investments. Example B shows the importance of time.

What You Don't Know Can Cost You

	ROSE			NANCY	
AGE	Investment	Total Value		Investment	Total Value
19	$2,000	$2,200		0	0
20	2,000	4,620		0	0
21	2,000	7,282		0	0
22	2,000	10,210		0	0
23	2,000	13,431		0	0
24	2,000	16,974		0	0
25	2,000	20,871		0	0
26	2,000	25,158		0	0
27	0	27,574		0	0
28	0	30,442		2,000	$2,200
29	0	33,486		2,000	4,620
30	0	36,834		2,000	7,210
31	0	40,518		2,000	13,431
32	0	44,570		2,000	16,974
33	0	49,027		2,000	20,971
34	0	53,929		2,000	25,158
35	0	59,322		2,000	29,874
36	0	65,255		2,000	35,062
37	0	71,780		2,000	40,768
38	0	78,958		2,000	47,045
39	0	86,854		2,000	53,949
40	0	95,540		2,000	61,544
41	0	105,094		2,000	69,899
42	0	115,603		2,000	79,089
43	0	127,163		2,000	89,198
44	0	139,880		2,000	100,318
45	0	153,868		2,000	112,550
46	0	169,255		2,000	125,005
47	0	186,180		2,000	140,805
48	0	204,798		2,000	157,086
49	0	225,278		2,000	174,994
50	0	247,806		2,000	194,694
51	0	272,586		2,000	216,363
52	0	299,845		2,000	240,199
53	0	329,830		2,000	266,419
54	0	362,813		2,000	295,261
55	0	399,094		2,000	326,988
56	0	439,003		2,000	351,886
57	0	482,904		2,000	400,275
58	0	531,194		2,000	442,503
59	0	584,314		2,000	488,953
60	0	642,745		2,000	540,048
61	0	707,020		2,000	596,253
62	0	777,722		2,000	658,078
63	0	855,494		2,000	726,086
64	0	941,043		2,000	800,995
65	0	1,035,148		2,000	883,185
TOTAL EARNINGS		**$1,019,148**			**$805,185**

Figure 6–23   Another example of the importance of getting started early and being consistent with your savings, this example is based on a 10 percent interest rate on your money.

when you sign on the dotted line, your brand new car depreciates in value by approximately 25 percent (the amount will vary depending on the type of automobile you buy). That means that if you have a note for $15,000 on a car, financed over three or four years, you will end up paying close to $18,000 for a car whose price dropped as soon as you bought it. And by the time you finish paying for it, it will be worth a lot less! Today, in our mobile society, a family needs to have a car. So, our suggestion is this: buy used and invest the difference![29]

What About the Repairs?   Many people express concern about making repairs on a used vehicle. You should never buy a used car without first allowing a reputable mechanic to check it out, and it is usually a good idea to purchase a used car maintenance warranty. It is also a positive selling factor if you trade in your car while there is still some time left on the warranty. Even so, repairs are much cheaper than on a new car. Remember this: an automobile is merely a means of getting you from point A to point B. The degree of style and luxury you can enjoy along the way depends on your commitment to avoid the debt trap!

Plan Three—The 25-plus-10 Plan

For the first job that you take at the age of 22, put $25.00 per month into an IRA or TSA during the first year. Each year thereafter, increase the monthly contributions by $10.00. In other words, the second year on the job, you would be putting $35.00 monthly into an IRA or TSA. Assuming a return of 12 percent, this will amount to a total of $1,830,000 by age 65. Most years, due to inflation, your raise will stay ahead of your initial $120.00 investment, even though you will be investing much money later in life. Thus, your percentage of investment of your salary will hardly ever be more than when you first started.

Plan Four—The 80-10-10 Plan

The beliefs and values of some people cause them to allot the first 10 percent of a check to charitable contributions. Another suggested strategy is an 80-20 or 90-10 plan. By following this plan, 10 percent goes to charity or as a gift, 10 percent is invested in an IRA or TSA, and 80 percent is used for living expenses. If 10 percent is not given to charity or as a gift, then a 90-10 plan could be followed. By following a budget and buying accordingly, one can retire as a multimillionaire by using this plan. This can be achieved with little disruption of your lifestyle.

You're probably saying: "I can't live off 100 percent of my income. How can I live off 80 percent?" By using the money management tips discussed earlier in the chapter, you can. Plan your work (budget), and work your plan. It is my belief that Americans are somehow programmed to spend 3 to 10 percent more than they make, no matter the income level. Simply budget as discussed earlier, and you can do it. Buy the next lower price automobile, get a house with a one-car garage rather than a two-car garage, or whatever else it takes to live within or even below your income level.

KEY FACTORS TO MAKING THE MILLIONAIRE PLANS WORK

A farmer does not simply plant a seed and let it grow, hoping for an abundant harvest. He or she has to work on it. Depending on their cropping method and the region of the country, the crop has to be cultivated, fertilized, sprayed for pesticides, and watered, either naturally or by irrigation. The same is true for dairy and livestock animals. Monthly, weekly—and for dairy, daily—care has to be taken. The same is true with the millionaire plans. They have to be worked at. You do not simply start something and let it grow without nurturing it. Knowledge of the following factors will help you reach the millionaire goal.

Manage Investments to Achieve a 12 Percent Interest Rate

Every plan is simply a mathematical process based on a 12 percent return on your investments. It is an achievable production goal. Ten pigs per litter, 100 bushels of corn per acre, and 18,000 pounds of milk per cow are viable production goals. They are challenges, but they can be achieved, just as the 12 percent interest can be achieved. It may not be achieved every year, but including very good and very bad years, this is our goal.

For a person who is able to save $2,000 per year for forty-three years (ages 22–65), a 1 percent increase in the interest rate would amount to $316,946.00. Look at the following figures:

Eight Percent Interest

🌿 Ages 22 to 65

🌿 $2,000 per year for 43 years at .08 interest compounded monthly

🌿 $839,054

Nine Percent Interest

🌿 Ages 22 to 65

🌿 $2,000 per year for 43 years at .09 interest compounded monthly

🌿 $1,156,000

Difference in 1 Percent (.01) Interest Rate. At $2,000 per year for forty-three years, a .01 (1 percent) interest rate difference equals $317,946. Also, refer to Figure 6–21 for a comparison of returns at 5 percent and 10 percent interest.

The Rule of 72

The Rule of 72 refers to the process of dividing the interest rate into 72 (72 ÷ 12 = 6) and determining how long it takes for one's money to double. Money that yields 12 percent interest will double in six years. For example, $1,000 yielding 18 percent interest would double in four years. If 36 percent were the interest rate, $1,000 would double to $2,000 in just two years. Figure 6–24 further illustrates this.

Invest in Mutual Funds to Achieve a 12 Percent Interest Rate

There are many types of investment strategies. Take courses, read, and seek advice from financial counselors. Do not simply take our word. However, we believe the novice (inexperienced) investor will also select **mutual funds**.

Mutual funds were invented to provide investment expertise for ordinary people. A mutual fund company makes investments solely on behalf of others. All the profits (or losses) go to the fund's shareholders. In existence in the United States since 1924, there are over 3,400 different mutual funds holding over $2 trillion for about 70 million shareholders. This amount rises yearly.

What you get from a mutual fund, at a relatively low cost, is professional management of your money by people who devote their full time and attention to investment decisions. Even a professional sometimes gets poor results, but the average fund does better than you probably can do for yourself, according to studies by Marshall Blume, chairman of the finance department at the University of Pennsylvania's Wharton School. With a mutual fund, you get a portfolio that usually consists of fifty or more stocks or other securities.

Select and monitor your mutual funds. There are many publications that give information on mutual funds. *Money* magazine publishes an annual yearbook that gives data on many of them. The *Wall Street Journal* and many local newspapers from larger cities list performances of the mutual funds daily. Most of the funds have toll-free numbers. Call them to get information. By law, you have to be given a **prospectus** before you can

Rule of 72				
ORIGINAL INVESTMENT	72	% INTEREST	YEARS TO DOUBLE	ENDING AMOUNT
$1,000	72	2%	36 YEARS	$2,000
$1,000	72	4%	18 YEARS	$2,000
$1,000	72	6%	12 YEARS	$2,000
$1,000	72	8%	9 YEARS	$2,000
$1,000	72	10%	7.2 YEARS	$2,000
$1,000	72	12%	6 YEARS	$2,000
$1,000	72	18%	4 YEARS	$2,000
$1,000	72	36%	2 YEARS	$2,000

Figure 6–24   The "Rule of 72" refers to the process of dividing the interest rate into 72 (72 divided by 12 equals 6) and determining how long it takes for one's money to double.

invest with any mutual fund. Consider the following as you select and monitor your funds:

ū *Select a mutual fund on the basis of whether the fund has a sales charge* (no-load fund). Many companies charge 4 percent or more just to invest. There are enough good mutual fund companies that do not charge anything.

ū *Select a mutual fund based on performance.* *Money* magazine, among a number of other magazines and newspapers, publishes mutual fund rankings each month, which show the funds that have gained the most during the past month, the past year, and the past five years. Do not select a mutual fund unless it has a lifetime track record of over 12 percent.

ū *Monitor your mutual fund closely.* Remember the earlier analogy of planting seeds. When you purchase a mutual fund, you are only planting the seed; then you have to nurture it.

ū *Diversify your mutual funds.* Do not put all your eggs into one basket. You may want to have three to five different mutual funds with different investment objectives. The minimum investment for many mutual funds for IRAs is only $500. If you have $2,000, for example, deposit $500 in each of four different mutual funds.[30]

Types of Mutual Funds.   As you select and monitor the investment, be aware of the types of mutual funds. Most funds carry a rating of high, medium, or low-risk. Furthermore, mutual funds carry this risk to some extent within each of the following investment strategy categories:

ū *Aggressive growth:* Strive for maximum capital gains, buying the **fidelity shares** of fast-growing companies.

ū *Long-term growth:* Own the least erratic issues of better-known and still rapidly growing firms.

❧ *Growth and income:* To achieve this combination, funds buy bonds and stocks of large companies that are big dividend payers and also offer some prospect for capital gains.

❧ *Income:* Funds with this objective attract investors seeking high yields.

❧ *Sector:* A sector fund specializes in stocks of one industrial sector or several related industries, such as health care, high tech, or energy.[31]

Are Mutual Funds Risky? The term **risk management** means the use of various ways to deal with potential personal or financial loss. To practice risk management means that an individual acknowledges the existence of potential losses and their magnitudes, and uses various methods to control them.[32] Even though mutual funds are more risky than some investments, they are less risky than others. In reality, mutual funds involve risk management since so many different stocks and securities are managed by a professional financial team. Even then, do not forget the old saying, "Nothing in life is guaranteed except death and taxes."

Are Mutuals the Only Investment? There are several types of investment, from low-risk bank savings (CDs) to high-risk commodities. Of course, stocks are also risky. Mark Twain summed it up well when he said, "October." This is one of the peculiarly dangerous months in which to speculate in stocks—the others are July, January, September, April, November, May, March, June, December, August, and February. Mutuals are in-between in risk. This chapter does not permit an explanation of each of these. However, Figure 6–25 shows various types of investment alternatives and their risk and earnings rankings.

More Information. For a directory of more than 3,400 funds, their addresses, phone numbers, investment objectives, fees charged, and other information, contact the Investment Company Institute, P.O. Box 66140, Washington, DC, 20035–6140. The institute is the national association of the investment company industry. Its mutual fund members represent approximately 95 percent of the industry's assets and have about 36 million shareholders.

Tax-Deferred Savings

The four millionaire plans are **tax-deferred savings** plans. Tax-deferred savings allow an individual to postpone taxes while building an investment. Examples of tax-deferred savings include Individual Retirement Accounts (IRAs), available to everyone who has earned income and contributions limited to $2,000 per year; 401(k) tax-sheltered plans, which are available to government employees and the self-employed; and 403(b) tax-sheltered plans, which are available to teachers and hospital employees. Contributions are allowed for a percentage of your income. In some situations, you can put up to 20 percent of your income into a 403(b). A relatively new IRA is the Roth IRA. With the Roth IRA, tax is paid on the front end, but neither the gains nor the amount you withdraw is fixed.

When Do I Get My Money? Withdrawals from such plans are designed to occur at retirement. As you withdraw your money, you pay taxes on your withdrawals according to your tax bracket. Actually, you can get it anytime, but you will pay taxes and heavy penalties on money withdrawn from these plans prior to age 59½.

FINANCIAL SECURITY

You are probably saying, "I don't care about retirement, I want to know how to get rich quick." But we cannot tell you how to do that by ethical, legal, or moral means. The best way to get rich quick, is

Financial Planning Pyramid

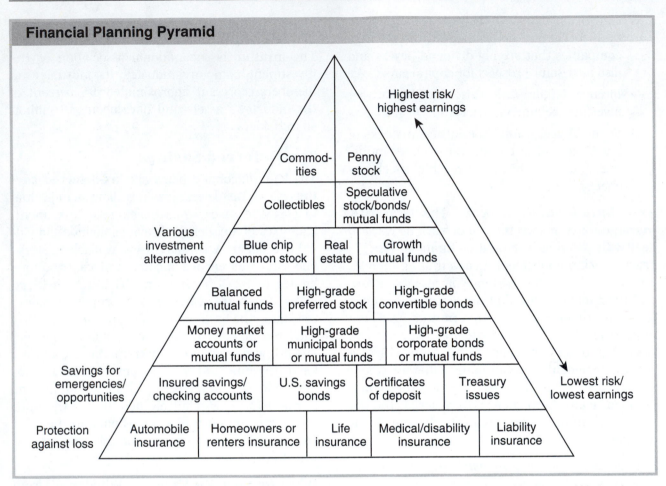

Figure 6–25 Financial planning pyramid showing the risk on various investment alternatives. (Courtesy of National Endowment for Financial Education)

to get rich slowly. We have heard it said that the safest way to double your money quickly is to fold it over once and put it into your pocket. However, you can reach financial security by the young adult years through knowledge gained in your agricultural education program and your supervised agricultural experience program (SAEP). However, you have to work and use "perspiration equity."

Financial Security Defined

Financial security involves being able to meet monthly expenses (house payments, car payments, utilities, groceries, medical, clothing, etc.) with your regular full-time job, plus accumulating (saving) approximately $100,000 in tax-free municipal or government bonds (with a return of 8 percent, which would be $8,000 tax-free earnings yearly). The interest on the savings is only to be used for travel, entertainment, and other non-budgeted items. This money is not to be used or confused with money from your full-time regular job intended for living expenses, house payments, and so on. Note that it is realized that you may have a different definition of financial security.

Assumptions for Reaching Financial Security

It is possible for an agribusiness manager or worker to become financially secure by age thirty to thirty-five. The age depends on whether a student continues his or her education after high school before pursuing other plans. For this to happen, there are some beliefs, assumptions, or criteria that must be taken into consideration.

1. *Part-time (avocational):* The key is **part-time (avocational) enterprises**.

2. *Marketing:* The key to a profit in avocational **enterprises** is marketing. As much emphasis must be placed on marketing as on production.

CAREER OPTION

Career Areas: Avocational (Part-Time) Production Agriculture, Agribusiness Recreation, Agricultural Mechanic, Nursery Production

Due to emerging technologies, many agriculture enterprises can now be done as an avocation or part-time endeavor. For example, sixty years ago, hay was harvested loose and put in hay lofts by a rope, spear, and pulley; forty years ago, almost every hay producer harvested with a square baler; but today, most hay producers use a round baler to harvest their hay. The invention of the round hay baler permits many cow-calf producers to manage a sixty- to seventy-cow herd on an avocational basis; sixty years ago, this would have been a full-time job. Refer to Figure 6–26 for several other avocational enterprises.

Another popular avocational enterprise near large metropolitan areas is strawberry production. Due to new strawberry varieties, using plastic and irrigation, some strawberry producers are able to gross $10,000 per acre. Besides using workers to pick the strawberries, some strawberry producers have pick-your-own operations.

The photo shows strawberry growing using special varieties, plastic, and drip irrigation.

Some strawberry producers are able to gross $10,000 per acre using new varieties, plastic, and drip irrigation. Raising strawberries as an avocational enterprise can be very advantageous if the market is available and the necessary skills can be mastered. (Courtesy of Cliff Ricketts)

3. *New way of thinking:* Traditional thinking and values will have to be put on the "back burner." For example, big may not be better. Enormous debt, big machinery, and large amounts of land are not necessary to make profits in agribusiness, production agriculture, and avocational enterprises.

4. *Manage investments:* As much thought and planning will have to be devoted to an **investment portfolio** as to that of an avocational enterprise.

5. *Get tough and be focused:* Self-discipline, hard work, and **perseverance** are prerequisites for anyone to achieve challenging goals. This is especially true for the goal to achieve financial security through avocational enterprises. Refer to the Career Option on page 155 for further explanation and examples of avocational enterprises.

Selecting Part-Time Agricultural Enterprises

The story is told about the farmer that sold his farm and went in search for diamonds. Many agriculture students go searching for diamonds and gold, and they do not realize they are already sitting on gold mines. Many agriculture students have resources or are available to use resources that most nonagriculture students do not have.

The $2,000 Work-Unit Approach. The "$2,000 work-unit approach" simply involves selecting avocational enterprises that will net you $2,000 on an acre or less. No more than five acres of land, including home and buildings, will be needed. There are many cases where no land, equipment, or supplies are needed, such as real estate investing. By selecting just four different avocational enterprises that will do this, you can achieve financial security if you are self-disciplined. Figure 6–26 lists over fifty examples. Remember, these are not automatic. However, if you study them, use all the

approved practices, visit people (working models) who have succeeded at that particular avocational enterprise, and become a "student of that enterprise," these fifty enterprises will meet the "$2,000 work-unit" criteria in most cases for most years. Remember, these can also be agricultural mechanic enterprises. In reality, they do not have to be agricultural all the time. Hobbies such as coin collecting, stamps, crafts, and so on can be used; as long as the $2,000 per year is net profit, it can be considered a $2,000 work-unit.

This approach is not needed if your family is established in agriculture production and you already have land, facilities, and equipment. Therefore, dairy, crop, and livestock producers (except in specialty situations) are excluded from this approach. However, students in many agricultural education programs in the United States do not come from farms, and if they do, their parents are generally not full-time production agriculturalists, or there is not enough economic opportunity for all their brothers and sisters to join the farming operation after graduation. Therefore, the $2,000 work-unit approach allows them to get into or stay with agribusiness and agriscience production. Also, by needing only five acres, few facilities, and little equipment, this approach addresses and avoids the factors that have caused farms to periodically endure hard times, which are:

❧ high interest payments

❧ high machinery costs

❧ depreciated land value

❧ inadequate market

Considerations for selecting part-time avocational enterprises that meet the criteria of the $2,000 work-unit approach include the following:

❧ Select enterprises that do not interfere with your full-time job. Remember, this is only your avocation. Your full-time job has priority.

Fifty $2,000 Work-Unit Approach

- Calves
- Feeder Pigs
- Boarding Horses
- Catfish
- Beekeeping
- Rabbits
- Hunting and Fishing Fees
- Bobwhite Gamebird Production
- Goats
- Burmese Python Production
- Artificial Insemination Technician
- Real Estate Sales
- Teacher, Specialty Classes
- Miniature Horses
- Llama
- Blueberries
- Apples
- Squash

- Lawn Maintenance
- Dogwoods
- Raising and Propagating Fruit Trees
- Raising Nursery Stock
- Watermelons
- Beans
- Specialty Crops
- Cucumbers
- Grapes
- Greenhouse Crops
- Burley Tobacco
- Peppers
- Selected Crafts
- Bedding Plants
- Ginseng
- Tomatoes
- Sweet Corn
- Cantaloupe

- Cabbage
- Asparagus
- Dried Flowers
- Herbs
- Shitake Mushrooms
- Mowing Lawns
- Interior Landscaping and Plant Maintenance
- War Games
- Real Estate Investing and Foreclosures
- Dogs and Cats
- Firewood
- Popcorn and Ornamental Corn
- Strawberries
- Blackberries
- Pumpkins
- Sweet Potatoes
- (There are others)

Each of these with an adequate market and management can earn $2,000 on an acre or less of land.

Figure 6–26   Fifty enterprises that potentially can net $2,000 per acre per year with an adequate market and management. However, market conditions can change rapidly.

🌿 Select an enterprise that you will not get **locked into**. This would mean you could not get out of what you were doing because your debts exceeded your assets.

🌿 Start small and capitalize on the **utilization of resources**.

🌿 Make sure that there is a market for the enterprise you selected. All else is in vain without a market. If there is no market, abandon your plan.

🌿 Select enterprises that are **compatible** with each other in relation to land, labor, and capital. For example, strawberry growing is intense during one particular month, so select an enterprise that is intense in another month, like blackberry growing.

🌿 Select avocational enterprises that will require one acre or less per enterprise.

🌿 Select enterprises that make a net profit of $2,000 yearly.

🌿 Select enterprises that are enjoyable and serve both as a leisure and **therapeutic activity**. Make this your hobby. Some people like animals; some people like plants; some like mechanical things.

Procedures for Potentially Achieving Financial Security by Age Thirty to Thirty-four

🌿 After graduation from high school or college, select one avocational enterprise that is compatible with your full-time job. Remember,

Four Enterprises at 8 Percent Interest
To Age 30

First Year Out of High School	Age 19	Strawberries	2,000.00
Second Year Out of High School	Age 20	Blackberries	2,000.00
		Strawberries	2,000.00
		Interest at 8%	320.00
			$4,320.00
Third Year Out of High School	Age 21	Crafts	2,000.00
		Blackberries	2,000.00
		Strawberries	2,000.00
		Previous Savings and Interest	4,320.00
			$10,320.00
Fourth Year Out of High School	Age 22	Catfish	2,000.00
		Crafts	2,000.00
		Blackberries	2,000.00
		Strawberries	2,000.00
		Previous Savings and Interest	11,134.80
			$19,134.80
	Age 23	Four Enterprises	8,000.00
		Previous Savings and Interest	20,642 58
			$28,642.58
	Age 24	Four Enterprises	8,000.00
		Previous Savings and Interest	30,933.99
			$38,933.99
	Age 25	Four Enterprises	8,000.00
		Previous Savings and Interest	42,048.71
			$50,048.71
	Age 26	Four Enterprises	8,000.00
		Previous Savings and Interest	54,052.61
			$62,052.61
	Age 27	Four Enterprises	8,000.00
		Previous Savings and Interest	67,016.82
			$75,016.82
	Age 28	Four Enterprises	8,000.00
		Previous Savings and Interest	81,018.17
			$89,018.17
	Age 29	Four Enterprises	8,000.00
		Previous Savings and Interest	96,139.62
			$104,139.62
	Age 30	Four Enterprises	8,000.00
		Previous Savings and Interest	112,470.78
			$120,470.78

Figure 6–27 Example of how selecting four enterprises that net $2,000 per year per acre each can potentially accumulate enough money to make you financially secure.

marketing is the key! The Business Research Center (or Small Business Development Center) at many universities can provide assistance in determining if there is a market for many avocational enterprises.

❧ Select an enterprise that will make a net profit of $2,000. Let us use strawberries as an example: one acre of properly managed strawberries will net $2,000.

❧ The second year after graduation, another enterprise can be added to net $2,000. Add an enterprise that will distribute your workload throughout the year. Since strawberries naturally produce in one month, an acre of blackberries can be added, since they naturally produce in another month.

❧ The third year after graduation, another enterprise can be added. The possibilities are unlimited. Some include blueberries, catfish, tomatoes, squash, nursery stock, bedding plants, floriculture crops, and agricultural mechanics projects. Refer to Figure 6–26 for further selections.

There are a thousand ways in which something can fail to work, but only one way it can. Select enterprises that will work for you. Part-time jobs and working overtime are other $2,000 work-unit alternatives.

❧ The fourth year another $2,000 enterprise can be added. (Another approach is to expand one of the first three enterprises. It may be best to add another acre of strawberries or blackberries.) Refer to Figure 6–27 for a summary of the generated income. Although our goal was $100,000 by age 30 to 34, Figure 6–27 shows over $120,000. As you can see, even with bad years, $100,000 can be achieved.

Invest Earnings in Tax-Free Bond Mutual Funds

There are many tax-free bonds. Many of these are mutual fund tax-free bonds. The goal of these funds is to achieve 8 percent interest.

Change in Strategy. Why is there a difference in strategy in the goal of 12 percent on a mutual fund investment for the millionaire plan compared to the goal of an 8 percent tax-free bond for the financial security plan? There are three reasons:

❧ Only $2,250 can be put into an IRA ($2,000 if your spouse also has an IRA).

❧ There is less paperwork and record-keeping, since interest income from tax-free bonds is, obviously, not taxable.

❧ There is less risk in tax-free bonds, especially utility bonds.

Limitations on Potentially Achieving Financial Security through Avocational Enterprises

Is this plan a sure thing? No, there are limitations that should be taken into consideration. Remember, if this were easy, everyone would do it. The limitations are as follows:

This is only a plan. Few things in life are absolute. However, if you do not try you surely will not succeed.

Market conditions change yearly. You will need to constantly seek and evaluate your markets.

Hard work and perseverance are taken for granted. Many people do not have the maturity or the ability to work as hard as this plan will require.

Life is a trade-off. You cannot hunt, fish, play ball, attend ball games every weekend, and still maintain four acres of avocational enterprises.

However, there will still be plenty of time for recreation.

Self-discipline is needed in order to not remove any of the principal or interest before the goal of $100,000 is achieved.

The stock market can crash, as it did in 1929, 1987, and 1998. This can drastically affect the value of one's investments.

Environmental conditions—agricultural enterprises, especially crop-related ones, are often beyond one's control environmentally, due to drought, freeze, flood, and pests.

LEVELS OF FINANCIAL MANAGEMENT

The more levels of financial security you can attain, the better off you will be toward retirement. Consider the following levels of financial management:

1. *Live within a budget:* Follow good money management practices to limit debt and live within your means.

2. *Social security:* This is required by law. Let us hope that the system is still stable when you retire.

3. *Retirement benefits through your job:* This is generally required by most places of employment.

4. *Life insurance:* **Universal life insurance** can also become a savings plan. If you do not reach retirement, your family will be protected financially if insured adequately. Some financial advisors advise using **term life insurance**. A yearly premium is paid, but the premium never accumulates toward a cash value. Since term insurance is much cheaper, many advise you to buy it and invest the difference in mutual funds.

5. *One of four plans to retire as a millionaire:* Select the plan that best fits your needs, as discussed earlier in the chapter.

6. *Financial security by age thirty to thirty-four, through part-time avocational enterprises:* This is the $2,000 work-unit approach.

7. *Stock market or futures (commodity) trading from interest only on principal from the previous level:* These were not discussed due to space, and they have a high risk. Refer to Chapter 14 for a discussion of Commodity (Futures) Marketing.

CONCLUSION

Few things in life are guaranteed. There are few absolutes. Thus, there is no promise that these financial ideas will work for you. However, we believe that we can say with 100 percent accuracy that if you do not try them or some other financial plan, you will not reach your financial goals. Many people have accumulated financial wealth, and they may not have been smarter than any of the rest of us. However, they have been educated to know how money works, to exercise restraint and self-control, to follow many of the suggestions made in this chapter, and to understand the power of compound interest. A part of being successful is learning to have power over our money rather than letting our money have power over us.

SUMMARY

The decisions about how to handle money will have a profound impact on your life. You must manage your money in an effective and responsible way. You must plan your work, and then work your plan.

As we earn money, we need to understand how to interpret our paychecks. There is a difference in gross versus net pay. Deductions include income taxes (federal, state, and local), social security tax (FICA), unemployment insurance, and worker's compensation. Employee benefits include life insurance, long-term disability, medical insurance, retirement savings plan, profit sharing, and several benefits that do not require deductions, such as paid holidays and vacations.

You must select a financial institution to help manage, protect, and increase your money. There are several types of institutions, including banks, savings and loans, credit unions, and brokerage firms. Some factors to consider in choosing a financial institution include types of service, convenience, cost, relationships, security, and customer qualifications.

Use caution with your checking account. Keep your checkbook properly recorded and balanced. Checks must be written accurately because banks need to understand exactly how much you want to give to whom. Remember to record your deposits. As you balance your checkbook, compare the bank statement with your checkbook register.

You must gain control over your money, or your money will gain control over you. Money is active. If you do not take action continually with your money, or lack of it, it will take action with you. You will have conflict, worry, shortages, and a general lack of fun and funds until you achieve some discipline in handling your money. You can discipline your spending with both fixed and variable expenses.

You need to plan and prepare a budget, which is a plan for spending and saving money. A good budget is an essential tool in every household. Budgets help us avoid unexpected expenses and provide greater freedom. A budget is not meant to complicate your life; on the contrary, when you begin to know where your cash is flowing, it will make life easier.

Some "sweet sixteen" financial management and security tips include: limit debt, do not use credit cards, save big dollars on home mortgages, reduce debt by using the debt snowball technique, develop power over purchases, live within and below your income, pay cash for a car rather than going into debt, negotiate and buy only big bargains, build an emergency fund, manage budget busters, use the envelope system, have enough insurance, master the magic of compound interest, use automatic saving plans, take advantage of matching stock purchases, and use your hobby to make money.

Is it possible for an agribusiness manager or worker to retire as a millionaire? Although nothing in life is guaranteed, things certainly will not happen unless you make them happen. If you think you can, you can. If you think you cannot, you cannot. Either way, you are correct. Your attitude determines your altitude. The four plans that hinge on a 12 percent interest rate are: Plan One, $2,000 a year for six years; Plan Two, buy a used car instead of a new car and invest the difference; Plan Three, the 25-plus-10 plan; and Plan Four, the 80-10-10 plan.

There are key factors to making the millionaire plan work. It has to be worked at. You do not simply start something and let it grow without nurturing it. You must manage investments to achieve a 12 percent interest rate; be aware of the Rule of 72; study, select, invest, and monitor the best mutual funds; and be aware of tax-deferred savings.

Financial security can be attained through knowledge gained in agricultural education and your supervised agricultural experience program. However, you have to work and use "perspiration equity." To do this, you need to select four avocational enterprises that will net $2,000 dollars each on an acre of land or less. This is called the "$2,000 work-unit approach." Invest the earnings in tax-free bond mutual funds. Of course, the plan has limitations due to changing markets, environmental conditions, and sudden downturns in the stock market. However, you can either make excuses or you can work hard, be persistent, stay focused, get tough, and just do it. Remember, if it were easy, everyone would do it.

END-OF-CHAPTER ACTIVITIES

Review Questions

1. What mandatory deductions are taken from your paycheck?
2. List ten employee benefits that may (some are optional) be paid by the employer.
3. What are six factors to consider in choosing a financial institution?
4. Name four reasons for writing checks.
5. List ten points to remember when writing checks.
6. What are six reasons for keeping a checkbook register?
7. What are five reasons for balancing your checking account monthly?
8. List eight common reasons why checkbook registers and account statements differ.
9. Give eight examples of fixed expenses.
10. Give ten examples of variable expenses.
11. What are six benefits of a budget?
12. List sixteen financial management and security tips.
13. What are three ways to save big money on home mortgages?
14. What are the "lucky seven" basic principles of negotiating?
15. What are five types of insurance that everyone should have?
16. Briefly discuss the four ways to potentially retire as a millionaire.
17. Why were mutual funds selected as the investment strategy most likely to get a 12 percent interest rate?
18. What are the five types of mutual funds?
19. Explain the "$2,000 work-unit approach."
20. List eight considerations for selecting part-time avocational enterprises.

Fill in the Blank

1. The math involved in keeping and balancing a checkbook is basic _____ and _____.
2. You must gain control over your money, or your money will gain _____ over you.
3. By failing to plan, we are planning to fail, and there is always too much _____ left at the end of the money.

4. Creating a _____ guards you from being blindsided by an unexpected expense or a forgotten bill.

5. Almost any person knows how to earn money, but not one in a million knows how to _____ it.

6. The _____ of money is the root of all evil.

7. Payments you make on a non- _____ basis can be budget busters if they are not planned.

8. You can get _____ percent interest on your money if your employer matches investments dollar for dollar.

9. If you do not see the money, you will hardly miss it—this is the benefit of a _____ savings plan.

10. The amount of interest that must be achieved to make the millionaire plans work is _____ percent.

11. Besides the interest rate, "Plan One," illustrated in Figure 6–22, shows the importance of _____ and _____.

12. If a person were to invest $2,000 per year for forty-three years, the difference in total yield between 8 percent and 9 percent would be _____.

Matching

a. savings and loan

b. credit cards

c. emergency fund

d. brokerage firms

e. envelope system

f. credit unions

g. banks

h. debt snowball

i. variable expense

j. fixed expenses

k. 8 percent

l. 12 percent

_____ 1. primary purpose is to buy and sell stocks and bonds on behalf of the customer

_____ 2. primary purpose is to offer checking accounts, savings accounts, and certificates of deposit (CDs)

_____ 3. primary purpose is savings accounts and home mortgages

_____ 4. primary purpose is to make low-interest loans available to members, who usually have an affiliation with the company

_____ 5. home mortgage

_____ 6. household maintenance and repair

_____ 7. can be very addictive

_____ 8. technique to reduce bills quickly

_____ 9. three to six months of income in savings

_____ 10. budgeting technique by using cash

_____ 11. interest goal of millionaire plan

_____ 12. interest goal of financial security plan

Activities

1. You have just accepted a position that pays an annual salary of $20,000. You will be paid in equal installments on the fifteenth and the last day of each month. Here is a summary of the deductions on your paycheck:

   ❦ federal income tax—10.60%

   ❦ state income tax—3.00%

   ❦ social security tax (FICA)—7.65% (employee portion; employer matches as required by law)

   ❦ medical insurance, family coverage (shared costs: employee pays $30.00 per pay period)

   ❦ dental insurance, family coverage (shared costs: employee pays $3.50 per pay period)

   What will be the amount of your check per pay period?

2. Contact a person at a financial institution that handles your money matters or a financial institution that you are considering and determine how many of the eighteen financial services are offered that are listed in Figure 6–3.

3. Write out the following check on a copy provided by the teacher. Use the following information.

Date:	Today's date
Payee:	Red Stafford's Feed Supply
Amount:	$132.47
Amount written:	One hundred thirty-two and 47/100 dollars
Memo:	Horse feed
Signature:	Your name

4. Fill out a checkbook register provided by the teacher.

Date	Payee	Amount	Check #
1/7	Carter Hardware	$50.00	#1731
1/10	Farmer's Co-op Supply	$23.52	#1732
1/12	Walmart	$8.48	#1733
1/14	Mires Rental	$35.00	#1734
1/18	Kroger Grocery	$18.50	#1735
1/21	Jones Supply	$12.36	#1736
1/6	Deposit	$69.35	
1/8	Deposit	$161.40	
1/15	Deposit	$337.80	
1/22	Deposit	$126.00	
Service Charge		$4.82	

5. At the end of the month, you will receive a statement from the bank. It will tell you how much money you have in your checking account. You must reconcile (balance) your bank statement every month. To complete this involvement activity, you will need to use the checkbook register you completed in activity 4. Reconcile your checkbook register with the bank statement illustrated here. Complete the reverse side of the bank statement, which follows.

BANK STATEMENT
Date of Statement: January 31, _____

DATE	#	CHECK	DEPOSITS	BALANCE
1-6			69.35	69.35
1-7	1731	50.00		19.35
1-8			61.40	80.75
1-10	1732	23.52		57.32
1-12	1733	8.48		48.75
1-15			337.80	386.55
1-21	1736	12.36		374.19
		4.82 SC		369.37
				LAST AMOUNT IN THIS COLUMN IS YOUR BALANCE

REVERSE SIDE OF BANK STATEMENT
January 31, _____

Subtract Service Charge (SC) from your checkbook balance		Bank balance amount shown on this statement	
		Add deposits made	
Subtract outstanding checks (checks written but not shown this statement)		Total	
Check No.	Amount	Subtract checks outstanding	
		Balance amount should agree with checkbook register	
Total			

6. Plan a budget for your present living and financial situation: living at home, living on your own (roommates), or family budget. You may write it out as Figure 6–12, or your teacher will provide you with a blank sheet of either Figure 6–13, 6–14, or 6–15.

7. Select one of the "sweet sixteen" financial management and security tips. Write a brief report on it and present it to the class. Use real-life examples or any other technique to enhance your presentation. Use today's prices and interest rate if different from the text.

8. Select a hobby that you feel would make you money. Research it and write a report, and present the report to the other class members.

9. Select the millionaire plan that you believe would work best for you. Write a short essay defending your selection and present the report to the other class members.

10. A person offers you a job in July. The conditions are that you work for a penny the first day, and your salary will be doubled each day for the next thirty-one days. If you take the job, how much money will you make in thirty-one days (considering you work thirty-one days without taking off for the weekends or Independence Day)?

11. Look in the advertisements of a financial magazine or newspaper. (Some cable TV stations advertise mutual funds.) Select three mutual funds. Call the toll-free numbers and have them send you a prospectus of the mutual fund that you have selected. Once you get the prospectus, determine their past performance and any other pertinent information, such as type and diversity of investments. Make a short presentation to your class of your findings. Note, the purpose of this exercise is to find mutual funds that have a track record of 12 percent.

12. Using the Rule of 72, determine the cash accumulated to age sixty-one, assuming a onetime investment of $1,000 beginning at age twenty-five.

    a. Age 25: $1,000, interest 4%

    b. Age 25: $1,000, interest 6%

    c. Age 25: $1,000, interest 12%

13. From the list in Figure 6–26, select ten "$2,000 work-units" that you believe would meet the criteria for selecting avocational enterprises.

14. Select one of the ten "$2,000 work-units" that you selected in activity 13, or come up with one that may not be on the list in Figure 6–26. Then write a paper justifying your selection. Remember, it has to meet the criteria for part-time enterprises (or the $2,000 work-unit approach). Convince the class why you believe your selection will work. If you do not believe it will work, make another selection. Budgets will enhance your defense of your selection.

NOTES

1. Dave Ramsey, *Financial Peace* (Nashville, Tenn.: Lampo Press, 1992), p. 43.

2. *Money Skills: A Resource Book for Teaching Personal Economics* (Atlanta, Ga.: NationsBank Corporation. *Money Skills* was produced for NationsBank by Martin Kidd Associates, Nashville, Tennessee, 1994), pp. 31–32.

3. Ibid., pp. 3.2–3.3.

4. Appreciation is extended to Dr. Joyce Harrison, Certified Financial Planner and faculty member of the Department of Human Sciences at Middle Tennessee State University, for input on the first portion of this chapter.

5. *Money Skills*, p. 5.2.

6. Ramsey, *Financial Peace*, p. 163.

7. Ibid., p. 164.

8. *Development of Financial Skills*. Successful Living Skills Series (Stillwater, Okla.: Oklahoma Department of Vocational and Technical Education, Curriculum and Instructional Materials Center, 1989), p. 238.

9. Ibid., p. 246.

10. Ibid., pp. 247–251.

11. *Money Skills*, p. 92.

12. Ibid.

13. *Development of Financial Skills*, pp. 251–255.

14. Ramsey, *Financial Peace*, pp. 16–17.
15. Ibid., p. 102.
16. Ibid., pp. 168–169.
17. Ibid., pp. 157–158.
18. *Money Skills*, p. 141.
19. *High School Financial Planning Program* (Denver, Col.: College for Financial Planning, 1988), pp. 54–57.
20. Ramsey, *Financial Peace*, p. 166.
21. Ibid.
22. Ibid., p. 103.
23. Samuel C. Ricketts, *Science IA (Agriscience)* (Nashville, Tenn.: Tennessee Department of Education, 1990), pp. 8–19.
24. Ramsey, *Financial Peace*, p. 118.
25. Ibid., p. 20.
26. Ibid., p. 70.
27. Ibid., p. 71.
28. Ibid., pp. 81–86.
29. Art L. Williams, *Common Sense* (Atlanta, Ga.: Parklake Publishers, 1986).
30. Ricketts, *Science IA (Agriscience)*, pp. 8–16.
31. Ibid.
32. *High School Financial Planning Program*, pp. 67–80.

CHAPTER 7

Agribusiness Record Keeping and Accounting

OBJECTIVES

After completing this chapter, the student should be able to:

- differentiate bookkeeping versus accounting
- complete budgets
- describe the single entry bookkeeping system
- describe the double entry bookkeeping system
- complete journals and ledgers
- complete a trial balance
- explain basic accounting considerations
- prepare an income statement, balance sheet, and statement of cash flow
- prepare a statement of owner equity
- describe the impact of computer technology on agribusiness record keeping
- explain the benefits of using computer software to complete tax returns

TERMS TO KNOW

accounting cycle	chronological	income statement
accounts payable	cost of sales	inventory
accounts receivable	credit	journals
accrued	creditor	ledger
assets	debit	liabilities
balance sheet	deficits	net profit
budget	depreciation	net worth
budgeting	double entry bookkeeping	note payable
capital	equities	operating budget
capital expenditures budget	erroneous	operating expenses
cash flow budget	incidental expenses	owner's equity

posting	salvage value	statement of cash flows
receivables	single entry bookkeeping	taxable income
reconciling	system	trial balance
revenue	solvency	write-off

INTRODUCTION

Planning is a big part of success. Part of planning in an agribusiness is the need to budget, plan, and keep good records. Keeping records takes time and effort, but it is essential. We should work as hard at our record keeping as at any other phase of the business.

Your attitude determines your success. Your agribusiness can achieve new heights if you develop a positive attitude toward record keeping and what it will do for your business. In reality, records save time and money. Records must be kept accurately and in an orderly fashion. We need records to:

❦ know our cash balance

❦ know who is in debt to our business

❦ keep track of our business debts

❦ file our tax reports correctly and promptly

❦ know our financial position

❦ make business decisions[1]

Role of the Financial Manager

Whether an agribusiness is large or small, one person is usually responsible for the bookkeeping, accounting, and financial management of the business. In a small agribusiness, the owner may be the person responsible for the finances. Therefore, the owner has to make sure the agribusiness can meet its financial obligations and payments. In order to do this, the financial manager determines how money comes to the business and how the agribusiness uses that money. Refer to Figure 7–1.

Sources of Income. Sources of income include revenues from the sale of products and services, profits reinvested in the business, loans and credit received from outside the business, and the **owner's equity**, which is the money the owner invests in the business. The owner or financial manager must decide the best way to use these sources of income.

Expenses. Operating expenses include wages and salaries, cost of maintaining the machinery and equipment needed to produce products and services, advertising, rent, electricity, and insurance. In order to meet unexpected expenses, the owner or financial manager needs to have a reserve or emergency fund of cash that can be made quickly available.[2]

ACCOUNTING VERSUS BOOKKEEPING

Often, the distinction between accounting and bookkeeping is confusing. Bookkeeping refers to the actual recording of business transactions. This clerical record-keeping function may be done manually or with the use of computers. Accounting goes far beyond bookkeeping. The purpose of an accounting system for a business is to prepare, review, and understand reports.[3]

Bookkeeping

If you were a bookkeeper, you would divide all the business's paperwork into meaningful categories, such as sales documents, purchasing receipts, and

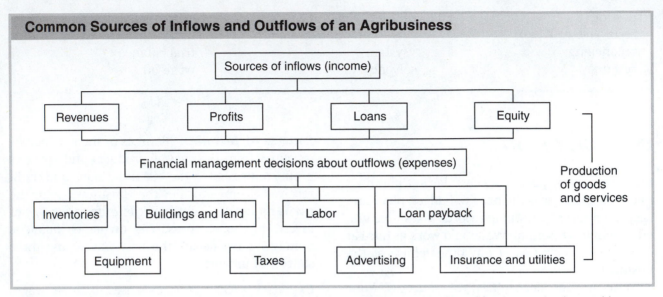

Figure 7–1   Agribusiness owners and managers try to balance the inflow and outflow of funds to maintain a stable business.

shipping documents. This information must then be recorded into record books called **journals**. Journals are the books where accounting data are first entered.

Accounting

Accountants classify and summarize the data provided by bookkeepers. They interpret the data and report them to the owners or management. Accountants also suggest strategies for improving the agribusiness's financial condition and progress.[4] Accounting is the basis for the financial statements and income tax preparation needed by an agribusiness. As you can see, an understanding of basic accounting is critical to the successful control of every agribusiness.

The Six-Step Accounting Cycle

The sequence of accounting procedures used to record, classify, and summarize accounting information is often termed the **accounting cycle**. The accounting cycle generally involves the work of both the bookkeeper and the accountant. It begins with the preparation of formal financial statements summarizing the effects of these transactions on the **assets**, **liabilities**, and owner's equity of the business. The objective of this chapter is to teach you each of the following six steps of the accounting cycle in order for you to perform them when the need arises:

Analyze and Categorize Documents.   This involves a system to get you organized. It may be as simple as putting incoming bills or documents in a shoebox or as complex as a very structured filing system.

Record the Information into Journals.   As each business transaction occurs, it is entered into journals, which creates a **chronological** record of events.

Post the Preceding Information into Ledgers. The **debit** and **credit** changes in account balances are posted from the journal to the **ledger** accounts. This procedure classifies the effects of the business transactions in terms of specific assets, liability, and owner's equity accounts.

Prepare a Trial Balance. A **trial balance** involves summarizing all the data in the account ledgers to see whether the figures are correct and balanced. If they are not correct, they must be corrected before the income statement and balance sheet are prepared.

Prepare an Income Statement, Balance Sheet, and Statement of Cash Flows. The **balance sheet** reports the business's financial condition on a specific date. The **income statement** reports revenues, costs, expenses, and profits (or losses) for a specific period of time showing the results of business operations during that period. It is good to make any notes disclosing any facts necessary for the proper interpretation of those statements.[5]

Analyze the Financial Statements. The sixth step is to analyze the financial statements and evaluate the financial condition of the firm. Bookkeeping and accounting are not done just as a formality or tradition. Besides being needed for income tax purposes, these financial statements can help you evaluate the success of your business so any adjustments can be made. These statements will also let you know whether you can expand or invest profits into other investments.

BUDGETS

If something will not work on paper, it will not work once the agribusiness opens. Once an agribusiness owner or manager has planned for its customers' needs and projected income and expenses, he or she must complete a budget. Proper planning can help deter business failure. Some owners of an agribusiness do their planning entirely in their head and have only a "rough idea" of where they stand. Others put their financial

plans on paper in an orderly fashion. This technique is called **budgeting,** and the plan is called a **budget**.[6] There are three types of budgets that must be prepared; an **operating budget**, a **cash flow budget**, and a **capital expenditures budget**. A discussion of each follows.

The Operating Budget

The operating budget summarizes the expected sales or production activities and related costs for the year. It is an estimate of sales and income plus the anticipated fixed and variable expenses for the year. The operating budget is a positive statement of what management expects to accomplish during the coming year in order to maximize profits.[7] Refer to Figure 7–2 for an example of an operating budget.

The Cash Flow Budget

The cash flow budget summarizes the amount and timing of income that will flow in and out of the business during the year. Income normally comes from sales and services, from borrowing, from sale of capital items, and from payments collected. Expenses can include payments for goods and services purchased from someone else, principal and interest payments on debt, taxes, salaries, and payments on capital purchases, such as a down payment on a truck. The cash flow budget shows:

🌾 when cash receipts should be available to the agribusiness during the year

🌾 when cash payments need to be made by the agribusiness

The cash flow budget is different from the operating budget since the cash flow budget leaves out such items as:

🌾 credit sales that do not generate a cash flow during the year

🌾 depreciation or other noncash expenses

🌾 noncash portions of any capital purchases or sales[8]

Operating Budget Green's Nursery & Landscaping Co. for the Year	
Income:	
Plants	50,000
Soil, Fertilizer, etc.	10,000
Accessories	6,000
Gross Sales Income	**66,000**
Cost of Goods	**33,000**
Gross Profit	33,000
Landscaping Fee Income	20,000
Lawn Care Fee Income	7,000
Total Income	**60,000**
Variable Expenses:	
Labor	30,000
Advertising	2,000
Office Expense	1,000
Total	**33,000**
Fixed Expenses:	
Equipment Depreciation and Lease	5,000
Rent	4,000
Insurance	3,000
Total	**12,000**
Total Expenses	**45,000**
Net Income before Income Tax	15,000
Income Tax Expense	5,000
Net Income	**$10,000**

Figure 7–2 The operating budget summarizes the expected sales or production activities and related costs for the year.

The cash flow budget may tell the owner little about the profitability of the business. Its purpose is to help the owner determine whether there will be adequate funds to meet obligations. If not, money can be borrowed. If there is excess cash during certain periods, it can be invested. Refer to Figure 7–3 for an example of a cash flow budget. Notice how the cash inflows and outflows are divided by quarter.

The Capital Expenditures Budget

Each agribusiness must make regular capital investments in equipment or depreciation due to wear and tear eventually taking its toll. The capital expenditures budget shows how money projected for capital expenditures is to be allocated among the various divisions or activities within the agribusiness. The capital expenditures budget is a list of projects (equipment, etc.) that management believes worthwhile, together with the essential cost of each.[9] Ranking the items or projects helps ensure that the money spent on capital expenditures will be allocated in an efficient manner. Refer to Figure 7–4 for an example of a capital expenditures budget.

Relationships between the Operating, Capital Expenditures, and Cash Flow Budgets

The sales part of the operating budget provides basic information for the cash flow budget. Only that part of sales that is actually expected to generate cash during the budget year is included in the cash flow budget. The capital expenditures budget provides information needed to identify big cash outflows.[10]

Cash Flow Budget
Green's Nursery & Landscaping Co.
for the Fiscal Year Ending December 31

	Quarter			
	1	2	3	4
Cash Balance, beginning of first quarter	10,000	10,075	27,250	16,725
Income:				
Plants	5,000	20,000	15,000	10,000
Soil, Fertilizer, etc.	1,000	4,000	3,000	2,000
Accessories	600	2,400	1,800	1,200
Gross Sales Income	6,600	26,400	19,800	13,200
Cost of Goods	3,300	13,200	9,900	6,600
Gross Profit	3,300	13,200	9,900	6,600
Landscaping Fee Income	2,000	8,000	6,000	4,000
Lawn Care Fee Income	700	2,800	2,100	1,400
Total Income	6,000	24,000	18,000	12,000
Variable Expenses:				
Labor	3,000	12,000	9,000	6,000
Advertising	200	800	600	400
Office Expense	100	400	300	200
Total Variable Expenses	3,300	13,200	9,900	6,600
Fixed Expenses:				
Lease Payments	625	625	625	625
Rent	1,000	1,000	1,000	1,000
Insurance	750	750	750	750
Total Fixed Expenses	2,375	2,375	2,375	2,375
Total Expenses	5,675	15,575	12,275	8,985
Net Income before Income Tax	325	8,425	5,725	3,025
Income Tax Expense	1,250	1,250	1,250	1,250
Net Inflow	(925)	7,175	4,475	1,775
Capital Expenditures			(15,000)	
Loan Repayments				(2,000)
Additional Borrowings		10,000		
Additional Investment	1,000			
	$10,075	$27,250	$16,725	$16,500

Figure 7–3   The cash flow budget summarizes the amount and timing of income that will flow in and out of the business during the year.

**Capital Expenditures Budget
Green's Nursery & Landscaping Co.
for the Year Ending December 31**

Capital Item	Priority	Estimated Cost	Total Cash Outlay
Computer Software and Printer	1	$3,000	$3,000
Trucks and Trailers	1	$15,000	$3,000
Bobcat Loader	2	$8,000	$2,000
John Deere Mower	1	$5,000	$5,000
(3) Small Trimmer Mowers	2	$1,000	$1,000
(4) Weedeater Trimmers	3	$1,000	$1,000

Figure 7–4   The capital expenditures budget is a list of projects and/or equipment needed for the agribusiness.

SINGLE ENTRY BOOKKEEPING SYSTEM

The simplest bookkeeping system is the **single entry bookkeeping system**. This is because the entry is made only in the checkbook or the checkbook and the journal. Therefore, one of the first activities when starting your agribusiness is to establish a checking or other bank account for your business.

Separate Business Accounts from Personal Accounts

Business accounts should always be separated from all personal accounts. This is best for tax records. Mixing personal accounts with business accounts will lead to problems. A sense of job security is attained, leading to unnecessary withdrawals for personal reasons from the business account. If the owner draws money from the business account for personal expenses, these funds should be paid back as soon as possible. Also, if funds are drawn from the business account, use a check to provide the appropriate documentation.

Sales Slips Plus Checking Account

Besides the checking account, a sales slip should be used in a small agribusiness. The checking account and use of sales slips will provide the information needed to file income taxes. Use a sales slip for every sale, and run all money through the checking account. Refer to Figures 7–5 and 7–6.

A checkbook and a sales slip book can document revenues and expenditures. The single entry system is basically a cash method of accounting, since the payments for goods and services are recorded when they are made. The single entry system must provide the owner of the agribusiness with:

❦ assurance that the business will pay all the taxes owed

❦ prevention of errors in **accounts receivable**

❦ prevention of **erroneous** payments for goods receivable

❦ an up-to-date cash position

❦ a possible projection for the future[11]

**Green's Nursery & Landscaping Co.
Rose Garden Drive
Murfreesboro, TN, 37000**

_____ ____

Customer:	
Name	_____
Address	_____
Phone #	_____

Quantity	Description	Unit Price	Extended Amount

For Office Use:
Date Completed: _____ Completed by: _____
Time Required: ___ Workers Used: _____

Figure 7–5 Use a sales slip for every sale to provide information needed to file income taxes.

DOUBLE ENTRY BOOKKEEPING SYSTEM

It is possible to make a mistake when recording a financial transaction. For example, you could easily write $12.97 as $12.79. Because of this, bookkeepers record all their transactions in two places. They can check one list against another list to make sure all transactions add up to the same amount. If not, the bookkeeper knows that a mistake was made.

Recording the Transaction in Two Places

The practice of writing every transaction in two places is called **double entry bookkeeping**. This process involves journals and ledgers. For example, if the owner of an agribusiness puts $10,000 of his or her own money into the business, the bookkeeper (which may be the owner) would debit (make an entry into) the cash account. Refer to Figure 7–7. To keep the transaction in balance, a corresponding credit entry would be recorded in the capital (owner's equity) account. Therefore, debits and credits are always in balance. A debit refers to a bookkeeping entry on the left side of the account.

Checking Account Register

Check Number	Date	Description of Transactions	G/L Acct #	Payment Amount	Deposit Amount	Balance $850.000
1001	6/15/xx	McMinnville Farm	501	500.00		350.00
	6/15/xx	Daily Deposit	401		1,200.00	1,550.00
1002	6/16/xx	Carl's Truck Repair Center	551	100.00		1,450.00
1003	6/16/xx	First Tennessee Bank—Payroll A/C	510	500.00		950.00
	6/16/xx	Daily Deposit	401		1,500.00	$2,450.00

Figure 7–6 By running all money through the checking account, information is recorded that will later be needed to file income taxes.

Simplified Example of Double Entry Bookkeeping System

Assets		=	Liabilities	+	Owner's Equity
Cash in Bank	25,000		20,000		80,000
Accounts Receivable	5,000				
Inventory	20,000				
Equipment	50,000				
	$100,000	=	$20,000	+	$80,000

Figure 7–7 The practice of writing every transaction in two places is called double entry bookkeeping. Debits and credits are always kept in balance.

A credit means the bookkeeping entry was recorded on the right side of the account. The preceding is illustrated by the accounting equation:

$$\text{Assets} = \text{Liabilities} + \text{Owner's Equity, or}$$

$$(\text{Resources} = \text{Claims})$$

Keeping the Books Balanced

Accountants insist on maintaining the balance of resources (assets) and claims or liabilities. Accountants want to identify and track each transaction and entry to ensure this. The balance sheet in Figure 7–8 is another example of double entry bookkeeping. Notice that the assets equal the liabilities, and the books balance.

JOURNALS AND LEDGERS

The Journal

The single entry and double entry systems of bookkeeping could be considered journals, but not in accountants' terms. In an actual accounting system, information is initially recorded in an account record called the journal, as shown in Figure 7–9. After the transaction has been recorded

in the journal, the debit and credit changes in the individual accounts are entered in the ledger. Since the journal is the accounting record in which transactions are first recorded, it is sometimes called the *book of original entry*.[12]

Use of the Journal. The journal is a chronological (day-by-day) record of business transactions. The information recorded about each transaction includes the date of the transaction, the debit and credit changes in specific ledger accounts, and a brief explanation. Again, refer to Figure 7–9.

Why Use a Journal? Since it is technically possible to record transactions directly in the ledger, why is the journal necessary? The unit of organization for the ledger is the transaction, but the unit of organization for the ledger is the account. By having both a journal and a ledger, three advantages are possible over using the ledger alone. They are as follows:

❧ The journal shows all the information about a transaction in one place and also provides an explanation.

❧ The journal provides a chronological record of all the events in the life of a business.

❧ The use of a journal helps to prevent errors.

**Green's Nursery & Landscaping Co.
Balance Sheet
as of December 31**

	Assets	Liabilities
Cash	25,000	
Accounts Receivable	5,000	
Inventory	20,000	
Equipment	50,000	
Total Assets	$100,000	
Accounts Payable		10,000
Bank Note		10,000
Surplus		80,000
Total Liabilities		$100,000

Figure 7–8 The balance sheet is another example of double entry bookkeeping. The assets equal the liabilities.

The Ledger

At convenient intervals, the debit and credit amounts recorded in the journal are transferred (posted) to the accounts in the ledger. The updated ledger accounts can then be used to prepare the balance sheet and other financial statements.

The ledger is an accounting system that includes a separate record for each item. For example, a separate record is kept for asset cash showing all the increases and decreases in cash that result from the many transactions in which cash is received or paid. A similar record is kept for every other asset, for every liability (such as telephone, fuel, office supplies, etc.), and for the owner's equity.

Therefore, the form of record used to show increases and decreases is a single balance sheet item called an account or a ledger account. All the separate accounts grouped together are called a ledger.

Parts of the Ledger. Very simply, a ledger account page has only three parts:

1. a title, consisting of the name of the particular asset, liability, or owner's equity
2. a left side, which is called the debit side
3. a right side, which is called the credit side

Students and those people unfamiliar with accounting often have erroneous notions about the meaning of the terms *debit* and *credit*. For example, many people unacquainted with accounting give the word *credit* a more favorable connotation than the word *debit*. This is not correct. *Debit* and *credit* simply mean *left* and *right*, without any hidden meaning.[13]

Posting. The process of transferring the debit and credits from the general journal to the ledger account is called **posting**. Each amount listed in the debit column (left side) of the journal is posted by entering it on the debit side of an account in the ledger, and each amount listed in the credit column of the journal is posted to the credit (right) side of a ledger account. Posting methods may vary with the preference of the individual, but the following sequence, as shown in Figure 7–10, is commonly used:

1. Locate in the ledger the first account named in the journal entry.
2. Enter in the debit column of the ledger account the amount of the debit as shown in the journal.
3. Enter the date of the transaction in the ledger account.
4. Enter in the reference column of the ledger account the number of the journal page from which the entry is being posted.

General Journal

Date		Account Titles and Explanation	LP	Debit	Credit
Jan	2	Cash		70,000.00	
		Owner's Equity			70,000.00
		Original Investment			
Jan	20	Cash		20,000.00	
		Bank Note			20,000.00
		Loan from bank			
Jan	31	Inventory		10,000.00	
		Cash			10,000.00
		Purchase of inventory			
Feb	10	Equipment		25,000.00	
		Cash			25,000.00
		Purchase of equipment			
Feb	15	Accounts Receivable		2,500.00	
		Sales Income			2,500.00
		Income from landscaping for Ideal Realty			
Feb	20	Cash		1,500.00	
		Accounts Receivable			1,500.00
		Partial payment, Ideal Realty			
Feb	25	Cash		2,750.00	
		Sales Income			2,750.00
		Cash sales of plants			

Figure 7–9 This is an example of a general journal showing several entries. Each of these entries will later be recorded in a ledger account.

Posting from General Journal to Ledger Account

General Journal

General Journal

Date		Account Titles and Explanation	LP	Debit	Credit
Jan	2	Cash	1	70,000.00	
		Owner's Equity	2		70,000.00
		Original Investment			

Ledger

Cash						Debit
Date		Explanation	LP	Debit	Credit	Balance
Jan	2	Original Investment	1	70,000.00		70,000.00

Owner's Equity						Credit
Date		Explanation	LP	Debit	Credit	Balance
Jan	2	Original Investment	2		70,000.00	70,000.00

Figure 7–10   An example of posting (transferring) the debits and credits from the general journal to the ledger account.

5. The recording of the debit in the ledger account is now complete; as evidence of this fact, return to the journal and enter in the ledger page (LP) column the number of the ledger account and page to which the debit was posted.

6. Repeat the posting process described in the preceding five steps for the credit side of the journal entry.[14]

TRIAL BALANCE

The proof of the equality of debit and credit balances is called a trial balance. Since equal dollar amounts of debits and credits are entered in the accounts for every transaction recorded, the sum of all the debits in the ledger must be equal to the sum of all the credits. Refer to Figure 7–11.

Before the account balances are used to prepare a balance sheet, the bookkeeper should check the figures to show that each account's debit balance equals the credit balance. This is a trial balance. A trial balance is a two-column schedule listing the names and balances of all the accounts in the order in which they appear in the ledger; the debit balances are listed in the left-hand column and the credit balances are listed in the right-hand column. The totals of the two columns should agree, as shown in Figure 7–11. Notice the correlation between the trial balance in Figure 7–11 and the entries in the ledger accounts in Figure 7–12 and the balance sheet in Figure 7–8.

Limitations of the Trial Balance

A trial balance does not prove that transactions have been recorded in the proper accounts, recorded in the correct account, or completely omitted from the ledger. In reality, the trial balance proves only that debits and credits are equal. Nevertheless, the trial balance is very helpful. It not only shows that

Green's Nursery & Landscaping Co. Trial Balance as of February 28		
Cash	59,250	
Accounts Receivable	1,000	
Inventory	10,000	
Equipment	25,000	
Prepaid Expenses	4,750	
Accounts Payable		10,000
Bank Note		10,000
Surplus		80,000
Total	$100,000	$100,000

Figure 7–11   The sum of all the debits in the ledger must be equal to the sum of all the credits. These sums will then be used to prepare a trial balance.

the ledger is in balance, it also serves as a convenient stepping-stone for the preparation of financial statements that will be required.

BASIC ACCOUNTING CONSIDERATIONS

Before we proceed to the next step (balance sheet, income statement, and statement of cash flows), let us discuss a few basic accounting considerations: assets and liabilities; capital and owners' equity; sales, cost of sales, and net profit; operating and incidental expenses; depreciation; inventory; and profit and loss.

Ledger

Ledger

Cash

Date		Explanation	LP	Debit	Credit	Debit Balance
Jan	2	Original investment		70,000.00		70,000.00
Jan	20	Bank note		20,000.00		90,000.00
Jan	31	Purchase inventory			10,000.00	80,000.00
Feb	10	Purchase equipment			25,000.00	55,000.00
Feb	20	Payment, Ideal Realty		1,500.00		56,500.00
Feb	25	Cash sales		2,750.00		59,250.00

Accounts Receivable

Date		Explanation	LP	Debit	Credit	Debit Balance
Jan	2	Sales to Ideal Realty		2,500.00		2,500.00
Feb	20	Payment, Ideal Realty			1,500.00	1,000.00

Inventory

Date		Explanation	LP	Debit	Credit	Debit Balance
Jan	31	Purchase inventory		10,000.00		10,000.00

Equipment

Date		Explanation	LP	Debit	Credit	Debit Balance
Feb	10	Purchase equipment		25,000.00		25,000.00

Figure 7–12   A trial balance is a two-column schedule listing the names and balances of all the accounts in the order in which they appear in the ledger. The totals of the two columns should agree.

Assets and Liabilities

Assets are something of value that is owned. Assets are economic resources that are owned by a business and are expected to benefit future operations. Some examples of assets are cash, **receivables**, inventory, investments, equipment, buildings, and prepaid accounts.

Liabilities are amounts that are owed, or debt. Examples of liabilities include accounts payable, notes payable, and mortgages. Even the larger agribusinesses find it convenient to purchase merchandise and supplies on credit rather than to pay cash at the time of each purchase. The liability arising from the purchase of goods and services on credit is called an **accounts payable**. The person or company to whom the account payable is owed is called a **creditor**.

When an agribusiness borrows money for any reason, a liability is incurred and the lender becomes a creditor of the business. The form of liability when money is borrowed is usually a **note payable**, which is a formal written promise to pay a certain amount of money, plus interest, at a definite future time.

Capital and Owner's Equity

Capital is the investment that the owner has put in the business. For example, money used to start the agribusiness is capital. **Capital** also includes machines, tools, and buildings. Owner's equity is the difference between total assets and total liabilities. The claims of the creditors legally come first. If you are the owner of an agribusiness, you are entitled to whatever remains after the claims of the creditors have been fully satisfied.

Revenue, Cost of Sales, and Net Profit

Revenue is the value of what is received from goods sold, services rendered, and other financial sources. Be careful not to confuse the terms, *revenue, sales,* and *income.* Most revenue (money coming into the agribusiness) comes from sales. However, there could be other sources of revenue, such as rents received, interest earned, or income from other sources.

Cost of sales (also known as *cost of goods sold*) is the expense and labor that are directly associated with the production or acquisition of products held for sale. This includes the purchase price plus any packaging or freight charges paid to bring in the goods, plus the cost associated with storing the goods, less the cost of any item returned.[15]

Net profits is revenue minus the cost of sales. Net profit is also called the *profit margin.* In an agribusiness service, there may be no cost of goods sold; therefore, net sales could equal gross margin.

Operating and Incidental Expenses

Operating expenses are any regular expenses associated with the operation of the business. Obvious ones include rent, salaries, supplies, utilities, insurance, and depreciation of equipment. **Incidental expenses** are any other expenses that seldom occur but are business related. Examples would be unexpected emergencies due to accidents, or weather damage.

Inventory

An important aspect of financial accounting is the development of the **inventory** for the agribusiness. An inventory is a physical count of all assets in a business along with their estimated worth.

When to Inventory. Although inventories can be done at any time, January 1 is a good date because it corresponds with the tax year. The inventory date must be the same each year to allow accurate comparisons of financial performance. The ending inventory for the year is always the beginning inventory for the next year.

Determining Inventory Value. There are two methods of putting a dollar value on agribusiness inventories. They are book value and market value. *Book value* is the original cost of an item less depreciation and any other adjustments. Typically, book value is used for income tax purposes and for estate planning. *Market value* is book value less any variable cost involving such things as transportation or sales commissions. This method is typically used for establishing credit and determining insurance needs.

There are many varieties of the two methods. No matter which method a manager selects to value inventory, it is important to use the same one consistently.

Depreciation

Depreciation is the systematic **write-off** of the value of a tangible asset over its estimated useful life. Agribusinesses are permitted to use **depreciation** as an expense of business operations. This loss is covered by normal use, weathering, and/or changes in technology. Depreciation of machinery, buildings, equipment, and storage structures constitutes a major agribusiness expense.

Calculating. There are various methods of calculating depreciation. Straight-line depreciation is the simplest and easiest to calculate. For calculating straight-line depreciation, compute the annual depreciation by dividing the original cost of the asset, less any **salvage value**, by the expected useful life. For example, a $25,000 asset with a $1,000 salvage value is determined to have an expected life of six years. The depreciable value would be $24,000. The depreciation amount is $4,000. The depreciation amount remains the same over the next five years. Refer to Figure 7–13.

Profit and Loss

Profit and loss are the difference between sales minus all expenses. In an income statement, which will be discussed later in the chapter, profits can be three different amounts: gross profit, operating income, and net income.

Straight-Line Depreciation Example

	Annual Depreciation	Accumulated Depreciation	Book Value
Year 1	$4,000	$4,000	$21,000
Year 2	4,000	8,000	17,000
Year 3	4,000	12,000	13,000
Year 4	4,000	16,000	9,000
Year 5	4,000	20,000	5,000
Year 6	$4,000	$24,000	$1,000

Figure 7–13 An asset costing $25,000 with a salvage value of $1,000 and a useful life of six years will depreciate by $4,000 yearly.

Gross profits is the term used when the cost of goods sold is deducted from sales revenue. The gross profit is a subtotal. Operating income results when operating expenses are deducted from gross profit. This is also a subtotal. Some call this *income from operations*. Net income is determined by subtracting "nonoperating" costs from operating income. Many call this the "bottom line."

PREPARING AN INCOME STATEMENT, BALANCE SHEET, AND STATEMENT OF CASH FLOWS

Agribusiness owners and managers need up-to-date financial information to make decisions about future operations. As discussed earlier, an agribusiness's financial information is recorded in ledger accounts. Transactions occurring during a fiscal period change the balance of ledger accounts. These changes should be reported periodically in the form of financial statements.

Financial statements summarize the financial information contained in ledger accounts. These financial statements provide adequate disclosure of the agribusiness. A financial statement showing the revenue, expenses, and net income or net loss of a business is called an income statement. A financial statement showing all assets and **equities** of a business on a specific date is known as a balance sheet.

A balance sheet and an income statement can be prepared any time that the financial information is needed. However, agribusinesses should always prepare these two statements at the end of a fiscal period. An income statement shows financial information over a specific period of time. A balance sheet shows financial information on a specific date. The financial progress of an agribusiness in earning a net income or a net loss is shown by an income statement. The financial condition of an agribusiness on a specific date is shown by a balance sheet. Refer to Figure 7–14.

Income Statement

An income statement measures the profit for an agribusiness over a given time period. This time period is usually one year; however, monthly or quarterly reports may be desired. Other names for the income statement are *operating statement* or *profit and loss statement.* The four major components of the income statement are revenue, expenses, taxes, and "other." Refer to Figure 7–15.

Comparison of Income Statement and Balance Sheet

The income statement reports financial progress from the beginning of the month until the end; from January 1st to January 31st.

For Month of January

S	M	T	W	T	F	S
	1	2	3	4	5	6
7	8	9	10	11	12	13
14	15	16	17	18	19	20
21	22	23	24	25	26	27
28	29	30	31			

The balance sheet reports the balances carried as of the end of a day; for example, January 31st.

As of January 31

S	M	T	W	T	F	S
	1	2	3	4	5	6
7	8	9	10	11	12	13
14	15	16	17	18	19	20
21	22	23	24	25	26	27
28	29	30	**31**			

Figure 7–14 Agribusiness owners should prepare these two statements at the end of every fiscal period.

Green's Nursery & Landscaping Co.
Income Statement
for the Year

Revenue			
Plants	200,000		
Soil, Fertilizer, etc.	50,000		
Accessories	30,000		
Gross Sales Income		280,000	
Cost of Goods		140,000	
Gross Profit			140,000
Landscaping Fee Income			20,000
Lawn Care Fee Income			7,000
Total Income			167,000
Variable Expenses			
Labor	30,000		
Advertising	4,000		
Office Expense	1,000		
Total Variable Expenses		35,000	
Fixed Expenses:			
Equipment Depreciation and Lease	5,000		
Rent	4,000		
Insurance	3,000		
Total Fixed Expenses		12,000	
Total Expenses			47,000
Net Income before Income Tax			120,000
Income Tax Expense			36,000
Net Income			**$84,000**

Figure 7–15 An income statement measures the profit for an agribusiness over a given time period.

Revenue. Many agribusinesses earn revenue from a variety of sources. Revenue comes in the form of cash receipts from product sales, services, and interest and dividend income. The gain or loss on the sale of intermediate and long-term assets also effect revenue. Wages earned from a job unrelated to the business and noncash adjustments in the value of inventories and **accrued** revenue due from others for products already sold are counted as revenue.[16]

Expenses. Total agribusiness expenses include cash operating expenses, interest expenses, plus depreciation and other noncash adjustments. Depreciation of property and equipment is a business expense even though there is no actual cash payment at the time. Accounts payable accrued expenses, cash investments in the business, and prepayments during the year may also require expense adjustment.[17]

Taxes. The difference between revenue and expenses represents the **taxable income,** or loss, of an agribusiness. The income statement typically uses a separate category to account for payment of income and self-employment taxes. This category does not include property taxes or employee taxes since they count as business expenses.[18]

Other. The final component of an income statement captures the gain or loss from unusual events. For example, a freeze that destroys an insured crop results in an insurance payment that would go into the "other" category.

Balance Sheet

The three major components of a balance sheet are assets, liabilities, and **net worth**. The name *balance sheet* comes from the fundamental accounting operation that states that "assets equal liabilities plus net worth." This equation always holds true. Any transaction changing one side of this equation results in an equal change to the other.

The balance sheet normally has two columns. The left side lists the assets, and liabilities and net worth appear on the right side. Assets and liabilities have direct values, whereas net worth has an indirect value because it is calculated from total assets minus total liabilities. The three classifications of both assets and liabilities are current, intermediate, and long term. Refer to Figure 7–16.

Current Assets and Liabilities. Current assets are those assets that an agribusiness manager could convert to cash within a twelve-month period without disrupting business operations. Current assets include such items as cash, accounts receivable, inventories, and prepaid expenses.

Current liabilities are debt payments due within twelve months of the balance sheet date. Those liabilities include such items as accounts payable, accrued expenses, and any payments due on loans. Any tax due (contingent) on the sale of current assets is also a current liability.

Intermediate Assets and Liabilities. Intermediate assets are those that normally yield service for longer than one year but less than ten. During this time frame, management either depreciates, liquidates, or replaces them. This group includes machinery, equipment, vehicles, and business furnishings.

Intermediate liabilities range in time frame from one to ten years. They primarily involve notes that mature during this time and taxes contingent on the sale of intermediate assets.

Long-term Assets and Liabilities. Long-term or fixed assets contain those assets that have an expected life or maturity exceeding ten years. These assets include land, buildings, and other permanent improvements.

Long-term or fixed liabilities have maturity periods exceeding ten years. They include mortgages on land and buildings and other long-term obligations.

Net Worth. Net worth is the difference between total assets and total liabilities. If this value is positive, the business has **solvency**. This means that if management converted all the assets to cash, there would be more than enough to cover all liabilities.

Statement of Cash Flows

In 1988, the Financial Accounting Standards Board required the statement of cash flows to replace the statement of changes in financial position.[19] The **statement of cash flows** reports

Green's Nursery & Landscaping Co.
Balance Sheet
December 31

Assets

Current Assets:		Current Liabilities:	
Cash	69,000	Accounts Payable	10,000
Short-Term Investments	20,000	Short-Term Notes Payable	2,000
Accounts Receivable	20,000	Current Portion of Long-Term Debt	2,000
Inventory	34,000	Accrued Rent	500
Supplies	5,000	Accrued Payroll	1,500
Prepaid Expenses	2,000	Accrued Taxes	1,000
Total Current Assets	150,000	Total Current Liabilities	17,000
Property and Equipment:		**Long-Term Liabilities:**	
Land	10,000	Mortgages net of current portion	20,000
Buildings	25,000	Notes Payable net of current portion	15,000
Machinery and Equipment	20,000	Total Long-Term Liabilities	35,000
Total Original Cost	55,000		
Accumulated Depreciation	5,000	**Total Liabilities**	52,000
Total Property and Equipment	50,000		
Long-Term Assets:		**Owner's Equity:**	
Long-Term Investments	15,000	Original Investment	70,000
Notes Receivable	4,000	Retained Earnings—Jan. 1st	14,000
Deposits	1,000	Current Year Net Income/Loss	84,000
Total Long-Term Assets	20,000	**Total Owner's Equity (net worth)**	168,000
Total Assets	**$220,000**	**Total Liabilities & Owner's Equity**	**$220,000**

Figure 7–16 The three components of a balance sheet are assets, liabilities, and net worth. The balance sheet has two columns. The left side lists the assets and liabilities and net worth appears on the right side.

cash receipts and disbursements related to the agribusiness's major activities: operations, investments, and financing. Refer to Figure 7–17.

The Generally Accepted Accounting Principles (GAAP) requires the inclusion of the statement of cash flows as one of the four basic financial statements. The other three required financial state-

ments are the income statement, balance sheet, and statement of owner's equity.

Comparison of Cash Flow Statement and Statement of Cash Flows. The statement of cash flows is an analysis and planning tool. The cash flow statement is a projection tool.

Green's Nursery & Landscaping Co.
Statement of Cash Flows
for the Year Ended December 31

Cash flows from operations:		
Sales	307,000	
Cash payments from Accounts Receivable	15,000	
Inventory	(140,000)	
Cash expenses for operations	(42,000)	
Income taxes paid	(36,000)	
Net cash from operations		104,000
Cash flows from activities other than operations:		
Investments	(24,000)	
Notes receivable	2,000	
Mortgage payments	(40,000)	
Proceeds from notes payable	15,000	
Net cash from nonoperations activities		(47,000)
Net change in cash		57,000
Cash at beginning of year		12,000
Cash at end of year		$69,000

Figure 7–17   The statement of cash flow reports cash receipts and disbursements related to operations, investments, and financing.

A *cash flow statement* highlights the financing arrangements necessary to cover cash requirements of the agribusiness. The previous year's cash flow statement shows when cash surpluses and deficits occurred. The surpluses and deficits largely determine the credit needed and a payback schedule. For example, a loan may be due in March yet your best month of sales may be April. Bankruptcy could occur because improper planning has not allowed for the fluctuation in cash flow, even though within the fiscal year, money would be available. Refer to Figure 7–18.

A *statement of cash flows* involves evaluation and planning. Questions that are answered by the statement of cash flows include:

◆ Where did the cash flows come from?

◆ How were dollars spent?

◆ Is the change in cash flow consistent with the balance sheet's cash position?

◆ Was cash used to buy stocks, bonds, and other investments?

◆ Were some investments sold that brought in cash?

Sources of Cash Flows.   Cash and cash equivalents are the only types of entries allowed on the statement of cash flows. This includes checking account balances, actual cash on hand, savings accounts, time certificates, government payment

Green's Nursery & Landscaping Co.
Quarterly Cash Flow Statement
for the Year Ended December 31

Categories	1st Qtr.	2nd Qtr.	3rd Qtr.	4th Qtr.	Total
Cash inflows from operations:					
1. Beginning Cash	12,000	28,500	16,500	26,000	12,000
2. Plants Receipts	30,000	70,000	70,000	30,000	200,000
3. Soil and Fertilizer Receipts	10,000	25,000	30,000	15,000	80,000
4. Landscaping Fees	2,000	6,000	8,000	4,000	20,000
5. Lawn Care Fees	500	2,000	3,500	1,000	7,000
	54,500	131,500	128,000	76,000	319,000
Cash outflows from operations:					
6. Inventory	5,000	90,000	40,000	5,000	140,000
7. Operating Expense	8,000	13,000	13,000	8,000	42,000
8. Income Tax Payments	9,000	9,000	9,000	9,000	36,000
	22,000	112,000	62,000	22,000	218,000
Cash Flows from Operations	32,500	19,500	66,000	54,000	101,000
Cash flows from other activities:					
9. Accounts Receivable	10,000	10,000	(30,000)	25,000	15,000
10. Investments	(4,000)	(20,000)			(24,000)
11. Notes Receivable		2,000			2,000
12. Mortgage Payments	(10,000)	(10,000)	(10,000)	(10,000)	(40,000)
13. Proceeds from Notes Payable		15,000			15,000
	(4,000)	(3,000)	(40,000)	15,000	(32,000)
Ending Cash Balances	$28,500	$16,500	$26,000	$69,000	$69,000

Figure 7–18 A quarterly cash flow statement highlights the financing arrangements necessary to cover the cash requirements of the agribusiness.

certificates, and so on. As stated earlier, three areas are used to categorize entries to the statement of cash flows: operating, investing, and financing activities.[20]

The values within the *Operating Income and Expenses* section should correspond with the accrual income statement. An example would be entries related to the production of goods and services that include consumable supplies. Another example would be withdrawals for family/personal living and investments. This step occurs if the agribusiness does not pay a wage or salary as a business expense for labor and management contributions of family members.[21]

Information as to why purchases and sales of capital assets occurred is provided in the *Investments* section. It is important to know if assets are being sold to account for operating cash deficiencies. Nonbusiness investments are included within this category. A negative cash flow requires either the operating or financing area to cover it, whereas a positive cash flow requires managers to scrutinize the records to determine if it is in the best interest of the business. Examples of investments include land, buildings, machinery/equipment, personal loans to others, retirement accounts, and so on.[22]

Borrowing funds or other equity injections comprise the primary cash inflows. Loan repayments or capital lease repayments are the major cash outflow entries. The *Financial Activities* section provides the operator(s) with a summary of debt financing. Both operating and term debt are accounted for; this information is an invaluable aid when making decisions concerning acquiring and repaying loans.[23]

STATEMENT OF OWNER EQUITY

Conducting a final check on the accuracy of the change in owner equity between the beginning and ending balance sheets with net income and other possible changes in owner equity is the primary role of the statement of owner equity. Owner equity is the most fundamental financial measure of the firm's financial position and performance. Verifying that the base financial statements, balance sheet, income statement, and statement of cash flows are in agreement is the goal of the statement of owner equity. Establishing and understanding what occurred financially within the business during the financial year is accomplished as well. This is a powerful addition to the basic financial statements, and it offers incredible insight to the financial status of a business.

As previously mentioned, the statement of owner equity utilizes information from the other three, basic financial statements. Beginning and ending owner equity amounts come from the balance sheets. Net income is calculated in the income statement, and other values are derived from the statement of cash flows.

Explanation of the Statement of Owner Equity Financial Statement

Figure 7–19 shows nine entries within the statement of owner equity financial statements. An explanation follows for the nine entries.

1. Beginning Owner Equity. This is the amount of equity at the beginning of the year. It is the same amount that would be taken from the balance sheet ("Total Owner Equity") of the previous year on December 31.

2. Net Income (Accrual). This amount is gross revenue minus expenses. It is the same amount as the "Net Business Income" taken from the income statement.

3. Gifts and Inheritances. This is any amount given to you outside the revenue producing ability of the agribusiness. An example would be a $5,000 inheritance that was invested in the agribusiness.

Green's Nursery & Landscaping Co. Statement of Owner's Equity for the Year Ended December 31		
1. Beginning owner's equity		84,000
2. Net income	84,000	
3. Gifts and inheritances	5,000	
4. Additions to paid-in capital, including investments of personal assets into the business	10,000	
5. Distributions of dividends, capital or gifts made (cash or property)	(5,000)	
6. Withdrawals for family living, gifts made, and investments into personal assets	(10,000)	
7. Total change in contributed capital and retained earnings (Line 2 + Line 3 + Line 4 − Line 5 − Line 6 = Line 7)		84,000
8. Change in valuation of equity		0
9. Ending owner's equity		$168,000

Figure 7–19   The primary role of the statement of owner equity is to conduct a final check on the accuracy of the change in owner equity. The statement of owner equity utilizes information from the other three basic financial statements: balance sheets, income statement, and statement of cash flows.

4. Additions to Paid-in Capital, Including Investments of Personal Assets into the Business.   If money is taken from investments in personal accounts, stocks, or any other personal assets and put into the agribusiness, this is where it is recorded. These contributions to the agribusiness are helpful, but their origin must be accurately noted to more accurately reflect the stability of the business.

5. Distributions of Dividends, Capital, or Gifts (Cash or Property).   These values are usually found directly on corporation or partnership financial distribution forms. These values are recorded in the statement of cash flows. However, few beginning agribusiness sole (single) proprietorships would utilize this entry.

6. Withdrawals for Living Family, Gifts Made, and Investments into Personal Assets.   As indicated, this entry represents money taken from the agribusiness for family living, gifts, and investments into personal assets. The amount is taken from the statement of cash flows.

7. Total Change in Contributed Capital and Retained Earnings.   Retained earnings is the amount of money put back into the business. This practice of reinvesting income back in the business is common in most firms where growth of the business is a goal. Also, in Figure 7–12 notice that Line 2 plus Line 3 plus Line 4 minus Line 5 minus Line 6 equals Line 7.

8. Change in Valuation Equity.   If the beginning balance sheet and ending balance sheet are done correctly, this amount is found by subtracting the amount in "Owner equity: retained earnings, contributed capital, personal net worth and valuation equity" in the beginning balance sheet from the same amount in the category in the ending balance sheet.

9. Ending Owner Equity. This amount is calculated by the following: Line 1 plus or minus Line 7 plus or minus Line 8 equals Line 9 ("Ending owner equity"). Line 9 and owner equity on the ending balance sheet should be the same.[24]

ANALYZING FINANCIAL STATEMENTS

The last step in the accounting cycle is to analyze the financial statements. Balance sheets, income statements, and cash flow statements provide information to agribusiness owners and managers on how well their business is doing. An agribusiness manager can obtain a clear picture of the financial position and performance of the business only by analyzing the components of all three financial statements together.

Compare All Financial Statements with Each Other

Agribusiness managers can expect problems evaluating financial position and performance when only one or two of the statements are available. For example, a solid net income in one year does not particularly mean the business is doing well, as profits might result due to the sale of intermediate or long-term productive assets. Examining the bal-

ance sheet would tell if total assets were increasing or decreasing and what was happening to net worth. Also, a strong cash position on the balance sheet does not necessarily mean that the business is in good shape.[25]

Furthermore, neither the balance sheet nor the income statement reveal critical financial periods during the year. Only the cash flow statements show when financial stress occurs during the year and the steps taken to improve the situation.

Objective of Financial Analysis

The objective of financial analysis is to estimate financial position and monitor financial performance. A financial analysis concentrates on the solvency, liquidity, and changes in the net worth of the agribusiness.

Methods of Financial Analysis

There are several methods available for analyzing financial statements: ratio analysis, historical analysis, and comparative analysis.

Ratio analysis expresses key financial relationships in percentage terms. This allows management to assess liquidity ratios, leverage (debt) ratios, profitability (performance) ratios, and activity ratios as compared to previous years or similar operations. These ratios provide valuable information to agribusiness owners, managers, and lenders. A brief description of each follows.

Liquidity ratios measure the business's ability to pay its short-term debts. These ratios are of particular importance to the firm's creditors, who expect to be paid on time. The two primary indicators of a company's liquidity are the current ratio and the acid-test ratio.[26] The current ratio is the ratio of a firm's current assets to its current liabilities. This information can be found on the firm's balance sheet. Let us say that Hall's Auto Repair has $25,000 of current assets and $10,000 of current liabilities. Its current ratio is thus 2.5. This means the company has $2.50 of current assets for every $1 of current liabilities.

$$Current\ ratio = \frac{Current\ liabilities}{Current\ assets} = \frac{\$25,000}{\$10,000} = 2.5$$

Leverage (debt) ratios refer to the degree to which a firm relies on borrowed funds in its operations. A firm that takes on too much debt could have problems repaying its lenders or meeting promises to stockholders. These two groups are generally very interested in a firm's debt ratios. The ratio of debt to owner's equity measures the degree to which the company is financed by borrowed funds that must be repaid:

$$\text{Debt of owners' equity ratio} = \frac{\text{Total liabilities}}{\text{Owners' equity}} = \frac{\$150,000}{\$275,000} = 54.5\%$$

A ratio above 1 (or 100 percent) shows that a firm actually has more debt than equity, which could be quite risky to both lenders and investors. However, it is always important to compare ratios among other firms in the same industry because debt financing is more acceptable in some industries than others. Comparisons with past years' ratios can also identify trends within the firm or industry.[27]

Profitability (performance) ratios measure how effectively the firm is using its various resources to achieve profits. Management's performance is often carefully measured by using profitability ratios.[28] Three of the more important ratios used are earnings per share, return on sales, and return on equity. An example of each follows:

$$\text{Earnings per share} = \frac{\text{Net income}}{\text{Number of common shares outstanding}} = \frac{\$120,000}{\$100,000} = \$0.12 \text{ earnings per share}$$

$$\text{Return on sales} = \frac{\text{Net income}}{\text{Net sales}} = \frac{\$50,000}{\$500,000} = 110\% \text{ return on sales}$$

$$\text{Return on equity} = \frac{\text{Net income}}{\text{Total owners' equity}} = \frac{\$50,000}{\$150,000} = 33\% \text{ return on equity}$$

Activity ratios measure the effectiveness of the firm's management in using its available assets. Converting the firm's resources to profits is a key function of management.

The *inventory turnover ratio* measures the speed of inventory moving through the firm and its conversion into sales. Inventory sitting idly in a business costs money. The more efficiently the firm manages its inventory, the higher the return. We measure the inventory turnover ratio as:

$$\text{Inventory turnover ratio} = \frac{\text{Cost of goods sold}}{\text{Average inventory}} = \frac{\$320,000}{\$116,000} = 2.75 \text{ turns}$$

A lower than average inventory turnover ratio often indicates obsolete merchandise or poor buying practices. A higher than average ratio may note an understocked condition where sales can be lost because of inadequate stock.[29]

Historical analysis makes comparisons over time. *Comparative analysis* compares events within the year with other similar firms.

COMPUTER TECHNOLOGY AND AGRIBUSINESS RECORDS

Many industries, such as tractor companies, have computer software programs for their dealers to use. Certified representatives can train the workers at the dealership to execute most of the business transactions. Many things can be done with these computer programs. The following are some examples:

- ❧ sales and billing
- ❧ customer accounts
- ❧ weekly payroll
- ❧ receive a product into inventory
- ❧ print monthly statements
- ❧ financial records and reports
- ❧ calculate quarterly taxes
- ❧ place monthly bills in correct accounts
- ❧ fully balance the journal

Besides financial records, these same computer software programs can assist with ordering parts, locating part numbers, inventory, and various other things to assist the agribusiness. Hopefully, the information in this chapter will give you a better understanding of the information computer programs can provide.

FILING TAXES WITH COMPUTER SOFTWARE

There are numerous computer software programs to file tax returns. These tax return programs have reached a high state of proficiency. If the taxpayer has a personal computer, any one of the leading software packages will be beneficial to use in filing taxes. The tax program assists in the calculation of sums, percentages, and tax liabilities. It is advised

that the computer program not be relied on solely. Mistakes can occur during the input phase, causing the output to be incorrect. It is suggested that the taxpayer prepare a handwritten copy first before using the software program.[30]

The software programs are designed to be easy to use, even for a beginner. These programs will lead you through all the necessary steps to prepare and file the return. The programs are accurate and fast, and all the leading programs provide technical support with the operation of the computer program. The support people are not trained in accounting and tax laws and will refer the taxpayer to the Internal Revenue Service for questions concerning taxes. The leading software programs have checks after the return is complete to check the accuracy of the return. An audit report that shows any possible errors in the return can be accessed. Data from a tax return can be transmitted electronically rather than sent by mail. (The taxpayer must have an operating modem to do this.) There are benefits to filing electronically, including acknowledgment that the tax return was received, and, for a fee, some programs even allow for the refund to be deposited in a checking account.[31] Many careers are available in agribusiness accounting and recordkeeping. Refer to the Career Option on page 195 for a further explanation of careers in this area.

CONCLUSION

Planning is a big part of success. Proper budgets, record keeping, reports, and financial analysis are a part of good business planning. Proper and accurate record keeping lets you know the financial status of your business. Financial reports let you know whether you can reinvest into the business, make other investments, or help you time your loans to match your cash flow. Almost all record

CAREER OPTION

Career Areas: Accountant, Bookkeeper, Financial Manager, Management Analyst

An important aspect of managing a sound business is the ability to plan ahead financially. Production agriculturalists today must spend their time managing crops and livestock, so the task of preparing financial paperwork must fall to someone other than the farm operator. A bookkeeper can handle most of the simple, day-to-day recording of business transactions. A bookkeeper position requires good math skills along with accounting classes at the high school level, and often a postsecondary education at a community college enhances employment opportunities.

An accountant performs a broad range of accounting, auditing, tax preparation, and consulting activities for their clients. Accountants have gained professional recognition through a certification of licensure test (Certified Public Accountant, or CPA). This is a four-part, two-day test, which only about one-fourth of all test takers pass the first time. Each of the four sections of the test can be taken separately and passed within a specified time period. The minimum requirement for licensure is a bachelor's degree in accounting or a related field. A master's degree, along with familiarity of computer software for accounting and an area of accounting expertise, are an advantage in today's job market. Financial management is an area of specialization within the accounting field.

A management analyst interprets financial information so that a business owner can make sound financial decisions. A management analyst prepares stockholder reports, financial statements for banks and other creditors, and government paperwork and tax forms. He or she can help you to plan for future growth by creating a budget for the business to follow. The prerequisites for this occupation are similar to those for the accountant. A bachelor's degree in accounting or business management is preferred, although expertise in both areas is a plus. Management analysts are listed on the Department of Labor, Bureau of Statistics, 1994–2005 National Industry Occupation Employment Matrix as among the top twenty-five fastest growing occupations requiring a bachelor's degree or higher for the period 1994 through 2005.

Cashier, billing clerk, and accounting clerk are entry-level positions, which can be attained with good grades in math and a high school diploma.

Computer programs for daily, weekly, and monthly record keeping for accountants, bookkeepers, financial managers, and management analysts are abundant. Most businesses also file their taxes with the aid of computer programs. (Courtesy of Cliff Ricketts)

keeping can be done by computer in order to make the process much easier.

SUMMARY

An important, yet often overlooked, part of planning in an agribusiness involves budgeting, planning, and keeping good records. We need to work as hard at record keeping as at any other phase of the business.

Often, the distinction between accounting and bookkeeping is confusing. Bookkeeping refers to the actual recording of business transactions. An accounting system includes the preparation, review, and understanding of reports.

There are six steps in the accounting cycle: analyze and categorize documents; record the information into journals; post the preceding information into ledgers; prepare a trial balance; prepare an income statement, balance sheet, and statement of cash flows; and analyze the financial statements.

If the business will not succeed on paper, it will not succeed once the agribusiness opens. Once an agribusiness owner or manager has planned for its customers' needs and projected income and expenses, he or she must complete a budget. There are three types of budgets that need to be prepared: an operating budget, a cash flow budget, and a capital expenditures budget.

The simplest bookkeeping system is the single entry bookkeeping system, in which the entry is simply made in the checkbook or in the checkbook and the journal. Besides a checking account, a sales slip should be used in this bookkeeping system. These two things will provide the information needed to file income taxes.

Double entry bookkeeping refers to the practice of writing every transaction in two places. This process involves journals and ledgers. The two places (columns) must be in balance. To keep the transactions in balance, debits and credits must always be in balance. A debit is a bookkeeping entry on the left side of the account, and a credit is a bookkeeping entry on the right side of the account.

Information from sales or purchases is recorded in an account record called a journal. After the transaction has been recorded in the journal, the debit and credit changes in the individual accounts are posted in the ledger. The proof of the equality of debit and credit balances is called a trial balance. A trial balance is a two-column schedule listing the names and balances of all the accounts in the order in which they appear in the ledger.

Every agribusiness owner should be familiar with basic accounting terminology. Among others, the owner or manager should understand assets and liabilities; capital and owner's equity; sales, cost of sales, and net profit; operating and incidental expenses; depreciation; inventory; and profit and loss.

Agribusiness owners and managers need up-to-date financial information to make decisions about future operations. Financial statements summarize the financial information contained in ledger accounts. A financial statement showing the revenue, expenses, and net income or net loss of a business is called an income statement. A financial statement showing the asset and equities of a business on a specific date is known as a balance sheet. The statement of cash flows reports cash receipts and disbursements related to the agribusiness's major activities: operations, investment, and financing.

Conducting a final check on the accuracy of the change in owner equity between the beginning and ending balance sheets with net income and other possible changes in the owner equity is the primary role of the statement of owner equity. Owner equity is the most fundamental financial measure of the firm's financial performance.

The last step in the accounting cycle is to analyze the financial statements. Balance sheets, income statements, and cash flow statements provide information to agribusiness owners and managers on how the business is doing. However, all three financial statements must be analyzed together to obtain a clear picture of the financial position of the agribusiness.

After studying the reports needed for good financial records to run an agribusiness, one can appreciate the value of computers in the accounting process. Computers can record and analyze data and print out financial reports providing continuous financial information for the agribusiness.

END-OF-CHAPTER ACTIVITIES

Review Questions

1. What are six reasons for keeping records?
2. What is the role of a financial manager within an agribusiness?
3. What are four potential sources of income for an agribusiness?
4. What are six potential expenses of an agribusiness?
5. Explain the differences between bookkeeping and accounting.
6. What are four reasons why accountants or accounting are vital to an agribusiness?
7. What are the six steps in the accounting cycle?
8. Explain the relationship between the operating, capital expenditures, and cash flow budgets.
9. Why should business accounts and personal accounts be separated?
10. What are five things the single entry system must provide the agribusiness owner?
11. What information is recorded in a journal?
12. What are three advantages of using both a journal and ledger rather than the ledger alone?
13. What are three parts of a ledger account page?
14. What are the six steps of posting in a ledger?
15. What are three limitations of trial balances?
16. Do *revenue, sales,* and *income* describe the same thing? Explain.
17. What are six examples of operating expenses?
18. Explain the two methods of determining inventory values.
19. Explain and give an example of how to calculate straight-line depreciation.
20. Name and briefly discuss the four parts of an income statement.
21. What are the four basic financial statements required by the Generally Accepted Accounting Principles (GAAP)?
22. List three reasons why a cash flow statement is important.
23. What five questions are answered by a statement of cash flows?
24. Name five sources of cash flow within a statement of cash flows.

25. Name and briefly discuss the three areas used to categorize entries on the statement of cash flows.

26. What is the primary role of the statement of owner equity?

27. List four contributions of the statement of owner equity.

28. Name and briefly explain the nine entries on a statement of owner equity financial statement.

29. Explain why all the financial statements should be analyzed together.

30. What is the objective of financial analysis?

31. Name and briefly explain the three methods of financial analysis.

32. Why should an agribusiness owner consider using computer technology for the agribusiness records?

33. List six reasons for filing taxes with computer software.

34. Name and briefly explain the four types of ratio (financial) analysis.

Fill in the Blank

1. The purpose of the cash flow budget is to help the owner determine whether there will be adequate _____ to meet obligations.

2. In a single entry bookkeeping system, a checking account and _____ _____ would provide the information needed to file income taxes.

3. _____ means that a bookkeeping entry was made on the left side of the account.

4. _____ means that a bookkeeping entry was recorded on the right side of the account.

5. When the assets equal the liabilities, the books are said to be _____.

6. The unit of organization for the ledger is the _____, but the unit of organization for the journal is the _____.

7. The _____ is an accounting system that includes a separate record for each item.

8. The sums of all the debits in the ledger must be equal to the sum of all the _____.

9. Examples of incidental expenses would be unexpected emergencies due to _____ or _____ damage.

10. The ending inventory for the year is always the beginning _____ for the next year.

11. An _____ _____ shows financial information over a specific period of time.

12. A _____ _____ shows financial information on a specific date.

13. The three major components of a balance sheet are _____, _____, and _____.

14. _____ is the most fundamental financial measure of the firms' financial position and performance.

Matching

a. capital expenditures budget

b. double entry bookkeeping

c. gross profits

d. operating budget

e. single entry bookkeeping

f. net income

g. exceeding ten years

h. one to ten years

i. operating income

j. cash flow budget

k. one year or less

_____ 1. summarizes the expected sales or production activities and related costs

_____ 2. summarizes the amount and timing of income that will flow in and out of the business during the year

_____ 3. list of projects (equipment, etc.) that management believes are worthwhile, with the estimated cost of each

_____ 4. this bookkeeping system may only involve a checkbook and journal

_____ 5. practice of writing every transaction in two places

_____ 6. cost of goods sold is deducted from sales revenues

_____ 7. results when operating expenses are deducted from gross profits

_____ 8. many call this the "bottom line"

_____ 9. current assets and liabilities

_____ 10. intermediate assets and liabilities

_____ 11. long-term assets and liabilities

Activities

1. In one paragraph, discuss why keeping good financial records is essential to the success of an agribusiness.

2. Name and discuss the three types of budgets used in planning for a business.

3. Illustrate the fundamental accounting equation.

4. Describe journals and ledgers and explain the process of posting from journals to ledgers.

5. Assets and liabilities are grouped by length of time to maturity. Name the groups and discuss the time factors of each.

6. Using the following account balances, prepare a balance sheet for Sumner Nursery, December 31.

Current Assets:		Current Liabilities:	
Cash	$35,000	Accounts Payable	$5,000
Short-Term Investments	$10,000	Short-Term Notes Payable	$1,000
Accounts Receivable	$10,000	Current Portion of Long-Term Debt	$1,000
Inventory	$17,000	Accrued Rent	$500
Supplies	$ 2,500	Accrued Payroll	$750
Prepaid Expenses	$ 1,000	Accrued Taxes	$500
Property and Equipment:		**Long-Term Liabilities:**	
Land	$ 5,000	Mortgages Net of Current Portion	$10,000
Buildings	$12,500	Notes Payable Net of Current Portion	$7,500
Machinery and Equipment	$10,000		
Accumulated Depreciation	$2,500	**Owner's Equity:**	
		Original Investment	$35,250
Long-Term Assets:		Retained Earnings—Jan. 1st	$7,000
Long-Term Investments	$ 7,500	Current Year Net Income/Loss	$42,000
Notes Receivable	$2,000		
Deposits	$500		

7. Using the following account balances, prepare an income statement.

Income:		Fixed Expenses:	
Plants	$100,000	Equipment Depreciation and Lease	$2,500
Soil, Fertilizer, etc.	$25,000	Rent	$2,000
Accessories	$15,000	Insurance	$1,500
Cost of Goods Sold	$10,000		
		Income Tax Expense:	$18,000
Landscaping Fee Income	$10,000		
Lawn Care Fee Income	$7,000		
Variable Expenses:			
Labor	$15,000		
Advertising	$2,000		
Office Expense	$500		

NOTES

1. William H. Hamilton, Donald F. Connelly, and D. Howard Doster, *Agribusiness: An Entrepreneurial Approach* (Albany, N.Y.: Delmar Publishers, Inc., 1992), p. 104.

2. Betty J. Brown and John E. Clow, *Introduction to Business: Our Business and Economic World* (New York: Glencoe/McGraw-Hill, 1997), p. 144.

3. *Advanced Agribusiness Management and Marketing: Accounting* (College Station, Tex.: Instructional Materials Service, 1990), 8709-B.

4. William G. Nickels, James M. McHugh, and Susan M. McHugh, *Understanding Business* (Chicago: Richard D. Irwin, 1996), p. 554.

5. Ibid., p. 555.

6. James G. Beierlein, Kenneth C. Schneeberger, and Donald D. Osburn, *Principles of Agribusiness Management* (Englewood Cliffs, N.J.: Prentice Hall, Inc., 1993), p. 102.

7. Ibid., pp. 104–105.

8. Ibid., p. 106.

9. Ibid., p. 106.

10. Ibid., p. 107.

11. Hamilton, Connelly, and Doster, *Agribusiness: An Entrepreneurial Approach*, p. 118.

12. Robert F. Meigs and Walter B. Meigs, *Accounting: The Basis for Business Decisions* (New York: McGraw-Hill, Inc., 1993), p. 67.

13. Ibid., p. 59.

14. Ibid., p. 70.

15. Nickels, McHugh, and McHugh, *Understanding Business*, p. 560.

16. *Advanced Agribusiness Management and Marketing* (College Station, Tex.: Instructional Materials Service, 1998), 8709–D, p. 2.

17. Ibid.

18. Ibid.

19. Nickels, McHugh, and McHugh, *Understanding Business*, p. 565.

20. National Council for Agricultural Education, *Decisions and Dollars* (Alexandria, Va.: Author, 1995).

21. Ibid.

22. Ibid.

23. Ibid.

24. Ibid.

25. Ibid.

26. Nickels, McHugh, and McHugh, *Understanding Business*, pp. 567–568.

27. Ibid., p. 568.

28. Ibid., pp. 568–569.

29. Ibid., p. 569.

30. J. K. Lasser, *J. K. Lasser's Your Income Tax 1997* (New York: Simon and Schuster/Macmillan, 1997), p. 617.

31. Margaret Wagner, *Personal Tax Edge* (Hiawatha, Ia.: Parsons Technology, 1997), pp. 19–22.

UNIT

3

The Agribusiness Input Sector

CHAPTER

8

The Agribusiness Input (Supply) Sector

OBJECTIVES

After completing this chapter, the student should be able to:

🌾 describe the changes in the agribusiness input sector

🌾 discuss the size of the agribusiness input sector

🌾 name the major types of agribusiness inputs

🌾 explain the history, consumption, types, market structure, and careers in the feed industry

🌾 discuss the history, analysis, use, and career opportunities in the fertilizer industry

🌾 discuss the history, types, benefits, and drawbacks of pesticides

🌾 describe the role and types of farm supply stores

TERMS TO KNOW

base mixes	defoliants	nonruminant
biopesticides	desiccants	pesticide
bulk feed	formulation	premix
centralization	fumigants	roughages
complete feed	fungicide	ruminant
complete fertilizer	hatcheries	straight fertilizer
concentrates	herbicide	supplements
consolidation	incomplete fertilizer	synthetic chemicals
cost-effective	insecticide	target customers
custom grinding	metric ton	vertically integrated

INTRODUCTION

Until recent years, one of the most neglected areas of the agricultural industry in terms of recognition and understanding was the agribusiness input (supply) sector. Agricultural economists erroneously assumed that this sector was covered in general business courses and programs. Part of the reason is attributed to various national legislative acts. Agricultural education programs were originally funded with the Smith-Hughes Act of 1917. However, it was not until the Vocational Act of 1963 that agribusiness input (and output) was allowed as part of the instructional program of federally funded agricultural education programs. More specifically, the act of 1963 allowed students to take vocational agriculture courses (now called agricultural education) if they offered Supervised Agricultural Experience Programs (SAEP) in areas other than production agriculture. Refer to Figure 8–1.

Figure 8–1 Students working in garden centers or nursery supply businesses were not considered legitimate Supervised Agricultural Experience Programs (SAEP) until the Vocational Act of 1963. (Courtesy of USDA)

CHANGES IN THE AGRIBUSINESS INPUT SECTOR

Historically, the pattern of agribusiness input has changed. Many years ago farmers produced agricultural products for their own use. They produced their own energy, used family labor, and produced their own food, clothing, and shelter. A small percentage of agricultural products produced on the farm was sold to purchase supplies.

As America gradually changed into an industrial country, the production of agricultural products became more specialized. It also become more economical for farmers to purchase production inputs from others. Today, production agriculturalists rely almost completely on purchased inputs and very little on products that they themselves produce.[1]

To further illustrate the changes in the agribusiness input sector, consider the following: Labor in 1910 accounted for 75 percent of the total production expenses. Today, however, labor accounts for less than 10 percent of production expenses. However, capital items (which includes buildings, livestock, machinery, equipment, operating inputs, etc.) accounted for only 17 percent of input in 1910 for the production of agricultural products. Today, approximately 80 percent of production costs are for capital items.[2]

SIZE OF THE AGRIBUSINESS INPUT SECTOR

In 1996, over $180 billion was spent on agribusiness input by American production agriculturalists,[3] and in 1996 some 6 million people had jobs filling the demand in the agribusiness input sector.[4]

To be more specific, farm equipment, including tractors, other motor vehicles, and machinery, cost approximately $8.8 billion and required 140,000 employees to produce.[5] Fuel, lubricants, and maintenance costs for machinery and motor vehicles were $10.1 billion. Other expenditures for agricultural inputs include $6.1 billion for seed, $25.2 billion for feed, $11.1 billion for livestock, $10.9 billion for fertilizer and lime, $17.3 billion for hired labor, and $18.9 billion for depreciation and other consumption of farm capital.[6] Many other examples could be added, but it already can be clearly seen that the agribusiness input sector is big business. In fact, the agribusiness input sector handles a wide range of products manufactured or processed by hundreds of firms.

TYPES OF AGRIBUSINESS INPUTS

Major Agribusiness Inputs

The agribusiness input sector includes many products that are supplied to production agriculturalists. The four biggest inputs are feed, fertilizers, agricultural chemicals, and farm machinery and equipment. The last input, farm machinery and equipment, will be discussed separately in Chapter 9.

Minor Agribusiness Inputs

Many other agribusiness inputs are required by production agriculturalists. Petroleum and petroleum products, such as diesel fuel, gasoline, motor oil, and transmission and hydraulic oil, are needed to keep tractors, trucks, and self-propelled machinery going. Seed and lime are necessary for crop producers. Veterinarian supplies are needed to maintain the health of the farm animals. Containers, bags, sacks, cartons, and crates are needed in order to ship and transport agribusiness supplies to production agriculturalists. Lumber and building materials are necessary in order to produce shelter for humans, animals, and plants in the production sector.

Agribusiness Inputs That Are Often Overlooked

Many agribusiness input suppliers are taken for granted or not even considered significant, yet production agriculturalists could not function without them. Hardware, iron, steel, and related products are needed for the construction and daily maintenance of agricultural buildings. Seeds, plants, and trees are needed for crop and forestry production. Utilities, including electricity, water, gas, and telephone services, are integral parts of the production of agricultural products. Credit, insurance, and other private and government services (which will be discussed in Chapters 10, 11, and 12), are also crucial input services in the production sector.[7]

FEED INDUSTRY

Historical Development

The feed industry is relatively new, dating back less than one hundred years. The **formulation** of animal feeds began shortly before the turn of the century, and scientific animal agriculture had its real beginning in the first two decades of the twentieth century. Until that time the by-products of corn and other grain refineries were considered industrial wastes and were dumped in the rivers. The soybean, which today has become the most important source of feed protein, was a botanical curiosity grown only in greenhouses. Today, among the numerous expenditures of production agriculturalists, the purchase of livestock feed is the largest. This represents about 14 percent of total farm expenses.

Recently, much of the manufactured or processed feed previously produced at feed mills is being produced on the farm. About 60 percent of the feed used, based on its cost, can be attributed to the manufactured feed industry. The other 40 percent is stored, ground, mixed, and enhanced with supplements on-site at the particular livestock operation. This can be done because livestock operations have become larger, making it **cost-effective** to invest in feed-processing equipment.[8] A prime example is the poultry industry, which is **vertically integrated** through ownership or a contract with a parent company. The integrated firms, such as Tyson Foods, mix their own feed.[9]

Feed Production and Consumption

Production agriculturalists spent $27 billion on feed in 1996. This represents an eleven percent increase over 1995 and a 30 percent increase since 1990. Surveys by the USDA and *Feedstuffs: 1997 Reference* issue indicate the increase was due primarily to higher prices of feed. Manufacturers in the United States produced 116.6 million tons of primary feed in 1996 and 2.9 million tons of secondary feed which represents a total of 128.5 million tons of commercial feed.

Today, approximately 7,000 animal-feed processing plants are in existence, making this one of the largest manufacturing industries in the United States. The feed industry employs some 100,000 workers, nearly three times more than in 1970. This figure does not include local feed mill workers, custom grinders and mixers, at least 2,000 wholesale feed dealers, over 20,000 retailers, and several thousand **hatcheries** that sell feed.[10]

Types of Feed

Feed is not produced for livestock in general, but for a particular type of livestock. Four major types of feed are fed in the poultry industry: chick starter, broiler grower, turkey grower, and laying feed. There are also different types of feed for swine, beef, and dairy. **Ruminant** and **nonruminant** animals, in particular, consume different types of feed. In reality, there are some 1,500 different combinations of ingredients. Refer to Figure 8–2 for the amounts consumed by each type of animal.

Commercial feeds are classified into three major categories: complete feeds, supplements, and premixes. **Complete feed** accounts for 80 percent of feed produced. Complete feeds are fed to animals without any additional preparation. **Supplements**, accounting for 18 percent of feed produced, are formula feeds requiring the addition of grain to form a complete ration. They contain the nutrients needed to supplement the incomplete levels of protein, vitamins, and minerals in feed. A **premix** accounting for about 2 percent of total feed production contains only vitamins and minerals and is used at the rate of less than 100 pounds per ton.[11] Premixes are mixed with feed grains and a protein source such as cottonseed or soybean meal to provide a complete ration.

Market Structure

The production and market structures of the feed industry are complex and interrelated. The general production and marketing channel for commercial feed is shown in Figure 8–3. The process

Species	Percentage of Total
Broilers	30.9
Beef/sheep	15.7
Dairy	13.4
Hogs	12.5
Starter/grower/layer/breeder	12.7
Turkey	7.8
Other	6.9
Total	100.0
Feedstuffs: 1997 Reference Issue, Vol. 69, no. 30.	

Figure 8–2 It may surprise many to know that broilers (poultry) consume more purchased feed than any other animal. Broilers are followed by beef/sheep, dairy, hogs, laying hens and turkeys. (Source: *Feedstuffs: 1988 Reference Issue,* Vol. 60, No. 31, July, 1988)

begins with the ingredient and premix suppliers, who sell the feed inputs to formula feed manufacturers. The manufacturers produce complete feeds, premixes, base mixes, and/or supplements. As previously mentioned, about 80 percent of formula feed is manufactured as complete feed, with the remainder consisting of supplements, concentrates, **base mixes**, and premixes.

The manufacturers ship feed to warehouses for storage or send it to commercial feedlots or feed retail stores. According to Omri Rawlins in his book, *Introduction to Agribusiness*, about 20,000

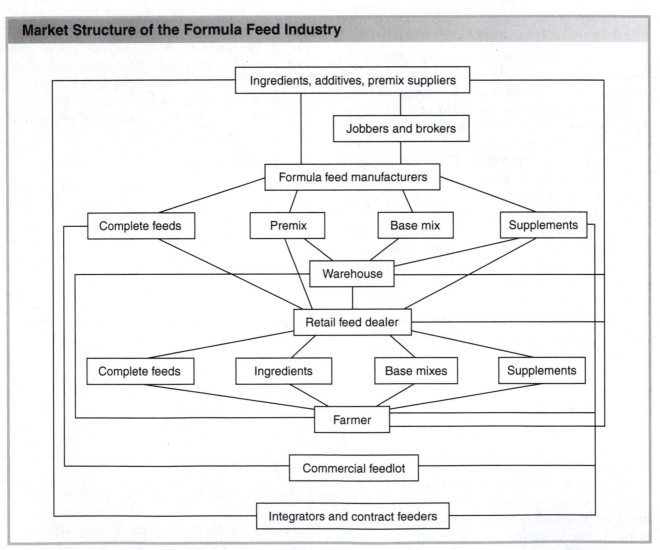

Figure 8–3 The flow of feed and market channel of the formula feed industry is complex and interrelated as shown by the above diagram. (Source: *Feedstuffs*)

feed dealers account for 60 percent of retail feed sales. The remaining 40 percent is handled by direct-selling feed firms or sold directly to large feeders. The retail feed stores vary considerably in the volume handled. About 25 percent of the stores handle 75 percent of the volume. Until recent years, the major volume of feed was sold in bags. However, the trend has reversed in favor of **bulk feed**, and now over 60 percent of formula feed is sold in bulk.

The feed industry is dominated primarily by the 10,000 manufacturing firms, although retail firms outnumber them two to one. This is due primarily to the large volume of feed produced by the manufacturers. The top four manufacturers are international in scope, whereas the remaining top twenty firms sell feed only on a national and regional basis. A listing of the top feed producers in the United States and the volume produced is shown in Figure 8–4.

An analysis of the top twenty feed manufacturing companies indicates that most of them are profit-type corporations, although four of the top twenty firms are cooperative corporations. Most of the firms are involved in the production of products other than feed, and the top four firms are highly diversified in the products they sell. Summary descriptions of the top four firms is shown in Figure 8–5.

Careers in the Feed Industry

There are many opportunities for employment in the feed industry. Graduates with a strong background in agribusiness and the animal sciences and a desire to work closely with production agriculturalists will find employment opportunities in the feed industry. The following are some specific job types: buyer for feed manufacturer; feed-manufacturing plant manager, assistant, or plant operator; sales representative; feed wholesaler; researcher; manager or employee of an agricultural cooperative; manager of a poultry hatchery; feed specialist

Estimated Annual Production of Top Feed Manufacturing Firms in the United States

Company	Volume (000 tons)
Purina Mills	7500
Cargill, Inc.	7000
PM Ag. Products	3500
Consolidated Nutrition LC	3000
Land O'Lakes, Inc.	2500
Kent Foods, Inc.	2000
Farmland Industries	1820
Continental Grain Co. (Allied Mills)	1500
Southern States Co-Op	1290
Cactus Feeders, Inc.	1200
Agway, Inc.	850
Hubbard Milling Co.	800
MFA, Inc.	796
Mark II Plan	770
Friona Industries, L.P.	680
Morenman Mfg. Co.	600
Quality Liquid Feeds, Inc.	600
SF Services Inc.	600
Star Milling	513
Pennfield Corp.	500

Figure 8–4 Feed manufacturing is a big part of agribusiness. By multiplying an average of $160.00 per ton to the tons produced, one can get an even more vivid view of the impact of the feed industry. (Source: N. Omri Rawlins, *Introduction to Agribusiness*, Murfreesboro, TN: Middle Tennessee State University, 1999)

with a chemical or pharmaceutical company; manager or employee of a retail feed and supply store; and positions in many related industries.[12] Refer to Figure 8–6 for the home office addresses of some of the major food companies.

A Summary of Selected Characteristics of the Four Major Feed Manufacturers

Company	Central Office	Scope	Major Feed Brand	Type of Feed	Other Products	Other Information
Ralston Purina	St. Louis, Missouri	International Profit Corporation	Checkerboard Chow	Beef, swine, dog, cat, dairy, poultry, horse, other pets, catfish, shrimp	Poultry products Cereals and snacks Sea foods (Chicken of the Sea) Chemical specialties Animal health products Protein products Candy and beverages Institutional foods Mushrooms, shrimp	Major feed producers in U.S. Owns and operates fast-food restaurant chain—Jack in the Box Owns and operates specialty dinner houses—Hungry Hunter, Boar's Head, Tortilla Flats Owns and operates Keystone Ski Resorts Owns and operates shrimp and mushroom farms
Allied Mills	Chicago, Illinois	International Profit Corporation	Wayne Feeds	Beef, swine, poultry, dog, horse, other pets, lab animal, zoo animal	Poultry products Pork products Soybean and oil meal Dehydrated alfalfa	Subsidiary of Continental Grain Company Second largest feed producer Sixth largest poultry processor Owns Baronet Corp., a major producer of leather goods
Central Soya	Fort Wayne, Indiana	International Profit Corporation	Master Mix and Provima	Beef, swine, dairy, poultry	Animal health products Poultry products Consumer foods: Mrs. Filbert's margarine Super beef mix Grain marketing	Third largest feed producer Fourth largest broiler processor Expanding rapidly into major food lines to promote "food-power"
Cargill	Minneapolis, Minnesota	International Profit Corporation	Nutrina	Beef, swine, dairy, other animals	Poultry products Food and feed grains Food and oil seeds Chemical products Farm seeds P-A-G Brand Fertilizers, metals, plastics, and resins	Fourth largest feed producer Second largest grain exporter Major transportation and warehousing system Over 250 plants and offices in North America and 60 in 36 other countries

Figure 8–5 The four major feed companies are involved in more than just feed production as shown above.

Feed Companies

A. E. Staley Manufacturing Co.
2200 East Eldorado Street
P.O. Box 151
Decatur, IL 62525
Tel.: 217-423-4411/Fax: 217-421-2936

Agway Grain Marketing
P.O. Box 4933
Syracuse, NY 13221-4933
Tel.: 315-449-6272/Fax: 315-449-5499

Archer Daniels Midland Co.
P.O. Box 1470, 4666 Faries Parkway
Decatur, IL 62425-1820
Tel.: 217-424-5450/Fax: 217-424-5447

Bluebonnet Milling Co.
100 South Mill, P.O. Box 2006
Ardmore, OK 73402
Tel.: 405-223-3010/Fax: 405-223-3546

Cargill
9300-MS48
Minneapolis, MN 55440
Tel.: 612-742-6212

Central Soya Co., Inc.
P.O. Box 1400
Fort Wayne, IN 46801
Tel.: 219-425-5100, 1-800-422-7692
Fax: 219-425-5254

Conagra Feed Co., Inc.
980 Molly Pond Road
Augusta, GA 30901
Tel.: 706-722-6681/Fax: 706-722-4561

Dekalb Feeds, Inc.
105 Dixon Avenue, P.O. Box 111
Rock Falls, IL 61071
Tel.: 815-625-4546/Fax: 815-625-4896

Dickey-John Corp.
P.O. Box 10
Auburn, IL 62615
Tel.: 217-438-6012, 1-800-637-2964
Fax: 217-438-6012

Doane Products Company
West 20th Street & State Line Road,
P.O. Box 879
Joplin, MO 64802
Tel.: 417-624-6166

Farmland Industries
P.O. Box 7305
Kansas City, MO 64116-0005
Tel.: 816-459-6000/Fax: 816-459-3980

General Mills, Inc.
54 S. Michigan Avenue
Buffalo, NY 14203-3086
Tel.: 716-856-6100/Fax: 716-857-3689

Hubbard Milling Company
P.O. Box 338
Ft. Pierre, SD 57532
Tel.: 605-223-2557

Kent Feeds Inc.
1600 Oregon Street
Muscatine, IA 52761
Tel.: 319-264-4211

Land O'Lakes Inc.
P.O. Box 64089
St. Paul, IN 55164
Tel.: 612-451-5412/Fax: 612-451-5407

Monsanto (Agricultural Group)
800 North Lindbergh Boulevard
St. Louis, MO 63167
Tel.: 314-694-1000/Fax: 314-694-3900

Moorman Manufacturing Co.
1000 N. 30th Street
Quincy, IL 62305-3115
Tel.: 217-222-7100/Fax: 217-222-4069

Murphy Farms Inc.
P.O. Box 759, U.S. Highway 117 S.
Rose Hill, NC 28458-0759
Tel.: 919-289-2111/Fax: 919-289-6400

Nutrena Feed-Nutrena Feed Division
15407 McGinty Road, P.O. Box 5614
Minneapolis, MN 55440-5614
Tel.: 612-475-7495/Fax: 612-475-6391

Purina Mills, Incorporated
1401 S. Hanley Road
St. Louis, MO 63144
Tel.: 314-768-4100/Fax: 314-768-4188

Walnut Grove
201 Linn Street
Atlantic, IA 50022
Tel.: 712-243-1030/Fax: 712-243-6360

Wayne Feed
10 S. Riverside Plaza
Chicago, IL 60606
Tel.: 312-930-1050

Gold Kist, Inc.
244 Perimeter Center Parkway, N.E.,
P.O. Box 2210
Atlanta, GA 30301
Tel.: 404-393-5000/Fax: 404-393-5347

Figure 8–6   Feed companies are a source of potential employment. The list of companies above gives the addresses of the home offices.

FERTILIZERS

The first mixed fertilizers sold in the United States for commercial purposes were manufactured in Maryland in 1849. Factories were soon built in other areas and production expanded rapidly, primarily in the eastern states. Some of the major events and dates in the fertilizer industry in the United States are summarized below:

1851—first record of fertilizer analysis by state officials

1856—first state fertilizer control law (Massachusetts)

1865—potash first imported into the United States

1928—first direct use of anhydrous ammonia in mixed fertilizers

1929—first shipment of synthetic nitrate of soda, from Hopewell, Virginia

1931—first commercial shipment of potash, from Carlsbad, New Mexico

1939—first experiments with the use of anhydrous ammonia on nonirrigated land (direct)

1942—first commercial use of anhydrous ammonia directly in the soil

1950–centennial of the first manufacture of mixed fertilizers in the United States[13]

Fertilizer Analysis

Commercial fertilizers are made by combining selected plant food materials to obtain specific ratios and quantities of plant nutrients. The three major plant food nutrients found in fertilizers are nitrogen, phosphorus, and potassium. Since these elements are rarely found in a pure, elemental form, they must be combined with other substances to be used as fertilizers. The mining, manufacture, and distribution of these materials serve as a basis for the fertilizer industry.

A **complete fertilizer** is one that contains nitrogen (N), phosphoric acid (P_2O_5), and potash (K_2O). Fertilizers containing only two of the three materials are known as **incomplete fertilizers**. A fertilizer containing only one of the primary materials is sometimes called a **straight fertilizer**. When the fertilizer industry distributed fertilizer in the past it was done primarily in 100-pound bags, so the analyses were given in the same order and the numbers represented the weight percentage of nitrogen, phosphoric acid, and potash. For example, a 100-pound bag of fertilizer labeled 4-12-12 contained 4 pounds of nitrogen, 12 pounds of phosphoric acid, and 12 pounds of potash, a total of 28 pounds of plant food. The remaining 72 pounds in the bag was "filler" or "conditioner." Federal law requires that each bag of fertilizer contain a tag stating the minimum amount of N, P_2O_5, and K_2O contained in the bag. However, a major portion of fertilizer today is sold in bulk by the manufacturers and mixed at the retail level through a process known as *bulk blend.*

Nitrogen. Nitrogen fertilizer manufacturing involves the process of combining nitrogen from the atmosphere with hydrogen from some natural source to produce synthetic ammonia. Natural gas is the primary source of hydrogen, although fuel oil, coke, naphtha, refinery gas, and other sources are also used. Nitrogen may be applied as a liquid, gas, or solid, although solids and liquids are the most common forms. The major gas source of nitrogen is anhydrous ammonia, whereas aqua-ammonia is the major liquid source. The three major sources of nitrogen available in solid form are ammonium nitrate, ammonium sulfate, and urea. The amount of nitrogen currently produced for fertilizer is about 10 million tons per year.

Phosphate. Phosphate fertilizers are made by treating rock deposits, certain iron ores, or bones with sulfuric or phosphoric acid. The phosphate content of fertilizer is expressed as phosphoric acid (P_2O_5). Phosphorous as a pure element cannot be used as a plant nutrient. The major sources of phosphates include normal superphosphate, concentrated superphosphate, ammonium phosphate, and phosphoric acid (H_3PO_4). Currently, over 5.7 million tons are being produced each year. The major phosphate deposits in the United States are found in Florida, North Carolina, Tennessee, Idaho, Montana, Utah, and Wyoming.

Potassium. Potassium in its pure state is a light gray element that bursts into flame when exposed to air. Thus, it must be combined with other elements before it can be used as fertilizer. Potassium in fertilizer is commonly called *potash* (K_2O). The major sources of potash in commercial fertilizers are muriate of potash (KCL), sulfate of potash (K_2SO_4), and nitrate of potash (KNO_3). The major potash deposits in the United States are found in New Mexico, Utah, and California, which currently produce over 2 million tons of potash each year. In addition, about 3.5 million tons are imported, primarily from Canada.

Fertilizer Use

Production of basic fertilizer products in the United States is carried out by some 40 to 50 manufacturers. The fertilizer is further processed by over 10,000 local marketing centers, which supply granular mixed fertilizer and liquid mixed fertilizer bulk blends.[14] The average size fertilizer plant has a storage capacity of 6,041 tons and an annual distribution volume of 15,400 tons. Almost half the fertilizer plants are single (sole) proprietorships, with approximately 42 percent cooperatives and about 9 percent corporations.[15]

Production agriculturalists in the United States spend approximately $11 billion yearly on fertilizers. Fertilizer usage increased substantially during the 1960s and 1970s, and peaked in 1981. Fertilizer use has been somewhat stable since the mid-1980s, ranging from 20.1 to 23.8 million tons through 1997. The amount of fertilizer used is affected by the number of acres in production, rates of application due to expenses, and exports, which were approximately $3 billion.[16]

Environmental Policy. Environmental policy will likely affect the fertilizer industry in the future. Environmental regulations on certain aspects of fertilizer production and use will become increasingly monitored. Local, state, and federal regulations concerning groundwater contamination by nitrogen fertilizers will certainly affect the industry. If contamination of underground water supplies becomes worse, regulations will increase.

Interrelationships among Fertilizers, Pesticides, and Other Chemicals

The market structures of pesticide (discussed in the next section) and fertilizer producers and distributors are very closely interrelated. Most of the major pesticide producers also produce some type of fertilizer, and vice versa. The chemical companies are very closely associated with the petroleum industry since farm chemicals are usually derivatives of oil products. The chemical industry is highly concentrated, with the processors or chemical manufacturers in control. The market is also highly integrated, with the processors normally owning the mines, processing plants, and wholesale operations. The retailers are primarily cooperative or independent farm supply stores.

The chemical industry is also highly diversified. Chemical processors handle a wide variety of products, including plastics, metals, and pharmaceuticals. Many handle numerous brands of food products. In the farm pesticide market, Novartis ranks first in chemical sales, and Monsanto and Zeneca rank second and third, respectively. Some of the major fertilizer producers include DuPont, Agrevo, Bayer, Rhone-Poulenc, Dow Agrosciences,

Cyanamid, and BASF. A list of the top 10 chemical companies is shown in Figure 8–7.

Rank	Company	Sales (Million $)
1	Novartis	4,199
2	Monsanto	3,126
3	Zeneca	2,674
4	DuPont	2,518
5	Agrevo	2,352
6	Bayer	2,254
7	Rhone-Poulenc	2,202
8	Dow Agrosciences	2,200
9	Cyanamid	2,119
10	BASF	1,855

Figure 8–7 The top ten agricultural chemical companies based on sales, 1997. (Courtesy of AEROW. World Crop Protection News, 1998.)

Careers in the Fertilizer Industry

All parts of this supply section need conventional business services, such as accounting and auditing, control of production cost, distribution cost, selling costs, data processing and management, office management, sales and sales promotion, credit and collections, communications, publicity, advertising, and customer relations. Manufacturer's representatives work with chemists, botanists, and other agricultural scientists. Manufacturing plants hire workers, as do wholesale distribution operations. Field service contracting includes applying fertilizer and spreading lime for farm customers. Refer to Figure 8–8 for the names of a few of the nations' fertilizer and pesticide companies.

PESTICIDES

Since the beginning of time, humankind has competed with insects and other animals for the fruits of the plant world. There has also been a constant battle between desirable and undesirable plants. Until the twentieth century, we used natural means and simple machines to control weeds, insects, and diseases. However, due to the successful efforts in chemical and biological research, weeds, insects, and diseases are now controlled primarily by **synthetic chemicals**. It is estimated that there are 10,000 species of harmful insects, over 1,500 diseases caused by fungi, 1,800 different weeds, and approximately 1,500 different types of nematodes that cause damage to crops and livestock.

Types of Pesticides

Pesticide is a broad term used to include all chemicals used to control weeds, insects, and diseases that affect crops, livestock, or people. The three major types of pesticides are herbicides, insecticides, and fungicides. Other types of pesticides include **fumigants**, **defoliants**, **desiccants**, and growth regulators. An **herbicide** (80 percent of pesticides sold) is one or more chemicals used to control undesirable plants, whereas an **insecticide** (15 percent) is used to control various insects. A **fungicide** (5 percent) is one or more chemicals used to control fungi that attack plants and animals.

Historical Background

The foundation for modern pesticides can be traced to the mid-nineteenth century. In 1859, Julius Sacks, a German botanist, wrote articles on the translocation of growth regulating substances in plants.

Fertilizer and Pesticide Companies

Abbott Laboratories, Chemical & Agricultural Products Division
1400 Sheridan Road
North Chicago, IL 60064
Tel.: 708-937-6737/Fax: 708-937-3697

Allied Procuts Corporation
10 South Riverside Plaza
Chicago, IL 60606
Tel.: 312-454-1020/Fax: 312-454-9608

American Cyanamid Co.
1 Cyanamid Plaza
Wayne, NJ 07470
Tel.: 201-831-2000/Fax: 201-831-4470

BASF Corporation
P.O. Box 13528
Research Triangle Park, NC 27709-3528
Tel.: 919-361-5300/Fax: 919-361-5722

Cargill Fertilizer Division
P.O. Box 9300
Minneapolis, MN 55440
Tel.: 612-475-7167/Fax: 612-475-7313

Ciba-Geigy Corporation
P.O. Box 18300
Greensboro, NC 27419-8300
Tel.: 910-632-6000

Conoco/Du Pont
P.O. Box 4784-TA2136
Houston, TX 77210-4784
Tel.: 713-293-3987

Dow Elanco
U.S. Crop Protection, 9330 Zionsville Road
Indianapolis, IN 4668
Tel.: 317-337-3000

E. I. du Pont de Nemours & Co.
Barley Mill Plaza, EA/WM5-167
Wilmington, DE 19880
Tel.: 302-992-3022

Monsanto (Agricultural Group)
800 North Lindbergh Boulevard
St. Louis, MO 63167
Tel.: 314-694-1000/Fax: 314-694-3900

Occidental Chemical Corporation, Ag. Products—Feed Products Division
5005 LBJ Freeway
Dallas, TX 75244
Tel.: 214-404-3800/Fax: 214-404-3200

Potash & Phosphate Institute
655 Engineering Drive, Suite 110
Norcross, GA 30092
Tel.: 404-447-0335

The Upjohn Company
700 Portage Road
Kalamazoo, MI 49001
Tel.: 616-323-4000/Fax: 616-323-4000

The Vigoro Corporation
225 N. Michigan Avenue
Chicago, IL 60601
Tel.: 312-819-2020/Fax: 312-819-2027

Farmland Industries
P.O. Box 7305
Kansas City, MO 64116-0005
Tel.: 816-459-6000/Fax: 816-459-3980

Fermenta Animal Health Company
10150 North Executive Hills Boulevard
Kansas City, MO 64153-2314
Tel.: 816-891-5500/Fax: 816-891-0663

Gold Kist Inc.
244 Perimeter Center Parkway, N.E.,
P.O. Box 2210
Atlanta, GA 30301
Tel.: 404-393-5000/Fax: 404-393-5347

Mobil Oil Canada
300-5th Avenue S.W.
Calgary, Alberta Canada T2P 2J7
Tel.: 403-260-7910/Fax: 403-260-4277

Figure 8–8   Fertilizer and pesticide companies are a source of potential employment. The home office addresses of the major companies are listed.

Around 1900, three scientists—Bonnet of France, Schultz of Germany, and Bolley of the United States—found that copper salt solutions would kill broad-leaved plants. These scientists were all working independently about the same time. In 1907, a German scientist, Hans Fitting, found that the swelling of the ovary and the fading of the flower after pollination were caused by a substance that developed during pollination. In 1908, Bolley of North Dakota reported effective weed control in wheat using sodium chloride, iron sulfate, copper sulfate, and sodium arsenite. Bolley predicted that chemical weed control would be one of the most significant developments in agricultural research, and a current analysis indicates he was correct.

Discovery of Herbicides.   During 1941, Pokorny discovered the compound 2,4-D. Other researchers later tried it as a fungicide and insecticide and found it ineffective. However, a year later Zimmerman and Hitchcock of the United States discovered that 2,4-D was a plant growth substance. In 1944, Garth and Mitchell established the selectivity of 2,4-D to broad leaved plants and Hamner and Tukey first used 2,4-D successfully in field weed control. In 1945, Templeman of England established the preemergence principle of soil treatment for selective weed control. Since that time, the development of chemical herbicides has expanded rapidly.

Discovery of Fungicides and Insecticides. The earliest fungicides and insecticides were inorganic compounds developed from copper, mercury, lead, and arsenic, many of which are still being used. The natural organic compounds of nicotine, pyrethrum, and rotenone were also used in limited amounts during the nineteenth century. The modern pesticide industry is based primarily on synthetic organic compounds, which made their first major impact during the 1940s. The compounds that had the greatest impact were DDT, an insecticide, and the herbicide 2,4-D.

DDT. The value of DDT as an insecticide was discovered in 1939 by Dr. Paul Muller, for which he received the 1948 Nobel prize in physiology and medicine. DDT was the major compound used to control malaria mosquitos during World War II, and the effectiveness of this compound encouraged researchers to develop other, similar compounds, such as chlordane, aldrin, dieldrin, and other valuable insecticides. Ironically, many of the most successful pesticides, which provided so much promise in earlier years, have become viewed as enemies of the environment. Currently, DDT, aldrin, and dieldrin are illegal, and numerous others are being evaluated.

Benefits of Pesticides

Pesticides represent over 4 percent of expenditures of production agriculturalists, who spend about $8.5 billion yearly on herbicides, insecticides, and fungicides to assure that crop production is not lost to pests.[17] Nevertheless, each year losses of about $3 billion are caused by hordes of insect pests that chew, suck, bite, and bore away at food, food crops, and livestock. Bacteria, viruses, fungi, nematodes, and weeds destroy another $7 billion, making for a total production loss of about $10 billion.[18]

Several thousand pesticide formulas are now registered with the government for sale to production agriculturalists. Pesticides have brought remarkable benefits to the agricultural industry: they can help control insects, kill weeds, fumigate and sterilize the soil, treat seed, fight plant diseases, regulate the time of harvesting certain crops, and help clear brush.[19]

Concerns with Pesticides

Although pesticides are very beneficial, they do cause some environmental damage. Research is being conducted to find effective pesticides that will not be harmful to water, air, soil, wildlife, humans, animals, or the food we consume. One example is the development of **biopesticides**, which will be safer to use.

Because of potential harm to the total environment, pesticide production and use are closely regulated by the Office of Pesticide Programs of the U.S. Environmental Protection Agency. The first major effort to control pesticides was the Federal Insecticide, Fungicide, and Rodenticide Act (FIFRA) of 1947. The USDA was responsible for administering FIFRA until 1970, when the authority was transferred to the Environmental Protection Agency.

Government restrictions make it extremely difficult to develop new and more effective pesticides. The registration of a new pesticide takes an average of five to ten years and costs approximately from $5 to $10 million. Future projections indicate that new pesticide development will take even longer to receive approval and be much more expensive. However, scholars contend that if the world's population is to be fed, more and better pesticides must be developed.

Animal Health Companies

A. L. Laboratories Inc.
1 Executive Drive
Fort Lee, NJ 07024
Tel.: 201-947-7774

American Cyanamid Co.
1 Cyanamid Plaza
Wayne, NJ 07470
Tel.: 201-831-2000/Fax: 201-831-4470

Animal Health Institute
P.O. Box 1417-D50
Alexandria, VA 22313-1480
Tel.: 703-684-0011

Elanco Animal Health
Lilly Corporate Center
Indianapolis, IN 46285
Tel.: 317-276-3000/Fax: 317-276-9430

Eli Lilly and Company
Lilly Corporate Center
Indianapolis, IN 46285
Tel.: 317-276-2000/Fax: 317-276-2000

Fermenta Animal Health Company
10150 North Executive Hills Boulevard
Kansas City, MO 64153-2314
Tel.: 816-891-5500/Fax: 816-891-0663

Merck Ag. Vet, Division of Merck &
Co., Inc.
P.O. Box 2000
Rahway, NJ 07065-0912
Tel.: 908-855-6709/Fax: 908-855-9366

Pfizer, Incorporated
235 E. 42nd Street
New York, NY 10017
Tel.: 212-573-2323/Fax: 212-573-7851

SmithKline Beecham Animal Health
Whiteland Business Park, 812 Spring-
dale Drive
Exton, PA 19341
Tel.: 215-363-3100/Fax: 215-363-3285

The Upjohn Company
7000 Portage Road
Kalamazoo, MI 49001
Tel.: 616-323-4000/Fax: 616-323-5251

Figure 8–9   The animal health companies which offer many careers also sell products beyond the agricultural industry. The home office addresses of several of these companies are listed above.

Careers

Just as in the fertilizer industry, many conventional business services are needed. Jobs are available in manufacturing, with wholesale distributors, with researchers, and in sales. Many new jobs are opening in the area of environmental policy, regulations, and control. Many colleges are now offering college degrees in environmental science and technology or programs with a similar name. Refer to Figure 8–8 for a list of the some of the major pesticide manufacturers which are listed along with the fertilizer companies. Figure 8–9 provides a list of animal health companies, which offer many career opportunities. Refer to the Career Option on page 218 for a further explanation of the careers in the agribusiness supply industry.

FARM SUPPLY STORES

Most of the agribusiness input supplies get to the production agriculturalist by the way of farm sup-

ply stores. There are approximately 36,000 farm supply stores, according to the U.S. Census of 1990 Business.[20] They sell seed, feed, fertilizer, pesticides, animal health products, farm machinery and equipment, petroleum products, and hardware. In the spring, many flower and garden supplies are also sold. The majority of small stores are organized as either sole (single) proprietorships or partnerships. The larger, chain-type farm supply stores are either corporations or cooperatives.

The pattern of many farm supply stores is **consolidation** and **centralization**, which means a complete line of farm production supplies and services are available. Some of the supplies and services include bagged and bulk fertilizers, bulk blending, delivery and spreading of fertilizers, use of fertilizer spreaders when their fertilizer is used, use of a feed mill to grind and mix **roughages** and/or **concentrates**, **custom grinding** and custom mixing, and soil testing. Many of these businesses offer technical information and furnish credit. Products for tractors, trucks, and automobiles

CAREER OPTION

Career Areas: Animal Nutritionist, Feed Formulator, Pesticide Applicator, Pesticide Specialist, Seed Analyst, Fertilizer Chemist

There are many jobs that make up the agribusiness input (supply) sector. Careers in animal nutrition are varied and interesting. One may work essentially as an organic chemist or technician in a laboratory, where complex equipment is used for research and analysis on animal needs and the nutritional values of feedstuffs. One also may work selling feeds or at an agribusiness producing feedstuffs. Careers in nutrition may be in the basic or applied sciences. They may focus on fish, small animals, pets, equines, poultry, livestock, dairy, or wild animals.

Tens of thousands of chemical formulations have been developed in recent years to control insects, diseases, weeds, nematodes, rodents, and other pests. We must use chemicals to help control pests. However, if not properly used, these materials are likely to be hazardous to the operator, other people, plants, animals, or the environment.

Pesticide applicators are employed by production agriculturalists, lawn service companies, farm and garden supply firms, termite control companies, highway departments, and railroads, and self-employed persons. Special training and licensing are required for handling most pesticides.

A person who chooses a career as a seed analyst works with seeds to make sure that their quality is high. Seed analysts require various degrees of technical or college-level training prior to employment. They must be patient individuals able to perform many repetitions of the same task.

Most of the major agricultural food and fiber crops are grown from seeds. Because large amounts of quality seeds are needed each growing season, a giant seed industry has developed. Many seed analysts work for large seed companies. They test the seed the company plans to sell to determine how well the seed can be expected to perform in the field and ensure high quality. Some seed analysts work with government agencies that are responsible for making sure that commercial seed meets the standard of quality defined by federal laws.

There are many jobs with fertilizer companies. Almost all production agriculturalists use some type of fertilizer. All inorganic fertilizers are manufactured from some raw material. A fertilizer chemist then tests, analyzes, and makes recommendations for the appropriate mixes. This position requires various degrees of technical or college-level training prior to employment.

At the Yakima Agricultural Research Laboratory in Wapato, Washington, technician Tom Treat applies a test pesticide to a rapeseed variety being grown for canola oil production. (Courtesy of USDA)

are also available, such as diesel, gasoline, propane, oil, tires, batteries, and accessories. Some stores even perform truck and car service and repair.

Types of Farm Supply Store

The types of farm supply stores can be divided into three major categories: farmer cooperatives, chain store corporations, and small single (sole) proprietorship feed and supply stores. The legal makeup of each of these businesses was discussed in Chapter 5. The following is a brief description citing examples of each.

Farmer Cooperatives. Approximately 28 percent of the supplies purchased by production agriculturalists come from farmer cooperatives. Farmer cooperatives have an even greater proportion of sales of fertilizer and lime, feed manufacturing and distribution, seed, agricultural chemicals, and petroleum. Production agriculturalists purchase about 45 percent of the fertilizer and lime, 23 percent of the feed, 13 percent of the seed, and 32 percent of the agricultural chemicals used in their production of animal and plant products from farmer cooperatives.[21]

Farmer Supply Chain Stores. Another major source of supplies for production agriculturalists is farmer supply chain stores. These are corporations and operate similar to WalMart and K-Mart except that their **target customers** are full-time and part-time production agriculturalists. Many other consumers also shop at these chain stores because of the hardware, lawn and garden, and work clothing available there.

There are several farm supply chain stores that are popular in different regions of the country. They sell small tractors, fertilizer, feed, seed, tractor supplies, auto and truck supplies, lawn and garden equipment and supplies, hardware, shop supplies, clothing, and various other products. Three of the larger farm supply chain stores are Tractor Supply Company (TSC), Central Tractor

Farm and Country Center, and Quality Farm and Fleet. Each of these also has catalog sales. Tractor Supply Company has 244 stores throughout the country, with national headquarters in Nashville, Tennessee. Central Tractor has 230 stores throughout the country, with the corporate headquarters in Des Moines, Iowa. Quality Farm and Fleet has 100 stores, concentrating in the northern and northeastern United States. Each of these stores employs approximately ten to twenty full-time and part-time workers, depending on the store size and hours of operation. The home offices of each also employs approximately seventy-five to one hundred people.

Single (Sole) Proprietorship Feed and Supply Stores. There are thousands of these operations throughout the United States. They tend to market a name-brand feed, such as Purina, Master Mix, Nutrena, Moorman, or Wayne, along with other, locally grown feeds. Seed, fertilizer, and miscellaneous farm supplies are also sold. The smaller stores are usually family managed, with some extra help during the rush seasons.

CONCLUSION

The agricultural input (supply) industries have developed mostly since the end of World War II. The main reason why the agribusiness input (supply) sector has flourished is because the supplies it sells are cheaper and more productive than what production agriculturalists could provide for themselves. The development of the agribusiness input (supply) sector has permitted production agriculturalists to concentrate more on production and made the supply firms a major force in agribusiness today. The agribusiness input sector provides approximately 70 percent of the total inputs used by production agriculturalists. The remaining 30 percent includes items, such as hay, livestock, and grains, that are produced on the farm for use as an input to other production of agricultural products.[22]

SUMMARY

Historically, the pattern of agribusiness input has changed. As America gradually changed to an industrial country, the production of agricultural products became more specialized and it became more economical for farmers to purchase production inputs from others. The agribusiness input sector is big business. It handles many products, which are manufactured or processed by hundreds of firms. In 1996, over $180 billion was spent on agribusiness inputs by American production agriculturalists, and some 6 million people have jobs filling the demand in the agribusiness input sector.

The agribusiness input sector includes many products that are supplied to production agriculturalists. The four biggest inputs are feed, fertilizer, agricultural chemicals, and farm machinery and equipment. Many other inputs are required for production agriculturalists, such as seed, lime, diesel fuel, gasoline, motor oil, transmission oil, hydraulic oil, veterinarian supplies, and many other products.

Among the numerous agribusiness input (supply) expenditures of production agriculturalists, feed for livestock is the largest, including about 14 percent of total farm expenses. Production agriculturalists spend over $27 billion per year on purchased feed. There are over 1,500 different combinations of feed ingredients for poultry, swine, beef, and dairy animals. The production and market structure of the feed industry is complex and interrelated, starting with premix suppliers, going through several processes, and ending

at the retail store for consumer purchase. There are many opportunities for employment in the feed industry.

Fertilizer was first manufactured for sale in Maryland in 1849. The three major plant food nutrients found in fertilizers are nitrogen, phosphorus, and potassium. Production of basic fertilizer products in the United States is carried out by some 40 to 50 manufacturers and further processed in over 10,000 local marketing centers. Production agriculturalists in the United States spend approximately $11 billion yearly on fertilizers. Environmental policy will likely affect the fertilizer industry in the future, adding to the already abundant career opportunities.

Due to the successful efforts in chemical and biological research, weeds, insects, and diseases are now controlled in the United States, primarily by synthetic chemicals. It is estimated that there are 10,000 species of harmful insects, over 1,500 diseases caused by fungi, 1,800 different weeds, and approximately 1,500 different types of nematodes that cause damage to crops and livestock. Production agriculturalists spend about $6.7 billion yearly on pesticides. Although pesticides are very beneficial, they do cause some environmental damage. Research is being conducted to find effective pesticides that will not be harmful to water, air, soil, wildlife, humans, or the food we consume.

Almost all agribusiness input supplies get to the production agriculturalist by way of farm supply stores. There are approximately 36,000 farm supply stores, which sell the items discussed in this chapter. The types of farm supply stores can be divided into three major categories: farmer cooperatives, chain store corporations, and small single (sole) proprietorship feed and supply stores. Many job opportunities are available with farm supply stores.

END-OF-CHAPTER ACTIVITIES

Review Questions

1. Define the Terms to Know.
2. What changes have occurred in the agribusiness input sector since the beginning of the century?
3. Name the four major agribusiness input products.
4. Name ten minor agribusiness inputs.
5. Name seven agribusiness inputs that are often overlooked.
6. What is unique about the feed industry among the industries that provide supplies to production agriculturalists?
7. Name and explain the three major categories of commercial feeds.
8. Name ten potential careers available within the feed industry.
9. What are the three major plant food nutrients found in fertilizers?
10. What is contained in a 100-pound bag of fertilizer labeled 4-12-12?
11. How is nitrogen fertilizer manufactured?
12. What are the three major sources of nitrogen in solid form?
13. How are phosphates made?
14. What are the four major sources of phosphates?
15. Where are the major sources of phosphates found in the United States?
16. What are the major sources of potassium in commercial fertilizer?
17. Where are the major potassium deposits in North America?
18. What three things affect the amount of fertilizer used?
19. Describe the interrelationships among fertilizer, pesticides, and other chemicals.
20. Briefly discuss the discovery of herbicides.
21. Briefly discuss the discovery of fungicides and insecticides.
22. What are seven benefits of pesticides?
23. What are fifteen types of supplies and services available at large consolidated and centralized farm supply stores?
24. Name ten items sold by farmer supply chain stores.

Fill in the Blank

1. In 1910, labor accounted for _____ percent of the total production expenses.
2. Today, capital items such as buildings, livestock, machinery, equipment, and operating inputs account for _____ percent of production costs.

3. In 1995, over _____ billion dollars was spent on agribusiness inputs by American production agriculturalists.

4. Production agriculturalists' largest expenditure of an input is livestock feed, representing about _____ percent of farm expenses.

5. About _____ percent of the feed used can be attributed to the manufactured feed industry.

6. The United States produces about _____ percent and consumes about _____ percent of the world's feed grains.

7. About 25 percent of the feed retail stores sell _____ percent of the feed.

8. Over _____ percent of formula feed is sold in bulk.

9. The first mixed fertilizers sold in the United States for commercial purposes were manufactured in Maryland in the year _____.

10. _____ will likely affect the fertilizer industry in the future.

11. One of the first and most popular insecticides, which is now illegal, is _____.

12. There are approximately _____ farm supply stores.

13. Approximately _____ percent of the supplies purchased by production agriculturalists come from farm cooperatives.

14. Production agriculturalists purchase about _____ percent of the fertilizer and lime, _____ percent of the feed, _____ percent of the seed, and _____ percent of the agricultural chemicals from farmer cooperatives.

Matching

a. 7,000

b. 5.7

c. 1,500

d. 5.5

e. 10 billion

f. 6 million

g. 100,000–150,000

h. 10 million

i. 10,000

j. 21 billion

k. 1,800

l. 6.7 billion

_____ 1. number of jobs in the agribusiness input sector

_____ 2. amount production agriculturalists spend yearly on purchased feed

_____ 3. number of animal feed–processing plants

_____ 4. number of people the feed industry employs

_____ 5. number of different combinations of feed ingredients

_____ 6. number of tons of nitrogen produced in the United States yearly

_____ 7. number of tons of phosphate produced in the United States yearly

_____ 8. number of tons of potassium produced yearly in North America

_____ 9. amount spent yearly by production agriculturalists on fertilizer

_____ 10. estimated number of harmful insects causing damage to crops and livestock

_____ 11. number of different weeds causing damage to crops and livestock

_____ 12. amount spent yearly on pesticides

Activities

1. Suppose that someone "puts down" or "makes light" of the agricultural industry. Given some facts and figures from this chapter, what can you tell them about the significance of agribusiness in the U.S. economy?

2. Visit the largest farm feed and support store in your community. List as many products as you can that they sell. Perhaps your teacher may wish to set a limit on the number or assign product categories to various students (example: animal health, pesticides, etc.).

3. Visit your local feed store. Find out the parent company (Purina, Cargill, etc.) and locate the home office address in Figure 8–6. Write this down and keep it for future reference if you desire to seek employment with the company.

4. Follow the same process as in Activity 3 for fertilizers, pesticides, and animal health products. Refer to figures 8–8 and 8–9 in this chapter.

5. Suppose you are working the sales desk of a local farm supply store and a customer orders a ton of 15-15-15 fertilizer. She wants to know how many actual pounds of nitrogen he or she is buying. What is your answer?

6. The agricultural industry is constantly monitoring the use of fertilizers, pesticides, and other agricultural chemicals. Write a paper or locate an article in a newspaper or magazine about a particular environmental concern. Present this to the class.

7. When it pertains to matters of health or the environment, many times the agricultural industry is criticized unjustly because someone is "crying wolf." Write a report or locate an article citing an example. Present this to the class.

8. Several agribusiness input supply companies were listed in this chapter. Select five companies that interest you as potential future employers. Give reasons for your selection of each and share these with your class.

NOTES

1. Randall D. Little, *Economics: Applications to Agriculture and Agribusiness* (Danville, Ill.:, Interstate Publishers, 1997).
2. Ibid.
3. United States Department of Agriculture, *Agricultural Statistics 1998* (Washington, D.C.: U.S. Government Printing Office, 1998).
4. Ibid.
5. Ibid.
6. Ibid.
7. Ewell P. Roy, *Exploring Agribusiness* (Danville, Ill.: Interstate Printers and Publishers, 1980).
8. Kevin Kimbe and Marvin Hayenga, *Agricultural Input and Processing Industries*, Report RD-05 (Ames, Ia.: Iowa State University, Department of Economics, 1992).
9. Little, *Economics*.
10. Marcella Smith, Jean M. Underwood, and Mark Bultmann, *Careers in Agribusiness and Industry* (Danville, Ill.: Interstate Publishers, 1991), p. 119.

11. Little, *Economics*, pp. 291–292.
12. Roy, *Exploring Agribusiness*.
13. Malcolm McVickar, *Using Commercial Fertilizer* (Danville, Ill.: Interstate Printers and Publishers, 1952), pp. 17–20.
14. N. Omri Rawlins, *Introduction to Agribusiness*, 3rd ed. (Murfreesboro, Tenn.: Middle Tennessee State University, 1999), p. 93.
15. Little, *Economics*, p. 292.
16. United States Department of Agriculture, *Agricultural Statistics 1998* (Washington, D.C.: U.S. Government Printing Office, 1998).
17. Ibid.
18. Ibid., p. 220.
19. Smith, Underwood, and Bultmann, *Careers in Agribusiness and Industry*, p. 218.
20. Ibid., p. 222.
21. Little, *Economics*, p. 296.
22. James G. Beierlein and Michael W. Woolverton, *Agribusiness Marketing: The Management Perspective* (Englewood Cliffs, N.J.: Prentice-Hall, 1991), pp. 105–106.

CHAPTER 9

The Agribusiness Input Sector: Farm Machinery and Equipment

OBJECTIVES

After completion of this unit, the student should be able to:

- describe the historical development of farm machinery and equipment
- describe the steam era
- discuss the historical development of the internal combustion engine
- discuss the historical development of farm tractors
- explain the market structure of the farm machinery and equipment industry
- analyze the diversification of the farm machinery and equipment industry
- evaluate the impact of foreign trade
- describe the trends in the farm machinery and equipment industry
- numerate various farm machinery and equipment products
- discuss various careers in the farm machinery and equipment industry
- describe the role of farm machinery and equipment wholesalers
- discuss farm machinery and equipment dealership
- explain the types of jobs in farm custom and contracting services

TERMS TO KNOW

broaches
castings
compression
crawler-type
cylinder
economy of scale
farm contracting
forge shop
foundry

full-line companies
hydraulic lifts
hydrostatic transmission
lathes
long-line companies
machine shop
mechanical power
milling machines
pneumatic

power take-off (PTO)
short-line companies
threshing machines
torque amplification
tricycle-type
turbocharger
turpentine
wholegoods

INTRODUCTION

One of the most significant developments of the agricultural revolution was the shifting from animal power to a highly mechanized agriculture. This power shift began in the latter half of the nineteenth century and the first half of the twentieth century. The shift from human and animal power to **mechanical power** originated in the industrial revolution, which began in the eighteenth century in England with the discovery of the steam engine. It was first used for pumping water out of coal mines, and by 1800 steam engines were used to power sawmills, textile mills, and numerous other industrial firms.

Through mechanization, animals such as horses, mules, and oxen were replaced with the power of machines. As a result, the hours of human labor required in production agriculture has been reduced greatly. Before 1830, it took nearly fifty-six hours of human labor to produce one acre of wheat. Today, according to the USDA, with modern farm machinery and equipment, less than two hours of human labor are needed to produce that same acre.[1]

HISTORICAL DEVELOPMENT

The development of modern farm equipment began before tractors made their impact. One of the first agricultural machines that had a significant impact on farming was the cotton gin, invented in 1784 by Eli Whitney. Three years later the cast-iron plow was patented by Jethro Wood. This plow worked very well in eastern soils but not the hard soils of the Midwest. In 1837 John Deere, founder of John Deere Tractor Company, made the first successful steel plow from a saw blade, and by 1846 he was building 1,000 steel plows per year. The steel did not wear out as fast as the cast iron, and the soil did not stick to the new plow as it did with the old one, so farmers were very happy with it.

Until about 1850 production agriculture changed very little from family to family. Most farm jobs, including sowing, tilling, and harvesting, were performed using human muscle power. Agriculture practices were handed down from father to son and from mother to daughter. As late as the year 1800, approximately 90 percent of all people in the United States lived on farms. Most production agriculturalists were nearly totally self-sufficient. They produced nearly everything they needed, including clothing, tools, feed, and farm equipment. They had to do so because the typical production agriculturalist had to struggle each year just to produce enough food to feed the family. There was little surplus production left to exchange.[2]

Beginning of Change

The nineteenth century brought the beginnings of change. The century began with the Louisiana Purchase in 1807 and the opening of the farmland west of the Allegheny Mountains.

The first cotton planter was patented in 1825, and a corn planter was developed three years later. In 1831, Cyrus McCormick, one of the founders of International Harvester Company, developed a successful grain reaper. Although this was not the first reaper invented, it was the first to achieve wide acceptance, and by 1860, 20,000 reapers were being sold each year. The reaper not only cut labor time and cost, it also reduced the risk of weather damage by more than 50 percent by reducing harvest time. In 1885 over 250,000 harvesting machines were produced and sold. The earlier farm machines were built for horsepower rather than tractors, but they adjusted very quickly when the source of power changed.

Between 1850 and 1880 the amount of land devoted to farms increased 82 percent, to 536 million acres from 294 million acres, while the number of farms rose 167 percent, to 4 million.[3] The rise of the Industrial Age near the middle of the century brought about shortages of farm workers as they flocked to northern industrial cities. The Civil War brought increased interest in the development of labor-saving tillage and harvesting tools by simultaneously raising farm commodity prices and reducing the supply of farm laborers.

From Manpower to Horsepower

At the end of the century, only 50 percent of Americans lived on farms and the transition from manpower to horsepower was nearly complete. Production agriculturalists were still largely self-sufficient, but they had raised their productivity to a level where they could afford to purchase some horse-drawn farm equipment and a few other inputs produced off the farm.[4]

THE STEAM ERA

Although steam power had its major impact on the industrial sector of the economy, it also played a major role on the farm between 1850 and 1900, which is generally called the "Steam Era." It is estimated that over 70,000 steam engines were produced for farm use. The first steam engines were stationary, and farm jobs had to be brought to the engine. The second phase produced a portable steam engine, which was pulled to the fields for specific jobs such as powering wheat **threshing machines** or sawmills. The third type of farm steam engines were called "traction engines" and were much more useful as a source of farm power. They pulled themselves as well as large equipment. The first steam traction engine for farm use was made in 1869 by J.I. Case Company.[5]

The steam traction engine ushered in a new era in agriculture by providing an alternate mobile source of power on the farm. However, steam engines had numerous problems, and it was obvi-ous that they were not the ideal source of power that farmers needed. The engines were extremely heavy, bulky, dangerous, expensive, and adaptable to very few chores.

INTERNAL COMBUSTION ENGINE

The development of the internal combustion engine occurred in stages similar to that of the steam engine. First, small, stationary one-**cylinder** engines were made for small jobs around the home and on the farm. Then, larger, two-cylinder motors were mounted on wheels or sleds and taken to the fields, shops, or wherever they were needed. Finally, two-cylinder engines were mounted on frames with wheels and a transmission so they could pull themselves as well as some type of equipment.

Fuels Used

A wide variety of fuels were used in the early internal combustion engines. Some of the major types were gun powder, **turpentine**, coal dust, and kerosene, which was commonly called coal oil. Even though early tractors were called gasoline tractors, the major source of fuel was kerosene. Most early tractors were made with a small tank for gasoline and a large one for kerosene. The farmers started the engine with gasoline and then switched to kerosene since it was cheaper and more efficient to use.

The Beginning of Internal Combustion Engines

The exact date when internal combustion engines were first made is not clear. However, some engines have been traced to the latter part of the seventeenth century. The first practical power unit was developed in 1876 by two German inventors, N. A. Otto and Eugene Langen. This engine was very popular and became known as the Otto engine.

In 1899 there were over 100 firms making internal combustion engines in the United States, not counting automobile engines, and by 1914 there were over 500 companies in operation.[6] Small engines continued to be a popular source of power through the 1940s. Currently, these small engines are experiencing a comeback through restoration by private collectors. However, they are used for fun, not for work.

FARM TRACTORS

The record is not clear as to when the first tractor was made and who made it. However, historian R. B. Gray reports that the Charter Gas Engine Company built six gasoline tractors in 1889 and that in 1890 George Taylor applied for a patent on a walking-type motor plow.[7]

The First Gasoline-Powered Tractor

In 1892 John Froehlich built what is sometimes called the first successful gasoline-powered tractor. The Froehlich tractor was the forerunner of the Waterloo Boy and the modern John Deere line of tractors. The Case Threshing Machine Company and the Dissinger and Bros. Company also built tractors in 1892 but they were experimental in nature and were not put to practical use until several years later. Several other companies began experimenting with gasoline tractors in the late 1890s, but the switch from horsepower and steam power to tractor occurred primarily during the first thirty years of the twentieth century. The term *tractor* was first coined in 1906 by a salesman for the Hart-Parr Tractor Company (a predecessor of Oliver, White Farm Equipment Company and today's AGCO). Previously, they were called "gasoline traction engines." Although tractors have been around for nearly one hundred years, mules are still needed for certain jobs. Refer to Figure 9–1.

Figure 9–1 These loggers still need a mule to drag these cedars from the woods in order for modern-day equipment to complete the loading process. (Courtesy of Cliff Ricketts)

Effect of World War I on Tractor Production

Tractor production expanded rapidly in the early 1900s. In 1910, 15 tractor companies sold 4,000 tractors. The onset of World War I marked another turning point in the development of agriculture and caused a rapid increase in tractor production, and in 1920, 166 tractor companies sold over 200,000 tractors. Spurred by higher incomes from feeding war-ravaged Europe, U.S. production agriculturalists began the process of replacing their horse-drawn equipment with gasoline-powered tractors and the larger tillage implements that they could pull.

Effect of the Depression Years on Tractor Production

After 1921, the number of tractor companies decreased about as fast as they had increased. This was due to the Depression of the 1920s. In 1921, 186 companies sold only 68,000 tractors, and by 1925 only 58 companies had survived, although the number of tractors sold increased. In 1935, 20 companies sold over 1 million tractors with 90 percent of sales coming from 9 major companies: International Harvester, John Deere, J.I. Case, Massey-Harris, Oliver, Minneapolis Moline, Allis Chalmers, Cleveland Tractor Company, and Caterpillar Tractor Company.[8]

Henry Ford and the Tractor

The first gasoline-powered tractors had the same problems as the steam engines: they were expensive, big, bulky, hard to drive, and very limited in their application. Tractor companies soon began experimenting with smaller tractors that were more suitable for small farms and less expensive. One of the most successful small tractors was the "Fordson" tractor made by Henry Ford. It was also the first mass-produced tractor on the market. The Fordson accounted for about 70 percent of the total tractor market by 1925. Refer to Figure 9–2.

Power Take-off Units, Tricycle-type Tractors, and Rubber-tired Tractors

In 1918 International Harvester announced a **power take-off (PTO)** unit, which allowed the operator to control mounted and drawn equipment with the tractor's engine. The **tricycle-type**

Figure 9–2 The Fordson tractor had a major impact on American agriculture as production agriculturists switched from horse to tractor power. (Courtesy of New Holland North America, Inc.)

tractor was introduced by International Harvester in 1924 and was very popular for cultivating as well as plowing. In 1932 Allis Chalmers, in cooperation with Firestone Rubber Company, introduced a **pneumatic** rubber–tired tractor, which completed the basic design of a light, versatile tractor that could handle most farm jobs. This finalized the basic transition from horses and mules to tractors with internal combustion engines. Animals were no longer needed as a major source of power although many smaller farmers continued to use horses and mules through the 1950s.

Effect of the Shift from Animal Power to Tractor Power

The shift of animal power to tractor power affected American farming in two major ways.

Decreased Demand for Animal Feed. A large portion of the land that had been used to produce animal feed was shifted to food production. It is estimated that if horses were still the major source of power on the farm, they would consume the output of approximately 25 percent of grain acreage.[9] This would be a major drain on food production capabilities. The conversion to mechanical power not only increased farmers' productivity but also increased the amount of food available for human consumption by lessening the amount of feed output needed by work animals. Refer to Figure 9–3.

Reduced Labor Time and Cost. In 1936, the Iowa State University Experiment Station reported that production agriculturalists with rubber-tired, two-plow tractors were producing 100 acres of corn with 51 days of fieldwork. The same operation with horses required 141 days.[10]

Advent of Various Fuels

Machine power continued to change and improve. In 1931, Caterpillar Tractor Company developed a

Figure 9–3   If horses were still the major source of power on the farm, they would consume the output of approximately 25 percent of grain acreage. (Courtesy of Massey-Ferguson Operations)

diesel-powered, **crawler-type** farm tractor. The crawler-type tractor did not fit most farm needs, but the diesel engine had a major impact a few years later. The gasoline engine itself has been improved through the development of high-**compression** engines. Also, in 1941 liquefied petroleum (LP) gas tractors were introduced by the Minneapolis Moline Company. This made it possible for farmers to use clean-burning, low-cost butane and propane fuels, especially in areas near these energy sources.

Modern Tractor Accessories

Today, **hydraulic lifts**, **torque amplification**, **hydrostatic transmission**, power steering, **turbochargers**, heated and air-conditioned cabs, and many other features provide an efficient and comfortable power unit for modern production agriculturalists. The production agriculturalist of today operating a 100-horsepower tractor can do the work of over 1,000 workers without machine or animal power. It is no wonder that the average American farmer produced enough for over 131 people. Refer to Figure 9–4.

Increased Size and Four-Wheel Drive

During the decades of the 1960s and 1970s the major changes included the shift to diesel as the major fuel, an increase in horsepower, and a shift to four-wheel drive power. Currently, over 80 percent of farm tractors use diesel and most major

Figure 9–4   A production agriculturalist today operating a 100-horsepower tractor can do the work of over 1,000 workers without machine or animal power. (Courtesy of New Holland North America, Inc.)

tractor companies offer tractors with a horsepower rating of 200 or more. The major change in the 1970s was the shift to four-wheel drive. The major advantages of four-wheel drive include the ability to use more power efficiently, better traction and flotation with less soil compaction, and increased safety. Four-wheel drive is now standard on extremely large models and optional on medium and small models. Refer to Figure 9–4.

MARKET STRUCTURE

Farm machinery companies may be classified into three types based on the type and variety of equipment sold: full line, long line, and short line. **Full-line companies** produce and sell tractors as well as a wide variety of equipment. **Long-line companies** produce and sell a wide variety of general farm equipment, including self-propelled combines, but no tractors. **Short-line companies** produce highly specialized equipment, such as planters and cultivators, forage equipment, and milking machines.

Full-Line Companies

The farm machinery industry is similar to the automobile industry in terms of market concen-

tration. According to (1997) Agricultural Dealer Surveys four full-line companies account for over 90 percent of total farm machinery sales in the United States. Deere, Inc., which is the largest producer, accounts for over one-third of total sales. New Holland is second, followed by Case and Agco. Full-line companies constitute the major basis of the farm equipment industry. They are the smallest in number, but they have the largest sales volume and are international in scope, with manufacturing plants and distribution centers throughout the world.

Long-Line Companies

The second major group is the long-line companies. They are larger in number, but they have less dollar volume than full-line companies. Most long-line companies operate on a national market. The major long-line companies include Gehl and Vermeer. Several previous long-line companies have expanded or merged into full-line companies. Refer to Figure 9–5.

Figure 9–5   Vermeer is an example of a long-line company. Vermeer produced the first round hay baler. (Courtesy of Vermeer Manufacturing Company)

Short-Line Companies

Short-line companies have the least amount of impact on the industry due to the lower volumes sold. Most short-line companies are regional in scope. They typically produce for full-line companies under some type of production agreement. Full-line companies acquire parts and assembled machines from short-line companies, foreign-based subsidiaries, and foreign markets.[11] Refer to Figure 9–6.

Size of the Farm Machinery and Equipment Market

There are around 1,500 firms involved in producing farm machinery in the United States.[12] These firms sell their products through approximately 7,000 retail outlets, which are franchised by major manufacturers. Presently, production agriculturalists spend about $4.6 billion each year on farm machinery and equipment. Average production expenditures for farm machinery in 1993 were just over $7,000 per farm. Tractors and self-propelled vehicles account for about 30 percent of total machinery and equipment sales.[13]

In many cases, a retail dealer may hold a franchise for more than one type of tractor or other types of equipment. This is a trend that has emerged as a result of the consolidation the farm equipment industry experienced throughout the 1980s. By having two to three franchises, sufficient volume is available to maintain full service and parts operation and to provide adequate income for the dealer. An average full-line dealer is required to have a minimum net worth of $250,000 to $500,000 before a franchise is awarded. Approximately two-thirds of the business volume comes from the sale of new and used equipment, with the remainder coming from parts, service, and small farm supply items.

DIVERSIFICATION OF FARM MACHINERY AND EQUIPMENT COMPANIES

The major farm equipment companies are highly diversified and do not limit their products to farm equipment. A major portion of recent expansion has been made in nonfarm product lines. For example, J.I. Case is a major producer of backhoes and other industrial equipment; Deutz-Allis, purchased by AGCO, was a major producer of electric generators; and White Motor Company, also purchased by AGCO, was a major producer of industrial trucks. In the early stage of development, Case and International Harvester made automobiles, but they discontinued production soon after they began. Ford is still a major producer of farm pickup trucks and cars. The Ford tractor was bought by the Fiat Company in Italy and is now sold under the New Holland name. Refer to Figure 9–7.

Most of the full-line companies have expanded their consumer line to attract urban home owners. The most attractive consumer line is riding lawn-mowers and garden equipment, but many other consumer products are offered, from chain saws and snowmobiles to bicycles. This phase of the industry is expected to expand even more as the

Figure 9–6 Bush Hog is an example of a short-line company. (Courtesy of Bush Hog Corporation)

Figure 9–7 Recent expansion of original farm machinery and equipment companies has been made in nonfarm product lines such as industrial equipment. (Courtesy of Cliff Ricketts)

Figure 9–8 Full-line companies have expanded to attract urban home owners in order to maintain a strong market for their products. (Courtesy of John Deere)

urban population increases and the number of farmers decreases. Refer to Figure 9–8.

FOREIGN TRADE

Due to the changes made in the farm machinery and equipment industry in the 1980s, it went from being primarily domestic to becoming competitive internationally. Large types of farm equipment and machinery were marketed primarily in the United States and Canada. Foreign countries were interested primarily in smaller pieces of equipment and machinery. Therefore, manufacturers in the United States and Canada first produced large machinery and equipment, but due to a growing market for smaller tractors in Western Europe and, eventually, the United States, farm machinery and equipment producers entered into agreements for the smaller tractors to be produced by foreign firms under domestic producer nameplates.[14]

Exports

Since most tractor manufacturers are now international in scope, foreign trade plays a major role in the economic success of the farm machinery equipment industry. From 1972 to 1981, the value of U.S. farm equipment exports increased at an annual rate of 24 percent. However, according to Omri Rawlins in his book *Introduction to Agribusiness* (1999) the export market peaked in 1981 and has declined by about 8 percent per year since that time. Most of the decrease was due to declining markets for high-priced machinery such as large tractors and self-propelled combines.

Imports

Because of the inflationary pressures in the farm machinery and equipment industry during the 1970s, American farmers became increasingly interested in the tractors made and imported from other countries. Imported tractors tend to be 10 to 20 percent cheaper, depending on the size. The major foreign tractors available to American farmers included the former Deutz-Allis (Germany), Long (Romania), Satra Belarus (Russia), Same (Italy), and numerous small to medium tractors from Japan such as the Kubota and Satoh.

While exports have declined since 1981, imports have continued to increase. The sharp increase in the demand for small tractors for nonagricultural uses has increased the market for small tractor imports. Low-power (under 40 horsepower) tractors account for about 30 percent of all farm machinery imports. The value of all tractors and parts imported amount to over one billion dollars. The United States imports about 56,000 tractors under 40 horsepower, and about 90 percent of these come from Japan.[15]

The major areas purchasing farm machinery from the United States are Canada, Western Europe, and Central America. The major areas selling farm machinery to the United States are Canada, Japan, the United Kingdom, and West Germany. Japan and Germany account for almost 80 percent of all foreign tractors sold in the United States.[16]

TRENDS IN THE FARM MACHINERY AND EQUIPMENT INDUSTRY

The farm equipment industry was the first resource group to experience the effect of the farm financial crisis of the 1980s. As incomes declined, farmers responded by repairing older farm equipment instead of buying new, thus reducing domestic demand. In addition, the increasing value of the dollar relative to other currencies reduced foreign demand. Increasing imports reduced demand even more for farm machinery made in the United States. As a result, farm expenditures for tractors and farm equipment peaked at $11.7 billion in 1970 and declined through the 1980s. In 1987 the industry experienced a slow recovery and sales volume increased to a peak in 1990 before experiencing a slight reduction. The Agricultural Dealer Survey conducted by Paine Webber in 1997 showed that since 1992 tractor sales have slowly increased.

Mergers

The large reduction in demand has resulted in several mergers of farm equipment manufacturers, a reduction in the number of dealerships, and a change in the types of tractors produced. Two major mergers of full-line companies occurred, and several long-line and short-line companies were purchased by full-line companies. In 1984, J.I. Case purchased the tractor division of International Harvester, creating Case-International, and in 1986 Steiger Equipment Company was purchased. Deutz-Allis was created in 1985 when Klockner-Humboldt-Deutz of West Germany purchased Allis-Chalmers. In 1980, a U.S. management team bought Deutz-Allis and formed AGCO. In 1991, AGCO acquired White Tractor Division, which earlier had acquired Oliver. AGCO recently purchased Massey-Ferguson.

Ford Tractor Company purchased New Holland Equipment Company from Sperry Rand Corporation in 1986 and became called Ford-New Holland for a while. Presently, Ford tractors are sold under the New Holland name. Ford also purchased Versatile in 1987 to expand its large tractor line. Allied Products, a newcomer to the farm equipment industry, purchased White Equipment Company and several short-line companies, including Bush Hog, Kewanee, and New Idea. Refer to Figure 9–9 for an illustration of other mergers.

PRODUCTS OF THE FARM MACHINERY AND EQUIPMENT MANUFACTURERS

Although we have concentrated on tractors and related tractor equipment up to this point, many other products are produced by the farm machinery and equipment industry. Since the agricultural industry includes the agribusiness input sector and agribusiness output (marketing) sector besides the production of agriculture sector, many types of farm machinery and equipment are produced. Besides tractors, field tillage and planting equipment, and harvesting equipment, the following products are sold:

- 🌾 farm structures and equipment
- 🌾 greenhouse, ornamental horticulture
- 🌾 forestry, logging equipment
- 🌾 lawn, garden, and landscaping equipment
- 🌾 light industrial power machines
- 🌾 grain-handling and storage equipment
- 🌾 livestock-feeding and -handling equipment
- 🌾 dairy-feeding and -milking equipment
- 🌾 cotton-ginning and -compressing equipment
- 🌾 fertilizer and pesticide equipment
- 🌾 irrigation equipment
- 🌾 prefabricated farm structures

As you can see, manufacturers produce a tremendous variety of equipment for production agriculturalists, contractors, gardeners, and landscape specialists. Billions of dollars are spent on farm machinery and equipment. Production agriculturalists now own over $2.5 billion worth of machinery and equipment.[17] Each year more nonfarm users purchase products from the farm machinery and equipment industry, which today also sell to landscaping firms, construction contractors, highway departments, parks, golf courses, rural and suburban home owners, and home gardeners.

CAREERS IN FARM MACHINERY AND EQUIPMENT MANUFACTURING

Careers are available in farm machinery and equipment manufacturing businesses for both high school and college graduates.

Factory Jobs

Factory jobs in farm machinery and equipment manufacturing businesses include: the **foundry**, where molten iron is poured into molds to make **castings**; the **forge shop**, where heavy hammers forge the red-hot steel billets into shape; the **machine shop**, with its **lathes**, **milling machines**, cutters, planes, **broaches**, and automatically controlled (and in many cases computer controlled) machine tools; the paint shop, the assembly line; and the shipping department.[18]

Office and Professional Jobs

There are several major divisions and departments conducting specialized, essential business services. Many of these jobs are for college graduates with a major in agribusiness or a related major. Refer to Figure 9–10. Although various manufacturing businesses vary, the following departments or divisions are potential sources of employment:

- 🌾 Purchasing Department
- 🌾 Sales and Marketing Division
- 🌾 Finance Department
- 🌾 Engineering Department
- 🌾 Consumer Relations Division
- 🌾 Human Resource Department
- 🌾 Parts Department
- 🌾 Distribution Department
- 🌾 Product Planning Division
- 🌾 Export Division
- 🌾 Administrative (or Executive) Department[19]

FARM MACHINERY AND EQUIPMENT WHOLESALERS

Once the farm machinery and equipment leaves the manufacturer, it goes to the wholesaler, which

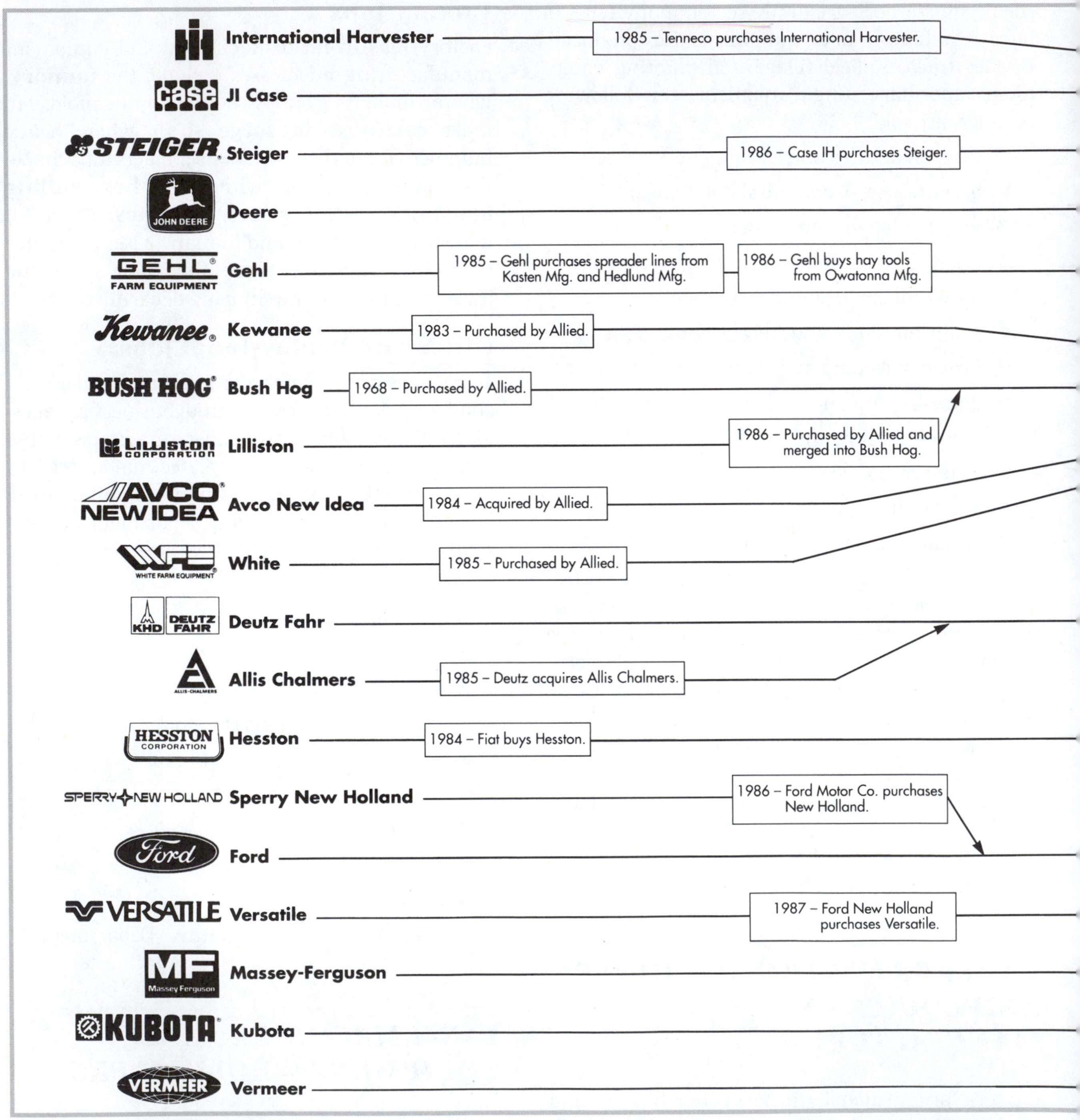

International Harvester —————————— | 1985 – Tenneco purchases International Harvester. |

JI Case ——————————————————————

Steiger ————————————— | 1986 – Case IH purchases Steiger. |

Deere ——————————————————————

Gehl ———————— | 1985 – Gehl purchases spreader lines from Kasten Mfg. and Hedlund Mfg. | | 1986 – Gehl buys hay tools from Owatonna Mfg. |

Kewanee ——————— | 1983 – Purchased by Allied. |

Bush Hog —— | 1968 – Purchased by Allied. |

Lilliston ————————————— | 1986 – Purchased by Allied and merged into Bush Hog. |

Avco New Idea ——————— | 1984 – Acquired by Allied. |

White ———————————— | 1985 – Purchased by Allied. |

Deutz Fahr ——————————————————————

Allis Chalmers ————————— | 1985 – Deutz acquires Allis Chalmers. |

Hesston ————————— | 1984 – Fiat buys Hesston. |

Sperry New Holland —————————— | 1986 – Ford Motor Co. purchases New Holland. |

Ford ——————————————————————

Versatile ———————————————— | 1987 – Ford New Holland purchases Versatile. |

Massey-Ferguson ——————————————————————

Kubota ——————————————————————

Vermeer ——————————————————————

Figure 9–9 Many mergers within the farm machinery and equipment industry have occurred in the last twenty years. More mergers have occurred since the printing of this book. (Courtesy of *Agri-Marketing Magazine,* October, 1991)

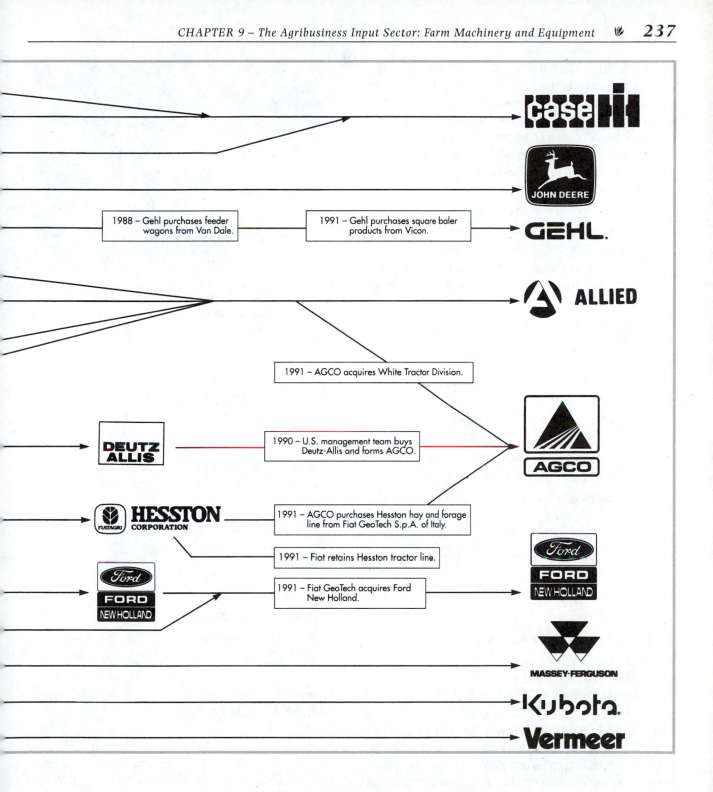

1988 – Gehl purchases feeder wagons from Van Dale.

1991 – Gehl purchases square baler products from Vicon.

1991 – AGCO acquires White Tractor Division.

1990 – U.S. management team buys Deutz-Allis and forms AGCO.

1991 – AGCO purchases Hesston hay and forage line from Fiat GeoTech S.p.A. of Italy.

1991 – Fiat retains Hesston tractor line.

1991 – Fiat GeoTech acquires Ford New Holland.

links the manufacturers to the dealerships. Because the manufacturers are often located far from the dealerships, the dealers cannot expect prompt delivery or quick service. Dealerships want to be close to a main source of supply to be assured of readily available **wholegoods**, repair parts, and prompt service. The following are some important wholesale services:

❧ assembling a relatively large inventory of equipment, machinery, and parts at a central point near dealerships in a region

❧ representing manufacturers and establishing good working relationships with retail dealers

❧ securing new retail dealers where necessary; contracting with them for the sale of the manufacturer's equipment and parts

❧ discharging the service and warranty obligations of the manufacturer

❧ assisting retailers in merchandising methods, displays, and showroom arrangements

❧ aiding retailers in sales training, sales promotion, advertising, and public demonstrations

❧ assisting dealers in service shop arrangement, training service mechanics, parts department arrangement, and parts inventory

❧ providing management counsel to retail dealers

❧ assisting dealers in their financing plans and programs[20]

Careers in Wholesaling

Many of the same jobs are available in wholesaling as in the manufacturing office and professional area. Postsecondary and college training will be an asset in getting many of these jobs. Regardless of the formal education, knowledge and experience should be attained in crop production methods, the principal farming enterprises, harvesting and storage of crops, animal production, and farm power

Figure 9–10 Many careers are available in farm machinery and equipment manufacturing. Specialized training is needed in many areas, and an essential skill for almost everyone is computer technology. (Courtesy of USDA)

and equipment. Knowledge and skills should also include marketing, economics, sales, accounting, sales promotion, communication, and public speaking. Especially important is a knowledge of the language of the production agriculturalist.

Specifically, the following areas offer potential employment opportunities in wholesaling:

❧ purchasing agent

❧ publications preparation

❧ territory or zone manager

❧ traffic and transportation manager

❧ liaison officer

❧ marketing surveying

❧ business services

❧ salesperson

DEALERSHIPS

There has been a major consolidation of farm machinery dealerships in recent years. The number of dealerships fell from 16,700 in 1967 to

9,500 in 1972, then to 7,900 in 1989. The number dropped even further over the following decade. However, average annual equipment sales per dealer increased from about $2 million in 1979 to over $3 million in 1988. USDA reports show farm equipment companies sold $4.6 billion worth of farm machinery during 1995.[21]

In order to survive, dealers are cutting costs in every way possible, increasing in size to gain **economy of scale**, expanding the variety of farm equipment and in some cases expanding into nonfarm equipment lines. Some analysts predict that future farm equipment dealers will resemble supermarkets, offering a wide variety of tractors and equipment.

A Complex Business

Dealerships are a very complex type of business because of the many different pieces of equipment handled, parts inventory, servicing, and repairing of equipment. Besides the capital required to start a business, funds will be unavailable should credit be extended. Manufacturers of farm equipment also provide consumers with financing, which is processed by the dealership. Challenges that add to the complexity of dealerships consist of the following:

- Sales of used machinery affect new machinery sales.
- There are greater demands for better service and repairs.
- The credit needs of farmers are increasing.
- The leasing of farm equipment is gaining popularity among farmers.
- There is a need for more automated livestock production equipment.
- Specialization in the handling and selling of equipment is increasing.
- Safety and environmental issues such as oil disposal are a growing concern.[22]

Dealership Careers

Although dealerships are complex, they offer a good place to start a career in the farm machinery and equipment industry. A wealth of experience can be gained from such an experience. By being on the ground floor, where you can see the whole business, much can be learned about farm markets, management, customer relations, finance, product promotion, and sales. Company training is also available. For example, Deere and Company has an excellent program, which includes marketing, parts, credit, services, and management. Specific jobs available at a dealership include:

- retail operations
- purchasing
- facility management
- sales
- parts department
- service department
- advertising
- finance, credit, collections
- accounting
- management[23]

Refer to the Career Option on page 241 for the skills needed to be an agricultural equipment dealer.

FARM CUSTOM AND CONTRACTING SERVICES

Custom and contracting services are another phase of the farm machinery and equipment industry. They are vital services of benefit to both the operator and the customer.

Custom Farming

Custom farmers perform a variety of jobs necessary for the production agriculturalists. They till, plant, cultivate, apply pesticides, and harvest crops. Production agriculturalists use custom farmers for several reasons. In harvesting, combines are very expensive to own, service, and operate. New combines can sell for over $100,000. Economically, it makes more sense to pay to get your crop custom-harvested since a combine might not be used for more than two months each year. Another reason for custom farming is that many custom farmers have multiple combines or other harvesting equipment. This provides the opportunity of getting the crop harvested quickly and preventing weather damage or loss. Besides harvesting grains, other custom-farming services may include grinding and mixing feed, baling hay and straw, picking cotton, spreading lime and fertilizer, spraying and dusting for pests, picking and packing fruits and vegetables for growers, hauling milk, and many others.[24] Refer to Figure 9–11.

The cost of the service varies according to the job. For example, grain harvesters are paid by the bushel; tillage operators charge by the acre; hay harvesters are paid by the bale. Whatever the cost, custom farming provides a service to production agriculturalists without the cost of owning the equipment, and it gets the job completed more quickly. On the other hand, the custom farmer can better afford ownership of equipment.[25]

Farm Contracting Services

Another career area in the farm machinery and equipment industry is **farm contracting** services. This includes special jobs done by individuals with specialized equipment. Special skills are needed such as mechanical operation, maintenance and repair; effective cost control through estimating, and bidding; and business management. There are several jobs that can be very profitable, such as:

- ❦ clearing land
- ❦ leveling and reforming land surface
- ❦ constructing irrigation ditches
- ❦ installing drainage systems
- ❦ terracing and constructing water courses
- ❦ developing watersheds
- ❦ aerial spraying[26]

CONCLUSION

The changes from manpower to horsepower and then to mechanical power each came in response to a felt economic need on the part of production agriculturalists. Each advancement was accepted by production agriculturalists because it helped them enhance their productivity and, more important, their profits. By helping production agriculturalists get what they wanted (higher output and profits), machinery and farm equipment manufacturers got what they wanted (higher sales and profits). In fact, Peter Drucker credits Cyrus McCormick with pioneering the development of marketing, particularly market research, pricing policies, parts, and servicing departments, and

Figure 9–11 Certain farmers perform a variety of jobs that are necessary for production agriculturalists, offering several economic advantages. (Courtesy of Holland North America, Inc.)

CAREER OPTION

Career Area: Agricultural Equipment Dealer

Acareer as an agricultural equipment dealer involves daily interactions with production agriculturists, mechanics, technicians, salespeople, customers, office staff, and the entire farm community. During years when production agriculture is profitable, the business is likely to do well, but during the low-profit years, an agricultural equipment dealership will often take financial losses. The success of the business is tied directly to economic conditions of those in production agriculture.

Many agricultural equipment dealers had entered the business as salespeople or mechanics and only later were able able to generate enough credit to buy into the ownership of the agribusiness. Large amounts of capital are needed to maintain the inventories of agricultural equipment and parts.

Success in an agricultural equipment dealership requires a good understanding of the machines that are used in producing crops. An agricultural background is useful in understanding the kinds of conditions in which agricultural machines are operated. An agricultural equipment dealer has a large investment in the business. In addition to the new and used farm equipment kept in the inventory, replacement parts must be purchased and stocked for resale.

Formal education and strong backgrounds in business practices are important in managing agricultural equipment dealerships. Good public relations skills and a real interest in providing services to people are important for success in this agricultural business.

An agricultural equipment dealer must follow up and assist the producer to make sure new equipment is functioning properly. (Courtesy of USDA)

also with the idea of installment credit purchases.[27] The purchasing patterns established for farm machinery and equipment soon spread to other agricultural inputs. Refer to Figure 9–12 for a list of selected farm machinery and equipment companies.

SUMMARY

One of the most significant developments of the agricultural revolution was the shift from animal power to a highly mechanized type of agriculture.

Farm Machinery and Equipment Manufacturers

A. O. Smith Harvestore Products Inc.
345 Harvestore Drive
DeKalb, IL 60115
Tel.: 815-756-1551/Fax: 815-756-7821

Acco Corporation
5295 Triangle Parkway
Norcross, GA 30092
Tel.: 404-447-5546/Fax: 404-246-6118

Badger Northland, Inc.
P.O. Box 1215
Kukauna, WI 54130
Tel.: 414-766-4603

Briggs & Stratton Corporation
P.O. Box 702
Milwaukee, WI 53201
Tel.: 414-259-5333/Fax: 414-259-5338

Bush Hog Corporation
P.O. Box 1039
Selma, AL 36702
Tel.: 205-872-6261/Fax: 205-872-0168

Butler Manufacturing Co.
7400 East 13th Street
Kansas City, MO 64126
Tel.: 816-968-6141

Case J.I.
700 State Street
Racine, WI 53404
Tel.: 414-636-6011, 636-7394
Fax: 414-636-7809

Caterpillar, Inc.
100 Northeast Adams Street
Peoria, IL 61629
Tel.: 309-675-1000/Fax: 309-675-6155

Cenex/Land O'Lakes Ag. Services
2827 Eighth Avenue South
Fort Dodge, IA 50501
Tel.: 515-576-7311/Fax: 515-576-2685

Cummins Engine Company, Inc.
500 Jackson Street, P.O. Box 3005
Columbus, IN 47201
Tel.: 812-377-3695

Deere & Co.
John Deere Road
Moline, IL 61265
Tel.: 309-765-8000/Fax: 309-765-5772

Detroit Diesel Corporation
13400 Outer Drive, West
Detroit, MI 48239
Tel.: 313-592-7642

Deutz Corporation
5295 Triangle Parkway
Norcross, GA 30092
Tel.: 404-564-7100

Ford New Holland Americas, Inc.
500 Diller Avenue
New Holland, PA 17557
Tel.: 717-355-1371/Fax: 717-355-3192

Vermer Manufacturing Company
One Mile E. New Sharon Road, P.O.
Box 200
Pella, IA 50219
Tel.: 515-628-3141/Fax: 515-628-7734

Gehl Co.
143 Water Street
West Bend, WI 50395
Tel.: 414-334-9461/Fax: 414-334-1565

Homelit Div. of Textron, Inc.
14401 Carowinds Boulevard
Charlotte, NC 28273
Tel.: 704-588-3200/Fax: 704-588-0926

J. I. Case-A Teneco Company
700 State Street
Racine, WI 53404
Tel.: 414-636-7843

John Blue Company
P.O. Box 1607
Huntsville, AL 35807
Tel.: 205-721-9090

Keystone Steel & Wire Co.
7000 S.W. Adams Street
Peoria, IL 61641
Tel.: 309-697-7020/Fax: 309-697-7422

Kohler Company/Power Systems
Group 444 Highland Drive
Kohler, WI 53044
Tel.: 414-457-4441

Kubota Tractor Corp.
13780 Benchmark Drive
Dallas, TX 75234
Tel.: 214-241-5900

Massey-Ferguson Operations
5295 Triangle Parkway
Norcross, GA 30092
Tel.: 404-447-5546/Fax: 404-409-6040

McCullough & Co.
4717 E. 119th Street, P.O. Box 9824
Kansas City, MO 64134
Tel.: 816-761-2857/Fax: 816-761-2845

Morton Buildings, Inc.
P.O. Box 399, 252 W. Adams Street
Morton, IL 61550
Tel.: 309-263-7474

New Holland, Incorporated
500 Diller Avenue, P.O. Box 1895
New Holland, PA 17557
Tel.: 717-355-1121/Fax: 717-355-1826

Patz Sales, Incorporated
Highway 14 S., P.O. Box 7
Pound, WI 54161
Tel.: 414-895-2251/Fax: 414-897-4312

Robison & Grisson/MT
Route 2, Box 36
Guntown, MS 38849
Tel.: 601-869-1028

White-New Idea Farm Equipment
Company
123 W. Sycamore Street
Coldwater, OH 45828
Tel.: 419-678-5311/Fax: 419-678-5496

Figure 9–12   There are many full-line, long-line, and short-line farm machinery and equipment manufacturers, which have retail dealerships throughout the country. The names and home office addresses of several of these companies are listed here.

This power shift began in the latter half of the nineteenth century and the first half of the twentieth century. The shift from human and animal power to mechanical power originated in the industrial revolution, which began in the eighteenth century in England with the discovery of the steam engine.

The exact date when internal combustion engines were first made is not clear. However, some engines have been traced to the latter part of the seventeenth century. In 1899 there were over 100 firms making internal combustion engines in the United States, not counting automobile engines, and by 1914 there were over 500 companies in operation.

In 1892 John Frolich built what is sometimes called the first successful gasoline tractor. The term *tractor* was first coined in 1906 by a salesman for the Hart-Parr Tractor Company. Tractor production expanded rapidly in the early 1900s. In 1910, 15 tractor companies sold 4,000 tractors. The onset of World War I caused a rapid increase in tractor production, and in 1920, 166 tractor companies sold over 200,000 tractors.

In 1932 Allis Chalmers, in cooperation with Firestone Rubber Company, introduced a pneumatic rubber–tired tractor, which completed the basic design of a light, versatile tractor that could handle most jobs. In 1931, Caterpillar Tractor Company developed a diesel-powered, crawler-type farm tractor. Today, hydraulic lifts, torque amplification, hydrostatic transmission, power steering, turbo chargers, heated and air-conditioned cabs, and four-wheel drives provide an efficient and comfortable power unit for modern-day production agriculturalists.

Farm machinery companies are usually classified as full-line, long-line and short-line. The full-line companies dominate the industry and are highly concentrated in number. The top two companies account for 57 percent, and the top six companies, for 93 percent, of all sales. Most of the full-line companies are international in scope, with manufacturing plants and distribution centers throughout the world. Most long-line companies are national in scope, whereas short-line companies generally serve regional and local markets. There are around 1,500 firms involved in producing farm machinery in the United States. These firms sell their products through approximately 7,000 retail outlets, which are franchised by the major manufacturers.

The major farm equipment companies are highly diversified and do not limit their products to farm equipment. A major portion of recent expansion has been made in nonfarm product lines. Due to the changes which were made in the farm machinery and equipment industry in the 1980s, it also changed from being primarily domestic to being competitive internationally. While exports have declined since 1981, imports have continued to increase.

The trend since the 1980s has been the merger of farm equipment manufacturers, a reduction in the number of dealerships, and a change in the types of tractors produced. Each year more nonfarm users purchase products from the farm machinery and equipment industry, which now sells to landscaping firms, construction contractors, highway departments, parks, golf courses, rural and suburban home owners, and home gardeners. As a result, many careers are available in the farm machinery and equipment manufacturing industry.

Once the farm machinery and equipment leaves the manufacturer, it goes to the wholesale companies. These companies link the manufacturers to the dealerships. Dealerships are a very complex type of business because of the need to handle many different pieces of equipment, parts inventory, and the servicing and repairing of equipment. Many dealerships have consolidated, resulting in many fewer dealerships, but the annual equipment sales per dealer has increased substantially.

Farm custom and contracting services form another phase of the farm machinery and equipment industry. These are vital services of benefit to both the operator and the customer. The farm machinery and equipment industry has helped the productivity and profit of the American production agriculturalist. Furthermore, purchasing patterns established in the farm machinery and equipment industry soon spread to other agricultural inputs and American business in general.

END-OF-CHAPTER ACTIVITIES

Review Questions

1. Describe the three types of steam engines used in the "Steam Era."
2. List five reasons why steam engines were not the ideal source of power that farmers needed.
3. Briefly discuss the three phases of use in the development of the internal combustion engine.
4. Name four fuels used in the early internal combustion engines.
5. What effect did World War I have on tractor production?
6. What effect did the Depression years have on tractor production?
7. What were the nine major tractor companies in 1985?
8. What were the two effects of shifting from animal power to tractor power?
9. What were the three major changes in tractors in the 1960s and 1970s?
10. What are the three major advantages of four-wheel-drive tractors?
11. What are the four major full-line tractor companies?
12. Compare the trend for tractor size manufacturing in both the United States and foreign countries.
13. What has been the trend for the exporting of tractors?
14. What has been the trend for the importing of tractors?
15. Explain four mergers of major tractor companies that have occurred in the last twenty years.
16. List sixteen products produced by the farm machinery and equipment industry.
17. Name eight nonfarm user groups of products produced by the farm machinery and equipment industries.
18. What are eight factory-type jobs that could lead to a career in the farm machinery and equipment industry?
19. What are eleven office and professional type jobs that could lead to a career in the farm machinery and equipment industry?
20. List nine important services of wholesalers.
21. What are twelve areas of knowledge, experience, and skills that should be attained for people preparing for careers in farm machinery and equipment wholesaling?

22. List eight areas that offer potential employment opportunities in farm machinery and equipment wholesaling.

23. Name five factors that make dealerships complex business.

24. What are seven challenges that add to the complexity of dealerships?

25. List ten specific jobs that are available at dealerships.

26. What are the advantages of a production agriculturalist using a custom farmer?

27. List seven types of jobs that custom farmers perform.

28. List seven types of jobs of those who do farm contracting services.

Fill in the Blank

1. Before 1830, it took nearly _____ hours of human labor to produce one acre of wheat.

2. Today, with modern farm machinery and equipment, less than _____ hours are needed to produce one acre of wheat.

3. In 1800, approximately _____ percent of all the people in the United States lived on farms.

4. In 1900, approximately _____ percent of Americans lived on farms.

5. Even though early tractors were called gasoline tractors, the major source of fuel was _____.

6. The _____ tractor accounted for about 70 percent of the total tractor market by 1925.

7. If horses were still the major source of power on the farm, they would consume the output of approximately _____ percent of grain acreage.

8. In 1936, production agriculturalists with rubber-tired, two plow tractors were producing 100 acres of corn with _____ days of fieldwork, while the same operation with horses required _____ days.

9. Today, the production agriculturalist operating a 100-horsepower tractor can do the work of over _____ workers without machine or animal power.

10. Currently, over _____ percent of tractors use diesel.

11. Four full-line tractor companies account for _____ percent of all farm machinery sold in the United States.

12. There are approximately _____ firms producing farm machinery through approximately _____ retail outlets.

13. Japan and Germany account for almost _____ percent of all foreign tractors sold in the United States.

Matching

a. Cyrus McCormick

b. Allis Chalmers

c. Minneapolis Moline Company

d. Fordson

e. J.I. Case Company

f. Eli Whitney

g. John Froelich

h. White Motor Company

i. Caterpillar Tractor Company

j. John Deere

k. International Harvester

l. Oliver

_____ 1. invented the cotton gin in 1794

_____ 2. made the first successful steel plow from a saw blade

_____ 3. developed a successful grain reaper in 1831

_____ 4. developed the first steam tractor engine for farm use in 1869

_____ 5. built, in 1892, what is sometimes called the first successful gasoline tractor

_____ 6. first mass-produced tractor on the market

_____ 7. first tractor with power take-off (PTO)

_____ 8. first pneumatic rubber–tired tractor

_____ 9. in 1931, developed a diesel-powered, crawler type farm tractor

_____ 10. introduced liquefied petroleum (LP) gas tractors in 1941

_____ 11. major producer of backhoes and other industrial equipment

_____ 12. major producer of industrial trucks

_____ 13. once made automobiles

_____ 14. the term *tractor* was coined by a salesman in 1906 for a company that evolved into this company

Activities

1. Select a particular farm machinery and equipment product. Go to the library and use other resources to prepare a two- to three-page paper on the historical development of the product. Share this with the class by giving an oral report.

2. Interview an older relative, friend, or farmer and get his or her impression of the changes in production agriculture over the years with the advances in farm machinery and equipment. Share these findings with your class.

3. Are four-wheel-drive tractors a luxury or a convenience? Defend your answer.

4. Refer to Figure 9–12 for a list of the various farm machinery and equipment manufacturers. Select each as a full-line, long-line, or short-line company.

5. Identify three different tractors owned by you, a neighbor, or someone in your community. Match each one with one of the four major tractor companies that exist today. Refer to Figure 9–9 for help.

6. List (if any) the farm machinery and equipment businesses in your community that are either manufacturers, wholesalers, or dealerships. Getting help from this chapter in the book, list the potential jobs available.

7. Defend whether you think it is better to buy equipment for all farm jobs or better to hire custom farmers for certain jobs. (Note, you might want to debate this in front of the class.)

8. Identify five farm contracting services in your community. Select one of these and give a report as to the profit potential.

NOTES

1. Jasper S. Lee and Max L. Amberson, *Working in Agricultural Industry* (New York: Gregg Division, McGraw-Hill Book Company, 1978), p. 50.
2. James G. Beierlein and Michael W. Woolverton, *Agribusiness Marketing: The Management Perspective* (Englewood Cliffs, N.J.: Prentice-Hall, 1991), p. 106.
3. Ibid.
4. Ibid.
5. Farm and Industrial Equipment Institute, *Men, Machines, and Land* (Chicago, Ill.: Author, 1974), pp. 19–21.
6. R. B. Gray, *The Agricultural Tractor, 1855–1950* (St. Joseph, Mich.: American Society of Agricultural Engineers, 1975).
7. Ibid.
8. Ibid.
9. Beierlein and Woolverton, *Agribusiness Marketing: The Management Perspective*, p. 107.
10. Farm and Industrial Equipment Institute, *Men, Machines, and Land.*
11. Randall D. Little, *Economics: Applications to Agriculture and Agribusiness* (Danville, Ill.: Interstate Publishers, 1997), p. 293.
12. N. Omri Rawlins, *Introduction to Agribusiness* (Murfreesboro, Tenn.: Middle Tennessee State University, 1998) p. 75.
13. Little, *Economics: Application to Agriculture and Agribusiness*, p. 292.
14. Ibid., p. 292.
15. Rawlins, *Introduction to Agribusiness* (Murfreesboro, Tenn.: Middle Tennessee State University, 1998) p. 77.
16. Ibid., p. 77.
17. United States Department of Agriculture (1997) *Census of Agriculture* (U.S. Government Printing Office, Washington, D.C.: 1997).
18. Marcella Smith, Jean M. Underwood, and Mark Bultmann, *Careers in Agribusiness and Industry* (Danville, Ill.: Interstate Publishers, 1991), p. 128.
19. Ibid., pp. 190–192.
20. Ibid., pp. 193–194.
21. USDA, *Agricultural Statistics, 1997.*
22. Ewell P. Roy, *Exploring Agribusiness* (Danville, Ill.: Interstate Printers and Publishers, 1980), p. 33.
23. Smith, Underwood, and Bultmann, *Careers in Agribusiness and Industry*, pp. 192–198.
24. Jasper S. Lee, James G. Leising, and David E. Lawrence, *Agrimarketing Technology* (Danville, Ill.: Interstate Publishers, 1994), pp. 162–163.
25. Ibid., p. 164.
26. Smith, Underwood, and Bultmann, *Careers in Agribusiness and Industry*, pp. 202–205.
27. Quoted in Beierlein and Woolverton, *Agribusiness Marketing*, p. 108.

10 Private Agribusiness Services

OBJECTIVES

After completing this chapter, the student should be able to:

- describe the general farm organizations
- discuss various agricultural commodity organizations
- explain the importance of agricultural research
- explain the importance of agricultural consultants
- describe the artificial insemination industry
- discuss the history, scope, and duties of veterinarians
- explain the importance of agricultural communications
- describe the history and impact of agricultural cooperatives

TERMS TO KNOW

agriservices
artificial insemination
collective bargaining
farm commodity
fraternal
glycerol
holding action

homogenous
husbandry
inseminate
physiologist
policy statements
political power
prominent

regulatory medicine
resource specialization
rural sociology
semen
sodium citrate

INTRODUCTION

In the early colonial days of American agriculture, when farmers were basically self-sufficient in the production and distribution of food and fiber, there was very little need for specialized services. Each farm family produced and processed the food needed, and shelters were built by the family members with trees from local forests. However, as the trend in farming shifted toward more specialized commercial production agriculture, in which the major production emphasis was to produce food and fiber for sale, the demand for specialized services began to develop. During the earliest phase of this development, special services were provided on an informal basis by trading services. When a major farm task needed to be done, farmers would ask their neighbors to help, and each one was given the specific task for which he or she was best qualified. These services often served a social function as well as an economic one.

Resource Specialization

Adam Smith, commonly called the father of economics, pointed out during the 1700s the economic advantages of **resource specialization**. However, for resource specialization to be feasible, a minimum size unit is required. The small family farm of early America was not very conducive to the specialization of labor. But in recent years, the increasing complexity of modern farming and the trend toward large farm operations have caused the demand for specialized services offered by the private agribusiness services input sector of agriculture to virtually explode.

Private Agribusiness Services

It would be a challenge to discuss all the types of private services offered to production agriculturalists today. Nevertheless, an understanding of the major agencies that provide them is imperative to anyone seeking a minimum level of knowledge about modern agriculture, especially the agribusiness phase. This chapter is designed to provide a basic understanding of the major **agriservices** currently offered to production agriculturalists by private firms, their economic importance to the total agricultural industry, and the major firms that provide these services.

GENERAL FARM ORGANIZATIONS

Farmers learned a long time ago that some problems could be solved better by group rather than individual action. As a result, thousands of farm organizations have been organized during the history of American agriculture. Numerous books have been written on the rise and fall of farm organizations and their impact on the society of their period. History shows that organizations that were formed in response to specific problems usually dissolved after the problems were solved, whereas organizations that sought to deal with a variety of issues and provide flexible current services tended to last for longer periods of time.

Currently, there are five major general farm organizations that are active in representing farmers' interests at the national level: the National Grange, the American Farm Bureau Federation, the National Farmers Union, the National Farmers Organization, and the American Agricultural Movement. There are several other organizations that also support various agricultural issues such as American Agri-Women (AAW), Women Involved in Farm Economics (WIFE), Concerned Farm Wives, Partners in Action for Agriculture, the National Catholic Rural Life Conference (NCRLC), and the Interreligious Task Force on U.S. Food Policy.

The National Grange

The National Grange is the oldest and second largest of the major general farm organizations. It was organized in Washington, D.C., during 1867 as the Patrons of **Husbandry**, but soon became known as the National Grange. It was organized as a **fraternal** organization with special emphasis on strengthening farm families and rural communities. Its activities included a wide range of social, educational, recreational, and economic functions. Although Oliver Hudson Kelley, a former USDA employee, is usually called the founder of the Grange, several other people played a major role in setting up the organization. William Saunders developed the plan of organization from the local to the national order. He suggested the name of "Grange" for the meeting place and the title of "Patrons of Husbandry" for the order.[1] John R. Thompson was primarily responsible for the rituals of the Grange and William Ireland developed the framework of the constitution and bylaws. Aaron Grosh, John Trimble, and Francis McDowell were also major contributors to early Grange development.

Accomplishments. The National Grange has played a major role in almost everything that has happened to improve agriculture and rural America during the last century. Some of the major accomplishments supported by the Grange include rural free delivery of mail, the founding of land grant colleges and experiment stations, the Agricultural Extension Service, vocational agriculture (now called agricultural education) and home economics education, antitrust legislation, the farm credit system, agricultural cooperatives, rural electrification, farm price supports, and numerous others. Today, Grange **policy statements** deal with issues that affect modern agriculture both directly and indirectly, including such areas as transportation, national welfare, conservation, foreign affairs, taxation, education, health, and safety.

Membership. Grange membership approaches 1 million in approximately 6,000 Subordinate Grange units. Although about thirty-five states have Grange organizations, the membership is concentrated primarily in the north-central and northwestern states. The National Grange goals and objectives are summarized in the national policy manual as follows:

> The Grange is more than a farm organization. Its purpose is to serve the total interest of the rural community and the Nation. Thus policies and programs of the Grange pertain to a broad array of circumstances affecting the lives of rural Americans; they result from member action generated by total community and national interest—not agricultural interest alone.[2]

National Farmers Union (NFU)

The Farmers Union was first organized in the Southwest with an emphasis on improving the education level of rural people and promoting agricultural cooperatives. Newt Greshman, a farmer and newspaper editor, along with ten other farmers in Paint, Texas, in 1902 organized the Farmers Educational Cooperative Union, which later became the Farmers Union. By 1903 the organization had spread to Arkansas, Georgia, Louisiana, and Oklahoma, and a national organization was established in 1905.

Intended Purposes. Although one of the early purposes of the Farmers Union was to provide a fraternal organization, the major purpose was to improve the financial conditions for farmers. First the union tried to control prices by setting a price and refusing to sell for less, but this tactic did not work. Second, they tried to limit output in order to improve prices, but this attempt also failed. As one NFU leader declared, "Whenever we tell the farmers to plant less cotton, they plant more."[3] Around 1920 the Farmers Union had over 140,000 members, although bankers, merchants, and lawyers were barred.

Even though the NFU was not very successful in controlling farm prices, it was very influential in other efforts to improve the agricultural cooperative movement, and it also provided farm credit to members. Other farm-marketing programs were developed, especially for cotton, and the NFU was also very influential in programs for agricultural education.

American Farm Bureau Federation

The first county Farm Bureaus were organized between 1912 and 1914, primarily to act as sponsoring agencies for county agricultural agents. As the Farm Bureau movement expanded, local units organized into state groups and finally into a national organization. The first efforts to establish a national Farm Bureau occurred in February 1919, when representatives of twelve state groups met in Ithaca, New York to discuss organizational plans. A second meeting was held, in Chicago in November 1919, to develop a constitution and bylaws. Adoption of the constitution and ratification of the bylaws occurred in March 1920, in Chicago, and the first annual meeting took place in Indianapolis in December 1920. Farm Bureau is the largest of the five major general farm organizations. Membership is based on farm families, and only one membership fee per family is required. Currently, approximately 4.8 million families from fifty states and Puerto Rico are members of Farm Bureau.[4]

Insurance. Although the major effort of Farm Bureau activities has been the representation of American agriculture in the political arena, Farm Bureau provides numerous other services as well. A broad-based insurance company is operated for Farm Bureau members at the state level, and the American Farm Bureau Service Company provides group purchasing of farm supplies under the Safe-mark brand. Distribution is handled locally through about 3,100 outlets in forty-one states and Puerto Rico. Since this service began in ten states in 1965, the volume of purchases has increased over 700 percent.

Marketing. The American Agricultural Marketing Association (AAMA) is the marketing arm for Farm Bureau and was established in 1960 to provide cooperative marketing services for members. This organization is especially active in marketing livestock. Numerous other services are provided by state and local associations depending on the specific needs of the area.

Accomplishments. Farm Bureau has given support, and in many cases has led the battle, for a better life for American farmers. Some of the major areas supported by Farm Bureau include the establishment of rural electricity and telephone services, agricultural education, research and extension services, the federal farm gasoline tax refund, rural development, the establishment of the Farm Credit Administration, increases in the amount of inheritance allowed before taxes are imposed, and numerous others.

National Farmers Organization (NFO)

The most controversial general farm organization is the National Farmers Organization which was organized in Corning, Iowa, in 1955. Jay Loghry, who was a feed salesman, met with thirty-five farmers at Carl, Iowa, to discuss the idea of a new farm organization that would have some input into determining the prices received for farm products. A week later, a meeting was held at Corning, Iowa, at which about 1,200 farmers attended with the idea of "unionizing" farmers. However, the term *union* received a negative response, and instead the group adopted the name National Farmers Organization.[5] Loghry was hired to assist in the development of the organization, and Dan

Turner, a former Iowa governor, was selected as an unpaid advisor.

Strategy to Improve Farm Prices. The first national meeting of the National Farmers Organization was held at Corning, Iowa, in December 1955, with about 750 members attending. Over 55,000 members were listed at that time, and Oren Lee Staley was elected president. By April 1956, the organization had expanded to 140,000 members. The advisor to NFO, former governor Dan Turner, recommended that they seek action on improving farm prices through political means, and they followed his advice in the beginning with several visits to Washington and a major effort to elect NFO members to national office. However, by convention time in 1957, the members had decided that this technique was not successful, so they turned to **collective bargaining** to improve farm prices.

Holding Action. The first **holding action** occurred in 1959 to test market reaction and to give other farmers the opportunity to join the organization. This action involved livestock only and lasted only seven days because it was confined to a small area around St. Joseph, Missouri. Additional livestock holding actions were conducted in 1960, 1961, and 1962. Prices increased for a short period of time after each holding action, but very few, if any, long-term gains were realized. The major impact of the holding actions resulted from the violence that occurred during this period. Rifle shots were fired at trucks hauling livestock to markets, tires were slashed, and nails were spread on highways.

In 1967, NFO members decided to apply their collective bargaining techniques to milk with a twenty-five-state holding action. Thousands of gallons of milk were poured on the ground, but the action had less effect on the market than anticipated. Again numerous acts of violence were reported. Since that time, several other minor holding actions were tried, with soybeans, grain, pork, and dairy products, which had similar results.

Price Improvement Challenge. Currently, NFO is seeking new members and encouraging the participation of previously inactive members in their bargaining programs. NFO estimates that 30 percent of the total production in any commodity would be sufficient to allow farmers to set their own prices, just as nonfarm businesses do. NFO is attempting to improve farm prices by cooperative marketing programs, with a major emphasis on transporting farm commodities in bulk quantities from surplus to shortage areas. NFO officials contend that:

> When the volume of farm commodities moving through and committed to NFO programs reaches 30 percent, meetings will be called in no larger than ten-county areas and members will vote on their price based on cost of production plus a reasonable profit.
>
> These prices will be announced and contracts will be presented to the companies who handle and process farm production. If the companies won't agree to sign contracts to maintain prices based on cost of production plus a reasonable profit, NFO will advise members to keep their products on their farms until they do get their prices and their contracts.[6]

American Agriculture Movement (AAM)

The American Agriculture Movement (AAM) began as a movement rather than an organization and operated several years without any formal organization. AAM members were primarily disenchanted with NFO and National Farmers Union members, which were grain and cattle producers from Texas, Colorado, Kansas, North and South Dakota, and Georgia. They were generally young farmers who had expanded production during the early 1970s and were now experiencing difficulty

General Farm and Service Organizations

Agriculture Council of America (ACA)
1625 I Street, NW
Washington, DC 20006

Agricultural Trade Council
1028 Connecticut Avenue, NW
Washington, DC 20036

American Farm Bureau Federation
225 West Touhy Avenue
Park Ridge, IL 60068

Farm and Land Institute
430 N. Michigan Avenue
Chicago, IL 60611

National Agricultural Transportation League
Box 960
Umatilla, FL 32784

National Farm Workers of America
La Paz
Keene, California 93531

National Farmers Organization
720 Davis Avenue
Corning, IA 50841

National Farmers Union
Box 2251
Denver, CO 80201

National Grange
1616 H Street, NW
Washington, DC 20006

Figure 10–1   Home office addresses of the major farm organizations and general agricultural service associations.

because of low prices for their products. When the cost-price squeeze became severe in 1976 and 1977, these farmers turned to Washington for relief.

The major technique for attracting attention was a "tractorcade" to Washington, where they camped for several weeks on the Capitol Mall and disrupted traffic with their tractors. The goal of the AAM in Washington was to obtain price supports for farm products above the level guaranteed by the 1976 farm bill.

The early efforts of AAM obtained considerable support in rural America, but the tactics soon lost favor with other farm organizations. In recent years, AAM has officially organized as a national farm organization, opened an office in Washington, and operates through acceptable channels like the other organizations. Refer to Figure 10–1 for the addresses of the five major organizations and six other associations providing farm general agricultural services.

AGRICULTURAL COMMODITY ORGANIZATIONS

Although most of the earlier farm organizations were general farm organizations, production agri-

culturalists soon learned that certain types of farmers had more in common than others. For example, wheat farmers had more problems in common than livestock producers, who had other problems. As a result, production agriculture organizations began to develop around farm commodity lines. The reasoning was that a specialized commodity organization could represent special commodity interest better than a general farm organization. As a result, specialized farm commodity organizations began to organize rapidly during the early part of the twentieth century.

Services Performed

The agricultural commodity organizations render a wide variety of services to their members in areas such as public relations, promotions, legislative lobbying, communications, sales training, auditing and record keeping, publicity, transportation, conducting research, and providing legislative and marketing information.[7] The staff of the agricultural commodity organization keeps in close touch with legislation directly affecting the industry. Special reports are made to special groups. A weekly or monthly newsletter or magazine is published to give news of general importance to the particular commodity group. Other things that may be done by commodity groups are as follows:

- 🌾 efforts to increase prices
- 🌾 train producers on how to handle, store, produce, and market commodity
- 🌾 help wholesalers with problems of transportation, trucking, materials handling, and warehousing
- 🌾 foster and improve trade relations among producers, distributors, and consumers
- 🌾 keep up-to-date on environmental issues
- 🌾 promote development of new products
- 🌾 render all services possible to members of their group
- 🌾 organize research
- 🌾 study legislation and government policy concerning it
- 🌾 aid in developing domestic and export markets
- 🌾 demonstrate and secure adoption of special accounting systems
- 🌾 conduct market surveys
- 🌾 provide government regulatory information[8]

Names of Selected Commodity Groups

Farm Commodity Groups. Today, almost every major farm commodity has one or more organizations operating at the state and/or national levels. Some of the larger ones include the National Cattle Beef Association, the American Dairy Association, the National Association of Wheat Growers, the National Cotton Council, the National Potato Council, and numerous others. These organizations keep their members informed on information relevant to the production and marketing of their products. In some cases, the **political power** of the commodity organization exceeds the power of some of the general farm organizations.

Agribusiness Commodity Groups. The organization of special commodity groups is not limited to production agriculture organizations. Some of the more **prominent** ones are the Farm and Industrial Equipment Institute, the Fertilizer Institute, National Independent Meat Packers, the National Canners Association, the National Association of Food Chains, Grocery Manufacturers of America, and numerous others. These serve basically the same function as the special farm organizations. The trend toward the founding of special group organizations is expected to continue for production agriculture and agribusiness groups. Refer to Figures 10–2 through 10–10 for names and addresses of many of the selected commodity groups and breed associations.

AGRICULTURAL RESEARCH

Traditionally, farmers have not been involved to a large degree in basic agricultural research, primarily because they cannot afford to be involved. Farm profits are not sufficient to invest in research because of the high degree of competition between production agriculturalists. Also, since the products produced by production agriculturalists are **homogeneous**, the benefits of new products cannot be controlled by those who invest the funds. As a result, production agriculturalists have depended on public funds for farm research since the long-term benefits help primarily the agribusiness firms and the consumer.

Agribusiness firms have benefited to a high degree from public agricultural research, and they have also invested huge amounts in private research. Profits for agribusiness firms have been high enough to justify research expenditures, and patent laws allow them to restrict the use of the new products thus developed. Private agribusiness firms are continuously looking for new and better

General Crop Associations

American Corn Millers Federation
1030 Fifteenth Street, NW,
Suite 912
Washington, DC 20005

American Cotton Shippers Association
318 Cotton Exchange Building
P. O. Box 3366
Memphis, TN 38103

American Rice Growers Cooperative
Association
211 Pioneer Building
Lake Charles, LA 70601

American Soybean Association
Box 158
Hudson, IA 50643

American Sugar Cane League of the USA,
Inc.
416 Whitney Building
New Orleans, LA 70130

American Sugar Beet Growers Association
1776 K Street, NW, Suite 900
Washington, DC 20006

Burley Tobacco Growers Cooperative
Association
620 South Broadway
Lexington, KY 40508

Durum Wheat Institute
710 North Rush Street
Chicago, IL 60611

Federated Pecan Growers Association of
the U.S.
Drawer AX, University Station
Baton Rouge, LA 70803

Flue-Cured Tobacco Cooperative
Stabilization Corporation
Box 12300
Raleigh, NC 27605

Grain Sorghum Producers Association
1708-A Fifteenth Street
Lubbock, TX 79401

Leaf Tobacco Exporters Association
P.O. Box 1288
Raleigh, NC 27602

National Association of Tobacco Distributors
58 East 79th Street
New York, NY 10021

National Association of Wheat Growers
1030 Fifteenth Street, NW, Suite 1030
Washington, DC 20005

National Cigar Leaf Tobacco Association
1100 Seventeenth Street, NW
Washington, DC 20036

National Corn Growers Association
P.O Box 358
Boone, IA 50036

National Cotton Council of America
1918 North Parkway, P.O. Box 12285
Memphis, TN 38112

National Cottonseed Products Association
P.O. Box 12023
Memphis, TN 38112

National Council of Commercial Plant
Breeders
1030 Fifteenth Street, NW, Suite 964
Washington, DC 20005

National Grain Trade Council
725 Fifteenth Street, NW, Suite 604
Washington, DC 20005

National Hay Association, Inc.
P.O. Box 1059
Jackson, MS 49204

National Onion Association
5701 East Evans Avenue, Suite 26
Denver, CO 80222 C

National Pecan Shellers and Processors
Association
111 East Wacker Drive, Room 600
Chicago, IL 60601

National Potato Council
Montbello Office Campus,
Suite 301
45th and Peoria
Denver, CO 80239

National Potato Promotion Board
1385 S. Colorado Boulevard
Denver, CO 80222

National Rice Growers Association
Box 683
Jennings, LA 70546

National Soybean Processors Association
1800 M Street, NW
Washington, DC 20036

National Sugar Beet Growers Federation
Greeley National Plaza, Suite 740
Greeley, CO 80631

North American Export Grain Association,
Inc.
1800 M Street, NW, Suite 610 N
Washington, DC 20036

Potato Association of America, The
114 Deering Hall
University of Maine
Orono, ME 04473

Rice Council of America
Box 22802
Houston, TX 77027

Shade Tobacco Growers Agricultural
Association, Inc., The
River Street, Box 38
Windsor, CT 06095

The Tobacco Institute
1776 K Street, NW
Washington, DC 20006

Tobacco Association of the U.S.
Box 1288
Raleigh, NC 27602

Tobacco Growers' Information Committee,
Inc.
Cameron Village Station Box 12046
Raleigh, NC 27605

Wheat Flour Institute
1776 F Street, NW
Washington, DC 20006

Wild Rice Growers Association
Box 366 Highway 210 West
Aitkin, MN 56431

Figure 10–2 Home office addresses of general crop associations.

General Dairy Associations

American Butter Institute, Inc.
110 N. Franklin Street
Chicago, IL 60606

American Dairy Association
6300 N. River Road
Rosemont, IL 60018

American Dairy Science Association
113 N. Neil Street
Champaign, IL 61820

American Dry Milk Institute, Inc.
130 N. Franklin Street
Chicago, IL 60606

Milk Industry Foundation
1105 Barr Building
910 Seventeenth Street, NW
Washington, DC 20006

National Dairy Council
6300 N. River Road
Rosemont, IL 60018

National Independent Dairies
Association
1730 K Street, NW, Suite 1302
Washington, DC 20006

National Milk Producers Federation
30 F Street, NW
Washington, DC 20001

Purebred Dairy Cattle Association,
Inc.
Box 126
Peterborough, NH 03458

United Dairy Industry Association
6300 N River Road
Rosemont, IL 60018

Whey Products Institute
130 N. Franklin Street
Chicago, IL 60606

Figure 10–3   Home office addresses of general dairy associations.

Dairy Breed Associations

American Guernsey Cattle Club
Peterborough, NH 03458

American Jersey Cattle Club
2105-J S. Hamilton Road
Columbus, OH 43227

American Milking Shorthorn Society
1722JJ S. Glenstone Avenue
Springfield, MO 65802

Ayrshire Breeders' Association
Brandon, VT 05733

Brown Swiss Cattle Breeders' Association
Box 1038
Beloit, WI 53511

Dutch Belted Cattle Association of America
Box 358
Venus, FL 33960

Holstein-Friesian Association of America
Box 809
Brattleboro, VT 05301

International Illawarra Association
313 S. Glenstone Avenue
Springfield, MO 65802

Figure 10–4   Home office addresses of dairy breed associations.

products, and many firms maintain extensive research facilities.

Areas of Agricultural Research

Research is an essential part of agribusiness. Research is conducted on all agricultural commodities and many other areas of the agricultural industry, including conservation, machinery, equipment, forestry, management, marketing, process-ing, **rural sociology**, and soils. The amount spent on private research is comparable to that spent by the USDA.

CONSULTING

Similar to research is agricultural consulting, since much of a consultant's job is research based.

Light Horse Breed Associations

American Bashkir Curly Registry
Box 453
Ely, NV 89301

American Albino Association
Box 79
Crabtree, OR 97335

American Mustang Association, Inc.
Box 338
Yucaipa, CA 92399

American Saddle Horse Breeders
Association, Inc.
929 S Fourth Street
Louisville, KY 40203

American Albino Association
Box 79
Crabtree, OR 97335

American Andalusian Association
Box 1290
Silver City, NM 88061

Appaloosa Horse Club, Inc.
Box 8403
Moscow, ID 83843

Arabian Horse Registry of America, Inc.
3435 S. Yosemite
Denver, CO 80231

American Buckskin Registry Association,
Inc.
Box 1125
Anderson, CA 96007

International Buckskin Horse Association,
Inc.
Box 357
St. John, IN 46373

Chickasaw Horse Association, Inc.
Box 8
Love Valley, NC 28677

National Chickasaw Horse Association
Route 2
Clarinda, IA 51232

Cleveland Bay Association of America
Middleburg, VA 22117

Galaceno Horse Breeders Association, Inc.
111 East Elm Street
Tyler, TX 75701

American Hackney Horse Society
Box 174
Pittsfield, IL 62363

American Hanoverian Society
809 W 106th Street
Carmel, IN 46032

Hungarian Horse Association
Bitterroot Stock Farm
Hamilton, MT 59840

American Lipizzan Horse Registry
Box 415
Platteville, WI 53818

Royal International Lipizzaner Club of
America
Route 7
Columbia, TN 38401

Missouri Fox Trotting Horse Breed
Association, Inc.
Box 637
Ava, MO 65608

Morab Horse Registry of America
Box 413
Clovis, CA 93613

American Morgan Horse Association, Inc.
Box 1
Westmoreland, NY 13490

Morocco Spotted Horse Cooperative
Association of America
Route 1
Ridott, IL 61067

American Paint Horse Association
Box 13486
Fort Worth, TX 76118

Palomino Horse Association, Inc.
Box 324
Jefferson City, MO 65101

Palomino Horse Breeders of America
Box 249
Mineral Wells, TX 76067

American Paso Fino Horse Association Inc.
Mellon Bank Building, Room 3018
525 William Penn Place
Pittsburgh, PA 15219

Paso Fino Owners and Breeders
Association, Inc.
Box 764
Columbus, NC 28722

American Association of Owners and
Breeders of Peruvian Paso Horses
Box 2035
California City, CA 93505

Peruvian Paso Horse Registry of North
America
Box 816
Guerneville, CA 95446

Pinto Horse Association of America, Inc.
Box 3984
San Diego, CA 92103

American Quarter Horse Association
Box 200
Amarillo, TX 79105

Standard Quarter Horse Association
4390 Fenton
Denver, CO 80212

Colorado Ranger Horse Association, Inc.
7023 Eden Mill Road
Woodbine, MD 21797

Spanish-Barb Breeders Association
Box 7479
Colorado Springs, CO 80907

Spanish Mustang Registry, Inc.
Route 2, Box 74
Marshall, TX 75670

United States Trotting Association
750 Michigan Avenue
Columbus, OH 43215

National Trotting and Pacing Association,
Inc.
575 Broadway
Hanover, PA 17331

Tennessee Walking Horse Breeders and
Exhibitors Association of America
Box 286
Lewisburg, TN 37091

The Jockey Club
300 Park Avenue
New York, NY 10022

American Trakehner Association, Inc.
Box 268
Norman, OK 73069

Ysabella Saddle Horse Association, Inc.
Prairie Edge Farm
Route 3
Williamsport, IN 47993

Figure 10–5 Home office addresses of light horse breed associations.

Draft Horses, Ponies, and Jacks, Donkeys, and Mules

Draft Horses

Belgian Draft Horse Corporation of
America
Box 335
Wabash, IN 46992

Clydesdale Breeders Association of
the U.S.
Route 3
Waverly, IA 50677

Percheran Horse Association of
America
Route 1
Belmont, OH 43718

American Shire Horse Association
6960 Northwest Drive
Ferndale, WA 98248

American Suffolk Horse Association,
Inc.
672 Polk Boulevard
Des Moines, IA 50312

Ponies

American Gatland Horse Association
Route 2, Box 181
Elkland, MO 65644

American Walking Pony Association
Route 5, Box 88
Upper River Road
Macon, GA 31201

American Cannemara Pony Society
Route 1, Hoshiekon Farm
Goshen, CT 06756

National Appaloosa Pony, Inc.
Box 296
Gaston, IN 47342

Pony of the Americas Club
Box 1447
Mason City, IA 50401

American Shetland Pony Club
Box 435
Fowler, IN 47944

Welsh Cob Society of America
Grazing Fields Farm
225 Head of the Bay Road
Buzzards Bay, MA 02532

Welsh Pony Society of America
Box A
White Post, VA 22663

Jacks, Donkeys, and Mules

Standard Jack and Jennet Registry of
America
300 Todds Road
Lexington, KY 40511

Miniature Donkey Registry of the U.S.,
Inc.
1108 Jackson Street
Omaha, NB 68102

American Donkey and Mule Society,
Inc.
2410 Executive Drive
Indianapolis, IN 46241

Figure 10–6 Home office addresses of draft horse, pony, and jack, donkey, and mule associations.

During recent years, private companies have emerged that offer technical services to production agriculturalists. Technical services are provided in a variety of areas. Crop and livestock consulting, laboratory analysis, and research services are offered to production agriculturalists. Some companies offer consulting in a specialized area, such as crop consulting, while others offer a full range of services.[9]

Crop Consulting

Many consulting jobs are performed in the crop area. Soil sampling and fertilizer recommendations; field checks including insect, weed and disease control recommendations; and equipment adjustments are offered to production agriculturalists. Tests conducted in a laboratory by consultants may include soil, feed, water, or plant tissue tests.[10]

Beef Breed Associations

Amerispain Corporation
Box 4029
San Angelo, TX 76902

American Angus Association
3201 Frederick Boulevard
St. Joseph, MO 64501

International Ankina Association
Route 1, Box 323
Liberty, KY 42539

Barzolia Breeders Association of
America
Box 142 1
Carefree, AZ 85331

American Beefalo Association
200 Semanin Building
4812 U.S. Highway 42
Louisville, KY 40222

Beef Friesian Society
210 Livestock Exchange Building
Denver, CO 80216

Beefmaster Breeders Universal
G.P.M. South Tower, Suite 720
800 NW Loop 410
San Antonio, TX 78216

Foundation Beefmaster Association
Livestock Exchange Building, Suite
200
4701 Marion Street
Denver, CO 80216

National Beefmaster Association
817 Sinclair Building
Fort Worth, TX 76102

Belted Galloway Society, Inc.
Box 5
Summitville, OH 43962

American Blonde d'Aquitaine
Association
217 Livestock Exchange Building
Denver, CO 80216

International Braford Association, Inc.
Box 1030
Fort Pierce, FL 33450

American Brahman Breeders
Association
1313 LaConcha Lane
Houston, TX 77054

International Brangus Breeders
Association, Inc.
9500 Tiaga Drive
San Antonio, TX 78230

American-International Charalois
Association
1610 Old Spanish Trail
Houston, TX 77025

American Chianina Association
Box 11537
Kansas City, MO 64138

Devon Cattle Association, Inc.
Box 628
Walde, TX 78801

American Dexter Cattle Association
707 W. Water Street
Decorah, IA 52101

American Galloway Breeders
Association
302 Livestock Exchange Building
Denver, CO 80216

Galloway Cattle Society of America
Springville, IA 52336

American Gelbvieh Association
202 Livestock Exchange Building
Denver, CO 80216

American Hereford Association
715 Hereford Drive
Kansas City, MO 64105

International Illawarra Association
313 S Glenstone Avenue
Springfield, MO 65802

North American Limonsin Foundation
100 Livestock Exchange Building
Denver, CO 80216

American Maine-Anjou Association
564 Livestock Exchange Building
Kansas City, MO 64102

American International Marchigiana
Society
Route 2, Box 65
Lindale, TX 75771

American Murray Grey Association
1222 N. 27th Street
Billings, MT 59102

American Normande Association
Box 350
Kearney, MO 64060

North American Norwegian Red
Association
Route 1, Box 346
Burns, TN 37029

American Pinzgauer Association
1415 Main Street
Alamosa, CO 81101

American Polled Hereford Association
4700 East 63rd Street
Kansas City, MO 64130

American Polled Shorthorn Society
8288 Hascall Street
Omaha, NB 68124

American Simbrah Association
1 Simmental Way
Bozeman, MT 59715

American Shorthorn Association
8288 Hascall Street
Omaha, NB 68124

(continued on following page)

Figure 10–7   Home office addresses of beef breed associations.

Beef Breed Associations (concluded)

American Simmental Association
1 Simmental Way
Bozeman, MT 59715

American South Devon Association
Box 248
Stillwater, MN 55082

International South Devon Association
Lynnville, IA 50153

Sussex Cattle Association of America
Box AA
Refugio, IX 78377

American Tarentaise Association
Box 1844
Fort Collins, CO 80522

Texas Longhorn Breeders Association
of America
204 Alamo Plaza
San Antonio, TX 78205

United States Welsh Black Cattle
Association
Route 1
Wahkon, MN 56386

National Buffalo Association
Custer State Park
Hermosa, SD 57744

Ranger Cattle Company
N. Pecos Station, Box 21300
Denver, CO 80221

Red Angus Association of America
Box 776
Denton, TX 76201

American Brangus Association
Box 1326
Austin, TX 78767

American Red Poll Association
3275 Holdrege Street
Lincoln, NB 68503

Red Poll Beef Breeders International
Dry Run Farm
Ross, OH 45061

American Romagnola Association
Box 8172
St. Paul, MN 55113

American Salers Association
Box 30
Weiser, ID 83672

Santa Gertrudis Breeders
International
Box 1257
Kingsville, TX 78363

American Scotch Highland Breeders
Association
Box 240
Walsenburg, CO 81089

Livestock Consulting

Livestock consulting may include the formulation of feed rations, computerized budgeting, computerized feed systems, automatic silage feeding, and testing of feed supplies. Agricultural consultants visit the farm or feedlot to check on livestock performance, feed ingredients, analyze their findings, and make recommendations with the goal of improving livestock production.[11] Consultants also screen and evaluate new products, evaluate irrigation systems, and work with production agriculturalists in many other areas.

ARTIFICIAL INSEMINATION

One of the major special services offered to dairy and livestock producers today is **artificial insem-** **ination**. Although some production agriculturalists **inseminate** their own livestock, most animals are bred by local technicians who are trained specifically for this service. For those dairy and livestock producers who prefer to do their own breeding, specialized equipment and **semen** are available.

Historical Developments

Although the use of artificial insemination was reported in several instances in the eighteenth century, it was first developed for practical application in livestock breeding in 1922 by a Russian **physiologist**. The first large-scale use of artificial insemination in the United States began in the 1930s. Two of the major developments that stimulated the use of artificial insemination were the discoveries that egg yolk was valuable as a semen

Sheep Breed Associations

American Cheviot Sheep Society
Route 1
Carlisle, IA 50047

Columbia Sheep Breeders
Association of America
Box 272
Upper Sandusky, OH 43351

American Corriedale Association, Inc.
Box 29C
Seneca, IL 61360

American Cotswold Record
Association
11903 Milwaukee Road
Britton, MI 49229

Debouillet Sheep Breeders
Association
2662 University
San Angelo, TX 76901

American and Delaine Merino Record
Association
Aleppa, PA 15310

Black Top and National Delaine
Merino Sheep Association
290 Beech Street
Muse, PA 15350

Texas Delaine Sheep Association
Route 1
Burnet, TX 78611

Continental Dorset Club, Inc.
Box 577
Hudson, IA 50643

Finnsheep Breeders Association
Route 3
Pipestone, MN 56164

American Hampshire Sheep Association
Route 10, Box 199
Columbia, MO 65201

Empire Karakul Registry
Route 1
Fabius, NY 13063

National Lincoln Sheep Breeders
Association
Route 6, Box 24
Decatur, IL 62521

Montadale Sheep Breeders Association.,
Inc.
Box 44300
Indianapolis, IN 46244

American North County Cheviot Sheep
Association
Aitchner Hall, University of Maine
Orono, ME 04473

American Oxford Down Record
Association
Box 401
Burlington, WI 53105

American Panama Registry Association
Route 2
Jerome, IN 83338

American Rambouillet Sheep
Breeders Association
2709 Sherwood Way
San Angelo, TX 76901

American Romney Breeders
Association
212 Withycombe Hall
Corvallis, OR 97331

American Shropshire Registry
Association, Inc.
Box 1970
Monticello, IL 61856

American Southdown Breeders
Association
Route 4, Box 14B
Bellefonte, PA 16283

American Suffolk Sheep Society
55 East 100 North
Logan, UT 84321

National Suffolk Sheep Association
Box 324
Columbia, MO 65201

U.S. Targhee Sheep Association
Box 40
Absarokee, MT 59001

National Tunis Sheep Registry
1826 Nesbitt Road
Attica, NY 14011

Goat Breed Associations

American Dairy Goat Association*
P O Box 865
Spindale, NC 28160

American Angora Goat Breeders
Association
P.O. Box 195
Rock Springs, TX 78880

International Nubian Breeders
Association
4496 Main Road West
Emmaus, PA 18094

*Note: Information on the following breeds can be attained here: Alpine, La Mancha, Oberhasli, Saanow, Toggenburg

Figure 10–8   Home office addresses of sheep and goat breed associations.

Swine Breed Associations

American Berkshire Association
601 W Monroe Street
Springfield, IL 62704

Chester White Swine Record
Association
Box 228
Rochester, IN 46975

Conner Prairie Swine, Inc.
14320 River Avenue
Noblesville, IN 46060

United Duroc Swine Registry
1803 W. Detweiller Drive
Peoria, IL 61614

National Large Black Swine Breeders
Association
Route 1, Box 44
Midland, NC 28107

Hampshire Swine Registry
1111 Main Street
Peoria, IL 61606

National Hereford Hog Record
Association
Route 2
Lena, IL 61048

Inbred Livestock Registry Association
14320 River Avenue
Noblesville, IN 46060

American Landrace Association, Inc.
Box 111
Culver, IN 46511

OIC Swine Breeders Association
Route 2, Box 166
Cloverdale, IN 46129

Poland China Record Association
Box 71
Galesburg, IL 61401

National Spotted Swine Record, Inc.
110 W. Main Street
Bainbridge, IN 46105

Tamworth Swine Association
Route 2, Box 126A
Killsboro, OH 45133

Wessex Saddleback Swine
Association
4010 Clinton Avenue
Des Moines, IA 50310

American Yorkshire Club, Inc.
Box 2417
West Lafayette, IN 47906

Figure 10–9 Home offices addresses of swine breed associations.

extender and that semen could be frozen without damage in a solution of **glycerol**, **sodium citrate**, and egg yolk. Artificial insemination is used predominantly with dairy cows, but beef and horse producers also use this service.

Makeup of the Artificial Insemination Industry

The artificial insemination industry is made up of semen collectors, distributors of semen and insemination supplies, and local service technicians. In the early years of the industry, most major artificial insemination firms were both collectors and distributors. However, through mergers and affiliations, the industry is now made up of a small number of semen collectors and a large number of distributors and technicians.

Artificial-Breeding Companies

The major collectors are Curtis Breeding Service, a division of Searle Agriculture, Inc.; Carnation Genetics, a division of Carnation Milk Company; American Breeding Service, a subsidiary of W. R. Grace Chemical Company; and Select Sires, a cooperative corporation owned and controlled by several regional and local service cooperatives. Each of these companies collects and sells semen from a wide variety of superior sires from all beef and dairy breeds. Many people believe artificial insemination to be one of this century's greatest technologies. Refer to Figure 10–11.

Miscellaneous Agribusiness Service Organizations

Agricultural Research Institute, Inc.
2100 Pennsylvania Avenue, NW
Washington, DC 22037

American Beekeeping Federation
Route 1, Box 68
Cannon Falls, MN 55009

American Egg Board
205 Touhy Avenue
Park Ridge, IL 60068

American Mushroom Institute
Box 373
Kennett Square, PA 19348

American Poultry Association, Inc.
Box 70
Cushing, OK 74023

American Rabbit Breeders
Association
1925 S Main Street, Box 426
Bloomington, IL 61701

American Seed Trade Association
1030 Fifteenth Street, NW
Washington, DC 20005

National Association of Animal Breeders
Box 1033
Columbia, MO 65201

National Livestock and Meat Board
444 N. Michigan Avenue
Chicago, IL 60611

National Livestock Producers Association
307 Livestock Exchange Building
Denver, CO 80216

National Livestock Exchange
Box 552
Memphis, TN 38101

National Association of Swine Records
112 1/2 N Main Street
Culver, IN 46511

National Pork Producers Council
4715 Grand Avenue
Des Moines, IA 50312

Agricultural Research Institute, Inc.
2100 Pennsylvania Avenue, NW
Washington, DC 22037

Poultry and Egg Institute of America
425 Thirteenth Street, NW
Washington, DC 20004

Poultry Science Association, Inc., The
University of Missouri
Columbia, MO 65201

United Egg Producers
3951 Snapfinger Parkway, Suite 580
Decatur, GA 30035

Figure 10–10 Home office addresses of miscellaneous agribusiness service organizations.

VETERINARY SERVICES

Although veterinary services have been in demand since the domestication of animals, the veterinary profession is not that old. The term *veterinarian* was first introduced in English writing in 1646 by Sir Thomas Browne, but the term was not generally accepted until 1846. Previously, veterinarians were called "cow doctors," "horse doctors," and similar names. One of the first indications of veterinary services in America was recorded by a medical historian named Blanton who referred to a William Carter as an expert "cow doctor" who lived in James City, Virginia in 1625. Richard Peters, a prominent patriot, lawyer, and country gentleman, was the first public figure to request the establishment of the veterinary profession in America.[12]

Father of Veterinary Medicine

John Busteed is often regarded as the "Father of Veterinary Medicine." He chartered the New York College of Veterinary surgeons in 1857 and operated it until 1899. He was also the major supporter of the veterinary profession at that time. The profession also received a boost when the American Veterinary Medical Association was organized in 1863. The first major goal of the association was to expand the number of veterinary colleges in the United States.

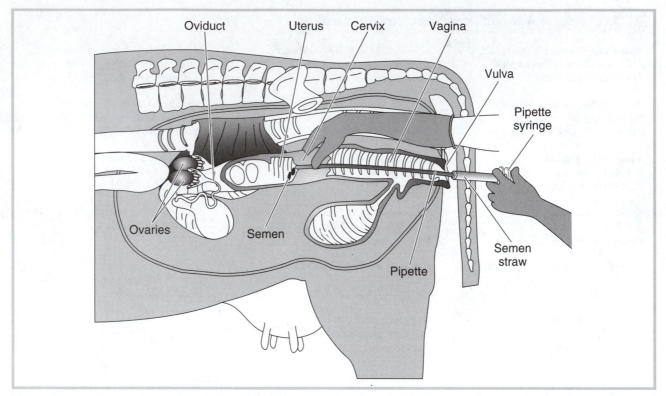

Figure 10–11 The diagram above shows the artificial insemination process. Semen is deposited from the tube just inside the uterus.

First Veterinary Schools

The first veterinary schools were private ventures, and during the latter part of the nineteenth century, the private schools flourished. However, they experienced numerous problems, and when pressures for veterinary research became strong, private schools could not adjust to the extra cost and were replaced by public schools. The last private veterinary school closed in 1927.

First College of Veterinary Medicine

The first public support in veterinary education was initiated in the form of special courses and programs, the first one being offered by Cornell University. The first college of veterinary medicine was established in 1879 by Iowa State University,

and the second was formally established at the University of Pennsylvania in 1884. Other veterinary schools then developed rapidly throughout the United States.

Present-Day Veterinarians

Currently, there are about 38,000 veterinarians providing a wide variety of services in private practice, teaching, research, public health, military service, private industry, and other special settings. The federal government employs about 2,000 veterinarians, primarily in the U.S. Department of Agriculture and the Public Health Service.[13] Veterinarians provide a valuable service to production agriculturalists. Among their many duties are diagnosis of the animals' medical prob-

Figure 10–12 Most veterinarians operate small animal clinics, but they are also employed in teaching, research, public health, private industry, and many other special services. (Courtesy of Cliff Ricketts)

lems, performing surgery, and prescribing and administering medicines and drugs. Veterinarians may specialize in large animals, small animals, or even exotic or zoo animals. However, most veterinarians today operate small animal clinics.[14] Other veterinarians engage in research, food safety inspection, or education. They also work in **regulatory medicine** and public health in the areas of food inspection, investigations of the outbreak of disease, or research work in scientific laboratories. All states require their veterinarians to be licensed. Applicants for licensing must have a Doctor of Veterinary Medicine degree and pass written and oral state board examinations.[15] Refer to Figure 10–12. In addition to careers in the field of veterinary medicine, there are many careers available as breed association representatives and artificial insemination technicians. Refer to the Career Option on page 266 for a further explanation of these careers.

AGRICULTURAL COMMUNICATIONS

Communications through the print media, computers, Internet, pictures, radio, television, videos, exhibits, and so forth are a part of all major segments of modern business and industry. Agribusiness is no exception. Many opportunities await those who have a good agricultural background and can combine it with the skills needed in communication. This part of the agribusiness private input sector also includes advertising; photography; books, magazines, and newspapers; trade publications; computer/Internet-related services; and public relations as well as many other areas.

Few things offer more excitement, challenge, and opportunity to keep up-to-date with every segment of agriculture and all its related activities than agricultural communications. A person working in agricultural communications may interview production agriculturalists, scientists, and executives; attend conventions, demonstrations, and legislative sessions; report new developments, ideas, and trends; and report in general on anything that is of interest to the agricultural industry. Refer to Figure 10–13 for a list of the major publications in the agricultural industry.

AGRICULTURAL COOPERATIVES

A cooperative is a business that is financed, owned, and controlled by the people who use it. Cooperatives are made up of individuals who have similar interests, desires, and problems. By working together, they combine their investments and influence. This gives them financial strength, greater independence, and a stronger voice in their own business affairs.

History and Development of Cooperatives

During America's frontier days, friends and neighbors traveled from miles around to enjoy the fellowship of others while they helped a family

CAREER OPTION

Career Areas: Breed Association Representative, Artificial Insemination Technician, Veterinarian, Animal Health Technician, Agricultural Communications

There are many career opportunities as field representatives for breed associations, feed companies, marketing cooperatives, supply companies, animal health products, herd improvement associations, and financial management firms. Also, since the greatest improvements in animal production and performance have come about as a result of selection and breeding, many jobs are available, such as artificial insemination technician. Similarly, many individuals utilize a background in the agricultural industry by pursuing writing, publishing, and telecasting careers.

In order to sell what is produced in this country every year, it is imperative that the buying public be aware of the available products. Communication is the key to accomplishing this task, and specialists in the field of communications are in high demand, ranking twenty-fifth on the U.S. Department of Labor, Bureau of Labor Statistics 1994–1995 National Industry–Occupation Employment Matrix.

In the agriculture industry, communications specialists span a wide area of careers. An agriculture communications expert is knowledgeable about all aspects of production and can relate information in a positive manner. This position requires a bachelor's degree in public relations or agricultural communications and a good working understanding of agricultural production. Agricultural industry councils, such as those representing the beef, milk, or pork industries, employ many communications specialists as spokespersons, advertising salespeople, marketers, and advertising copywriters.

Veterinarians and their associates work constantly to treat animals with disorders and to improve animal health. The desire to become a veterinarian has been high among youth in recent years. This interest has permitted the supply of veterinarians to increase despite the rigor of the college curriculum and stiff competition to get into veterinarian schools. Colleges of animal sciences offer curriculum in other career areas in animal sciences, such as nutrition, breeding, education, and production.

In urban and suburban areas, animal shelters, hospitals, kennels, and pet stores provide many career opportunities for people interested in becoming an animal health technician. In rural areas, large-animal veterinarian, veterinary assistant, laboratory veterinarian, and laboratory technician are typical positions in the animal health industry.

A career as a veterinarian is both challenging and rewarding for those willing to accept the rigor of the college curriculum and competition to get into veterinarian schools. (Courtesy of USDA)

General Farm Magazines

Publication	Address
Agribusiness Report	The Miller Publishing Company 2501 Wayzata Boulevard Minneapolis, MN 55440
Agri-Fieldman	Meister Publishing Company 37841 Euclid Avenue Willoughby, OH 44099
Agri-Finance	5520 Touhy Avenue, Suite G Skokie, IL 60076
Big Farmer	Big Farmer, Inc. 131 Lincoln Highway Frankfort, IL 60423
Doan's Agricultural Report	8900 Manchester Road St. Louis, MO 63144
Farm Journal	230 W. Washington Square Philadelphia, PA 19105
Progressive Farmer	P O Box 2581 Birmingham, AL 35202
Successful Farming	1716 Locust Des Moines, IA 50336
The Drovers Journal	1 Gateway Center Fifth and State Kansas City, KS 66101
Feedstuffs	Box 67 Minneapolis, MN 55440

Note: Numerous other agricultural magazines are offered at the regional level, especially in the livestock industry. In addition, almost all livestock breeder associations publish one or more breeder magazines which can be obtained from the various breeder associations.

Agricultural Communications Groups

American Agricultural Editors Association
2001 Spring Road
Oak Brook, IL 60521

American Association of Agricultural College Editors
23 Agricultural Administration Building
Ohio State University
2120 Fyffe Road
Columbus, OH 43210

National Association of Farm Broadcasters
Box 119
Topeka, KS 66601

Newspaper Farm Editors of America
4200 Twelfth Street
Des Moines, IA 50313

Figure 10–13 Home office addresses of general farm magazines and agricultural communications groups.

"raise" a barn or build a house. There were also afternoons when women gathered at quilting bees, each sewing a patch to contribute to the project. Although these were not formal cooperative associations by any means, they were some of the nation's first cooperative efforts.

First Agricultural Cooperative. The first agricultural cooperative was established in Babylon around 2,000 B.C. Its objective was similar to the goals of modern agricultural cooperatives—that is, to provide greater flexibility and more independence for member farmers.

Forerunner of Modern-day Cooperatives. The forerunner of modern-day cooperatives, the Rochdale Society of Equitable Pioneers, was founded in England in 1844. It started with twenty-eight members, each of whom purchased one share of stock worth about $150 in today's economy. Its membership consisted of craftsmen such as weavers and shoemakers. Working together they were able to sell their products under one roof and use a part of the earnings to purchase supplies in large quantities at economical prices. Another portion of the earnings was reinvested in the society so that the funds could continue to grow. The remainder was returned to the individual members in the form of refunds.

It was not until the Civil War and the coming of the industrial revolution that American agricultural cooperatives really took root. In the 1920s, the federal government and various state governments began to step in with laws designed to encourage the growth of agricultural cooperatives. Perhaps the most important new law was the Capper-Volstead Act of 1922, which allowed farmers to organize legally to buy and sell their products and is the law under which cooperatives operate today.

Present-day Cooperatives. Today's cooperatives, whether in agriculture, utilities, banking, retailing, insurance, or wholesaling, are organizations in which people of similar needs and interests can work together to help themselves and each other. By doing so, they strengthen themselves economically. They keep profits within their own control, and they stimulate free enterprise, while helping to protect the family farm.[16] Therefore, cooperatives are organized to:

❧ improve bargaining power

❧ reduce costs

❧ obtain products or services that are otherwise unavailable

❧ expand new and existing market opportunities

❧ improve product or service quality

❧ increase income

Impact of Cooperatives

Today, the life of every American is touched in some way by cooperative enterprises. Approximately 45,000 separate cooperative organizations, with more than 90 million members, currently operate in this country. Cooperatives serve one out of every four citizens. Statistics show that of these 45,000 cooperatives, approximately 5,000 are farmer-owned, providing marketing, purchasing, and related services for farmers and producers. These agricultural cooperatives operate at a business value of over $73 billion. Cooperatives market about 28 percent of all agricultural products and provide about 25 percent of all production supplies used on farms each year.

The cooperative business structure provides insurance, credit, health care, housing, telephone, electrical, transportation, child care, and utility services. Members use cooperatives to buy food, consumer goods, and business and production supplies. Farmers use cooperatives to market and process crops and livestock, purchase supplies and services, and to provide credit for their operations.[17]

Types of Cooperatives

In Chapter 5, three major types of agricultural cooperatives were briefly discussed: purchasing or supply, marketing, and service cooperatives. A further explanation of each of these three types of cooperatives follows.

Purchasing or supply cooperatives are organized to purchase farm supplies for members in large volumes, which allows production agriculturalists to buy resources at a lower price and to have some control over their quality. In some cases these cooperatives produce their own products rather than buy them from someone else. Recent reports by the USDA, Agricultural Chartbook (1997) indicate that about 25 percent of all farm supplies are purchased through cooperatives. The major resources production agriculturalists purchase through their cooperatives, in order of volume, are petroleum products, feed, and fertilizer, which account for about 75 percent of total purchases from cooperatives.

Marketing cooperatives provide one or more marketing services beyond the production process. Larger volume and consistency in quality give farmers more control in negotiating higher prices for their products. Of course, some products are more adaptable to cooperative marketing than others. Grain, soybeans, dairy products, and livestock products account for about 85 percent of all farm products marketed. Over 35 percent of all farm products are sold through more than 3,000 agricultural cooperatives, with an annual gross sale of $47 billion. Cooperatives market 78 percent of milk, 33 percent of cotton, and 17 percent of fruits, nuts, and vegetables.[18]

Service cooperatives provide special services to production agriculturalists, such as credit, artificial breeding, irrigation, telephone, electricity, insurance, import-export, and numerous other functions. In addition to separate cooperatives providing these services, many marketing and supply cooperatives provide special services as a part of their operation. Refer to Figure 10–14 for the addresses of several cooperative associations.

Service cooperatives account for about 2 percent of all agricultural cooperatives. The following provides a brief description of some service cooperatives (co-ops).

🌿 *Electric cooperatives*: Like other types of cooperatives, electric co-ops exist to serve members in providing a service that would otherwise be unavailable to them. Today, nearly 1,100 of these cooperatives provide electricity to 11 million consumers on farms and in many small towns and communities. Electric co-ops began to light the countryside in 1935, with the creation of the Rural Electrification Administration (REA) by President Franklin Roosevelt's executive order. Co-ops were needed because it was not economically feasible for existing power companies to build lines in the sparsely populated corners of rural America. REA still provides loans on a continuing basis for rural electrification.

🌿 *Telephone cooperatives*: Congress amended the Rural Electrification Administration Act in 1949 to include financing for telephone service as well as electric service in rural areas. Cooperatives were organized by citizens whose only interest was ensuring good, dependable telephone service in areas where there was little or no service and it was not economically feasible for commercial (for-profit) companies to do so. There are 240 telephone cooperatives in the United States serving over 2 million members.

🌿 *Artificial breeders' cooperatives*: The purpose of artificial breeders' cooperatives is to offer members superior genetics in dairy and beef

Cooperative Associations

American Institute of Cooperatives
1129 Twentieth Street, NW
Washington, DC 20036

American Rice Growers Cooperative
Association
211 Pioneer Building
Lake Charles, LA 70601

Burley Tobacco Grower Cooperative
Association
620 South Broadway
Lexington, KY 40508

Cooperative League of the USA
1828 L. Street, NW, Suite 1100
Washington, DC 20036

Farm Educational and Cooperative
Union of America
12025 East 45th Avenue
Denver, CO 80251

Flue-Cured Tobacco Cooperative
Stabilization Corporation
Box 12300
Raleigh, NC 27605

National Association Supply Cooperative
Box 303
New Philadelphia, OH 44663

National Council of Farmer Cooperatives
1129 Twentieth Street, NW
Washington, DC 20036

National Federation of Grain
Cooperatives
1129 Twentieth Street, NW, Suite 512
Washington, DC 20036

National Livestock Producers Association
307 Livestock Exchange Building
Denver, CO 80216

National Milk Producers Federation
30 F Street, NW
Washington, DC 20001

National Rural Electric Cooperative
Association
2000 Florida Avenue, NW
Washington, DC 20009

National Telephone Cooperative
Association
2100 M Street, NW
Washington, DC 20037

Pacific Agricultural Cooperative for
Export, Inc.
465 California Street, Suite 414
San Francisco, CA 94104

Figure 10–14 Home office addresses of cooperative associations.

sires. They serve their cooperative members by conducting training seminars to teach the techniques of artificial insemination and the proper thawing and handling of semen; providing herd counseling in sire selection and equipment needs, and making deliveries of semen, nitrogen, and other supplies to farmer members on regular intervals.

❦ *Insurance:* Earlier we discussed the Farm Bureau as a general farm organization. Farm Bureau also provides insurance in most states. Farm Bureau is also a cooperative. The various service companies of Farm Bureau insurance include home, automobile, life insurance, and group health insurance through the Rural Health Program.

❦ *Credit:* Organizations such as Farm Credit Services are farmer-owned cooperative businesses that exclusively serve rural communities. The banks' and associations' primary goal is to provide dependable and constructive credit and related services to production agriculturalists, agricultural cooperatives, and rural residents in the area they serve.

❦ *Irrigation cooperatives:* Production agriculturalists, especially in the western states, formed mutual irrigation companies which now supply water to one-quarter of the irrigated land. They acquire or develop sources of water supply and distribute the water to users. Production agriculturalists in dry lands could produce little if they lacked irrigation.[19]

❦ *DHIA:* The Dairy Herd Improvement Associations are cooperative oganizations of about twenty-five milk producers each. Each association employs a trained supervisor to keep monthly milk and butterfat production records on association herds. The milk producers then decide which cows in their herds are underproducing milk and should be removed.

❦ *Export-Import cooperatives:* Export-import cooperatives play an important part in international trade of farm products. A large portion of the $54 billion plus annual exports include companies that are cooperatives. The export business makes up to 50 percent of the total volume of some cooperatives.[20] Some cooperatives are importers as well as exporters. Refer to Figure 10–15 for some examples of export-import cooperatives.

In summary, agricultural cooperatives play a significant role in the private agriservices sector, as well as the input and marketing sector of agriculture. The economic importance of agricultural

Export-Import Cooperatives

Agway Inc. 333 Butternut Drive DeWitt, NY 13214	Ocean Spray Cranberries One Ocean Spray Drive Lakeville-Middleboro, MA 02349	Tree Top 220 East 2nd Avenue P.O. Box 248 Selah, WA 98942	Sun Diamond Growers P.O. Box 1727 Stockton, CA 95201
Sunkist Growers, Inc. P.O. Box 7888 Van Nuys, CA 91409-7888	Texas Citrus Exchange Expressway 83 and Mayberry Rd. P.O. Box 793 Mission, TX 78573-0793	California Almond Exchange 1802 C Street P.O. Box 1768 Sacramento, CA 95812	Harvest States 1667 North Snelling Avenue P.O. Box 64594 St. Paul, MN 55164-0594

Figure 10–15 Home office addresses of export-import cooperatives.

cooperatives is indicated by the fact that agricultural cooperatives are included in *Fortune* magazine's annual listing of the nation's 500 largest industrial companies, based on sales volume. In addition, several others are listed in the second 500 largest corporations. The top agricultural cooperative by sales volume is Farmland Industries with $9.2 billion in sales.[21]

Careers.   Careers in cooperatives are as many and as varied as their diverse activities and enterprises. It is not possible to list here all the types of jobs and opportunities available with cooperatives. A few of the major categories of jobs include managers and assistant managers, public relations and communication specialists, laboratory and field technicians, product specialists, and providers of operations and business services, professional services, and finance, credit, and insurance services.

CONCLUSION

There are many other jobs in the private agribusiness services. Those providing the most opportunities were discussed in this chapter. The following is a list of the other jobs in the private agribusiness input sector that involve service to production agriculturalists: auctioneer, farm tax accountant, blacksmith, land appraiser, agricultural engineer, farm record service, livestock buyer, agricultural scientist, rural sociologist, and agricultural law. There are many other jobs in the agribusiness input sector, which will be discussed in Chapter 11, along with public agriservices.

SUMMARY

The private agriservices sector of agribusiness is increasing dramatically, and this trend is expected to continue. The major reasons for this increase include the increasing complexity of production agriculture and the expansion of farm units to the point that specialized services are economically feasible. Some of the major services offered by private agriservice groups include political representation, advertising and promotion by specialized commodity organizations, agricultural research, consulting, artificial insemination, veterinary services, communications, and purchasing, supplying, marketing, and servicing production agriculturalists through cooperatives.

General farm organizations play a major role in the private agriservices group. Although there are numerous organizations of this type in operation, the four major ones include the National Grange, the American Farm Bureau Federation, the National Farmers Union, and the National Farmers Organization. Although the general goals of all these organizations are basically the same, they differ considerably in specific goals and the procedures for implementing them. Almost all production agriculturalists are members of at least one of these organizations, and many production agriculturalists are members of more than one.

In addition to general farm organizations, numerous specialized organizations have developed around specific farm commodities. The major purpose of these organizations is to promote the development of specific products through research, education, and market expansion efforts. Almost every major farm commodity has one or more organizations to assist in its development. Some examples of these specialized organizations include the National Cattle Beef Association, the American Dairy Association, the National Association of Wheat Growers, the National Cotton Council, the National Potato Council, and numerous others.

Another major function of the private agriservices sector is to develop new technology through constant research on new ways of making agriculture more productive. Agribusiness firms provide a tremendous amount of funds for agricultural research in two major ways: providing grants to public research institutions and conducting applied research in their own research facilities. Similar to research is agricultural consulting, since much of a consultant's job is research based.

Another major service offered to dairy and livestock producers today is artificial insemination. Although some production agriculturalists inseminate their own livestock, most animals are bred by local technicians who are trained specifically for this service.

There are about 38,000 veterinarians providing a wide variety of services in private practice, teaching, research, public health, military service, private industry, and other special services. Veterinarians provide a valuable service to production agriculturalists.

Communications through the print media, computers, Internet, pictures, radio, television, videos, exhibiting, and so on are a part of all major segments of modern business and industry.

Agribusiness is no exception. This private agriservice offers many opportunities to those who have a good agricultural background and combine it with good communication skills.

Agricultural cooperatives also play a major role in providing a wide variety of services and products for production agriculturalists. Today, the life of every American is touched in some way by cooperative enterprises. Approximately 45,000 separate cooperative organizations, with more than 90 million members, currently operate in this country. Cooperatives serve one out of every four citizens. These agricultural cooperatives operate at a business value of over $73 billion. Three types of cooperatives are purchasing and supply, marketing, and service.

There are many other jobs in the private agribusiness input sector. These include auctioneer, farm tax accountant, blacksmith, land appraiser, agricultural engineer, farm record service, livestock buyer, agricultural scientist, rural sociologist, and agricultural law.

END-OF-CHAPTER ACTIVITIES

Review Questions

1. Why has the importance of private agribusiness services become more important in recent years?
2. What are the names of the five major general farm organizations?
3. What are ten things in which the National Grange played a major role that served to improve agriculture and help rural America during the last century?
4. What is the Grange doing today with issues that affect modern agriculture?
5. What was the major purpose of the National Farmer's Union when it was first organized?
6. What are four major efforts and/or services of the American Farm Bureau Federation?
7. What are seven accomplishments of the Farm Bureau that gave support to, or led the battle for, a better life for American farmers?
8. How does the National Farmer's Organization attempt to improve farm prices in today's market?
9. Why did agricultural commodity organizations become popular?

10. List fifteen services performed by agricultural commodity organizations.
11. Name five of the larger farm commodity groups.
12. Name six of the more prominent agribusiness commodity groups.
13. Name ten subjects or areas in which agricultural research is conducted.
14. What are five jobs conducted by crop consultants?
15. What are five jobs conducted by livestock consultants?
16. What were two major developments that stimulated the use of artificial insemination?
17. What are the three major parts of the artificial insemination industry?
18. Name six places where veterinarians can work.
19. What are eight duties of veterinarians?
20. What are twelve things (areas) that jobs in communications would include?
21. What are four major areas in the job of agricultural communications specialist?
22. What are six things that today's cooperatives are organized to do?
23. What are three major types of cooperatives?
24. What are seven examples of services cooperatives?
25. What four types of insurance does Farm Bureau sell?
26. List nine other jobs from the private agribusiness input sector not previously mentioned.

Fill in the Blank

1. NFO estimates that _____ percent of the total production in any commodity would be sufficient to allow farmers to set their own prices.
2. In some cases, the political power of commodity organizations exceeds the power of some _____ _____ organizations.
3. Production agriculturalists have depended on public funds for farm research since the long-term benefits help primarily _____ firms and the _____.
4. Artificial insemination was first developed for practical application in livestock breeding in 1922 by a _____ physiologist.
5. Artificial insemination is used predominantly with dairy cows, but beef and _____ producers also use this service.
6. _____ _____ is often regarded as the "Father of Veterinary Medicine."
7. The first College of Veterinary Medicine was established in 1879 by _____ _____ University.
8. Currently, there are about _____ veterinarians providing a wide variety of services.
9. A _____ is a business that is financed, owned, and controlled by the people who use it.
10. Approximately _____ separate cooperative organizations with more than _____ million members currently operate in the United States.

11. There are approximately _____ farmer-owned cooperatives providing marketing, purchasing, and related services at a business value of over _____ billion dollars.

12. Cooperatives market about _____ percent of all agricultural products and provide about _____ percent of all production supplies used on farms each year.

13. Cooperatives market _____ percent of milk, _____ percent of cotton, and _____ percent of fruits, nuts, and vegetables.

14. Today, nearly _____ cooperatives provide electricity to _____ million consumers.

Matching

a. American Farm Bureau Federation
b. National Farmers Union
c. American Agriculture Movement
d. National Farmer's Organization
e. The National Grange

_____ 1. oldest and second largest of the major general farm organizations
_____ 2. first organized in the Southwest in 1902
_____ 3. largest of the five general farm organizations
_____ 4. organized in Washington, D.C.
_____ 5. originally called the Farmers Educational Cooperative
_____ 6. first organized between 1912 and 1914 to sponsor county agricultural agents
_____ 7. has as members approximately 2.5 million families from fifty states and Puerto Rico
_____ 8. organized in Corning, Iowa, in 1955
_____ 9. originally called Patrons of Husbandry
_____ 10. most controversial of general farm organizations due to holding action of livestock and pouring out of milk
_____ 11. originally, members were primarily disenchanted NFO and NFU members
_____ 12. primarily young farmers, who expanded the organization until prices dropped dramatically in the mid-1970s.

Activities

1. Determine which general farm organization is the strongest in your community. Prepare a brief report of its mission and the things that it has done recently to help improve the lives of production agriculturalists.

2. Determine which agricultural commodity organizations are strongest in your community. Make a list of the potential jobs that are available with this organization.

3. Prepare a one- to two-page paper on a particular commodity and report some research that has been done on the commodity. Share this report with your class. Indicate whether the research was conducted with private or public (government) funds.

4. Name three agricultural consultants in your community. If there are not three, give examples of three agricultural consultants.

5. Some believe that artificial insemination is one of the most important agricultural technological developments in this century. Give reasons as to whether you agree or disagree with this statement.

6. Prepare a list of courses (classes) needed by a high school student who desires to be a veterinarian. Next, find out from your teacher, guidance counselor, or a college catalog what college courses are needed to be a veterinarian. Also, make a brief statement on the importance of grades.

7. Make a list of the potential jobs that are available within the agricultural communications profession in your community.

8. Make a list of the various cooperatives in your community. Identify for each whether it is a purchasing (supply), marketing, or service cooperative.

NOTES

1. W. L. Robinson, *The Grange, 1867–1967* (Washington, D.C.: The National Grange, 1966), p. 3.

2. The National Grange, *Legislative Policies, 1977* (Washington, D.C.: Author, 1977).

3. Gladys T. Edwards, *The Farmer's Union Triangle* (Denver, Colo.: The Farmer's Union Education Service, 1941), p. 19.

4. N. Omri Rawlins, *Introduction to Agribusiness* (Murfreesboro, Tenn.: Middle Tennessee State University, 1999), p. 140.

5. Luther Tweeten, *Foundations of Farm Policy* (Lincoln: University of Nebraska Press, 1970).

6. National Farmers Organization, *NFO: Historical and Background Information* (Corning, Ia.: NFA, Public Information Department, 1977).

7. Marcella Smith, Jean M. Underwood, and Mark Bultmann, *Careers in Agribusiness and Industry* (Danville, Ill.: Interstate Publishers, 1991), p. 287.

8. Ibid., pp. 288–292.

9. Ibid., p. 297.

10. Ibid.

11. Ibid.

12. J. F. Smithcors, *The American Veterinary Profession* (Ames: Iowa State University, 1963).

13. Bureau of Labor Statistics (1998) *Occupational Outlook Handbook, 1998–1999: Veterinarians.* Available Online: http://stats/bls.govlocolocos076.htm

14. Ibid.

15. Smith, Underwood, and Bultmann, *Careers in Agribusiness and Industry,* p. 300.

16. Tennessee Council of Cooperatives, *What Is a Cooperative?* (Lavergne, Tenn.: Author, n.d.).

17. USDA, *What Is a Cooperative?* (Washington, D.C.: USDA, Rural Business and Cooperative Development Service, March 1995).

18. United States Department of Agriculture (1993) *Farmer Cooperative Statistics: CS Service Report 43* (U.S. Government Printing Office, Washington, D.C.)

19. Smith, Underwood, and Bultman, *Careers in Agribusiness and Industry,* p. 310.

20. United States Department of Agriculture, *Agricultural Statistics, 1997* (Washington, D.C.: National Agricultural Statistics Service).

21. Source: http://www.pathfinder.com/fortune/global500/companies/499.html (1999) *Fortune* magazine: Available Online.

Public Agribusiness Services

OBJECTIVES

After completing this chapter, the student should be able to:

- discuss the historical perspective of public agribusiness services
- explain reasons for government services
- describe the organizational structure of the USDA
- describe the following divisions of the USDA and each of the major agencies:
  food, nutrition, and consumer services
  marketing and regulatory services
  farm and foreign agricultural services
  rural development
  natural resources and environment
  research, education, and economics
  food safety
- discuss the role of the state departments of agriculture
- explain the land grant system
- explain non–land grant agricultural programs
- describe the sea grant program
- discuss agricultural education programs

TERMS TO KNOW

agricultural statistics	flex time	interdisciplinary
attachés	germ plasm	liaison
consortium	grants	policy analysis
distance learning	income stabilization	price support
economic indicators	inflation	production controls
eradicate	initiatives	teleconferencing

INTRODUCTION

American agriculture has not become the world's greatest food producer by accident, but by planning and hard work on the part of agriculturalists and nonagriculturalists within a conducive social, economic, and political environment. American leaders have long recognized the importance of a strong agriculture, possibly because many of our political leaders had their roots in farming. According to Vivian Whitehead, twenty-three presidents have had close farm ties.[1] Included in this list are George Washington, Thomas Jefferson, Andrew Jackson, Abraham Lincoln, Theodore Roosevelt, Harry Truman, Lyndon Johnson, and Jimmy Carter. In addition, numerous members of Congress were part-time farmers before and during their terms of service.

The major role of government in agriculture has come from the federal level. However, state and local governments also play a major role in agricultural development. Agriculturalists obtain information from teachers of agricultural education, their county agents, and their state colleges of agriculture. American governments have influenced agriculture directly in the following types of programs: research, education, and extension; health regulations; **price support** laws and **production controls**; and the collection and distribution of **agricultural statistics**. Agriculture is also affected by **inflation**, unemployment, foreign policy, and all other general policy areas of the nation as a whole. There is very little that agriculturalists do today that is not affected, either directly or indirectly, by government action.

The government's role in agriculture has always been controversial, some aspects more than others. For example, price supports and production controls have been extremely controversial. Many production agriculturalists who are supposed to be helped by price supports and production controls dislike the government interference in the free market system, while other production agriculturalists and farm organizations demand higher price supports and either more or fewer production controls. The major controversy over research, education, and regulation is how much should be provided. Agriculturalists want more research and educational services and less regulation, while consumer groups want more regulation and less funds spent on research and education.

HISTORICAL PERSPECTIVE

When the term government is used without qualification, we usually assume it means federal government, and the major role of government in agriculture has come from the federal level. However, state and local governments also play a major role in agricultural development.

The first federal aid to agriculture in the United States came in 1839 when Congress appropriated $1,000 to collect farm statistics and to collect and distribute seeds of various types. An agricultural division was established in the Patent Office to carry out these objectives. Ten years later the Patent Office was shifted from the Department of State to the newly created Department of Interior, taking the agriculture division with it. On May 15, 1862, President Lincoln signed the act creating the U.S. Department of Agriculture. The purpose of the department was "to acquire and diffuse among people of the United States useful information on subjects connected with agriculture."[2] The administrators of the newly created Department of Agriculture were called commissioners until 1889, when Congress raised the head of the department to the rank of cabinet secretary. Isaac Newton was the first of seven commissioners, and Norman Colman was the first secretary of agriculture.

REASONS FOR GOVERNMENTAL SERVICES

Production

Various governmental agencies and services are provided to aid production agriculturalists in solving some of their farm problems. Many believe that governmental assistance to production agriculturalists is a recent development. However, various forms of service and assistance have occurred for over 150 years. One of the first services provided by the government was the Homestead Act of Congress, which aided farmers in securing land to help feed America. In 1862, land grant colleges were made possible by an act of Congress and President Abraham Lincoln. This enabled better education on farming technology in order to increase production. In 1890, an act of Congress provided support for land grant colleges for blacks.[3]

Education

The land grant colleges, as well as other colleges offering agriculture (non–land grant colleges) provide education to students about agribusiness inputs, agribusiness services, production agriculture, agriscience, and the agribusiness output sector. Extension specialists are graduates of these universities who provide agricultural information to production agriculturalists, processors, handlers, farm families, communities, and consumers. The Extension Service also administers the 4-H program. The government provides funds to state and local districts to help finance instruction in agricultural education. These funds are administered through the U.S. Department of Education.[4] The agricultural education teachers provide instruction for production agriculturalists of all ages, as well as for high school students. Agricultural education teachers are also advisors to the Future Farmers of America.

Services

The government offers many services that we take for granted. It is ironic that many of those who criticize government spending in agriculture also criticize some lack of service, product, or consumer production. The following is a list of services provided by the government. Most of these are administered through programs of the U.S. Department of Agriculture (USDA), which will be discussed later in this chapter:

🌾 manages national forest system

🌾 provides loans to farmers and cooperatives

🌾 makes loans to rural electric system

🌾 improves marketability of crops

🌾 encourages fair trade practices

🌾 requires truthful labeling

🌾 extends patent-type protection to developers of plants

🌾 prevents discrimination against producers

🌾 helps prevent loss of stored products

🌾 maintains fair and open competition in marketing, livestock, poultry, and meat

🌾 inspects commodity exports and imports

🌾 expands foreign markets for agricultural products

🌾 provides food assistance for those in need

🌾 inspects food to assure safety and quality

🌾 assists states in controlling insects, weeds, and plant diseases that are a serious threat to the nation

🌾 keeps foreign diseases from entering the country

🌾 eradicates diseases that do get into the country

🌾 fights major domestic animal disease outbreaks

❧ provides for humane treatment of animals

❧ regulates all futures trading

❧ protects natural resources

❧ provides comprehensive information services for food and agricultural sciences

❧ conducts research in the fields of livestock, plants, soil, water, air quality, energy, lead safety and quality, food processing and marketing, nonfood agricultural products, and international development

❧ sets aside land not suitable for farming to be used for forests and reservations[5]

To carry out and accomplish all these services, thousands of workers are needed. In the sections that follow, the names of departments within the USDA will be discussed that perform most of these services.

ORGANIZATIONAL STRUCTURE OF THE USDA

Since the USDA plays such an important role in the agricultural industry, a student of agribusiness should have a basic understanding of the agencies involved and the organizational structure. The USDA is not one agency but a rather complex organization of individual agricultural agencies, in addition to several administrative ones. The specific number changes frequently as agencies are reorganized and new ones added, especially with the change of each new administration. However, the USDA normally has fifteen to twenty agencies. Many divisions and agencies within the USDA changed names and functions in response to the Department of Agriculture Reorganization Act of 1994.

The USDA is administered by a secretary, deputy secretary, and several assistant secretaries and undersecretaries. Each assistant secretary or under-

secretary is responsible for administering several agencies that relate to specific areas of agriculture. The major divisions currently organized include Food, Nutrition and Consumer Services; Marketing and Regulatory Services; Farm and Foreign Agricultural Services; Rural Development; Natural Resources and Environment; Research, Education, and Economics; and Food Safety. In addition, there are numerous offices that provide administrative services. According to the USDA's Agricultural Fact Book (1997) the USDA currently operates with a budget of over $65 billion and employs over 100,000 people, including part-time employees. Figure 11–1 shows the current organization of the USDA. The following sections discuss the major divisions and most of the agencies that provide a specific service for the clients of the USDA. (The administrative agencies will not be included).

FOOD, NUTRITION, AND CONSUMER SERVICES

Until 1977 the Food and Consumer Services function was carried out through Agricultural Marketing Services, but it has now been given separate division status in order to coordinate the largest budget of any division in the USDA. It administers food assistance and nutrition programs that account for 40 to 70 percent of the USDA budget, depending on the size of farm program expenditures.[6] The major responsibilities of this division stem back to the 1960s, when the USDA had tons of surplus food with no markets. Under Public Law 480, programs were developed to distribute this food to needy individuals as a part of the war on poverty. Currently, emphasis is also being directed toward consumer education on all aspects of nutrition. The agency currently assigned to this division is the Food and Consumer Service.

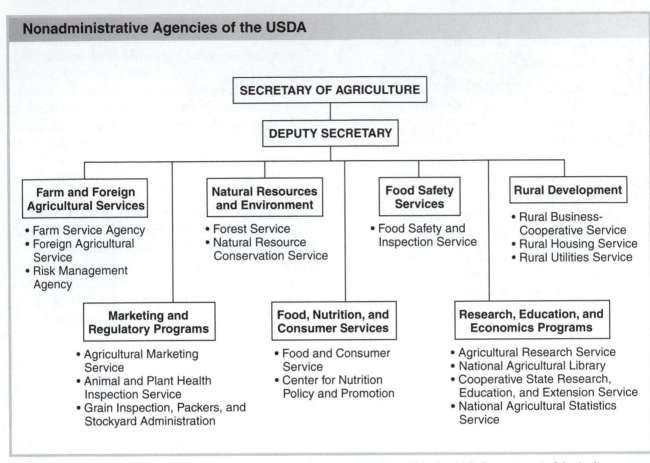

Nonadministrative Agencies of the USDA

SECRETARY OF AGRICULTURE

DEPUTY SECRETARY

Farm and Foreign Agricultural Services
- Farm Service Agency
- Foreign Agricultural Service
- Risk Management Agency

Natural Resources and Environment
- Forest Service
- Natural Resource Conservation Service

Food Safety Services
- Food Safety and Inspection Service

Rural Development
- Rural Business-Cooperative Service
- Rural Housing Service
- Rural Utilities Service

Marketing and Regulatory Programs
- Agricultural Marketing Service
- Animal and Plant Health Inspection Service
- Grain Inspection, Packers, and Stockyard Administration

Food, Nutrition, and Consumer Services
- Food and Consumer Service
- Center for Nutrition Policy and Promotion

Research, Education, and Economics Programs
- Agricultural Research Service
- National Agricultural Library
- Cooperative State Research, Education, and Extension Service
- National Agricultural Statistics Service

Figure 11–1 Presently, there are seven divisions and nineteen agencies within the U.S. Department of Agriculture.

Food and Consumer Service

The Food and Consumer Service administers a number of programs that provide food assistance to individuals through state and local agencies. These programs are designed to meet the food needs of families, individuals with special nutritional requirements, and persons in certain institutions. The larger of these programs, in terms of people served, is the school lunch program, which provides more than 25 million students in the nation's schools with subsidized lunches.[7]

According to the USDA's *Agricultural Fact Book* (1997), other programs administered by the Food and Consumer Service include the Food Stamp Program, which includes about 1 of every 10 Americans, the most expensive of the nutrition programs; Nutrition Assistance for Puerto Rico; Child Nutrition Programs; Special Milk Programs; Special Supplemental Food Programs; Cash and Commodities for Selected Groups; and Temporary Emergency Food Assistance Programs. Although the Food Stamp Program is the largest program relative to the amount of funds dispersed and the number of people assisted, more than 5 million people are reached by the other programs as well. Refer to Figure 11–2.

Figure 11–2   The USDA employees shown here are reviewing the National Nutrient Database for the Child Nutrition Program. (Courtesy of USDA)

MARKETING AND REGULATORY SERVICES

This division administers one of the largest number of agencies and has the overall responsibility of coordinating the activities of firms involved in moving food and fibrous products from the farm to the final consumer. These agencies provide protection for farmers, on one hand, and consumers, on the other. The specific agencies involved in this division include the Agricultural Marketing Service, the Animal and Plant Health Inspection Service, the Grain Inspection Service, and the Packers and Stockyards Administration.

The Agricultural Marketing Service

The Agricultural Marketing Service promotes the orderly and efficient marketing and distribution of agricultural commodities. The Federal-State Marketing Improvement Program, under the Agricultural Marketing Service, studies problems at the local and state levels. It is involved in improving marketability of crops, finding new marketings, improving market efficiency, and improving mar-

ket information.[8] Specific activities administered by the Agricultural Marketing Service include a market news service, egg products inspection, development of quality standards for commodity grading, and administering marketing orders and agreements for certain commodities. All of these programs are producer and/or marketing firm initiated.

Animal and Plant Health and Inspection Service

The major purpose of the Animal and Plant Health and Inspection Service is to protect plants and animals from economically dangerous diseases and pests. The agency initiates survey and diagnostic activities and, with the affected states, conducts programs to control or **eradicate** these hazards. This agency conducts an inspection program at numerous ports of entry to prevent the entry and spread of potentially damaging pests and diseases. In addition, the agency deals with emergency outbreaks of plant pests and animal diseases such as the gypsy moth outbreak in Oregon and the citrus canker infestation in Florida during 1985. The Animal and Plant Health and Inspection Service is also designed to provide the humane treatment of animals.

Grain Inspection, Packers, and Stockyards Administration

The Federal Grain Inspection Service was established in 1976 in response to international scandals in foreign grain sales. This agency establishes official standards for grain, conducts weighing and inspection activities, and inspects other agricultural products under the authority of the Grain Standard Act and the Agricultural Marketing Act of 1946. Inspection and weighing costs are recovered by user fees; they are required at all export ports and optional at interior locations. It is also the regulatory agency that monitors livestock markets for potential unfair trade practices.

FARM AND FOREIGN AGRICULTURAL SERVICES

The Farm and Foreign Agricultural Services division deals with national efforts to keep the supply and demand of agricultural products in balance. During the 1960s the primary efforts were directed toward reducing the supply of agricultural products through massive storage programs and production control at the farm level. During the 1970s and 1980s, the major effort had been directed toward increasing demand through the development of new foreign markets for agricultural products. However, a major effort is still being concentrated on supply control as well. This division currently includes the Farm Service Agency, Foreign Agricultural Services, and Risk Management.

Farm Service Agency (FSA)

This agency represents the biggest change in the USDA structure in recent years. FSA is now a one-stop shopping area where the production agriculturalist can sign up for farm programs, obtain loans, and obtain assistance on soil conservation programs. The Farm Service Agency resulted from the consolidation of the former Agriculture Stabilization and Conservation Service (ASCS) and the Farmers Home Administration (FmHA).

History. This agency has the most direct contact with the modern production agriculturalist. During the 1950s, the old Agricultural Stabilization and Conservation Service (ASCS) received about 65 percent of the total USDA budget, but the FSA currently administers less than 50 percent due to major increases in the food assistance programs. The ASCS began in 1933 as the Agricultural Adjustment Administration (AAA), commonly known as the Triple A. In 1945, it became known as the Production and Marketing Administration,

which survived until 1953. It was then divided into two agencies, known as the Agricultural Marketing Service and the Commodity Stabilization Service. The Agricultural Marketing Service (discussed earlier) continued its status as a separate agency, and the Commodity Stabilization Service became the ASCS until its merger into the Farm Service Agency.

The Farmers Home Administration (FmHA) was established in 1946 to provide financial assistance to farmers who were unable to obtain loans from private agencies. In these high-risk loans and in housing loans, the borrower is expected to refinance the loan with a private lender as soon as possible. Since the beginning of FmHA, it has added numerous other responsibilities. With the creation of rural development programs in the 1970s, FmHA made a higher percentage of farm loans because of the farm financial crisis. Rural development efforts are directed toward loans for rural home repairs, development of rural water and waste disposal systems, and other development projects. With the FmHA being merged with FSA, loans for land are still made with the agency. However, loans for rural housing are made through the Rural Housing Service.

Responsibilities. The responsibilities of the FSA are to:

- 🌿 stabilize the nation's agricultural economy through price support loans and purchases and crop and **income stabilization**
- 🌿 conserve the nation's farm resources through a variety of conservation measures and land retirement programs
- 🌿 protect the nation's food and food reserves through commodity storage programs and inventory management programs
- 🌿 assist in the nation's defense through civil defense and defense mobilization activities as assigned

❦ assist with farm loans

❦ furnish information on commodities

Size. In addition to the national office in Washington, D.C., the FSA has three regional commodity offices, 50 state offices, and about 2,800 county offices that administer the various programs. It is officially directed at the local level by elected farmer committee members and at the state level by committee members appointed by the secretary of agriculture. FSA employs over 4,000 workers.

The FSA deals on a day-to-day basis with hundreds of thousands of producers, carriers, exporters, handlers, warehouse workers, and others. FSA responsibilities include the handling of millions of documents yearly.[9] Refer to Figure 11–3.

Commodity Credit Corporation

The Commodity Credit Corporation (CCC) was organized in 1933 to act as a purchasing agent for the federal government. The CCC loan program was the tool that acted as a price support mechanism. It was not until 1939 that the CCC was transferred to the Department of Agriculture. The CCC was created to stabilize, support, and protect farm income and prices; to help maintain a balanced supply of agricultural commodities and their products (foods, feeds, and fiber); and to assist in their orderly distribution. Basically, CCC is the financial arm of the FSA, but it also finances other agricultural programs that are closely related, such as farm export activities of the Foreign Agricultural Service.

Foreign Agricultural Service

The Foreign Agricultural Service has played a major role in making the United States the world's largest exporter of agricultural products. This agency is the export promotion and service agency for American agriculture, and **attachés** in 65 foreign countries maintain a constant flow of infor-

Figure 11–3 This Farm Service Agency employee is advising a local production agriculturist with one of the agency's farm programs. (Courtesy of USDA)

mation on the international production and marketing of food and fiber. Import and export regulations are administered by the Foreign Agricultural Service in cooperation with numerous local, state, regional, national, and international trade groups. The Foreign Agricultural Service is assisted in its mission by the Office of International Cooperation and Development. It is responsible for maintaining an export credit sales program and financing the sale of agricultural commodities under Public Law 480. The Foreign Agricultural Service employs about 75 people and operates with an annual budget of about $75 million.

The Office of International Cooperation and Development is responsible for cooperative international research, scientific and technical exchanges, and **liaison** with international agricultural organizations. It directs training and technical assistance programs in many developing countries.

Risk Management

The Risk Management Agency is the new name of the former Federal Crop Insurance Corporation (FCIC). This agency provides farmers with insurance that repays crop production losses due to extreme weather, insect damage, diseases, and other uncontrollable natural conditions. The Risk Management Agency provides the only all-risk crop investment protection available to farmers in many major agricultural counties. An all-risk crop insurance bill was introduced in Congress in 1922 but was not enacted until 1938. Coverage was first available to wheat but soon expanded to cotton and other crops. Currently, insurance is available on twenty-three crops in thirty-nine different states.

Since the enactment of the Federal Crop Insurance Act of 1980, substantial progress has been made in shifting the crop insurance's program to the private sector. Currently, government insurance companies account for about 75 percent of crop insurance sales, and the remainder is handled by private agents through sales and service agreements.

A part of the mission of these agencies is to research, develop, and pilot new crop programs, insurance plans, and risk management strategies; to evaluate and make recommendations for improvement of existing crop programs, insurance plans, and risk management strategies; and to coordinate the development of and support for specialty crop programs.

RURAL DEVELOPMENT

The emphasis on rural development began in the late 1960s and early 1970s when policy makers realized that a tremendous economic gap had developed between the urban and rural sectors of the U.S. economy. It developed as part of the war on poverty initiated during the Lyndon Johnson administration.

A new and sharper focus on rural development took shape with passage of the Department of Agriculture Reorganization Act of 1994. Rural development work was focused in three new organizations reporting to the undersecretary for rural development: the Rural Utilities Service (RUS) offers telecommunications and electric programs along with water and sewer programs; the Rural Housing Service (RHS) includes rural housing programs as well as rural community loan programs; and the Rural Business-Cooperative Service (RBS) includes cooperative development and technical assistance, plus other business development programs and the Alternative Agricultural Research and Commercialization Center.

Rural Utilities Service

The Rural Utilities Service (RUS) was established as the Rural Electrification Administration (REA) in the early 1930s to help rural inhabitants obtain electric and telephone services by making long-term loans to rural electric cooperatives and other power suppliers to build and operate rural electric systems. This agency also extends loans, through the Rural Telephone Bank, to independent companies and cooperatives for extending and improving telephone service in rural areas. Over 1,000 cooperatives and other organizations have borrowed RUS funds to provide electric service. Loans are also made for water and waste disposal programs.

Over $2 billion has been loaned to provide electric services to over 20 million rural inhabitants. In addition, over 900 firms have borrowed over $2 billion to provide modern telephone service to about 10 million rural people. These funds were loans, not grants, and must be repaid with interest. The repayment on these loans has totaled over $1.5 billion since the REA (now RUS) was established as a part of President Franklin D. Roosevelt's "New Deal." This system, more than any other, has allowed rural people to enjoy the same modern conveniences as their urban neighbors.

Today, the mission of the Rural Utilities Service is to serve a leading role in improving the quality of life in rural America by administering its Electrification, Telecommunications, and Water and Waste Disposal Programs in a service-oriented, forward-looking, and financially responsible manner.

Rural Housing Service

The Rural Housing Service operates credit programs for rural communities and rural housing. A challenge of this agency is to make sure loans are for housing in truly rural America rather than counties adjacent to metropolitan areas. Previously, rural housing loans were handled by the Farmers Home Administration.

Rural Business-Cooperative Service

This agency, originally called the Agricultural Cooperative Service, was first organized in the 1930s as part of the New Deal. Its major functions are to provide technical assistance and conduct research on economic, financial, organizational, managerial, legal, social, and other issues that affect agricultural cooperatives.

Today, the mission of RBS is to enhance the quality of life for all rural Americans by providing leadership in building competitive businesses, including cooperatives that can build sustainable economic communities, compete successfully in the domestic arena, and develop emerging market opportunities in the international arena. RBS objectives are to invest its financial resources and technical assistance in businesses and communities and to build partnerships that leverage public and private resources to stimulate rural economic activity.

NATURAL RESOURCES AND ENVIRONMENT

The Natural Resources and Environment division is a relatively new organization of agencies from other divisions. Their major objectives are to conserve our natural resources and improve the environment in which we live and work. The financial support of this division has been reduced considerably, and user fees have been increased.

Forest Service

The U.S. Forest Service was created in 1905 and is responsible for the conservation and management of the nation's forest and other wildlife resources. There are 154 national forests and 19 national grasslands, containing 187 million acres in forty-one states and Puerto Rico. The major resources obtained from these forests and rangelands are wood, water, wildlife, forages, and recreation. The Forest Service conducts research in timber, range, water, and recreation management and in forest product utilization in order to ensure maximum benefits for the greatest number of people. The Forest Service also cooperates with state forestry agencies and private landowners to provide technical assistance for the protection and management of 480 million acres of state and private forest lands.

In addition to trees, the national forests and grasslands provide a home for over 4 million big game animals and 79 species of threatened or endangered wildlife. The Forest Service provides services that benefit production agriculturalists and the general public as well. Refer to Figure 11–4.

Natural Resources Conservation Service

The nation's leaders have been concerned about water and soil conservation since the 1800s, and numerous conservation acts have been passed. However, it was not until 1933 that a specific agency was created to deal with this problem. The Soil Erosion Service was established in the Interior Department but transferred to the Department of Agriculture two years later and renamed the Soil

Figure 11–4 The U.S. Forest Service is involved with recreation management as well as having responsibility for the conservation and management of the nation's forest and other wildlife resources. (Courtesy of Donald Waldrop)

Conservation Service (SCS). In 1994, it was renamed the Natural Resources Conservation Service. The NRCS administers programs to conserve soil and water resources through providing technical assistance to production agriculturalists and other landowners in soil and water conservation districts. It also provides technical assistance to communities and local governments in rural development and flood prevention programs. Soil classification, mapping, and surveys are also conducted by 2,950 local conservation districts employing about 17,000 people.

Jobs. There are many interesting jobs in the NRCS, which include the following:

🌿 Soil conservationists work with production agriculturalists and other landowners on ways to conserve soil, build farm ponds, and reduce water pollution.

🌿 Soil conservation technicians assist production agriculturalists and others in installing needed conservation practices, and then supervise the installation.

🌿 Soil scientists map and classify soils; identify problem areas, such as wetness and erosion; identify soils on aerial photographs; write soil descriptions; and prepare other information about soils.

🌿 Range conservationists help plan grazing systems, suggest ways to control brush, and offer advice on water management, as well as many other tasks.

🌿 Engineers assist in erosion control, water management, structural design, construction, hydraulics, soil mechanics, and environmental protection.[10]

RESEARCH, EDUCATION, AND ECONOMICS

The Research, Education, and Economics Division evolved from an agency created in 1978 that encompassed the activities of several research and education agencies. This division provides support, coordination, and joint planning of food and agricultural research. It also provides extension and teaching efforts among federal departments, state extension services, and experiment stations, universities, and other private and public institutions. The specific agencies involved include the Agricultural Research Service and National Agricultural Library; the Cooperative State Research, Education, and Extension Service; the Economic Research Service; and the National Agricultural Statistics Service.

Agricultural Research Service

The Agricultural Research Service (ARS) is the major agency that does basic research on plants and animals.

Animal and Plant Research. The agency deals with disease prevention and cure as well as research related to improving the productivity of plants and animals. Current agricultural research priorities include improving animal reproductive efficiency; improving the use of plant **germ plasm** (genes) and *preservation*; removing barriers to crop productivity; conserving soil, water, and air; reducing the effects of erosion on soil productivity; using energy efficiently; developing new pest control technology, and so on.[11]

Marketing Research. ARS also maintains marketing research programs related to developing new and improved products. Other goals are to expand domestic and foreign markets for agricultural products, protect and maintain the quality of agricultural products in marketing channels, and improve the efficiency of market facilities. Although much of the work of the Agricultural Research Service is in cooperation with the states, the service also sends many researchers to foreign countries.

National Agricultural Library. The library contains the most extensive source of agricultural information in the world with about two million books, and over 25,000 periodicals. The National Agricultural Library is located in Beltsville, Maryland, and its major users are USDA employees. Any individual can use the materials, but you must request what you want through a public or university library.

Cooperative State Research, Education, and Extension Service (CSRES)

The Cooperative State Research Service. This service administers federal grant funds for research in the production phase of agriculture, agricultural marketing, rural development, and forestry. Funds are provided to state agricultural experiment stations and other designated institutions in the fifty states, Puerto Rico, Guam, the Virgin Islands, and the District of Columbia. Cooperative State Research Service staff members review research proposals, conduct reviews of research in progress, and encourage cooperation between states.

Extension Service. The Extension Service (CSRES) is one of the major educational agencies of the Department of Agriculture (USDA). It is the federal phase of the Cooperative Extension Service, which includes federal, state, and local governments. All three groups share in the financing, planning, and conducting of agricultural education programs. The Extension Service was created by the Smith-Lever Act of 1914, and its major purpose is to help the public learn about the latest technology developed through research by various agricultural institutions.

The Extension Service has educational faculty and facilities in 1862 and 1890 land grant universities, and it staffs at least 3,150 county offices representing local governments. Extension personnel offer education programs to production agriculturalists, families, individuals, and communities. **Distance learning**, **teleconferencing**, the Internet, and other communication technologies are used to deliver programs, research findings, science, technology, and other services. Recently, the Cooperative Extension Service identified eight **initiatives** for its work in the coming years. The areas of concentration are as follows:

- 🌿 competition and profitability of U.S. agriculture
- 🌿 alternative agricultural opportunities
- 🌿 water quality
- 🌿 conservation and management of natural resources
- 🌿 revitalizing rural America
- 🌿 improving nutrition, diet, and health

❦ family and economic well-being

❦ building human capital[12]

4-H Clubs. Depending on the size of the county or local district, the Agricultural Extension Service has one to five agents. One of these has the responsibility of the adults' programs and one has the responsibility of the youth program, the 4-H Club (Hands, Head, Heart, Health). Educators learned long ago that young people are more receptive to learning and adopting new ideas than are adults. Therefore, it was logical that the Agricultural Extension Service direct its attention to educating rural youth as well as adults, and the media developed to reach young people were 4-H Clubs. The purpose of the clubs is to give young people an opportunity to participate in problem-solving activities that affect themselves, their homes, and their communities. Members are encouraged to develop projects that will improve their knowledge in agriculture, science, home economics, personal development, community service, and numerous other areas. Membership in 4-H Clubs in 1984 was estimated at about 5,700,000 in approximately 76,572 organized clubs.[13] Members are active in all fifty states, Puerto Rico, the Virgin Islands, and numerous other countries. Youth organizations in over eighty countries have adopted the 4-H system. Approximately $115 million is appropriated annually to 4-H activities from federal, state, and local governments, in addition to about $25 million from private sources. Thirty-eight states now have 4-H foundations, and over 200 of the nation's major corporations contribute to 4-H annually. Most of this support is coordinated through the National 4-H Council. Refer to Figure 11–5.

Economic Research Service

The Economic Research Service provides timely and reliable agricultural economic data, including historical research, forecasts from major eco-

Figure 11–5 One of the objectives of 4-H is personal development. This member is participating in a 4-H-style show. (Courtesy of Marlene Graves)

nomic indicators, and **policy analysis** that serves as a basis for economic decision making by production agriculturalists, consumers, extension workers, private analysts, marketing firms, input suppliers, and public officials. It employs the largest single group of agricultural economists in the world.

National Agricultural Statistics Service

The National Agricultural Statistics Service has the unique responsibility of continuing the original mandate of the Department of Agriculture, which was to collect and distribute agricultural data. This agency provides agricultural data to fill in the gaps between agricultural census years. Crop and livestock estimates are made by a central office in Washington, D.C., and by state statistical offices, which are operated as joint state and federal services. The agency also conducts and coordinates research on statistical techniques and data collected by satellites. Refer to Figure 11–6.

Food Safety and Inspection Service

The Food Safety and Inspection Service is responsible for making sure that the nation's meat and poultry supplies are safe, wholesome, and properly labeled and packaged. This agency provides plant inspection of domestic meat- and poultry-processing operations, reviews inspection systems in foreign countries exporting meat and poultry products to the United States, and inspects products at ports of entry. These services are funded by state and federal programs.

USDA Summary

In summary, the U.S. Department of Agriculture plays a very important role in modern agriculture. The results of USDA activities affect, not only those in the agricultural industry, but the public at large. The USDA has expanded considerably since it was established in 1862, as the need for more and better services has increased. The USDA was established primarily as an agency to serve the needs of production agriculturalists but has now expanded to a group of agencies intended to serve the agricultural industry and the nation as a whole. Currently, over 50 percent of the USDA's budget is designed to see that the needy are at least provided a minimum level of food through the Food and Nutrition Service.

Currently, the USDA is funded at about $65 billion annually and maintains a payroll for about 100,000 people. Most of the positions are under the U.S. Civil Service Commission. To qualify, you have to pass a civil service examination. There is stiff competition for the agricultural research federal civil service jobs. However, a job with this agency is desirable, as the federal government

Figure 11–6 A Global Positioning System antenna on the top of the tractor operator's cab communicates with a satellite and a yield monitor to allow an onboard computer to plot corn yields about every six feet as the combine moves along. Data stored in the computer can later be used to produce color-coded yield maps for each field. (Courtesy of USDA)

offers many outstanding benefits to its employees, such as vacations, sick leave, **flex time**, alternative work schedules, child care arrangements, credit unions, and recreation and fitness centers.

STATE DEPARTMENT OF AGRICULTURE

Although the major public support and regulation of agricultural programs are initiated and funded at the federal level, state governments also play a major role in modern agriculture. Presently, all fifty states have departments of agriculture or other, comparable agencies responsible for administering agricultural programs at the state level. State departments of agriculture are administered by secretaries, commissioners, or directors, most of whom are appointed by the state governor. However, some of these administrators are elected rather than appointed.

Mission

The state departments of agriculture are primarily regulatory agencies. However, numerous other activities are funded at the state level, including disease prevention and treatment programs, regulating weights and measures, collecting and disseminating statistical data, supervising agricultural fairs, enforcing chemical fertilizer and seed quality laws, and agricultural promotion activities. The emphasis of the various activities vary with each state.

National Association of State Departments of Agriculture (NASDA)

In an effort to influence the agricultural industry beyond the state level, the administrators of state departments of agriculture organized the National Association of Commissioners, Secretaries and Directors of Agriculture in 1915. In 1955, the name of this group was changed to the National Association of State Departments of Agriculture (NASDA). Since 1968 NASDA has maintained an office in Washington, D.C., headed by an executive secretary with a small staff, to gather and disseminate data for the Congress and other federal agencies on agricultural matters.

Duties of NASDA

NASDA staff members do not perform lobbying functions, but they do assist individual members and their staffs in preparing reports for congressional committees. Policy positions on current agricultural issues are often prepared by NASDA and presented to the proper federal agencies for consideration. NASDA has been, and continues to be, a very effective means of keeping the federal agencies abreast of grassroots problems in agricul-

ture. "The interest of NASDA covers the full scope of agriculture from production through processing and transportation, consumer affairs, national regulatory programs and the fiscal policies of the Federal and State governments."[14]

LAND GRANT SYSTEM

One of the most significant developments in the history of American agriculture was the establishment of what we know today as the land grant system. This system was designed to develop new technology through agricultural research and to make this technology available to the agricultural industry through a comprehensive educational system. It consists of the land grant universities, agricultural experiment stations, and the Agricultural Extension Service.

History of Land Grant Systems

The first part of the system to develop was the land grant universities or colleges. In 1853, Justin S. Morrill, a representative from Vermont, introduced a bill in the House requesting that public lands be donated "to the several states which may provide colleges for the benefit of Agriculture and Mechanical Arts."[15] Each state not in rebellion was entitled to 30,000 acres of public land for each senator and representative in Congress. The bill finally passed by a narrow margin in 1858 but was vetoed by President James Buchanan.

Representative Morrill again introduced his bill in early 1862, but again it was rejected. However, the bill was later amended, passed by both houses, and signed by President Lincoln in July 1862. The final content of the bill required that 30,000 acres would be allotted to each state for each senator and representative, with a maximum of 1 million acres in any one state. The proceeds from the sale of such land were to be used to support agricultural and mechanical programs at the college level. Each state was directed to designate one or more colleges as land grant colleges.

Colleges of 1890

In 1890, a second Morrill Act was passed, which provided annual appropriations to support teaching at the land grant colleges and added a "separate but equal" provision authorizing the creation of colleges for blacks, who, at that time, were not allowed to attend many of the universities. Sixteen southern states took advantage of this act and created colleges that are currently referred to as the "colleges of 1890."

Agricultural Experiment Stations

The first Morrill Act was followed by the Hatch Act in 1887, which authorized federal support to each state that would establish an agricultural experiment station in conjunction with its land grant college. An annual appropriation of $15,000 was approved for each state for setting up the stations and conducting experiments in the agricultural and mechanical arts. An important factor in the development of the land grant system was the establishment of the National Association of State Universities and Land-Grant Colleges in 1887. This organization became very powerful politically and had a major influence on the future expansion of research and education in agriculture.

Development of the Cooperative Extension Service

It soon became obvious that the new technology developed through agricultural experiment stations was not reaching people in rural communities. In response to this need, the Cooperative Extension Service was established in 1914 by the Smith-Lever Act. This act provided an organizational structure through which the Cooperative Extension Service is funded cooperatively by the federal, state, and local governments and adminis-

tered by the land grant universities in each state. Refer to Figure 11–7 for a list of the land grant universities.

NON–LAND GRANT AGRICULTURAL PROGRAMS

During the early years of higher education in the United States, the major responsibility of teaching in agriculture was delegated to the land grant colleges and universities. However, it soon became evident that these institutions alone could not adequately serve all the young people who were interested in studying agriculture beyond the high school level. There are still some states today that offer programs in agriculture only at the land grant universities. However, at least thirty-four states now have one or more agricultural programs in non–land grant institutions. The non–land grant schools now play a major role in educating young people for agricultural occupations, especially at the undergraduate level. There are more programs available in non–land grant institutions, and they have about 50 percent of total agricultural enrollment. Many of the students in junior colleges transfer to land grant and non–land grant institutions, and many non–land grant graduates transfer to land grant institutions for graduate study.

Comparison of Land Grant and Non–land Grant Universities

Land grant and non–land grant institutions are different in terms of financial support and objectives. Land grant institutions are supported by state and federal funds on an annual basis. The only question is how much they receive. Non–land grant institutions receive state funds on a regular basis; they may receive federal funds for special projects, but not on a regular basis. Non–land

A List of the Land Grant Universities in All 50 States

State	Land-Grant Universities	State	Land-Grant Universities
Alabama	Auburn University	Missouri	University of Missouri
	Alabama A&M[a]		Lincoln University[a]
Alaska	University of Alaska	Montana	Montana State University
Arizona	University of Arizona	Nebraska	University of Nebraska
Arkansas	University of Arkansas	Nevada	University of Nevada
	Arkansas A&M[a]	New Hampshire	University of New Hampshire
California	University of California	New Jersey	Rutgers University
Colorado	Colorado State University	New Mexico	New Mexico State University
Connecticut	University of Connecticut	New York	Cornell University
Delaware	University of Delaware	North Carolina	North Carolina State University
	Delaware State University[a]		North Carolina A&M[a]
Florida	University of Florida	North Dakota	North Dakota State University
	Florida A&M[a]	Ohio	Ohio State University
Georgia	University of Georgia	Oklahoma	Oklahoma State University
	Fort Valley State College[a]		Langston University[a]
Hawaii	University of Hawaii	Oregon	Oregon State University
Idaho	University of Idaho	Pennsylvania	Pennsylvania State University
Illinois	University of Illinois	Puerto Rico	University of Puerto Rico
Indiana	Purdue University	Rhode Island	University of Rhode Island
Iowa	Iowa State University	South Carolina	Clemson University
Kansas	Kansas State University		South Carolina State University[a]
Kentucky	University of Kentucky	South Dakota	South Dakota State University
	Kentucky State University	Tennessee	University of Tennessee
Louisiana	Louisiana State University		Tennessee State University[a]
	Louisiana Tech Southern University	Texas	Texas A&M University
Maine	University of Maine		Stephen F. Austin
Maryland	University of Maryland, College Park		Prairie View A&M[a]
	University of Maryland, Eastern Shore[a]	Utah	Utah State University
Massachusetts	University of Massachusetts	Vermont	University of Vermont
Michigan	Michigan State University	Virginia	Virginia Polytech Institute
Minnesota	University of Minnesota		Virginia State University[a]
Mississippi	Mississippi State University	Washington	Washington State University
	Alcorn A&M[a]	West Virginia	University of West Virginia
		Wisconsin	University of Wisconsin
		Wyoming	University of Wyoming

[a]Colleges of 1890.

Figure 11–7   Land grant universities were created in 1862 and 1890.

grant schools are primarily, and in some cases strictly, concerned with teaching, although most teachers are encouraged to do some research and public service. Land grant institution faculties are responsible for divided duties in teaching, research, and/or extension activities.

SEA GRANT PROGRAM

Historical Development

In conjunction with increased efforts in the early 1960s to expand our knowledge about the availability of usable resources located in our major bodies of water, Athelstan Spilhaus, chair of the Committee on Oceanography of the Natural Academy of Sciences, proposed the creation of sea grant universities, similar to the land grant universities established under the Morrill Act of 1862. The idea was further developed at a sea grant conference in 1965, which was sponsored by the University of Rhode Island. As a result, Senators Clairborne Pell of Rhode Island and Paul Rogers of Florida introduced the Pell-Rogers Sea Grant Bill. The National Sea Grant College and Program Act was signed into law by President Johnson in 1966, and the National Science Foundation was given the administrative responsibility for the program. In 1970, the administrative responsibility of the sea grant program was transferred to the National Oceanic and Atmosphere Administration, an organizational unit of the U.S. Department of Commerce.

Mission

The national sea grant program is concerned with the exploration and development of the resources of the oceans, the Great Lakes, and the coastal waters. It was established to promote research, education, and advisory services in marine resources, including conservation, and socially and economically sound management.

Reason for the Sea Grant Name

The term *sea grant* was chosen to emphasize its parallel with the land grant program, with which it shares a basic organizational pattern of research, teaching, and extension functions.

Funding

Through a matching fund program, the sea grant program provides grants to colleges and universities, as well as private industry, state agencies, private foundations, consumer interest groups, and private individuals, to conduct research and educational projects in marine development. Usually, two-thirds of the funds come from federal appropriations and the remaining one-third is supplied by state, local, or private funds. Normally, 60 percent of the sea grant funds is allocated to research, and the remaining 40 percent is used for education and advisory services programs. Grants awarded by the Office of Sea Grant are classified into four types: project awards, coherent project programs, sea grant institutional programs, and sea grant college programs.

Project Awards. **Grants** are awarded for specific projects in research, education, or advisory services that are appropriate to sea grant objectives. These awards are based on proposals submitted by competing individuals or institutions, with specific goals and completion schedules spelled out. Emphasis is placed on innovative and current projects designed to meet the needs of a specific area.

Coherent Project Programs. The purpose of the coherent project programs is to provide opportunities for academic and nonacademic institutions to use their resources in solving marine-related problems. This program allows for work on broad-based problems and allows more flexibility in time and methods of project completion than in the grant program. These projects are usually **interdisciplinary** in scope and may include research, education, and/or advisory services, depending on the institution's resource capabilities.

Sea Grant Institutional Programs. Status as a sea grant institutional program is given to an institution, or **consortium** of institutions, with a broad competence in marine affairs and a comprehensive marine program in research, education, and advisory services. An institution with coherent project status may advance its project to the classification of Sea Grant Institution Program through the growth and development of a broad-based program over a period of time.

Sea Grant Colleges. The highest level of federal support is given to those institutions of higher learning that have been designated as sea grant colleges. This status is given to those institutions that have clearly demonstrated outstanding leadership in the quality, quantity, and productivity of performance in research, education, and advisory services for a period of at least three years. The designation of sea grant college is a highly coveted honor. It is sought by many but granted to only a limited number of universities. Only twenty-nine sea grant college designations have been awarded. The location of these programs is shown in Figure 11–8.[16]

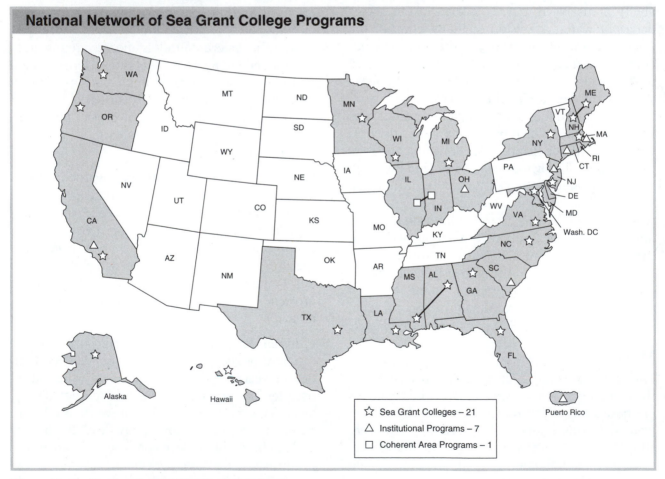

National Network of Sea Grant College Programs

☆ Sea Grant Colleges – 21
△ Institutional Programs – 7
☐ Coherent Area Programs – 1

Figure 11–8 The locations of 21 sea grant colleges.

International Mission of the Sea Grant Programs

In 1973, the Sea Grant Program Act was amended to provide a means for sharing the results of marine research with other nations through cooperative research and educational programs. The sea grant program has made a significant contribution, although it is only about three decades old. It seems evident that this program will play an even greater role in the future in providing food and nonfood resources from the sea and major lakes for the United States and the rest of the world.[17]

AGRICULTURAL EDUCATION

To complement agricultural research at the university level, agricultural education was developed to train young men and women at the high school level who want to pursue a vocation of production agriculture or agribusiness. Modern agricultural education is based on a cooperative federal, state, and local program set up under the Smith-Hughes Act of 1917. The act was named after Senator Hoke Smith and Representative Dudley Hughes, both of Georgia, who were the major sponsors. The program is administered by the U.S. Office of Education through state staff for agricultural education. Each state is responsible for presenting an agricultural education plan to the U.S. Office of Education. When this plan is approved, it becomes the responsibility of the states to implement the program through cooperation of local schools.

Agricultural education was originally designed to prepare high school students for farming, primarily through agricultural instruction planned around supervised, on-farm programs (Supervised Agricultural Experience Programs) of rural youth. Many current programs are still production oriented, but others have expanded into numerous agribusiness areas, depending on the particular needs of each community. Besides agricultural

education teacher, there are many careers in the public agribusiness services. Refer to the Career Option on page 296 for an explanation of some of the careers in this area.

Future Farmers of America (FFA)

The Future Farmers of America organization evolved from "agriculture clubs" which began forming shortly after the Smith-Hughes Act of 1917 was passed which provided federal support of vocational agriculture programs. However, it was not until 1928 that a national organization was formed, when thirty-three delegates from eighteen states met in Kansas City, Missouri, and formulated a constitution based on the organization of the Future Farmers of Virginia. Leslie Applegate of New Jersey was elected national president, Dr. C. H. Lane of Washington, D.C., became the first national advisor, and the first executive secretary-treasurer was Henry Groseclose of Virginia.[18]

Size

During the first ten years of FFA, membership increased to 100,000, representing 4,000 chapters and forty-eight state associations. In 1939, the organization purchased 28 1/2 acres of the George Washington estate in Alexandria, Virginia, which became the national headquarters and an FFA camp. In two years' time, FFA membership had increased to over 245,000, with about 7,500 chapters. Currently, the national FFA organization serves more than 400,000 active members and about 7,200 chapters located throughout the United States. In 1997, over 42,000 FFA members, advisors, and guests attended the national convention. The impact of FFA as a national organization is emphasized by the fact that numerous presidents, including Presidents Truman, Eisenhower, Johnson, Nixon, Ford, Carter, and Bush, have addressed national FFA conventions. The mission statement of the National FFA Organization is to develop premier leadership, personal

CAREER OPTION

Career Areas: Natural Resources Conservation Service, Air Quality Control, Federal and State Regulatory Agencies, Agricultural Extension and/or 4-H Leader, Agricultural Education Teacher

Soil and water conservation career options provide extensive opportunities for indoor and outdoor work. Work may be found in the Natural Resources Conservation Service. One may work in the field, served by such offices in most counties of the United States. Typical job titles are conservation technician, farm planner, soil scientist, and soil mapper.

The U.S. Department of Interior, state departments of natural resources, city and county governments, industry, and private agencies hire people with soil and water conservation expertise. Their employees manage water resources for recreation, conservation, and consumption. They may also work as consultants, law enforcement officers, technicians, administrators, heavy equipment operators, and the like.

Air quality control employees monitor and help to improve air quality. Technicians collect, and chemists analyze, samples of air taken from various places in the atmosphere, buildings, and homes. Employees of environmental protection agencies and environmental advocacy groups strive to help maintain a healthful environment. Air quality specialists advise and assist industry in reducing harmful emissions from motor vehicles and industrial smokestacks.

Various federal and state governmental agencies provide services to aid production agriculturists in solving some of their farm problems.

Employment is available in the following areas: conducting food inspection, keeping foreign diseases from entering the country, protecting our natural resources, preventing disease, providing treatment programs, regulating weights and measures, supervising agricultural fairs, enforcing chemical fertilizer and seed quality laws, and sponsoring agricultural promotion activities.

Careers are also available as an agricultural extension agent/4-H leader or an agricultural education teacher/FFA advisor. Both these public agriservice jobs are very rewarding as they involve helping young people grow and learn.

Agricultural extension agents and 4-H leaders have very rewarding jobs, including working with elementary school students. (Courtesy of USDA)

Figure 11–9 The mission statement of the National Future Farmers of America is to develop premier leadership, personal development, and career success. These FFA members are attending the FFA Washington Leadership Conference and meeting governmental officials from their states. (Courtesy of the National FFA Organization)

development and career success. Refer to Figure 11–9.

Jobs in Agricultural Education

There are over 10,000 positions as agricultural education teachers throughout the United States. Most rural school systems and many urban systems have agricultural education programs. There is expected to be a shortage of agricultural education teachers in the future. Four years of college is needed for most programs to become a certified agricultural education teacher.

CONCLUSION

The federal and state governments have a major impact on the agribusiness service sector of the agricultural industry. It would be difficult for production agriculturalists to survive without the public agribusiness services. Many government services are taken for granted, but it would be difficult to imagine the accomplishments of the agricultural industry without the various USDA agencies, state departments of agriculture, the land grant and non–land grant universities, and agricultural education programs.

SUMMARY

American leaders have long recognized the importance of a strong agricultural sector, possibly because many U.S. political leaders had their roots in farming. At least twenty-three presidents were involved in production agriculture, and numerous members of Congress were part- or full-time production agriculturalists before and during their terms of service. The major role of government in agriculture has come from the federal level. However, state and local governments also play a vital role in agricultural development.

The first federal aid to agriculture in the United States came in 1839, when Congress appropriated $1,000 to collect farm statistics and to collect and distribute seeds of various types. In 1862, the Department of Agriculture was created, and in 1889 it was given a cabinet ranking. From this beginning, the USDA has expanded into a complex organization of several agencies, which is administered by a secretary, deputy secretary, assistant secretaries, and numerous administrative directors.

Various governmental agencies and services are provided to aid production agriculturalists in solving some of their farm problems. The government offers many services that we take for granted, such as inspecting food to ensure safety and quality, keeping foreign diseases from entering the country, and protecting our natural resources as well as many others.

The U.S. Department of Agriculture plays an important role in the agricultural industry. There are several divisions and agencies with the USDA. The major divisions currently organized include: Food, Nutrition, and Consumer Services; Marketing and Regulatory Services; Farm and Foreign Agricultural Services; Rural Development; Natural Resources and Environment; Research, Education, and Economics; and Food Safety.

State governments also play a major role in modern agriculture. Presently, all fifty states have departments of agriculture or other, comparable agencies that are responsible for administering agricultural programs at the state level. The state departments of agriculture are primarily regulatory agencies. However, numerous other activities are funded at the state level, including disease prevention and treatment programs, regulation of weights and measures, collection and dissemination of statistical data, supervision of agricultural fairs, enforcement of chemical fertilizer and seed quality laws, and the sponsoring of agricultural promotion activities.

One of the most significant developments in the history of American agriculture was the establishment of what we know today as the land grant system. The system consists of land grant universities, agriculture experiment stations, and the Agricultural Extension Service. In 1862, the Morrill Act established land grant universities, and in 1887, the Hatch Act provided funds for establishing agricultural experiment stations. In 1914, the Agricultural Extension Service was created by the Smith-Lever Act and added to the system.

In addition to the land grant system, non–land grant schools play a major role in educating young people for agricultural occupations, especially at the undergraduate level. Approximately 50 percent of all undergraduate agricultural students are enrolled in non–land grant institutions. The major difference is that non–land grant schools are primarily concerned with teaching, whereas land grant schools are responsible for teaching, research, and extension activities.

In the 1960s, a sea grant program, similar to the land grant program, was developed to explore and develop the resources of the oceans, Great Lakes, and coastal waters. Through a matching fund, the sea grant program provides grants to colleges and universities as well as to private industries, state agencies, private foundations, and others to conduct research and educational projects in marine development.

The Smith-Hughes Act of 1917 provided funds for agricultural education programs at the high school level. These programs are designed primarily for young men and women who have selected some phase of agriculture as a vocation. The youth group associated with agricultural education is the Future Farmers of America. The mission statement of the National FFA Organization is to develop premier leadership, personal development, and career success. The 4-H Clubs are also organized and supervised by Agricultural Extension Service personnel.

END-OF-CHAPTER ACTIVITIES

Review Questions

1. Name eight American presidents who had close farm ties.
2. List twenty-four services provided by the federal government.
3. What are the seven divisions within the USDA?

4. What programs are administered by the Food, Nutrition, and Consumer Service?

5. What are three agencies within the Marketing and Regulatory Services Division?

6. What are three agencies within the Farm and Foreign Agricultural Services Division?

7. What are three responsibilities of the Farm Service Agency?

8. What was the Risk Management Agency called before it acquired its new name?

9. What are three reasons why the Commodity Credit Corporation was created?

10. What is the mission of the Rural Utilities Service?

11. What is the mission of the Rural Business-Cooperative Service?

12. What are the four agencies within the Rural Development Division?

13. List and briefly describe five types of jobs with the Natural Resources Conservation Service.

14. What are the four agencies within the Research, Education, and Economics Division of the USDA?

15. What are eight initiatives of the Cooperative Extension Service?

16. What is the purpose of 4-H Clubs?

17. What types of projects are 4-H members encouraged to develop?

18. Give seven examples of animal and plant research being conducted by the Agricultural Research Service.

19. List six activities conducted by State Departments of Agriculture.

20. What is the purpose and/or duties of the National Association of the State Departments of Agriculture (NASDA)?

21. What was the land grant system designed to do?

22. What three groups do the land grant system include at each designated university?

23. Many universities have *A&M* at the end of their name, such as Texas A&M. What does the A&M stand for?

24. What are the two major differences between land grant and non–land grant universities?

25. What is the mission of the national sea grant program?

26. Name and briefly explain the four types of grants awarded by the Office of Sea Grant.

27. What is the mission statement of the National Future Farmers of America Organization?

Fill in the Blank

1. Agriculturalists want more research and educational services and less regulation while consumer groups want more regulation and less funds spent on _____ and _____.

2. The first federal aid to agriculture in the United States came in _____, when Congress appropriated $1,000 to collect farm statistics and to collect and distribute seeds of various types.

3. On May 15, _____, President Lincoln signed the act creating the USDA.

4. _____ _____ was the first commissioner (now called secretary) of agriculture.

5. There are _____ national forests and _____ national grass lands, containing 187 million acres in forty-one states and Puerto Rico.

6. The national forests and grasslands provide a home for over _____ million big game animals and _____ species of threatened or endangered wildlife.

7. Soil classification, soil mapping, and soil surveys are conducted by _____ local conservation districts employing about _____ people.

8. The _____ Act of 1862 started land grant universities.

9. The _____ Act of 1887 started Agricultural Experiment Stations.

10. The Cooperative Extension Service was established in 1914 by the _____ Act.

11. Agricultural Education was established in 1917 by the _____ Act.

Matching

a. Farm Service Agency
b. Foreign Agricultural Service
c. Risk Management
d. Food and Consumer Service
e. Food Safety and Inspection Service
f. Agricultural Marketing Service
g. Animal and Plant Health Inspection Service
h. Grain Inspection, Packers and Stockyards Administration
i. Forest Service
j. National Resource Conservation Service
k. Agricultural Research Service
l. National Agricultural Library
m. Cooperative State Research, Education, and Extension Service
n. Economic Research Service
o. National Agricultural Statistics Service
p. Rural Utilities Service
q. Rural House Service
r. Rural Business-Cooperative Service
s. Commodity Credit Corporation (CCC)

_____ 1. provides timely and reliable agricultural economic data

_____ 2. collects and distributes agricultural data

_____ 3. administers the Food Stamp Program

_____ 4. provides technical assistance and conducts research on issues that affect agricultural cooperatives

_____ 5. promote the orderly and efficient marketing and distribution of agricultural commodities

_____ 6. protects plants and animals from economically dangerous diseases and pests

_____ 7. establishes official standards for grain, conducts weighing and inspection activities, and inspects other agricultural products

_____ 8. responsible for making sure that the nation's meat and poultry supplies are safe

_____ 9. maintains a good agricultural economy through price supports, makes loans and purchases, and conserves the nation's farm resources

_____ 10. This loan program was the tool that acted as a price support mechanism.

_____ 11. played a major role in making the United States the world's largest exporter of agricultural products

_____ 12. provides financial assistance to farmers who are unable to obtain loans from private agencies for farmland

_____ 13. provides farmers with insurance that repays crop production losses due to uncontrollable natural conditions

_____ 14. helps rural people obtain electric and telephone services by making long-term loans to rural cooperatives

_____ 15. responsible for the conservation and management of the nation's forest and other wildlife resources

_____ 16. administers programs to conserve soil and water resources through technical assistance to production agriculturalists and other landowners

_____ 17. helps the public learn about the latest technology developed through research by various agricultural institutions

_____ 18. contains the most extensive source of agricultural information in the world with about 2 million books and over 25,000 periodicals

_____ 19. major agency that does basic research on plants, animals, and marketing

_____ 20. administers federal grant funds for research in the production phase of agriculture, agricultural marketing, rural development, and forestry

_____ 21. makes loans for housing in rural areas

Activities

1. Make a list of all the federal and state government jobs in your community that are related to the agricultural industry. Share your findings with the class. Get a total of the agricultural-related federal and state jobs.

2. Suppose there were suddenly an act of Congress making taxes illegal. Explain if and how you would then pay for the agricultural-related federal and state jobs.

3. In the "Reasons for Government Services" section of this chapter, twenty-four services of the federal government were listed. Select and defend the five that you feel are the most important.

4. Select five government services that you feel are the least important and explain why.

5. There are approximately eighteen agencies within the USDA. Prepare a brief report on one of these agencies and present it to the class. If your school library does not have enough information, you may wish to try the Internet, contacting local agencies, or writing for additional information.

6. List the fringe benefits of working for either the federal or state government.

7. Compare the purpose and/or mission of agricultural education and the FFA to the purpose and/or mission of the Extension Service and the 4-H.

8. Of all the public agribusiness business services discussed in this chapter, which do you believe is the most important to you personally? Prepare a brief paper and share this with the class.

NOTES

1. Vivian Whitehead, "White House Farmers," in USDA, ed., *The American Farmer* (Washington, D.C.: USDA, Economic Research Services, 1976).

2. USDA, *A Guide to Understanding the United States Department of Agriculture* (Washington, D.C.: USDA, Office of Personnel, 1970).

3. Alfred H. Krebs and Michael E. Newman, *Agriscience in Our Lives* (Danville, Ill.: Interstate Publishers, 1994), p. 57.

4. Ibid., p. 58.

5. Ibid., pp. 58–60.

6. United States Department of Agriculture (1997). *Agriculture Fact Book, 1997* (Washington, D.C.: U.S. Government Printing Office, 1997).

7. Ibid.

8. Ibid.

9. Marcella Smith, Jean M. Underwood, and Mark Bultmann, *Careers in Agribusiness and Industry* (Danville, Ill.: Interstate Publishers, 1991), p. 253.

10. Ibid., pp. 253–255.

11. Ibid., p. 256.

12. Ibid., p. 263.

13. 4-H Club National Home Page (1999). Available Online: http://www.4-H.org.

14. National Association of State Departments of Agriculture (NASDA), *NASDA—What's It All About?* (Washington, D.C.: Author, 1977).

15. Benjamin A. Hubbard, *A History of the Public Land Policies* (New York: Peter Smith, 1939), p. 348.

16. N. Omri Rawlins, *Introduction to Agribusiness* (Murfreesboro, Tennessee: Middle Tennessee State University, 1999), p. 129.

17. Ibid.

18. Future Farmers of America (FFA), *Official FFA Manual* (Mt. Vernon, Va.: Author, January 1997), p. 6.

CHAPTER

12

Agribusiness Input Services: Credit

OBJECTIVES

After completing this chapter, the student should be able to:

- discuss the importance of farm credit
- explain three fundamentals of credit
- describe three areas of credit needs
- differentiate the three lengths of financing terms
- explain the components of a credit profile
- compute interest
- list the agricultural credit sources for real estate and non-real estate loans
- describe the Farm Credit System
- describe the Consolidated Farm Service Agency (FSA)
- describe the role of commercial banks in agribusiness credit
- describe the role of life insurance in agribusiness credit
- describe the Commodity Credit Corporation
- discuss the role of individuals and others in agribusiness credit
- explain the career opportunities in agribusiness credit input services

TERMS TO KNOW

actuarial interest rate
add-on interest
amortize
annual percentage rate (APR)
annuities
appreciate
buyer's fever
collateral
contractual interest rate
debt-equity ratio
default
delinquencies

depreciable
discount (prime) rates
dividends
equilibrium price
equity capital
farm assets
financial assets
fixed expenses
foreclosure
interest
interest rates
lien

long-term credit
nonrecourse loans
open account
operating expenses
principal
secured
securities markets
simple interest
speculative investments
start-up expenses
viable

INTRODUCTION

Decisions about credit often are the most important judgments that people in the agricultural industry must make. These decisions often determine whether individuals operating in the agribusiness input, production sector or in the agribusiness output sector will succeed in making a profit. Remember the old saying, "It takes money to make money." A business must have sufficient financial resources if it is to show a profit. Money is used in every area of the agricultural industry, including for land, buildings, equipment, livestock, crops, and operating expenses.

Questions about whether to use credit for specific purposes must be examined carefully. The ability to make sound business judgments comes from education, careful study, and practical business experience.[1]

IMPORTANCE OF FARM CREDIT

In the agricultural industry, credit is needed to overcome a shortage of **equity capital**. Limited capital, fluctuating interest rates, and lack of credit information are major problems facing many production agriculturalists. Credit is important to production agriculturalists in order to:

❦ increase production

❦ raise the quality of what is produced

❦ improve operations to make them more profitable[2]

The money that is borrowed must at least generate enough additional income to pay for the cost of the borrowed money (the **interest**) and to ensure that the **principal** is repaid within the specified period of the loan.[3]

Change in Credit Needs

One of the major changes occurring in production agriculture during recent years has been the substitution of capital for labor. Around 1900, capital (including land) contributed about 25 percent to the farm production process and labor, 75 percent. Currently, however, about 90 percent of the production of food and fiber is attributed to capital, while labor only accounts for 10 percent. Although the rate of substitution has been reduced since 1990, the trend still continues. The total volume of capital needed has become so large that farm credit agencies now provide a major supply of capital. As a result of increased borrowing and rising interest rates, interest payments have become one of the fastest growing costs of production agriculturalists during the last two decades.

Magnitude of Agricultural Credit

Agriculture and its products and services make-up also one-fourth of the U.S. gross domestic product. The value of all **farm assets** increased from about $839 billion in 1990 to $957 billion in 1995. In 1995, real estate (land and buildings) accounted for $726 billion, non–real estate assets (equipment, livestock, and crops) totaled $183 billion, and **financial assets** amounted to $48 billion. Debt in agriculture was about $151 billion, or 16 percent of assets.[4] Most of this debt is split between real estate and non–real estate loans. Refer to Figure 12–1.

Average Debt per Farm

The average farm in the United States has real estate worth $350,217, non–real estate assets worth $88,085, and financial assets amounting to $23,155. Total assets increased from $21,530 in 1950 to $461,457 in 1995. During this same time period, debt on these assets increased from $1,930 to $72,697 per farm. Refer to Figure 12–2. Thus, the equity of individuals working in production agriculture increased from $19,600 to $388,760 per farm.[5]

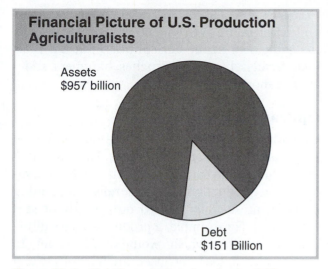

Figure 12–1 Debt in agriculture is only about 16 percent of total farm assets (1995).

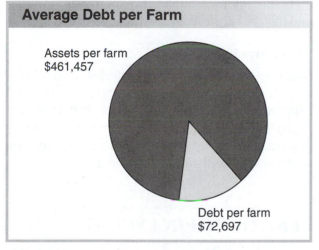

Figure 12–2 Average debt per farm is also only about 16 percent of total farm assets (1995).

THREE FUNDAMENTALS OF CREDIT

In education, you have traditionally heard of the three "R's." In credit, there are also three "R's": returns, repayment, and risk.

Returns

The main reason for borrowing money is to increase net returns and make a profit. Will the net returns or profit be greater with the use of added assets in the form of credit than if it were not used? A production agriculturalist must carefully choose among the best alternatives in considering credit. Money should not be borrowed based on quick, thoughtless decisions or **buyer's fever**. It is important to analyze each investment based on the ability to pay back the borrowed money.[6]

Repayment Ability

Agricultural lenders expect their money to be repaid in full plus interest. Production agriculturalists and others in the agricultural industry must determine their repayment ability on loans, just as the lenders must. Since borrowers usually pledge their farms or agribusinesses as collateral for their loans, inability to repay may lead to **foreclosure**. Therefore, priority should always be given to loans that have earning capacity. For example, borrowing for dairy cows would take priority over borrowing for automatic feeders to replace hand feeding. The automatic feeder does not have the repayment ability, since it generates no direct income, unlike the sale of milk from the dairy cows. Repayment ability is also tied to sacrifices in standard of living in favor of additional investments. For example, it may be better to continue to drive a good used pickup and apply your funds to productive investments, such as the dairy cows, rather than to purchase a new truck.[7]

Risk

Borrowers with strong assets can assume more risk than those with few assets. It is only reasonable that lenders tend to favor individuals in the agricultural industry who have enough stability to absorb a potential loss. Borrowers do not like **speculative investments** due to their high risks.[8] However, they do like to make loans for enterprises that can produce enough profit to ensure income stability or that **appreciate** in value.

THREE AREAS
OF CREDIT NEEDS

In the agricultural industry, finances are needed in three areas: fixed expenses, operating expenses, and start-up expenses.

Fixed Expenses

Fixed expenses refers to items that can be used over and over for a long period of time, resulting in the same price (expense) each year. Examples of fixed expenses include land, buildings, machinery and equipment, tools, and fixtures. Fixed expenses usually result in a large amount of money borrowed, but the repayment amount is fixed monthly or yearly for several years.[9] A brief discussion of the major fixed expenses follows.

Land. Land is a large, onetime expense item. Land is used, not only by production agriculturalists, but also by those in the agribusiness input, output, and service sectors. Improvements to land can also be expensive. Examples of improvements include fencing, clearing land, leveling and ditching farmland, and preparing building sites.

Buildings. Buildings represent a onetime, fixed expense that can be used for twenty-five or more years. Examples include dairy barns, hay sheds, grain elevators, livestock houses, and various other buildings for agribusiness.

Machinery and Equipment. Machinery and equipment usually have a useful life that is less than that of buildings. The useful lives of machinery and equipment depends on how often they are used, care and maintenance, quality, and how complicated they are.

Fixtures and Tools. Cash registers, shelves, and refrigerators are some of the fixtures used as the equipment in the agricultural industry. Tools include such things as wrenches, hammers, shovels, thermometers, and so on.

Operating Expenses

Operating expenses cover everything needed to run a farm, ranch, or agribusiness. The amount of money needed depends on the size of the operation. For production agriculturalists, operating expenses are usually figured on a yearly or seasonal basis. For example, a production agriculturalist who raises wheat would need money to purchase seed, feed, fertilizer, and chemicals and to pay workers. In the agribusiness input, output, and service sectors, operating expenses would include communication, transportation, utilities and fuel, taxes, insurance, and advertising.

Start-up Expenses

Start-up expenses are payable before the business begins its operation. Examples of start-up expenses include attorneys' fees, incorporation expenses, and costs for the development of a site or product.[10] Start-up expenses do not include the initial cost of land and other fixed items.

LENGTH OF FINANCING

Lending (financing, or credit) is important to any country's economic well being. It stimulates economic activity by providing a means of purchasing that which would otherwise be impossible to obtain. In reality, a loan is a contract between the borrower and lender. Loans usually fall into three borrowing time frames: short, intermediate, and long term. However, there are different types of loans within each category.

Short-Term Loans

Short-term loans are distinctive in that their terms are normally one year or less. The main use of the short-term loan is to finance operating inputs. Banks, individuals, and merchants and farm credit services are among some of the suppliers of short-term loans. Short-term operating loans are probably the most frequently used in this category. They usually range anywhere from one month to one year in length. A typical operating loan is for six months, with a single payment retiring the loan amount at the end of the period.

In business applications, short-term or operating credit assists in purchasing items such as fuel, fertilizer, chemicals, and seed, and for maintenance expenses. Short-term credit may be the most important type for an agribusiness's survival because it finances the everyday operations of the firm, which generate the cash flow in the business. In order to maintain good credit, these loans should be repaid on receipt of money at harvest or auction time.[11]

Intermediate-Term Loans

These loans vary in length from one to ten years. They finance assets that may be **depreciable** over their expected lives. Farm machinery and equipment, breeding livestock, irrigation systems, and any modernization of farm facilities are examples of these assets.

As technology increases, the importance of intermediate-term loans increases as well. Those in the agricultural industry must stay up with technology in order to stay competitive. Many of them will have to borrow money to be able to purchase the new technology. Since most of the technology is in the form of depreciable assets, intermediate-term loans assist in these purchases.

Commercial banks are responsible for most of the outstanding intermediate-term debt, followed by the Farm Credit Agency. Banks make both commercial and consumer loans on depreciable assets. They offer different types of loans according to the description of the asset, the borrower's needs, and loss security. They will want to see financial records and past credit history when considering a loan application.

Having good credit ratings and relationships with lending institutions will lower the cost of buying intermediate-term assets. A lender may require **collateral** when considering whether to grant a loan. A **lien** is the lender's right to take possession of the asset in collateral if the borrower fails in repayment of the loan.[12]

Long-Term Credit

Loans with lengths that extend over ten years are possible with long-term credit. Purchases of land, buildings, and housing create the need for this type of credit. Typically, the credit instrument used in long-term financing is a mortgage. Businesses and individuals both use mortgages for long-term financing needs.

Long-term credit is especially important when starting an agribusiness. The manager will have to determine how much land and what types of buildings are suitable to the scale of operations. The start-up of the agribusiness is, in effect, dependent on whether the lender is willing to extend this credit.

The growth of the agribusiness also depends on the ability to obtain long-term credit. As the agribusiness expands, it will need to purchase more land and storage space. Maintaining a good long-term credit source will lower the cost of borrowing in the expansion process. Repayment of loans of this type comes from funds left over after deducting all other expenses for the year.[13] Net income or retained earnings are examples of sources of repayment for long-term loans. Sources of loans for long-term credit include, Farm Credit Services, Farm Service Agency (formerly FmHA),

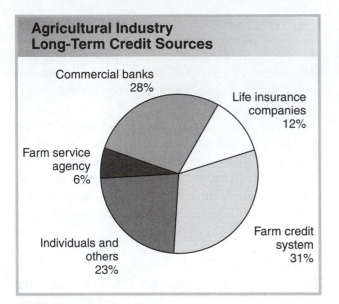

Figure 12–3 Loans for long-term credit come from several sources. Long-term credit is needed for land purchases, buildings, and housing.

life insurance companies, commercial banks, and individuals. Refer to Figure 12–3.

COMPONENTS OF A CREDIT PROFILE

When reviewing a loan application, lenders will require a credit profile. Lenders want positive answers to the following questions and more before extending credit:

❧ *Personal Characteristics*: What are the borrower's honesty, integrity, reputation, and judgment, as a person and manager?

❧ *Management Ability*: Is the applicant a capable decision maker, with the appropriate experience, education, and background?

❧ *Financial Position*: Does the applicant provide accurate and sufficient information with the financial statements? Do the firm's balance sheet, income statement, and cash flow statement show loan repayment ability?

❧ *Loan Purposes*: For what reason does the applicant want the loan? Does the length of the loan match the asset's life?

❧ *Loan Security*: Are there existing assets that would amount to sufficient collateral?[14]

To illustrate one way a lender may evaluate a borrower's credit capacity, consider the following example. Assume that José's real estate lender will extend credit up to a **debt-equity ratio** of 2. In other words, José can have two times as much debt as equity in the business. The non–real estate lender allows a debt-equity (D/E) ratio of 1. José can obtain additional operating credit of up to 75 percent of expected gross crop production and 85 percent of expected gross cattle sales. Figure 12–4 shows that José's business can support a total credit capacity of $216,400. Refer to Figure 12–5 for an example of a loan application.

COMPUTING INTEREST

Interest can be a major expense to agribusinesses that use debt to finance their operations. The **interest rates** that lending institutions charge actually represent the prices they charge for the use of their money.

Interest Rates

Normally, a lender will charge an interest rate tied to the current **discount (prime) rates** set by the Federal Reserve System. Supply and demand factors for money influence the movement of these rates. Interest rates can, and often do, fluctuate over time.

Debt-Equity Ratio

A. Asset Credit

Source of Credit	Asset Value (A)	Debt Outstanding (D)	Equity in Assets E=A-D	Max. D/E	Credit Capacity (E)(D/E)
Real Estate	$50,000	0	$50,000	2	$100,000
Non-Real Estate Assets	30,000	0	30,000	1	30,000
Total	$80,000	0	$80,000		$130,000

B. Income Credit

	Gross Value per Unit	Credit Rate	Credit per unit	No. of Units	Credit Capacity
Crops	$200/acre	75%	$150	150	$22,500
Cattle	$250/head	85%	$213	300	$63,900
Total					$86,400

Figure 12–4   Examining debt-equity ratio is one way in which a lender may evaluate a borrower's credit capacity.

It is often hard to think of money as a commodity that is exchanged for a certain **equilibrium price**. However, the interest rate charged on a loan is actually its selling price to the borrower. Figuring out the actual price, or interest rate, on loaned money is important to controlling the agribusiness's costs of capital.[15]

Cost of capital is the difference between the actual amount paid for the loan and the actual amount received from it. For example, if a person decides to borrow $100 at 9 percent interest, the cost of capital would be $100 x 9% = $9. Thus, the person would receive $100 and pay back $109.

The **contractual interest rate** will not always be equal to the actual rate charged on a loan. This creates an important need to be able to know what a lender is charging for the use of the money. Most borrowers will often choose a lender that offers the lowest **actuarial interest rate**.

The **annual percentage rate (APR)** is a common name referring to the actuarial interest rate. However, the lender with the lowest APR may not always suit your needs. You may need the lower monthly payments on a longer-term loan, which another lender may offer at higher rates.[16] The following are three ways to calculate the annual percentage rate:

Simple Interest.   **Simple interest** applies to loans with a single payment. An example would be a six-month, short-term operating note due in a single payment at the end of the period. The APR on the loan would be the rate charged if there were no down payment or borrowing fees required. In the example just presented, the APR is 9 percent.

Remaining Balance Method.   This method applies to installment-type loans where a series of installment payments go toward paying off the loan. Suppose that a person borrows $100 at 9

Form Approved - OMB No. 0560-0167

FSA 410-1
(03-31-97)

U.S. DEPARTMENT OF AGRICULTURE
Farm Service Agency

REQUEST FOR DIRECT LOAN ASSISTANCE

Federal Agencies may not conduct or sponsor, and a person is not required to respond to, a collection of information request unless it displays a currently valid OMB control number. Public reporting burden for this collection of information is estimated to average 60 minutes per response, including the time for reviewing instructions, searching existing data sources, gathering and maintaining the data needed, and completing and reviewing the collection of information. Send comments regarding this burden estimate or any other aspect of this collection of information, including suggestions for reducing this burden, to Department of Agriculture, Clearance Officer, OIRM (OMB No. 0560-0167), Stop 7630, Washington, D.C. 20250-7630. **RETURN THIS COMPLETED FORM TO YOUR FSA COUNTY OFFICE.**

INSTRUCTIONS TO APPLICANT: (For individuals, partnerships, or joint operations, show names, and trade names if any. Business entity applicants must provide additional information listed in Item 31. A husband and wife who want to apply for a loan together will be considered a joint operation. Either a husband or wife can apply as an individual.)

1. APPLICANT'S NAME	2. SPOUSE'S NAME	3. APPLICANT'S TELEPHONE NO.
4. APPLICANT'S ADDRESS		5. APPLICANT'S SOCIAL SECURITY NO. OR TAX IDENTIFICATION NO.

6. APPLICANT'S BIRTH DATE	7. SPOUSE'S BIRTH DATE	8. SPOUSE'S SOCIAL SECURITY NUMBER	9. TOTAL NUMBER OF HOUSEHOLD MEMBERS

10. TYPE OF OPERATION: ☐ INDIVIDUAL ☐ PARTNERSHIP ☐ JOINT OPERATION   ACRES OWNED _____
☐ CORPORATION ☐ COOPERATIVE   ACRES RENTED _____

11. MARITAL STATUS ☐ MARRIED ☐ SEPARATED ☐ UNMARRIED (INCLUDING SINGLE, DIVORCED, AND WIDOWED)

	YES	NO
12. Have you or any member of your organization ever been in receivership, been discharged in bankruptcy, or filed a petition for reorganization in bankruptcy? If YES, please provide details in Item 31.		
13. Are you, or any member of your organization, or the organization itself, involved in any pending litigation? If YES, provide details in Item 31.		
14. Do you now, or have you ever, conducted business under any other name? If YES, give name in Item 31.		
15a. Have you or any member of your organization ever obtained a direct or guaranteed farm loan from the Farm Service Agency (FSA) or Farmers Home Administration?		
15b. If Item 15a is YES, did the government ever forgive any debt through a write-off, debt settlement, compromise, write-down, charge-off, adjustment, reduction, or bankruptcy? If bankruptcy, please provide details in Item 31. If Item 15a is NO, leave blank.		
16. If you obtained a guaranteed loan, did the government pay the lender a loss claim? Leave blank if you did not obtain a guaranteed loan.		
17. Are you or any member of your organization delinquent on any federal debt? If YES, provide details in Item 31.		
18. Are you a citizen or permanent resident of the United States of America? If permanent resident, provide a copy of Form I-151 or I-551, "Alien Registration Receipt Card."		
19. Are you a veteran? If YES, please indicate Branch and Dates of Service in Item 31.		
20. Are you now, or have you ever farmed or ranched? If YES, provide the number of years and brief explanation in Item 31.		
21. Are you an FSA employee or are you related to or closely associated with any FSA employee? If YES, please explain in Item 31.		

22. PURPOSE OF LOAN	23. APPROXIMATE AMOUNT OF LOAN NEEDED
24. NAME AND ADDRESS OF APPLICANT'S EMPLOYER	25. NAME AND ADDRESS OF SPOUSE'S EMPLOYER
26. APPLICANT'S APPROXIMATE ANNUAL INCOME	27. SPOUSE'S APPROXIMATE ANNUAL INCOME

28. FSA USE ONLY

A. DATE FORM FSA 410-1 RECEIVED	B. DATE APPLICATION COMPLETE

C. CREDIT REPORT FEE $	D. DATE RECEIVED	E. INITIALS

F. TYPE OF ASSISTANCE: FO ☐ OL ☐ EM ☐ SUBORDINATION ☐ OTHER (SPECIFY) ☐

Figure 12–5 An example of a loan application provided by the USDA Farm Service Agency.

FSA-410-1 (03-31-97) Page 2

29.	**FOR BUSINESS ENTITY APPLICANTS ONLY** **(COOPERATIVES, CORPORATIONS, PARTNERSHIPS, OR JOINT OPERATIONS)**

    I.   MEMBER INFORMATION - Business entity applicants must attach the following information regarding all members, stockholders, partners, and joint operators (If there are no individually owned assets, then husband and wife joint operations may submit one consolidated balance sheet):

        A.   Name, address, social security number, birth date, principal occupation, and a balance sheet not more than 90 days old.

        B.   The full name of each non-U.S. citizen.

        C.   Each members' ownership interest (expressed as a percent).

    II.   BUSINESS ENTITY INFORMATION - The business entity must provide:

        A.   Any Organizational and Operational Documents (e.g. Charter, Articles of Incorporation, Bylaws, Partnership or Joint Operation Agreements, etc).

        B.   Any evidence of its current registration with relevant state regulatory agencies (good standing).

        C.   A duly adopted resolution to apply for and obtain financing.

        D.   Tax identification number.

NOTE: Individual liability will be required regardless of the type of business organization.

30. VOLUNTARY INFORMATION FOR MONITORING PURPOSES:
The following information is requested by the Federal Government in order to monitor FSA's compliance with federal laws prohibiting discrimination against loan applicants on the basis of race, color, national origin, religion, sex, marital status, handicapped condition or age (provided that the applicant has the capacity to enter into a binding contract). You are not required to furnish this information, but are encouraged to do so. This information will not be used in evaluating your application or to discriminate against you in any way. However, if you do not furnish it, FSA is required to note your race/national origin and sex on the basis of visual observation or surname.

A. RACE/NATIONALITY: ☐ White ☐ Black ☐ Hispanic

☐ Asian or Pacific Islander ☐ American Indian or Alaskan Native

B. SEX: ☐ Male ☐ Female

31.   ADDITIONAL SPACE FOR ANSWERS. Write the number to which each answer applies. If you need more space, use additional sheets of paper the same size as this page. On each sheet, write the applicant's name.

ITEM NUMBER	REMARKS

FSA-410-1 (03-31-97) Page 3

32. A signed and dated balance sheet not more than 90 days old is required. Business organizations must provide individual members' balance sheets. You may use this form or attach your own. If you have a balance sheet on file with FSA that is less than 90 days old, you need not complete this section at this time.

BALANCE SHEET — **AS OF** _____

CURRENT FARM ASSETS			$VALUE	CURRENT FARM LIABILITIES				$AMOUNT
Cash on hand	Checking	Savings		Farm Accounts and Notes Payable				
$	$	$		(Include Principal and Interest)				
Other Investments:				Creditor	Payment Due Date	Interest Rate	Monthly or Annual Installment ($)	
Time Certificates		Other						
$	$							
Accounts and Notes to be Received (Receivables)								
Crops and Feed	Units	Price Per Unit ($)						
Livestock to be Sold	No. / Unit Weight	Price Per Unit ($)		CCC Loan:				
				Type	Quantity		Due Date	
				Current Portion of Principal Due on:				
				Intermediate Liabilities				
				Long Term Liabilities				
				Accrued interest on:				
Growing Crops	Acres	Cost/Acre ($)		Intermediate Liabilities				
				Long Term Liabilities				
				Accrued Taxes on:				
				Real Estate, Personal Property and Assessments				
Supplies and Prepaid Expenses				Income Tax and Social Security				
Leases				Accrued Rent/Lease Payments				
Other				Other (judgments, liens, etc.)				
TOTAL CURRENT FARM ASSETS▶				**TOTAL CURRENT FARM LIABILITIES▶**				

INTERMEDIATE FARM ASSETS				INTERMEDIATE FARM LIABILITIES (portion due beyond 12 months)				
Accounts and Notes to be Received beyond 12 months (Receivables)				Creditor	Payment Due Date	Interest Rate	Amount Delinquent ($)	
Breeding Livestock	No.	Price Per Unit ($)						
Machinery, Equipment and Vehicles								
Co-op Stock								
Cash Value, Life Insurance	(Face Amount $)							
Farmer-Owned Reserve:								
Type	Quantity	Price/Unit ($)		Loans Secured by Life Insurance Policy(ies)				
				Farmer-Owned Reserve				
Other				Other				
TOTAL INTERMEDIATE FARM ASSETS▶				**TOTAL INTERMEDIATE FARM LIABILITIES▶**				

LONG TERM FARM ASSETS (Farm Real Estate)				$VALUE	LONG TERM FARM LIABILITIES (portion due beyond 12 months)				$AMOUNT
Acres	Date Bought	Annual Tax	Cost		Creditor	Payment Due Date	Interest Rate	Amount Delinquent ($)	
			$						
			$						
Co-op Stock									
Equity in Partnerships/Corporations/Joint Operations/Cooperatives									
Other					Other				
TOTAL LONG TERM FARM ASSETS▶					**TOTAL LONG TERM FARM LIABILITIES▶**				
TOTAL FARM ASSETS▶					**TOTAL FARM LIABILITIES▶**				

NONFARM ASSETS				NONFARM LIABILITIES				
Household Goods				Nonfarm Accounts and Notes Payable				
Car, Recreational Vehicles, etc.				Creditor	Payment Due Date	Interest Rate	Monthly or Annual Installment ($)	
Cash Value of Life Insurance								
Stocks, Bonds								
Nonfarm Business								
Other Nonfarm Assets								
Nonfarm Real Estate (Annual Tax $)								
				TOTAL NONFARM LIABILITIES▶				
				TOTAL FARM LIABILITIES▶				
TOTAL NONFARM ASSETS▶				**TOTAL LIABILITIES▶**				
TOTAL FARM ASSETS▶				**NET WORTH▶**				
TOTAL ASSETS▶				**TOTAL LIABILITIES AND NET WORTH▶**				

FSA-410-1 (03-31-97) Page 4

33. SPECIAL PROGRAM INFORMATION

Certain FSA programs are, by law, designed to reach targeted applicants. If you are interested in any of the programs described below, or have questions about these programs and whether you may qualify for a specific program, the FSA office processing your application will help you.

A. **SOCIALLY DISADVANTAGED APPLICANTS:** A portion of FSA farm ownership and operating loan funds are, by law, targeted to applicants who have been subjected to racial, ethnic or gender prejudice because of their identity as a member of a group, without regard to individual qualities. Under the applicable law, groups meeting this condition are: Women, Blacks, American Indians, Alaskan Natives, Hispanics, Asians, and Pacific Islanders.

B. **BEGINNING FARMER ASSISTANCE:** FSA has the authority to assist beginning farmers and ranchers through the farm operating and ownership loan programs. A portion of FSA farm ownership and operating loan funds are, by law, targeted to beginning farmers and ranchers. In addition, FSA has a beginning farmer down payment program, which receives special funding. In some States, FSA has agreements with State beginning farmer programs to help meet the credit needs of beginning farmers and ranchers.

C. **LIMITED RESOURCE LOANS:** Limited resource farm ownership and operating loans are available to qualified FSA applicants. This program provides loans at reduced interest rates to low-income farmers and ranchers whose farm operations and resources are so limited that they cannot pay the regular rates for FSA loans. The program is also intended to provide beginning farmers with an opportunity to start a successful farming operation.

34. STATEMENT REQUIRED BY THE PRIVACY ACT

The following statements are made in accordance with the Privacy Act of 1974 (5 U.S.C. 552a): The Farm Service Agency (FSA) is authorized by the Consolidated Farm and Rural Development Act, as amended (7 U.S.C. 1921 et seq.), or other Acts, and the regulations promulgated thereunder, to solicit the information requested on its application forms. The information requested is necessary for FSA to determine eligibility for credit or other financial assistance, service your loan, and conduct statistical analyses. Supplied information may be furnished to other Department of Agriculture agencies, the Internal Revenue Service, the Department of Justice or other law enforcement agencies, the Department of Defense, the Department of Housing and Urban Development, the Department of Labor, the United States Postal Service, or other Federal, State, or local agencies as required or permitted by law. In addition, information may be referred to interested parties under the Freedom of Information Act (FOIA), to financial consultants, advisors, lending institutions, packagers, agents, and private or commercial credit sources, to collection or servicing contractors, to credit reporting agencies, to private attorneys under contract with FSA or the Department of Justice, to business firms in the trade area that buy chattel or crops or sell them for commission, to Members of Congress or Congressional staff members, or to courts or adjudicative bodies. Disclosure of the information requested is voluntary. However, failure to disclose certain items of information requested, including your Social Security Number or Federal Tax Identification Number, may result in a delay in the processing of an application or its rejection.

35. GENERAL INFORMATION

A. **RIGHT TO FINANCIAL PRIVACY ACT OF 1978 and TITLE XI, 1113(h) OF PUB. L. 95-630:** FSA has a right of access to financial records held by financial institutions in connection with providing assistance to you, as well as collecting on loans made to you or guaranteed by the government. Financial records involving your transaction will be available to FSA without further notice or authorization but will not be disclosed or released by this institution to another government Agency or Department without your consent except as required by law.

B. **THE FEDERAL EQUAL OPPORTUNITY ACT** prohibits creditors from discriminating against borrowers on the basis of race, color, religion, sex, handicap, familial status, national origin, marital status, age (provided the borrower has the capacity to enter into a binding contract), because all or a part of the borrower's income derives from any public assistance program, or because the borrower has in good faith exercised any right under the Consumer Credit Protection Act.

C. **FEDERAL COLLECTION POLICIES FOR CONSUMER DEBTS:** Delinquencies, defaults, foreclosures and abuses of mortgage loans involving programs of the Federal Government can be costly and detrimental to your credit, now and in the future. The mortgage lender in this transaction, its agents and assigns as well as the Federal Government, its agencies, agents and assigns, are authorized to take any and all of the following actions in the event loan payments become delinquent on the mortgaged loan described in the attached application: (1) Report your name and account information to a credit bureau; (2) Assess additional interest and penalty charges for the period of time that payment is not made; (3) Assess charges to cover additional administrative costs incurred by the Government to service your account; (4) Offset amounts owed to you under other Federal programs; (5) Refer your account to a private attorney, collection agency or mortgage servicing agency to collect the amount due, foreclose the mortgage, sell the property and seek judgement against you for any deficiency; (6) Refer your account to the Department of Justice for litigation; (7) If you are a current or retired Federal employee, take action to offset your salary, or civil service retirement benefits; (8) Refer your debt to the Internal Revenue Service for offset against any amount owed to you as an income tax refund; and (9) Report any resulting written-off debt to the Internal Revenue service as taxable income. All of these actions can and will be used to recover debts owed to the Federal Government, when in its best interests.

36. CERTIFICATIONS

A. **RESTRICTIONS AND DISCLOSURE OF LOBBYING ACTIVITIES**

1. The loan applicant certifies that: if any funds, by or on behalf of the loan applicant, have been or will be paid to any person for influencing or attempting to influence an officer or employee of any agency, a Member, an officer or employee of Congress, or an employee of a Member of Congress in connection with the awarding of any Federal contract, the making of any Federal grant or Federal loan, and the extension, continuation, renewal, amendment, or modification of any Federal contract, grant, or loan, the loan applicant shall complete and submit Standard Form - LLL, "Disclosure of Lobbying Activities," in accordance with its instructions.

2. The loan applicant shall require that the language of this certification be included in the award documents for all sub-awards at all tiers (including contracts, subcontracts, and subgrants, under grants and loans) and that all subrecipients shall certify and disclose accordingly.

3. This certification is a material representation of fact upon which reliance was placed when this transaction was made or entered into. Submission of this statement is a prerequisite for making or entering into this transaction imposed by 31 U.S.C. 1352. Any person who fails to file the required statement shall be subject to a civil penalty.

B. **ABUSE OF CONTROLLED SUBSTANCES:**

The loan applicant certifies that he/she as an individual, or any member, stockholder, partner or joint operator of an entity applicant, has not been convicted under Federal or State law of planting, cultivating, growing, producing, harvesting, or storing a controlled substance since December 23, 1985, in accordance with the Food Security Act of 1985 (Public Law 99-198).

C. **TEST FOR CREDIT**

The individual or authorized party certifies that the needed credit, with or without a loan guarantee, cannot be obtained by the individual applicant, or in the case of a business entity, the needed credit cannot be obtained considering all assets owned by the business entity and all of the individual members.

D. **ACKNOWLEDGMENT**

I, THE UNDERSIGNED LOAN APPLICANT, UPON SIGNING THIS LOAN APPLICATION, CERTIFY THAT I HAVE RECEIVED THE ABOVE NOTIFICATIONS AND ACCEPT AND COMPLY WITH THE CONDITIONS STATED THEREON. I CERTIFY THAT THE STATEMENTS MADE BY ME IN THIS APPLICATION ARE TRUE, COMPLETE, AND CORRECT TO THE BEST OF MY KNOWLEDGE AND BELIEF AND ARE MADE IN GOOD FAITH TO OBTAIN A LOAN. I UNDERSTAND THAT THE 60-DAY PROMPT APPROVAL PERIOD WILL NOT BEGIN UNTIL A COMPLETE APPLICATION HAS BEEN FILED. (WARNING: SECTION 1001 OF TITLE 18, UNITED STATES CODE PROVIDES FOR CRIMINAL PENALTIES TO THOSE WHO PROVIDE FALSE STATEMENTS ON LOAN APPLICATIONS. IF ANY INFORMATION ON THIS APPLICATION IS FOUND TO BE FALSE OR INCOMPLETE, SUCH FINDING MAY BE GROUNDS FOR DENIAL OF THE REQUESTED CREDIT.)

SIGNATURE OF LOAN APPLICANT OR AUTHORIZED REPRESENTATIVE	DATE

percent interest and will pay the loan off in four payments. The payments will be annual. The payment of principal will be $25 per year. The interest payments in each year calculate as follows:

Year 1 $100 × .09 = $9 + $25 = $34.00
Year 2   75 × .09 = $6.84 + $25 = $31.84
Year 3   50 × .09 = $4.50 + $25 = $29.50
Year 4   25 × .09 = $2.25 + $25 = $27.25
Total $22.59 + 100 = $122.59

In this example, the remaining balance is the basis on which the borrower pays interest. The contractual rate and the APR in this case would be the same.

Add-on Method. Often, lenders use **add-on interest** when making loans on automobiles, tractors, or other types of machinery. In computing the interest, you take the total interest paid on a loan and add it to the loan amount. After adding in the interest, you divide the total amount by the number of payment periods to get the installment payment amount.

The type of interest will be different from the contracted rate; in most cases, it will be higher.[17] The following example helps explain this point.

Suppose Mr. Shinn needs $10,000 to finance the purchase of a used truck. The bank offers to finance the truck for four years at 7 percent interest using the add-on method. Mr. Shinn's financial advisor tells him that the stated rate on the loan will not be his actual rate, or APR. She computes the APR as follows:

❦ Interest calculation: $10,000 × 7% × 4 years = $2,800. This is the finance charge.

❦ Then, she adds the interest to the loan amount: $10,000 + $2,800 = $12,800. This is the total amount to be repaid.

❦ Calculation of the monthly payment would be: 4 years × 12 months = 48 monthly payments.

❦ Then, she divides $12,800/48 payments = $266.67 per month.

Use the following formula to calculate the APR for add-on interest.

$$R = \frac{2C}{L(P + A)} \times 100$$

where

R = annual percentage rate
C = total interest cost
L = length of the loan in years
P = beginning principal of the borrowed amount
A = payment in each period

The payments do not have to be monthly in this equation. The variable *A* could apply to annual, semiannual, or quarterly payments as well. Substituting in the numbers from the problem above, the advisor computes the following APR.

$$\frac{2 \times \$2,800}{4(\$10,000 + \$266.67)} \times 100 = 13.64\%$$

Mr. Shinn is actually paying 13.64 percent interest, instead of 7 percent, as stated on the loan contract.

Amortization Tables

An amortization table is a table that calculates constant payments needed to repay both the principal and interest on a sum of money. The term **amortize** refers to a loan that is set up with equal installment payments. These can be set up as annual, semiannual, quarterly, or monthly payments. The tables are set up most conveniently for yearly payments, but they can be used for the other payment schedules as well. There are many computer programs that can provide the same information.

With equal payments there is a larger amount of interest cost and smaller amount applied to the principal during the early stages of the loan. As the number of payments increases, the amount of interest decreases and the amount going to principal increases. Interest is only being paid on the actual amount still owed. Because the total size of the payments is always the same, managers can more accurately calculate their cash flow.[18]

The amortization table in Figure 12–6 is used by looking at the far left column. It will either be marked in years or time periods. Drop down to year five. Now cross over to 10 percent interest. The factor is .2637. So if $10,000 is borrowed at 10 percent for five years, the annual payments would be $10,000 times .2637, or $2,637. The payment schedule is illustrated in Figure 12–7.

If semiannual payments are made, the same table can be used. If the loan is for five years with semiannual payments, the time period, or number of payment periods, is ten. The interest must be divided by two because half the interest twice a year is the same as all the interest once a year. Find the factor for ten time periods (years) at 5 percent interest. That factor is .1295. The semiannual payment will be $1,295, or $1,000 times .1295. The total cost of the loan would be $12,950. It costs less than the other loan because the money is being paid back more frequently, which lowers the interest cost.

Feasibility

If an investment is feasible, it will generate enough after-tax income to pay for itself. The investment must have a positive cash flow during its economic life (the loan period). Many agricultural investments show a negative cash flow during the first year or two of activity. The difference must be made up from profits in other parts of the operation, in savings, or by short-term borrowing. Because this can be an excessive burden to the operation, it has become an important managerial concern.

There are certain steps that can help overcome this cash flow problem. Prepare an estimated cash flow statement for the investment item. Then, if the cash flow is negative in the first year or two,

determine whether the operation can absorb any added costs of the new investment. Try to match the economic life, or length of loan, to the physical life, or how long the item will last. Also, the more money that can be invested at first, the easier it will be for the cash flow to become positive because the payment will become smaller. An example would be a large machinery purchase.[19] Many times, the loan payment will be more at first than the money earned with the machine.

Profitability Index

The most practical way to determine if something is profitable is to determine if the *benefits* (returning in sales) outweigh the *costs* (inventory or investment). For a profitability index (PI), the *benefits* must be determined and divided by the *costs.* This provides a useful management tool for making investment decisions.

The profitability index allows for a comparison of investments with different rates of return and different terms. The most profitable investment can then be selected. The present value of each of the investments will be used through the life of that investment. This puts the value of the investment in terms of today's dollar, which can easily be compared. If a person was currently earning $110 from a $100 investment, the profitability index would be figured as follows:

$$\frac{\$110 \text{ benefit}}{\$100 \text{ cost}} = 1.10 \text{ profitability index}$$

As long as the profitability index is greater than 1, the investment is profitable. When deciding between two investments, simply choose the one with the highest profitability index. Remember, money is a tool—nothing more and nothing less. If the tool is used right, it will benefit the operation. If the tool is used with poor judgment, it might not be there to use tomorrow. Refer to Chapter 7 if further clarification is needed on financial (profitability) ratios.

Annual Payment per $1 of Loan at Given Interest Rates and Maturities (Amortization Table)

Year	4%	5%	6%	7%	8%	9%	10%	11%
1	1.0400	1.0500	1.0600	1.0700	1.0800	1.0900	1.1000	1.1100
2	.5301	.5378	.5454	.5530	.5607	.5684	.5761	.5839
3	.3603	.3672	.3741	.3810	.3880	.3950	.4021	.4092
4	.2754	.2820	.2885	.2952	.3019	.3086	.3154	.3223
5	.2246	.2309	.2373	.2438	.2504	.2570	.2637	.2705
6	.1907	.1970	.2033	.2097	.2163	.2229	.2296	.2363
7	.1666	.1728	.1791	.1855	.1920	.1986	.2054	.2122
8	.1485	.1547	.1610	.1674	.1740	.1806	.1874	.1943
9	.1344	.1406	.1470	.1534	.1600	.1667	.1736	.1806
10	.1232	.1295	.1358	.1423	.1490	.1558	.1627	.1698
11	.1141	.1203	.1267	.1333	.1400	.1469	.1539	.1611
12	.1065	.1128	.1192	.1259	.1326	.1396	.1467	.1540
13	.1001	.1064	.1129	.1196	.1265	.1335	.1407	.1481
14	.0946	.1010	.1075	.1143	.1212	.1284	.1357	.1432
15	.0899	.0963	.1029	.1097	.1168	.1240	.1314	.1390
16	.0858	.0922	.0989	.1058	.1129	.1203	.1278	.1355
17	.0821	.0886	.0954	.1024	.1096	.1170	.1246	.1324
18	.0789	.0855	.0923	.0994	.1067	.1142	.1219	.1298
19	.0761	.0827	.0896	.0967	.1041	.1117	.1195	.1275
20	.0735	.0802	.0871	.0943	.1018	.1095	.1174	.1255
25	.0640	.0709	.0782	.0858	.0936	.1018	.1101	.1187
30	.0578	.0650	.0726	.0805	.0888	.0973	.1060	.1150
35	.0535	.0610	.0689	.0772	.0858	.0946	.1036	.1129
40	.0505	.0505	.0664	.0750	.0838	.0929	.1022	.1117

12%	13%	14%	15%	16%	17%	18%	19%	20%
1.1200	1.1300	1.1400	1.1500	1.1600	1.1700	1.1800	1.1900	1.2000
.5916	.5994	.6072	.6151	.6229	.6308	.6387	.6466	.6545
.4163	.4235	.4307	.4379	.4452	.4525	.4599	.4673	.4747
.3292	.3361	.3432	.3502	.3573	.3645	.3717	.3789	.3862
.2774	.2843	.2912	.2983	.3054	.3125	.3197	.3270	.3343
.2432	.2501	.2571	.2642	.2713	.2786	.2859	.2932	.3007
.2191	.2261	.2331	.2403	.2476	.2549	.2623	.2698	.2774

(continued on following page)

Figure 12–6 An amortization table calculates the constant payments needed to repay both the principal and interest on a given sum of money.

Annual Payment per $1 of Loan at Given Interest Rates and Maturities (Amortization Table) *(concluded)*

12%	13%	14%	15%	16%	17%	18%	19%	20%
.2013	.2083	.2155	.2228	.2302	.2376	.2452	.2528	.2606
.1876	.1948	.2021	.2095	.2170	.2246	.2323	.2401	.2480
.1769	.1842	.1917	.1992	.2069	.2146	.2225	.2304	.2385
.1684	.1758	.1833	.1910	.1988	.2067	.2147	.2228	.2311
.1614	.1689	.1766	.1844	.1924	.2004	.2086	.2168	.2252
.1556	.1633	.1711	.1791	.1871	.1953	.2036	.2121	.2206
.1508	.1586	.1666	.1746	.1828	.1912	.1996	.2082	.2168
.1468	.1547	.1628	.1710	.1793	.1878	.1964	.2050	.2138
.1433	.1514	.1596	.1679	.1764	.1850	.1937	.2025	.2114
.1404	.1486	.1569	.1653	.1739	.1826	.1914	.2004	.2094
.1379	.1462	.1546	.1631	.1718	.1807	.1896	.1986	.2078
.1357	.1441	.1526	.1613	.1701	.1790	.1881	.1972	.2064
.1338	.1423	.1509	.1597	.1686	.1776	.1862	.1960	.2053
.1275	.1364	.1454	.1546	.1640	.1734	.1829	.1924	.2021
.1241	.1334	.1428	.1523	.1618	.1715	.1812	.1910	.2008
.1223	.1318	.1414	.1511	.1608	.1707	.1805	.1904	.2003
.1213	.1309	.1407	.1505	.1604	.1703	.1802	.1901	.2001

Payment Schedule

	Interest	Payment on Principal	Total payment
Year 1	$1,000	$1,637	$2,637
2	836	1,801	2,637
3	656	1,981	2,637
4	458	2,179	2,637
5	218	2,419	2,637
	$3,168 +	$10,017* =	$13,185

*Principal did not quite total $10,000 due to rounding.

Figure 12–7 Refer to Figure 12–6 and notice that a five-year loan for 10% interest has a factor of .2637. The payment schedule here shows that the yearly payments for a $10,000 loan are $10,000 × .2637 = $2,637.

AGRICULTURAL CREDIT SOURCES FOR REAL ESTATE AND NON–REAL ESTATE LOANS

Production agriculturalists have access to all credit sources that other businesses have plus some special sources set up strictly for them. Agricultural lending is big business. Many lenders recognize this and compete for a share. The major sources of credit for those in the agricultural industry are the Farm Credit System, commercial banks, Farm Service Agency (formerly Farmers Home Administration), Commodity Credit Corporation, life insurance companies, and individuals.

Non–Real Estate Loans

Most of the money borrowed by production agriculturalists has been used to finance farm expansion and higher production cost items, such as farm machinery and motor vehicles. In 1995, money borrowed for non–real estate loans **secured** by farm assets totaled $72 billion. Commercial banks supplied 50 percent; the Farm Credit System, 17 percent; the Farm Service Agency, 7 percent; the Commodity Credit Corporation, 5 percent; and "individuals and others," 21 percent.[20] Refer to Figure 12–8.

Real Estate Loans

In 1995 the Farm Credit System supplied 31 percent of all outstanding real estate debt; life insurance companies supplied 12 percent; commercial banks, 28 percent; the Farm Service Agency, 6 percent; and individuals, 23 percent.[21] Historically, individuals have been the major source of funds for land transfers. As can be seen, the Farm Credit System is the largest lender involved in the land mortgage field. The total real estate debt outstanding as of December 31, 1995, was $78.7 billion.[22] Refer to Figure 12–3.

FARM CREDIT SYSTEM

Historical Beginning of Federal Land Banks

The Federal Land Banks were organized by the Federal Farm Loan Act of 1916, which established twelve Federal Land Banks across the United States. Federal Land Banks made long-term loans by first mortgages on real estate through more than 490 local Federal Land Bank Associations. The federal government provided initial financial support to these banks ($9 million in 1916, which was repaid in 1932, plus $189 million obtained in 1933–1937.) The remaining money was then repaid to the federal government, and the Federal Land Banks became owned by their borrowers.[23]

Historical Beginning of Production Credit Associations

The production credit system was established under federal laws in 1923 and 1933 to provide short-term and intermediate-term loans to production agriculturalists and rural residents. The twelve Federal Intermediate Credit Banks were first authorized by the Farm Credit Act of 1923 to provide discounting services to banks for agricultural loans. In 1933, Congress authorized local Production Credit Associations to make direct short- and intermediate-term loans to individuals in the agricultural industry.[24] By 1986, there were 420 Production Credit Associations throughout the United States, with 1,500 full-time offices.

Historical Beginning of the Bank for Cooperatives

The Bank for Cooperatives was set up by the Farm Credit Act of 1933. It provides loan funds to

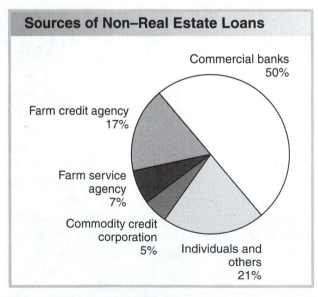

Sources of Non–Real Estate Loans

- Commercial banks 50%
- Farm credit agency 17%
- Farm service agency 7%
- Commodity credit corporation 5%
- Individuals and others 21%

Figure 12–8 Commercial banks are the major source of non–real estate loans (1999).

farmer-owned cooperatives that are involved in marketing, supply, and service. To be eligible to borrow, a cooperative has to have at least 80 percent of its voting control with agricultural producers, and the cooperative has to do at least 50 percent of its business with its members.[25] Refer to Figure 12–9 to see how the Farm Credit System was organized until 1986.

Restructured and Present-Day Farm Credit Systems

The nation's agricultural industry experienced rough times during much of the 1980s. After several years of rising land prices and increasing interest rates, plus a significant drop in the market for agricultural products, international competitiveness, and increased domestic production, farm product prices began to decline. As a result, production agriculturalists were less able to repay their debts.[26]

By 1985, the Farm Credit System reported that it had several billion dollars in loans that might go into **default** over the next few years. This turned out to be true. The Federal Land Banks lost about 66 percent of this total; the Federal Intermediate Credit Banks and Production Credit Associations about 33 percent; and the Bank for Cooperatives, about 1 percent.[27]

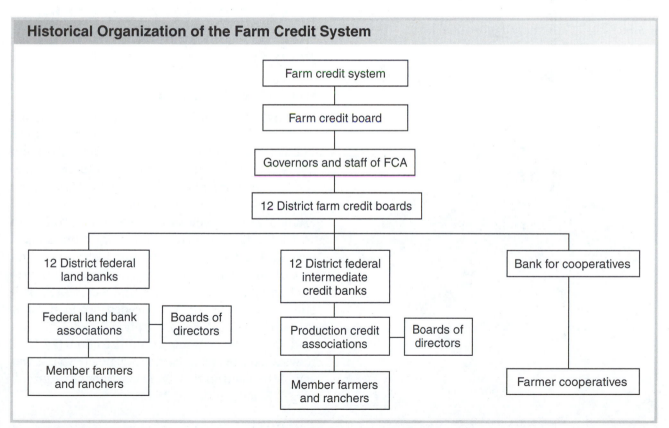

Historical Organization of the Farm Credit System

Figure 12–9 The organization chart shown here was effective until 1986. With the Agricultural Credit Act of 1987, the Federal Land Bank Associations and Production Credit Associations merged into what is called Farm Credit Services in most districts.

Agricultural Credit Act of 1987. The financial problems of the Farm Credit System prompted Congress to pass the Agricultural Credit Act of 1987. This act required the Federal Land Banks and Intermediate Credit Banks within the twelve Credit Districts to merge. There is now a single institution in each of the seven districts, called the Farm Credit Banks, which handles both long- and short-term credit needs of production agriculturalists.

In essence, the Federal Land Bank Associations and Production Credit Associations merged into what is called Farm Credit Services in most districts. The merger eliminated about half of the local offices of the Farm Credit System. In 1980, there were 915 associations, and by 1996 mergers and consolidations reduced the number of associations to 228.[28] Refer to Figure 12–10.

Farm Credit Services

Figure 12–10 The Farm Credit Service Agencies make short-term, intermediate-term, and long-term loans. Loans range from farm to educational loans. (Courtesy of Farm Credit Services)

Types of Farm Credit Service Loans. Besides real estate, farm, and land financing, the Farm Credit System loans money for the following: capital improvements, operating loans, family living expenses, revolving lines of credit, farm equipment, livestock, speciality loans (such as educational), farm-related business loans, farm and rural home loans, farm credit leasing, and related services.[29]

Banks for Cooperatives. Ten Banks for Cooperatives merged with the Central Bank for Cooperatives to form a single institution, which was named CoBank and headquartered in Denver, Colorado.

CONSOLIDATED FARM SERVICE AGENCY (FSA; FORMERLY FMHA)

Historical Beginning

Although the Farm Credit System was completed in 1933, it did not fill all the farm credit gaps. Therefore, the Farmers Home Administration (FmHA) was started originally in 1935 as an independent government agency called the Resettlement Administration and established under President Franklin D. Roosevelt's Executive Order 7027.[30] In 1937, the Resettlement Administration was taken over by the Farm Security Administration as an expansion of federal aid to agriculture. Young production agriculturalists still found it extremely difficult to provide the collateral needed for a loan. Thus, the Farmers Home Administration Act of 1946 created the Farmers Home Administration (FmHA), now called the Farm Service Agency (FSA). It was established within the U.S. Department of Agriculture.[31] Refer to Figure 12–11.

Purpose

The major purpose of FmHA (FSA) was to provide supervised credit to production agriculturalists

Figure 12–11 In order to save money at both the federal and state levels, many of the USDA agencies were consolidated. The old Agricultural Stabilization Conservation Service (ASCS) is now the Natural Resources Conservation Service. The Farm Service Agency (formerly FmHA) shares the facility shown here with other USDA and state agencies. (Courtesy of Marlene Graves)

who were unable to get adequate credit from other sources at reasonable rates. Production agriculturalists also had to agree to refinance their loan with another agency as soon as it became economically feasible. In other words, FmHA was strictly a hardship agency for high-risk loans. FmHA was also created to provide emergency credit to farmers affected by natural disasters such as extreme drought, hail, and floods. Fortunately, this has been only a minor portion of FmHA lending.

Change from FmHA to FSA

The Farmers Home Administration (FmHA) ceased to operate with the signing of the Federal Crop Insurance Reform and Department of Agriculture Reorganization Act of 1994. Farm loans are currently handled by the Farm Service Agency (FSA), including both direct and guaranteed loans. Being a lender of last resort, for farm loans, FmHA experienced a great amount of loan **delinquencies** and defaults. Loan losses averaged about $1.7

billion per year from 1988 to 1995.[32] Loans made by the FSA are of the following general types:

- 🌾 agriculture[33]
- 🌾 American Indian land acquisition
- 🌾 business and industrial
- 🌾 community facility
- 🌾 farm emergency
- 🌾 farm labor housing
- 🌾 farm operating
- 🌾 farm ownership
- 🌾 grazing association
- 🌾 home life development
- 🌾 individual home ownership
- 🌾 irrigation and drainage
- 🌾 recreation enterprise
- 🌾 rental and cooperative housing
- 🌾 resource conservation and development
- 🌾 self-help technical
- 🌾 soil and water conservation
- 🌾 watershed
- 🌾 youth

Nonfarm Portion of FmHA. The nonfarm portion of FmHA, the Rural Development Administration, Rural Electrification Administration, and Agricultural Cooperative Service, were merged into a new USDA organization in 1994 called the Rural Housing Service, which is an agency within the Rural Development division. (Note that before the restructuring of the FmHA, it had 1,700 field offices, normally in county seats.)

COMMERCIAL BANKS

Since early colonial days, commercial banks have been the major source of short- and intermediate-term loans. Banks are still the major non–real estate lender. There are about 13,000 commercial banks in the United States.[34] About 4,500 of these are classified as agricultural banks. To be designated as an agricultural bank, a bank needs to hold farm loans of 25 percent or more.[35]

Organization of Commercial Banks

Commercial banks are corporations created under federal or state law. They are owned by stockholders, who have invested in them. The stockholders elect a board of directors, which hires a staff to run the bank. In reality, a commercial bank is a private business, but it is regulated by federal and state agencies to ensure the safety of depositors' funds.

Agricultural Loans

Most commercial bank loans to agricultural producers are for short-term production expenses such as fertilizer, cattle, feed, seed, fuel, labor, and chemicals. Recently, commercial banks provided about 52 percent ($38 billion) in non–real estate loans to agriculture and about 28 percent ($22 billion) in real estate loans, representing a market share of 40 percent of total farm debt.[36]

Many banks have departments that specialize in agricultural credit needs. The banks employ skilled agricultural specialists, who can analyze farm business decisions and provide guidance in assisting production agriculturalists in using their capital wisely and meeting their repayment schedules.[37]

LIFE INSURANCE

Life insurance companies have played a strong role in financing real estate loans. Most of the major life insurance companies have agents whose primary responsibility is to invest company funds, and a prime target for these funds is farm real estate. Insurance companies also finance real estate indirectly, through commercial banks. The banks negotiate the loan with farmers and then discount the note with insurance companies.

Life insurance companies finance about 6 to 7 percent of the total farm debt and about 13 percent of the real estate loans, or about $10 billion.[38] However, life insurance loans vary greatly among different regions of the country. For example, although relatively inactive in the Northeast, they are very popular in the Corn Belt as well as the Southwest.[39]

Size of Loans

Life insurance companies tend to focus on large farm businesses. At least one company reportedly will not make a real estate loan for less than $250,000. The average-size life insurance company loan in recent years has been two to three times the average-size Farm Credit System loan. Farm mortgages traditionally account for 2 to 3 percent of the total investments made by life insurance companies.[40]

Source of Funds

Life insurance companies get their biggest source of funds for loans from premium payments from policyholders. Individual pension and retirement plans have **annuities** paid by companies for their employees. Life insurance companies receive **dividends** from the investments they have in the **securities markets**. These last two sources are also funds that insurance companies use to provide loans.[41]

Generally, life insurance companies are a **viable** source of real estate financing as long as the interest rate is reasonable, the repayment period is satisfactory, and there is no penalty for early repayment.

COMMODITY CREDIT CORPORATION (CCC)

The Commodity Credit Corporation (CCC) is the credit arm of the Farm Service Agency. The purpose of the CCC is to finance the loan and price support programs of the FSA. The CCC was created through an executive order from President Roosevelt in 1933 and was set up as a government corporation, with $100 million capital and authority to borrow additional funds if needed. In 1939, the CCC became a part of the Department of Agriculture, and in 1948 it received a permanent charter from Congress.

Nonrecourse Loans

A credit responsibility of the CCC is to make **nonrecourse loans** to farmers through their local FSA agency. These loans are designed to prevent them from having to sell their commodities immediately upon harvest, when prices are usually at their lowest seasonal level. Farmers who qualify for these loans use their stored crops as collateral for a short-term period. If the crops are sold during the period of the loan, the loan is paid back with interest, but if the farmer chooses not to sell the crop during this period, the crops become the property of the CCC and the loan is considered paid in full.

Limitations

The major limitations of CCC loans are that they are not available on all crops and that only production agriculturalists who qualify for the specific price support programs can obtain this type of loan. However, for those who can qualify, they provide an excellent means of storing products until the price improves while still obtaining needed capital to pay current operating expenses.

In recent years, the CCC has become a major source of short-term funds for production agriculturalists. Direct loans to farmers have amounted to about $7.8 billion per year.[42]

INDIVIDUAL AND OTHER LOANS

Individuals are a major source of credit for real estate loans. These are primarily farmers who sell their land, take partial payment, and hold a personal note on the remainder. Interest rates paid to individuals tend to be 1 to 2 percent lower than institutional rates, primarily because of lower overhead cost. Individuals play a very minor role in short- and intermediate-term loans, and their role in long-term loans is decreasing. The chief disadvantage for the borrower is that in the event of failure to make the payments, all preceding payments can be treated as rent, with the borrower losing any money that has been paid.

Merchants

Merchants represent a major part of the non–real estate loan market. They offer trade credit to managers of agribusinesses. The manager may establish an **open account** with the merchant to charge purchases. The balance on these purchases is normally due in thirty days.

Farm Implement Dealers

Farm implement dealers are responsible for a small part of the non–real estate debt. They provide financing on the machinery and equipment they sell, usually through credit institutions. The manufacturer sets up these institutions to provide credit at the dealerships. Dealer credit normally has longer terms but higher interest rates.

CAREER OPPORTUNITIES IN AGRIBUSINESS INPUT CREDIT SERVICES

There are many career opportunities in agribusiness credit or financial services for individuals with the needed qualifications. A knowledge of the economy in the agricultural industry, both locally and nationally, and the ability to understand and interpret the policies of agriculture are valuable assets for professionals in the field of finance. Refer to the Career Option on page 325 for an explanation of a career as a Farm Credit officer.

Experience

Agribusiness financial institutions making loans to production agriculturalists are often interested in hiring employees who have a basic knowledge in finance, accounting, risk management, and agricultural economics. A knowledge of production agriculture, land values, farm practices, production costs, and the marketing of agricultural products is required. Lending institutions depend on their agricultural specialists to keep their farm credit activities sound.

Education

Of the people currently working with agribusiness financial institutions, a high percentage have college degrees. Others have had specialized vocational or technical training. More important, agribusiness financial institutions are looking for those willing to learn. For people who go to college, courses should include agricultural economics, accounting, money and banking, farm management, marketing, future markets and trading, crop and livestock production, farm finance, soil management, business law, insurance and risk management, management information systems, and computer applications.

Company training is highly desirable and is usually required by agribusiness financial institutions. These courses provide understanding and knowledge of the methods, customs, and practices of the business. Such training programs can help you learn practical applications of any previous formal education.

Potential Available Jobs

The following are examples of jobs available with agribusiness financial institutions:

- ❦ credit officer for Farm Credit System
- ❦ credit officer for Farm Service Agency
- ❦ agricultural specialist for a rural bank
- ❦ loan appraiser for life insurance company
- ❦ salesperson for farm crop insurance company
- ❦ adjuster for farm crop insurance company
- ❦ credit officer for manufacturer of farm equipment or supplies
- ❦ collection officer for a manufacturer or supplier
- ❦ credit officer for a wholesaler or retail dealers
- ❦ loan representative for a federal or private lending agency
- ❦ appraiser of farmland and production agriculture enterprises
- ❦ staff member at any of the agribusiness financial institutions
- ❦ credit officer for an agricultural cooperative[43]

CONCLUSION

The agricultural credit area is big business. This input service is a vital component of the agricultural industry. Without it, production agriculturalists would encounter a broken link in the chain of their business.

Many jobs are available. Agricultural industry assets totaled $957 billion in 1995. Claims against these assets amount to only $151 billion.[44] This represented a debt-to-asset ratio of only 15.8 percent. This amount is borrowed from commercial

CAREER OPTION

Career Area: Farm Credit Officer

*T*he Farm Credit officer fulfills many roles in helping to supply needed financial resources to farmers, their families, and their businesses, and to the local communities, which depend on healthy economic agricultural productivity. Often the Farm Credit officer acts as a salesperson, operating at the state and local levels to locate new prospects for loans and financial services. At other times the credit officer acts as a farm management advisor and resource person for the local producer.

Farm Credit is a cooperative with a national focus, but whose districts are geographically divided so that loan officers can be more specialized in local business and farming trends. This local presence, coupled with a national interest and financial basis in agribusiness, allows the Farm Credit officer to provide broad-based yet personalized financial advice. Though some commercial banks offer loans to farm business enterprises, the benefit of Farm Credit agencies lies in their understanding of, and roots in, agriculture and their focus on the particular financial problems that farmers face. Farm Credit offers loans for homes in some nonagricultural areas, but the mission of the agency is to provide loans for agricultural development. This includes loans for farms, livestock, farm equipment, and farm operating expenditures. Repayment options are tailored to suit the specific agribusiness. Farm Credit officers often help small farms begin a new operation or rebuild or update operations, and they help part-time farmers become more efficient and productive by providing financial support and services. Often the Farm Credit officer can help steer family farm owners from a less productive enterprise to a more productive one as the agency encourages new production ideas and supports local educational efforts. These efforts include training and workshops on topics specific to the agribusiness community.

A Farm Credit officer should always thoroughly assess the potential borrower's financial position before making a loan. (Courtesy of USDA)

banks, life insurance companies, individuals, the Commodity Credit Corporation, the Farm Service Agency, and the Farm Credit System.

SUMMARY

Decisions about credit often are the most important judgments that people in the agricultural industry must make. In the agricultural industry, credit is needed to overcome the general shortage of equity capital. Limited capital, fluctuating interest rates, and lack of credit information are major problems facing many production agriculturalists.

The three "Rs" of credit are returns, repayment, and risk. The main reasons for borrowing money are to increase net returns and make a profit. Agricultural lenders expect their money to be repaid in full plus interest. Borrowers with strong assets can assume more risk than those with few assets.

In the agricultural industry, finances are needed in three areas: fixed expenses, operating expenses, and start-up expenses. Fixed expenses include items that have a fixed yearly price, such as land, buildings, machinery and equipment. Operating expenses cover everything needed to run a farm, ranch, or agribusiness. Start-up expenses include attorneys' fees, incorporation expenses, and costs for the development of a site or product.

Loans usually fall into three borrowing time frames: short, intermediate, and long term. Short-term loans usually range anywhere from one month to one year in length; intermediate-term loans vary from one to ten years; and long-term loans extend over ten years.

When reviewing a loan application, lenders require a credit profile. Lenders want positive answers to the following questions before extending credit: personal characteristics, management ability, financial position, loan purpose, and loan security.

Interest can be a major expense to agribusinesses that use debt to finance their operations. Normally, a lender will charge an interest rate tied to the current discount or prime rates set by the Federal Reserve System. An amortization table is a table that calculates constant payments needed to repay both the principal and interest on a sum of money. If an investment is feasible, it will generate enough after-tax income to pay for itself. The profitability index allows for a comparison of the same investment with different rates of return and different lengths of time.

There are several sources of credit for individuals in the agricultural industry. Production agriculturalists have access to all credit sources that other businesses have, plus some special sources set up strictly for their business. The major sources of credit are the Farm Credit System, commercial banks, the Farm Service Agency, the Commodity Credit Corporation, life insurance companies, and individuals.

There are many career opportunities in agribusiness credit or financial services for people with the needed qualifications. Experience, education, a knowledge of the economy in the agricultural industry both locally and nationally, and the ability to understand and interpret the policies of agriculture are valuable assets for a professional in the field of agribusiness finance.

END-OF-CHAPTER ACTIVITIES

Review Questions

1. What are three reasons why credit is important to production agriculturalists?
2. Name and briefly discuss the three fundamentals—"three R's"—of credit.
3. Name the three areas of credit needs.
4. What are four types of fixed expenses?
5. What items are typically purchased with short-term loans?
6. What is the key to maintaining good credit with short-term loans?
7. What items are typically purchased with intermediate-term loans?
8. What items are typically purchased with long-term loans?
9. Name and write down the questions asked by lenders when securing a credit profile on a borrower.
10. What would be the cost of capital for a $1,000 loan at 8 percent interest?
11. List and briefly describe the three ways to calculate the annual percentage rate.
12. What are six sources of credit for production agriculturalists?
13. When and why was the Federal Land Bank started?
14. When and why was the Production Credit Association started?
15. When and why was the Bank for Cooperatives started?
16. Why did the Farm Credit System have to be restructured?
17. List ten types of loans made by the Farm Credit Service.
18. Discuss briefly the historical beginning of the Farmers Home Administration.
19. What were three purposes of the Farmers Home Administration?
20. Name twenty general-type loans made by the Farm Service Agency.
21. What are the names of three agencies of the nonfarm portion of the old FmHA, which merged into a new USDA organization in 1994 called the Rural Housing Service?
22. Explain the organization of commercial banks.
23. What percentage of agricultural non–real estate loans are made by commercial banks?
24. What percentage of agricultural real estate loans are made by commercial banks?
25. What size of loans for agricultural purchases are made by insurance companies?
26. What are the sources of funds for insurance companies for money to loan to agricultural borrowers?
27. What is the purpose of the Commodity Credit Corporation (CCC)?
28. What are the limitations of CCC loans?
29. What are one advantage and one disadvantage of borrowing from individuals to purchase land?

30. Explain how merchants and farm implement dealers make loans.

31. What experiences are needed for individuals wanting to enter a career in agribusiness credit services?

32. List twelve examples of potential jobs available with agribusiness financial institutions.

Fill in the Blank

1. _____ _____ has become one of the fastest growing costs of production agriculturalists during the last two decades.

2. Agriculture and its products and services make up about _____ of the U.S. gross domestic product.

3. _____ _____ loans are usually for one year or less.

4. _____ _____ loans vary from one to ten years.

5. _____ _____ loans usually extend for over ten years.

6. A/an _____ table is a table that calculates payments needed to repay both the principal and interest on a sum of money.

7. If an investment is _____, it will generate enough after-tax income to pay for itself.

8. _____ _____ is determined by comparing the benefits with the costs.

9. For non–real estate loans, commercial banks supply _____ percent, Farm Credit Service supplies _____ percent, the Farm Service Agency _____ percent, the CCC supplies _____ percent, and individuals and others supply _____ percent.

10. For farm real estate loans, Farm Credit System supplies _____ percent, life insurance companies _____ percent, commercial banks _____ percent, the Farm Service Agency _____ percent, and individuals _____ percent.

Matching

a. $72 billion
b. Agricultural Credit Act of 1987
c. $88,085
d. CoBank
e. $957 billion
f. $151 billion
g. $72,697
h. Farm Service Agency
i. $78.7 billion
j. $350,217
k. $7.8 billion
l. Federal Crop Insurance Reform and Department of Agriculture Reorganization Act of 1994

_____ 1. value of all farm assets in 1995
_____ 2. debt in agriculture in 1995, or 16 percent of all assets
_____ 3. average farm real estate worth
_____ 4. average farm non–real estate assets

_____   5. average debt per farm in 1995

_____   6. money borrowed for non–real estate loans secured by farm assets in 1995

_____   7. farm real estate debt in 1995

_____   8. caused the Federal Land Banks and Production Credit Association to merge

_____   9. new name following the merge of the Bank of Cooperatives and the Central Bank for Cooperatives

_____   10. formerly called Farmers Home Administration

_____   11. amount of CCC loans to production agriculturalists

_____   12. legislation that terminated the FmHA

Activities

1. Make a list of all the financial institutions in your community that make loans to those in the agricultural industry.

2. Suppose a lender has a debt equity ratio of 2. How much money could you borrow for agricultural purposes if you already had $150,000 equity in land?

3. Suppose a lender has a debt equity ratio of 1. How much money could you borrow for agricultural purposes if you had $24,000 in equity?

4. You want to buy a commercial lawn mower for your lawn service business. The mower cost $2,700, and the interest rate is 11 percent. How much simple interest will you have paid at the end of one year?

5. Suppose you are buying a used small tractor for your lawn service business. The used tractor cost $5,000, and you take out a loan for 8.5 percent interest to be paid off in five years. You set this up on the "remaining balance" method of payment discussed in this chapter. Set up a payment plan showing the balance, principal, and interest paid yearly.

6. Calculate the price of the tractor described in Activity 5 by using the add-on method over five years at 8.5 percent interest. (a) What would be the total payment amount? (b) How much would the monthly payments be? (c) What would be the actual annual percentage rate (APR) of the loan?

7. Refer to the Amortization Table in Figure 12–6. You want to purchase a 100-acre farm for $3,000 per acre. The lender will charge a 9 percent interest rate to be paid off in twenty years. (a) How much will the payments be yearly? (b) How much will you pay for the farm over twenty years, including principal and interest?

8. What was the profitability index if a person earns $3,000 from a $2,500 investment?

9. Interview an agricultural loan officer in your community. List three things that the loan officer tells you will help you get a loan. List three things that will keep you from getting a loan.

10. Select a potential job that makes loans to those in the agricultural industry. Prepare a one-page paper describing the experience, qualifications, and education needed for that particular job. Share this information with your class.

NOTES

1. Jasper S. Lee and Max L. Amberson, *Working in Agricultural Industry* (New York: Gregg Division, McGraw-Hill Book Company, 1978), pp. 107–108.
2. Randall D. Little, *Economics: Applications to Agriculture and Agribusiness* (Danville, Ill.: Interstate Publishers, 1997), p. 131.
3. Ibid.
4. Gail L. Cramer, Clarence W. Hensen, and Douglas D. Southgate, Jr., *Agricultural Economics and Agribusiness*, 7th ed. (New York: John Wiley & Sons, 1997), p. 248.
5. Ibid.
6. Little, *Economics*, pp. 146–147.
7. Ibid.
8. Ibid.
9. Lee and Amberson, *Working in the Agricultural Industry*, pp. 108–109.
10. Ibid.
11. *Agribusiness Management and Marketing*, 8720B (College Station, Tex.: Instructional Materials Service, 1988).
12. Ibid.
13. Ibid.
14. Ibid.
15. Ibid.
16. Ibid.
17. Ibid.
18. Veronica Feilner, *Agricultural Management and Economics* (Columbia, Miss.: Instructional Materials Laboratory, 1988), pp. I:32–33.
19. Ibid., pp. I:33–34.
20. Cramer, Hensen, and Southgate, *Agricultural Economics and Agribusiness*, pp. 250–251.
21. Ibid., p. 251.
22. Ibid.
23. Ibid.
24. *Agribusiness Management and Marketing*, 8710-C, p. 1.
25. Cramer, Hensen and Southgate, *Agricultural Economics and Agribusiness*, p. 255.
26. Ibid.
27. Ibid., p. 258.
28. Ibid., p. 259.
29. Ibid.
30. Ibid., p. 255.
31. USDA, *A Brief History of the Farmers Home Administration* (Washington, D.C.: U.S. Department of Agriculture, Farmer's Home Administration, February 1990).
32. Cramer, Hensen and Southgate, *Agricultural Economics and Agribusiness*, pp. 255–256.
33. Ibid.
34. Ibid., p. 269.
35. Marcella Smith, Jean M. Underwood, and Mark Bultman, *Careers in Agribusiness and Industry* (Danville, Ill.: Interstate Publishers, 1991), p. 274.
36. Cramer, Hensen, and Southgate, *Agricultural Economics and Agribusiness*, p. 269.
37. Ibid.
38. Ibid.
39. John B. Parson, Jr., Ralph D. Pope, and Michael L. Cook, *Introduction to Agricultural Economics* (Englewood Cliffs, N.J.: Prentice-Hall, 1986), p. 454.
40. Ibid.
41. *Agribusiness Management and Marketing*, 8710-C, p. 4.
42. Cramer, Hensen, and Southgate, *Agricultural Economics and Agribusiness*, p. 263.
43. Smith, Underwood, and Bultman, *Careers in Agribusiness and Industry*, p. 281.
44. Cramer, Hensen, and Southgate, *Agricultural Economics and Agribusiness*, p. 272.

UNIT
4

The Agribusiness Output (Marketing) Sector

CHAPTER

13

Basic Principles of Agrimarketing

OBJECTIVES

After completing this chapter, the student should be able to:

- explain agribusiness marketing
- discuss how marketing developed
- explain the importance of supply and demand
- describe the prerequisites to an efficient economic system
- examine the factors to consider in a consumer-driven market
- discuss farm commodity marketing
- discuss the marketing of agribusiness products
- describe commodity research and promotion boards
- explain how to conduct a market analysis

TERMS TO KNOW

agrimarketing	futures contract	product advertising
allocation of resources	futures options	profitable
auctioneer	hedging	selling
basis	institutional advertising	spot price
blocked currency	market analysis	supply
check-off program	marketing	transformation
commission	marketing cooperatives	trichinae-safe pork
cotton gins	merchandising	value adding
demand	niche	vertical integration
efficient economic system	premium	yardage fee
emulsifier	prerequisite	
free market economy	pricing efficiency	

INTRODUCTION

Little is more important than marketing in the agricultural industry. In reality, production agriculturalists have been so efficient and effective in raising an abundance of food and fiber that many have produced themselves into bankruptcy. Production of food and fiber is not the problem; rather, it is marketing agricultural commodities in order to obtain a fair price that poses the challenge.

Agricultural marketing, or **agrimarketing**, is a large and important discipline in the agribusiness output sector. In 1996, the cost of marketing U.S. domestic food amounted to $421 billion, while the cost of producing farm food products was $123 billion. Of the $544 billion spent by consumers for food products, less than one-fourth was returned to producers, while over three-fourths of that amount was used for marketing costs.[1]

Marketing drives the economy. Without it a country will develop at a very slow rate. Many people do not realize the importance that the marketing function plays in all sectors of the economy. Marketing's role in agriculture assures producers of adequate profit while conveniently meeting consumers' needs. A good agribusiness manager realizes that a well-planned marketing strategy leads toward a maximizing of profits.[2]

Agribusiness is the nation's largest industry, and marketing is its largest segment. Over 80 percent of those involved in agribusiness are employed in marketing. Agribusiness-marketing activities generate more than 17 percent of America's annual gross domestic product, and about 75 cents of each consumer's food dollar goes to cover marketing expenses. Agribusiness marketing is clearly a major part of our national economy.[3]

Marketing is one of the four major decisions that a production agriculturalist faces, along with production, labor, and capital. Marketing's rank is

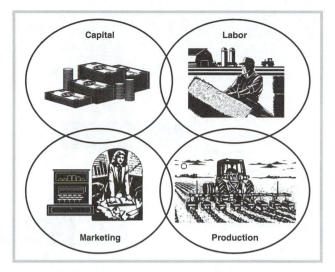

Figure 13–1   Decisions concerning marketing, capital, labor, and production are the four major decisions that a production agriculturist faces. Which do you believe is the most important?

equal to these other functions. Managers make decisions on production, labor, and capital in order to keep the business running smoothly and efficiently. Refer to Figure 13–1. However, these decisions may not always be enough to keep the business **profitable** in a changing economic environment.[4] Without adequate markets offering reasonable prices, profit will not be attained.

One factor that influences profits in a direct way is selling price. Profit is the difference between a product's selling price and its total cost to produce. Therefore, the higher the selling price, the greater profits.

WHAT IS AGRIMARKETING?

Agrimarketing is the sum of the processes, functions, and services performed in connection with food and fiber from the farms on which they are produced until their delivery into the hands of the consumer. Many steps are involved in agrimarketing,

Agrimarketing

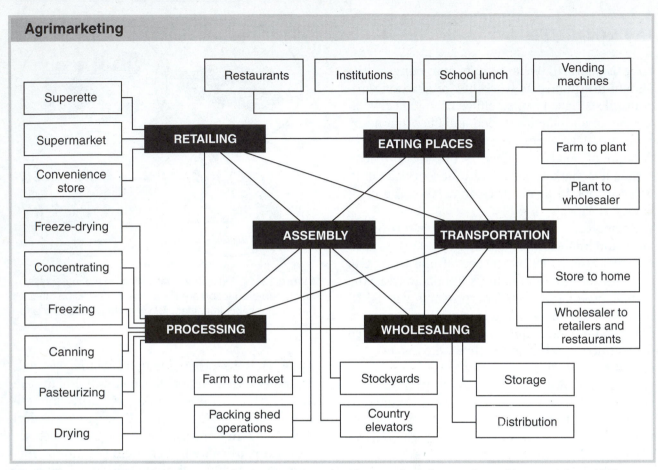

Figure 13–2   Agrimarketing is the name for the processes, functions, and services performed in connection with the movement of food and fiber from the farm on which they are produced until their delivery into the hands of the consumer.

such as buying, assembling, storing, packing, warehousing, communicating, advertising, financing, transporting, grading, sorting, processing, conditioning, packaging, selling, retailing, **merchandising**, risk taking, insuring, standardizing, regulating, inspecting, and gathering market information.[5] Refer to Figure 13–2.

The Difference between Selling and Marketing

Selling.   Historically, production agriculturalists have taken any products not used on the farm to town and sold them for whatever a local buyer would pay. If the price seemed too low to the production agriculturalists, they could try to bargain for a higher price. But buyers could usually buy all they wanted at the posted price. Unable to bargain for higher prices, production agriculturalists could only take the wheat, hogs, eggs, or cream back to the farm. Usually they did not; they took the price offered.

This is not marketing; this is **selling**.[6] When individual production agriculturalists were small and the amount of excess commodities they wanted to sell was also small, the difference between selling and marketing did not matter very much. In reality, marketing was not an option in the early days. The production agriculturalists might feel cheated and go away disgusted, but there was always the hope of getting higher prices next year.

Marketing. Many production agriculturalists now market their products instead of merely selling them. **Marketing** starts with analyzing the market to see what is needed before beginning production. It may be a different variety of wheat; hogs with longer, leaner body types; earlier maturing apples; or perhaps new or exotic crops. By being willing to shift production to a different crop or livestock animal, the production agriculturalist may be able to substantially increase revenues and profit.[7]

For example, many vegetable producers plant a crop only when they have a sales contract in hand. This takes the marketing uncertainty out of the picture for production agriculturalists, who would rather not speculate on price but do want to make a reasonable profit by being capable, efficient producers. The vegetable buyers are assured of a supply of product at a known price so that uncertainty is eliminated for them, too. Marketing contracts are becoming more common in production agriculture, particularly in cases where the quality of the product is important or where buyers require a specific raw material to use in a food-manufacturing process.[8]

HOW AGRIMARKETING DEVELOPED

In the formative years of this nation, most Americans worked as farmers, producing food and fiber mainly for domestic consumption. They did more than just produce food and fiber. They also processed farm products into a form that could be consumed: grain into flour, fibers into cloth, hides into leather. They cured meat and processed milk so it could be stored for future consumption. As time passed, production agriculturalists found that their particular geographic location, soil, and climate allowed them to produce some commodities better and more efficiently than others.

The food-marketing sector of agriculture developed primarily in response to the industrial and agricultural revolutions, which allowed, and in some cases caused, increasing numbers of people to shift to nonfarm occupations. Fewer people produced their own food, so they depended on farmers to provide food for them. Also, nonfarmers began to demand that food be brought to them in a different form than farmers normally sold their products. As a result, marketing firms began to develop to provide the services that consumers were demanding, and that farmers were reluctant to provide. Today, marketing is the largest phase of agriculture in terms of the number of employees as well as the sales value of food and fiber.[9]

SUPPLY AND DEMAND

It is a basic fact of life—anything you will ever want to buy or sell has a price. That price is determined by how much of a product people are willing and able to sell and how much other people are willing and able to buy; in other words, supply and demand.

Supply and demand have long been the factors that determine whether the production of a product can be profitable. **Supply** is the amount of a product available at a specific time and price. Some of the factors that may determine supply are how many people there are in the business of producing that particular product in the market area, how much of the product is coming into the market area from other areas, and the past history of profitability for that product in your area.

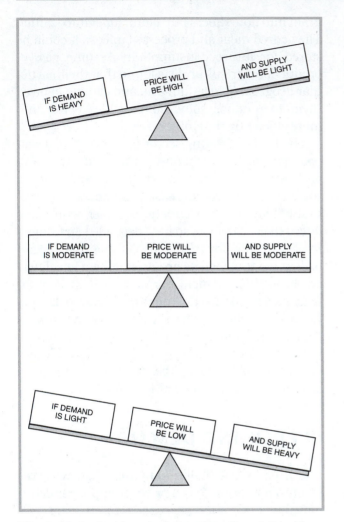

Figure 13–3 In a free market system, the relationship between supply and demand will determine price.

demand for a product. The amount of money available to consumers to buy that product is a factor that is often overlooked. Competition from similar products may reduce demand. Seasonal variations in demand also need to be considered.[10] Refer to Figure 13–3.

PREREQUISITES OF AN EFFICIENT ECONOMIC SYSTEM

When every **prerequisite** is met in the general economic system, the marketing system can lead, not only to an efficient satisfaction of producer and consumer needs, but also to an efficient allocation of society's limited resources. Three of the prerequisites are as follows.

A Free Market Economy

A **free market economy** must exist, where consumers provide the answers to the questions of what to produce, how much to produce, when to produce, who should produce, and for whom goods should be produced. The level of profits in such a system are a measure of how well producers answer these questions.

Prices That Reflect the Full Value of Market Resources

When prices reflect the full value of resources, resources are allocated to their highest and best use. How close an economic system comes to achieving this is a measure of **pricing efficiency**. A high degree of pricing efficiency is desirable because it results in:

Producers using the most efficient technology available and the lowest-cost combination of inputs to make their products.

Demand for a product is also determined by numerous factors. **Demand** is the amount of a product wanted at a specific time and price. It is often determined largely by price. The less expensive a product, the more of it will be wanted. However, there are other factors that also influence

---

The Conflicting Needs of Producers and Consumers That Agrimarketing Seeks to Resolve

Producers try to:	Consumers try to:
◆ Maximize long-run profit	◆ Maximize the satisfaction they receive from the products they consume with their limited incomes
◆ Sell large quantities of a few products	◆ Buy small quantities of many products
◆ Seek the highest prices	◆ Seek the lowest prices

Figure 13–4 Producers and consumers must work together to determine the needs in agrimarketing.

Consumers purchasing only those products that lead to the maximization of their total satisfaction.[11]

A High Degree of Interaction between Consumers and Production Agriculturalists

When consumers and production agriculturalists interact, both parties have a good knowledge of the variety and prices of goods available for sale and therefore can make informed choices.

The satisfaction of these three prerequisites is desirable because it will lead to the development of an **efficient economic system** where there will be:

❦ decision making by the consumers as to the who, what, and where of the economic system

❦ a pricing system that brings about an efficient **allocation of resources**

❦ the maximization of consumers' satisfaction

❦ the maximization of producers' long-run profits[12]

Refer to Figure 13–4.

FACTORS TO CONSIDER IN A CONSUMER-DRIVEN MARKET

Consumers are ultimately the determinant of agrimarketing success. Commodities or services may be available to consumers, but if the consumer chooses not to buy and does not submit to persuasion, there will be no sale. In a democratic system, we have the right to produce and market goods and services for personal profit (within the state and federal legal guidelines). Similarly, the right of refusal to buy goods and services is a basic right of consumers. The agrimarketing system involves the interactions among producers and consumers, and also among all the people, processes, and jobs in between.[13]

Consumers are the people, producers, agribusinesses, and other parties that use goods and services. They have certain needs and preferences. Consumers make decisions about what and how much they will consume. Success in agrimarketing involves understanding the consumer and the role of consumption. You want to offer a product that will be chosen over all others!

Agrimarketing is the link between producing and consuming. Products must be marketed in a variety of ways to satisfy all consumers. Understanding the many kinds of consumers and how their choices affect the marketing process is essential.[14]

Meeting Modern Demands

Today, modern food stores and delicatessens and a dazzling array of restaurants ranging from fast food to elite, white-tablecloth establishments demonstrate the complicated, dynamic nature of agrimarketing. Markets now offer a wide choice of products, various systems of distribution, and many built-in services, such as precooked meats and microwave meals. Much of the present market diversity results from the keen awareness by food processors, manufacturers, and retailers that the market is consumer driven. Consumers vote every day in the marketplace with their dollars, and the market hears what consumers are saying.[15]

Lifestyles are changing rapidly. Today, in many households both spouses work outside the home. Over one-half of all American women over the age of fifteen are in the workforce, leaving less time to prepare food. This has increased the demand for highly prepared foods and for meals at restaurants. Product requirements in the prepared food and food service markets are different from the requirements of retail grocery stores. The rising proportions of food sold through food service outlets or in a highly prepared form are creating demands for new food characteristics.

Predicting Consumer Demands Is Big Business

Finding out what consumers want and how they feel about various product characteristics has become big business. Management practices now involve **market analysis**, studying changes in consumer lifestyles and preferences and adjusting businesses to capitalize on those changes. The food-marketing industry spends billions of dollars each year on market research, new product development, and advertising. For example, the cost of advertising for 30 seconds in the 1998 Super Bowl was 1.3 million dollars.

Modern marketing involves product development, pricing strategies, efficient distribution, and product promotion. Marketing strategies integrate these components into an overall plan. Firms develop new products with desirable characteristics revealed through consumer research.[16]

Once the characteristics of consumer demand are evaluated, firms must develop desired products and pursue appropriate marketing techniques. Successful firms have strategies for accomplishing these tasks. Most new products fail in the marketplace within the first year after introduction, despite huge investments. Over 10,000 new products or product variations are put on grocery store shelves yearly.[17]

The Consumer Population

The earth's population is 5.6 billion people. North America including the United States, Canada and Mexico has 444 million people. Within the United States, there are approximately 260 million people. The world's population increases at the rate of 2.8 people every second. This translates to a 10,080 person increase in population every hour. At the current rate of population growth, the world's population will double every 30 years. By the year 2050, there will be nearly 9.4 billion people on earth.[18] They will need food and fiber. Agricultural production and marketing must keep pace if the needs of these people are to be met.

FARM COMMODITY MARKETING

When agrimarketing is discussed, it is discussed in two phases. The first phase is the marketing of commodities directly from the farm. After the product goes through the needed processes (*transformation*) to get it ready for the consumer, it goes through another market phase, the agribusiness product–marketing phase. The phases are different, yet they are interrelated. The agricultural products produced on the farm are sold, and marketed and consumed by people the way they like it.

For example, both beef and swine now are leaner due to consumer demand.

Production agriculturalists are not always in a position to set the price that they will receive for their products. Instead, they are accustomed to accepting the price that buyers of their products will pay. Production agriculturalists have historically been "price takers" rather than "price makers," and thus have been at the mercy of current market prices. Unfortunately, current market prices do not guarantee that the production of livestock or a particular crop will be profitable. Production agriculturalists can do little to control the profitability of their products unless they approach marketing as a vital component of producing agricultural commodities.[19]

In the agricultural industry, marketing is as important as the actual production of the commodities. Correct marketing decisions can mean the different between success and failure of the agricultural enterprise. Therefore, at least as much time and effort should be spent on marketing animals and crops as is spent on their production.

Marketing Strategies for Farm Commodities

In order to make the most money from a farm commodity, successful marketing is essential. Strategies that can be used to market farm commodities more profitably include:

❦ determine what types of markets are available to you

❦ determine the cost of various types of marketing

❦ determine transportation costs to market at each of the markets available to you, and sell where transportation costs are favorable

❦ determine the most profitable form in which to market your product (age, size, weight, degree of preparation)

❦ advertise to create markets where none existed before

❦ market your seasonal products at the peak of demand

❦ strive to shorten the marketing channels between producer and consumer[20]

Types of Livestock and Dairy Markets

There are various types of agricultural markets available to production agriculturalists. The one or more types chosen by an individual producer are often a matter of what is available and what the producer prefers. Care should be taken to carefully choose the type of marketing. Intelligent marketing practices may well make the difference between profit and loss in a very competitive business.

Terminal Markets. Terminal markets are usually located in a stockyard, which holds animals until they are sold. The terminal market never actually owns the animals. The animals that are delivered to the terminal market are consigned to a selling agent, who does the actual selling of the animals. The terminal market charges the seller a fee for caring for the animals until they are sold. This is called a **yardage fee**. The selling agent also receives a fee, called a **commission**, for selling the animals.

The use of terminal markets to market animals has always been mostly confined to the midwestern and western states. In recent years, the use of terminal markets to market animals has greatly decreased, and most livestock are currently being marketed by other methods.[21]

Auction Markets. Auction markets are places where animals are sold by public bidding on individual groups of animals. An **auctioneer** generally conducts the sales at an auction market. These markets are very widespread and are convenient to most local communities. They have grown in popularity and now represent the most common means of marketing farm animals. Auction markets are usually most practical for the smaller livestock producer. Video markets are a branch of this where buyers view videotapes of animals for sale.

As is true when selling in terminal markets, commissions are charged for selling animals at auction markets. The amount of commission charged varies with the type and size of the animal. Because many auction markets are small, there may not be the competition that is present at terminal markets to buy smaller numbers of animals.[22] At auction markets, you are really selling, not marketing.

Direct Sales. *Direct sales* refers to the producer selling animals (or crops) directly to processors. This method of marketing has several distinct advantages. There are no commission fees to be paid to selling agents, and there are no yardage fees. Transportation costs are kept to a minimum because the buyer generally comes to the farm to make purchases. The animals look their best for the buyer because they have not been exposed to the stresses of hauling and contact with strange facilities and animals.[23] Direct sales of animals, especially beef cattle, have increased significantly in recent years.

Cooperatives. Approximately 75 percent of the milk marketed in the United States is sold through farmer milk-marketing cooperatives. **Marketing cooperatives** are groups of producers who join together to market a commodity. The cooperatives then either process the milk and sell it directly to consumers or sell it to other, large processing plants. Marketing cooperatives have the ability to maintain product quality, arrange for transportation of the products from farm to market, balance the supply and demand of agricultural products, and plan advertising to increase sales of products. Refer to Figure 13–5.

Cooperatives are also used to market selected crops, although not to the same extent as with milk. An example is **cotton gins**. Some producers have formed ginning cooperatives to process their cotton at cost.

Government. The government purchase program also buys significant quantities of the milk. This milk either goes into the school lunch program or the Public Law 480 program. Public Law 480, also known as the Agricultural Trade Development and Assistance Act, was passed into law in 1954. The act provided for selling the surpluses to developing nations for **blocked currency**. This meant that money could be exchanged only within the nation making the purchase to boost the

Figure 13–5 Approximately 75 percent of milk marketed in the United States is sold through farmers' milk-marketing cooperatives. (Courtesy of USDA.)

Figure 13–6 With vertical integration producing, marketing, processing, and even inspection are joined. USDA inspectors work on-site in the processing plants. (Courtesy of USDA)

economies of underdeveloped nations. The program also allowed for donating surpluses to nations suffering from disasters.[24]

Vertical Integration. In the poultry market, almost 99 percent of the chickens produced for meat are grown under a system called **vertical integration** (several steps in the production, marketing, and processing of animals are joined together). Refer to Figure 13–6. The use of vertical integration in the production and marketing of farm products allows for extremely large systems of production that can be very efficient. Only the anticipated number of animals that will be needed by consumers are produced. There is less competition from other producers, and all phases of production can be controlled. Tyson Foods's poultry production system is an example of vertical integration. The swine industry is also fast becoming a vertical integration marketing system.

Futures Markets. **Hedging** is done by many livestock producers and feedlot operators. The Chicago Mercantile Exchange is the major exchange for livestock futures contracts. Refer to Chapter 14 for a comprehensive discussion of the **futures contract** and commodity (futures) marketing.

Types of Grain Markets

Several types of marketing are available for grain producers. These include futures contracts, forward contracting, harvest or spot pricing, and postharvest pricing. Futures contracts were mentioned previously as a type of marketing for livestock. Futures contracts will not be addressed in this section since all of Chapter 14 is devoted to commodity (futures) marketing.

Forward Contracting. Forward contracting allows the producer to establish a price in order to avoid a price decline. However, forward contracting also helps lock in a profit. Cash-forward contracting is another alternative for preharvest pricing. Like futures contracts, the producer attempts to establish a price in order to avoid a loss due to price decline. However, elevator or forward contracting also locks in a profit.

The main objectives of forward contracting are to capture pricing opportunities and reduce price risk. They also guarantee a market upon harvest and increase the firm's creditworthiness to lenders. Forward contracts have availability of up to a full year before delivery. The contract terms include quality, quantity, delivery time, grading procedures, and terms for settling a legal dispute should one arise. It also specifies a fixed price, less discounts for shrinkage and quality factors.

It is important to note that forward contracts do carry risk. While they do reduce price risk, they incur risks such as losses due to bad weather, droughts, fire, or other crop damage. Also, there are binding legal implications when entering into a forward contract. Producers should examine these factors carefully before using forward contracting.

Some advantages that forward contracts have over futures contracts are:

🌾 buyer and seller of forward contracts can negotiate the quantity, whereas futures contracts trade in 5,000 and 1,000 units

🌾 forward contracts do not have brokerage fees and commissions, as do futures contracts

🌾 forward contracts do not require as much knowledge of the market as trading in futures contracts[25]

A drawback to using a forward contract is that the seller cannot withdraw from it, whereas if a producer hedges with a futures contract, he or she may withdraw at any time by lifting the hedge.

Harvest Pricing. Harvest pricing is the traditional method of pricing used by producers. They simply harvest their grain, truck it to the local elevator, and sell it for the going price. Producers now use less harvest pricing because of depressed grain prices during harvest time. The **spot price** is another name for the prevailing cash price.

Producers can also price grain at harvest time but delay the delivery and payment for the commodity. They can do this by either forward contracting with an elevator or hedging with a futures contract. This strategy is most effective when there is a wide **basis**. In this situation, storage returns can be above normal by holding grain. The deferred income is helpful for tax purposes because realization of income does not occur until a later date. The drawback to this strategy is that it will not meet immediate cash flow needs.[26]

Postharvest Pricing. Postharvest pricing is the most frequently used strategy in pricing grain. Under this method, pricing can occur from right after harvest up to ten months or more later. The storage of grain is the important element in postharvest pricing.

Prices can rise or decline during the storage period. Producers who use postharvest pricing

need a tool to protect themselves against price risks. **Futures options** are a type of insurance to protect against price decline. The **premium** paid for the option is the cost of providing insurance for the stored crop. Use of the futures markets and forward contracts also protects against price decline.

Types of Fruit and Vegetable Markets

Consideration must be given to the product being produced, availability of markets, labor availability, and personal desires before a decision on which type of market will be used. Roadside markets, farmer's markets, pick your own, local brokers, wholesaling firms, and selling directly to processors are all viable options.

Roadside Markets. Roadside markets have long been a way of marketing fruits and vegetables for those who have grown them. Some producers grow fruits and vegetables for the express purpose of selling them in roadside stands. Some producers purchase fruits and vegetables from others to market in addition to the produce that was grown on their farms. Some roadside stands have become major businesses. Consumers like to purchase food from roadside stands because of the perceived freshness and wholesomeness of the fruit and vegetables. Some customers are actually willing to pay more for produce at these businesses because of the quality. Roadside stands are seasonal businesses in most parts of the country, unless they purchase produce from other sources that have access to produce grown in parts of the world with longer or different growing seasons.[27]

Farmers' Markets. Farmers' markets have appeared in small towns and large metropolitan areas to cater to the demands of urban consumers. They give urban and suburban consumers access to fresh products directly from the producers. Farmers' markets give the producer access to markets that would seldom be available otherwise. The producer also has the opportunity to educate the consumer concerning the value of good farm products. Refer to Figure 13–7.

Of course, there are several costs and inconveniences associated with marketing products at farmers' markets. There are often fees to be paid for the privilege of participating in farmers' markets. Competition may be higher, especially if several producers are selling the same type of product. Vehicles with heating and/or cooling may be necessary to get products to the market as fresh as possible. Products may be subject to certain food regulations and packaging requirements to meet state and federal standards. Facilities to hold and display products must also be available at the market.[28]

Figure 13–7 An advantage of farmers' markets is that the producer has the opportunity to educate the consumer concerning the value of good farm products. (Courtesy of USDA)

Pick Your Own. Pick your own is a simple concept. The producer grows a typical garden crop on a large scale, then allows customers to pick their own. Examples of fruit and vegetables that can be found at pick-your-own operations include strawberries, raspberries, peas, green beans, sweet corn, tomatoes, pecans, apples, peaches, plus just about any other fruit or vegetable.

Pick-your-own operations are pleasing people for several reasons. People like to eat fruits and vegetables that are as fresh as possible. Also, some view picking or gathering their food as a wholesome family activity. There appears to be a certain amount of entertainment value achieved through pick-your-own fruits and vegetables. Other people like to preserve fresh fruits and vegetables for consumption later on in the year.[29]

Food Brokers. Food brokers buy and sell fruits and vegetables and have them delivered to retail stores, restaurants, and institutions such as hospitals and public schools. The main characteristic of food brokers is that they may never actually handle the product.

Wholesalers. Wholesalers buy the product and hold it at a warehouse. It is then graded, boxed, or packaged, and shipped to retail stores, restaurants, and institutions.

Processors. Processors many times buy the products directly from the producer, as discussed earlier with livestock. The product goes directly to the processing plant, where the products are frozen, canned, or processed into some other form ready for the consumer.

Speciality Markets

There are many types of markets that are aimed at a particular **niche**. It works for some producers in a unique situation. Although there are several types, only five types will be briefly discussed: fee fishing, organically grown crops, hormone-free beef, on-farm restaurant, and wineries.

Fee Fishing. A variation of the pick-your-own concept is fish farms that allow anglers to fish for a fee. Catfish and rainbow trout are typical species of fish that are available this way. Some fish farm operations charge a fee to fish plus a per pound fee for any fish caught. Others simply charge a per pound fee. Some per pound fees include cleaning the fish, whereas other operations charge an additional fee if you want the fish cleaned.[30]

Organically Grown Crops. Organically grown crops include fruits and vegetables, as well as other crops, that are grown without chemical fertilizers, herbicides, or insecticides. There is a growing segment of the U.S. population that is interested in eating only foods that have been grown without the aid of chemicals. These people feel that food grown organically is more healthy and wholesome than food that is grown with chemicals.[31]

Hormone-Free Beef. Another specialty is beef that is grown without the use of growth stimulants and other medications. This type of beef is referred to as *hormone-free beef.* Some consumers are willing to pay a higher price for this product because it is perceived to be healthier.[32] Much concern has been raised about the use of hormones to stimulate growth in beef animals. Some producers of hormone-free beef have established a market by selling beef directly from the farm. There is also a market for beef that is low in fat and cholesterol.

On-Farm Restaurant. Opening an on-farm restaurant is another strategy for shortening the gap between the producer and consumer. The producer of fish has the advantage of removing several layers of middlemen. By doing this, the producer has the advantage of marketing a value-added product. Patrons of the restaurant enjoy dining on fresh fish that has never been frozen.

Wineries. Wineries represent another type of value-added marketing strategy. Many, if not most, wineries have their own vineyards for growing wine grapes. Some wineries also purchase grapes on contract from other producers. Wineries convert a relatively low-value crop into a high-value beverage.[33] Some wineries also bottle nonalcoholic juices to satisfy customers who do not consume alcohol. Wineries sell their product on-site as well as in other markets.

MARKETING AGRIBUSINESS PRODUCTS

Once farm commodities are marketed, most of the commodities are processed and marketed again by agribusinesses. Agrimarketing firms and commodity groups spend more than $10 billion annually to advertise and promote food.[34] Four key ingredients of marketing agribusiness products are advertising, price promotion, merchandising promotions, and public relations activities. Refer to the Career Option on page 345 for a further explanation of a career as a salesperson and agrimarketing specialist, just two of the many careers in agrimarketing.

Advertising

Advertising distributes messages about a product on television, radio, billboards, newspapers and newspaper supplements, and magazines. It aims to shape consumer perceptions and attitudes about the product or commodity and to establish a long-term market base.

Product advertising focuses on the product itself. This might include the product's usefulness, durability, price, value, and customer's need for the item. On the other hand, **institutional advertising** is designed to create a favorable image of the firm or institution offering the products or service.

Price Promotions

Price promotions are used to stimulate a quick sales response. Using "dealer price incentives," for example, manufacturers allow retailers and whole-

CAREER OPTION

Salesperson/Agrimarketing Specialist

Once farm commodities are marketed, most of the commodities are processed and marketed again by agribusinesses. Agrimarketing firms and commodity groups spend more than $10 billion annually to advertise and promote food. Effective salespersons and marketing specialists perform a variety of functions. These include creating an awareness of the product, motivating potential consumers, and reinforcing the value of purchases already made.

The educational level needed by a salesperson or marketing specialist depends on the company or the product. Although educational level is important, equally important are the ability to communicate, a positive attitude, knowledge of the product, belief in the product, and excellent human relations skills. Also, you must project a positive image of the company you are representing.

At this international food exposition in New Orleans, Louisiana, agrimarketing specialist Diana Tucker meets with industry representatives who have a potential interest in manufacturing products or developing technologies identified by the Agricultural Research Service. (Courtesy of USDA)

salers to purchase products for limited periods of time at discounts. Retailers may decide to pass all, part, or none of the incentive on to consumers. Price incentives are often aimed directly at consumers through cents-off coupons and rebates.

Merchandising Promotions

Merchandising promotions focus on the point of sale at the retail food store, restaurant, or fast-food outlet. Merchandising techniques include end-of-aisle displays, additional shelf facings, banners and signs, and in-store demonstrations. These are used to introduce new products and stimulate quick sales response.

Public Relations Activities

Public relations activities include support for local and national civic projects or sports events, support of public television and radio, educational programs and materials, meetings with food editors, and periodic news releases. The goal is to stimulate sales by accomplishing one or more of three objectives:

🌾 present the company in a favorable light

🌾 present the product in a favorable light

🌾 foster activities that call attention to the product[35]

Coordination

For maximum effectiveness, advertising must be coordinated with other marketing techniques, such as public relations, promotional programs, pricing, personal selling, product description and characteristics, merchandising promotions, warranties, and public relation activities. Effective marketing performs a variety of functions. These include creating an awareness of the product, motivating a customer to seek out the product, educating the potential customer, and reinforcing the value of purchases already made.[36]

The Four "P's" of Agribusiness Marketing and the Marketing Mix

As agribusiness marketing managers plan for success, four controllable variables should be included to make up the agrimarketing mix. They must be properly combined to meet the needs of the intended market. The agrimarketing mix is composed of four items each beginning with the letter "P": product, price, place, and promotion.

- product—the firm must develop the right product to give maximum satisfaction to the members of the target market

- price—the right product must carry the right price in light of market conditions

- place—the right product, at the right price, must be in the right place to be purchased by the members of the target market

- promotion—telling the members of the target market in the right way that the right product, at the right price is available at the right location[37]

Developing the proper agrimarketing mix of product, price, place, and promotion in order to satisfy consumer needs is what agrimarketing is all about, and what marketing managers do to earn their living. Agrimarketing managers use the four "P's" of marketing as tactical tools in the satisfac-

tion of consumer needs. By manipulating the marketing mix, they are able to implement the company's plans via its organization structure.[38]

Value Adding

Between the time the corn leaves the field and the corn flakes reach the consumer's table, many steps occur. At each step, value has been added to the commodity. At each step the value increases by more than the additional input, and **value adding** occurs. For example, when corn is ground into cornmeal, the value of the cornmeal is more than the value of the corn plus the cost of turning the corn to cornmeal. Furthermore, when the cornmeal is made into corn flakes, the value of the corn flakes is more than that of the cornmeal and other ingredients that go into them. By the time the corn flakes reach the consumer, the corn has been transported, ground, baked, packaged, marketed, displayed, and sold. Each of the processes adds value to the commodity.[39]

Value Adding Can Occur at Any Point During Marketing. Value is added if the product is transported to a more favorable marketplace. Likewise, value is added when marketing is timed to coincide with particular events. For the producer, however, most value is added when the product form is changed. In other words, if the producer can complete part of the processing before selling the commodity, value is added.

Packaging. A particularly important value-adding consideration in today's marketplace is packaging. Because many consumer buying decisions are made at the point of sale (in the store), packaging in eye-catching, appealing ways is very important. Some of the functions that packaging can perform in adding value to an agricultural product are:

- providing name brand identification

- advertising at the point of sale

🌿 transporting the product with the least damage

🌿 increasing home shelf life

🌿 convenience[40]

COMMODITY RESEARCH AND PROMOTION BOARDS

Food and fiber commodity research and promotion boards are the result of congressional legislation; most of them were created in the 1970s and 1980s. They are monitored by the USDA's Agricultural Marketing Service (AMS), and they assist the food and fiber industry in developing new products through their research programs. They know that their ability to increase demand for their products depends on their ability to work with industry, universities, and government to develop new products and new uses for their commodities. By staying tuned to changes in consumer needs and disseminating their research to food technologists and manufacturers, these boards contribute significantly to the development of new food and fiber products.[41]

Check-off Program

In this program the seller contributes a small amount for each unit sold to a money pool to be used to purchase advertising for the specific commodity. The program is referred to as the **check-off program** for the given commodity. It is estimated that about 90 percent of all United States farmers contribute to some 300 federal and state generic promotion programs covering approximately eighty commodities. The spending for research and promotion by these programs is reportedly near the half-billion-dollar mark annually. Media advertising for agricultural goods and services typically includes the use of newspapers, radio, television, outdoor billboards, signs, pamphlets, transportation advertising, direct mail, telephone book classified section, breed journals, product catalogs, and magazines.[42]

New Product Development by Commodity Boards

Recently, new products under development financed by the commodity research and promotion boards include a prototype new beef product, called beef surimi, that has a potential market of $70 million a year; a prototype all-cotton disposable diaper which, when produced and marketed, could be worth $240 million in additional annual sales to cotton growers; and an innovative dairy product with cultures that actually consume and remove cholesterol from fermented dairy products, such as sour cream.[43]

The following are examples of what the commodity boards are doing to monitor consumer trends and help industry introduce new products that can meet consumers' demands and manufacturers' needs.

The Potato Board. The Potato Board assists companies that develop new food products by providing them with statistical information about potato production and consumption trends. It also furnishes companies with information about consumer attitudes and the nutritional benefits of the potato.

This board explores new ways in which the potato can be used. For example, the potato can be used in food processing as a binding agent in cereal and pasta. Potato products, such as starch and flour, may be used by food technologists as an ice crystal retardant in ice cream, as an **emulsifier** in salad dressings, as a fat extender in dairy products, and as a nutritional supplement in processed foods.

Cattlemen's Beef Promotion and Research Board. According to the Cattlemen's Beef Promotion and Research Board, a new type of 100 percent beef product, called beef surimi, is being developed and processed into high protein surimi hot dogs, bologna, beef nuggets, finger food, even a new ready-to-eat snack beef chip. Beef surimi gets its name from a Japanese food process, now widely used, for making fish sticks and fish patties. Beef surimi has high-bonding qualities and will contain about 5 percent fat, much less than currently found in such beef products as hot dogs. Prototypes for the beef surimi hot dog and snack chip already have been developed.

The National Honey Board. In a typical supermarket, forty to fifty products contain honey. Honey is increasingly in demand because today's consumers prefer natural-tasting foods without added chemicals or preservatives. The National Honey Board has developed a list of new product applications. For example, honey attracts moisture and can reduce shrinkage in hams, cured meats, and baked goods. Honey also can clarify wine and is being tested as a seasoning for corn chips.

Commodity Research and Promotion Boards	Congressional Act and Year Started
The American Egg Board 1460 Renaissance Dr., Suite 301 Park Ridge, IL 60068	Congressionally enacted by the Egg Research and Consumer Information Act of 1974
Cattlemen's Beef Promotion and Research Board P.O. Box 3316 Englewood, CO 80155	Congressionally enacted by the Beef Promotion and Research Act of 1985
The Cotton Board 5350 Poplar Ave., Suite 210 Memphis, TN 38119	Congressionally enacted by the Cotton Research and Promotion Act of 1966
National Dairy Promotion and Research Board 2111 Wilson Blvd., Suite 600 Arlington, VA 22201	Congressionally enacted by the Dairy and Tobacco Adjustment Act of 1983
National Honey Board 9595 Nelson Rd., Box C Longmont, CO 80501	Congressionally enacted by the Honey Research, Promotion, and Consumer Information Act of 1984
National Pork Board P.O. Box 9114 Des Moines, IA 50306	Congressionally enacted by the Pork Promotion, Research, and Consumer Information Act of 1985
The Potato Board 1385 S. Colorado Blvd., Suite 512 Denver, CO 80222	Congressionally enacted by the Potato Research and Promotion Act of 1971

Figure 13–8 Food and fiber commodity research and promotion boards contribute significantly to the development of new food and fiber products through offering funding and conducting research.

The Cotton Board. Research funded by cotton growers has assisted the textile manufacturing industry in developing new cotton fabrics and products. Since 1984, research supported by the Cotton Board and conducted by its subsidiary, Cotton Incorporated, has produced new cotton products now used extensively in clothing and in the home: cotton-wool blends for year-round clothing; high-loft, resilient cotton for all-cotton pillows and mattress pads; and all-cotton stone-washed denim for jeans and jackets.

The American Egg Board. Through its research programs, the American Egg Board has developed several new products, such as frozen egg crust pizza, egg dip, egg sandwich loaf, and egg-filled burritos.

The National Pork Board. The most significant research currently being funded by the National Port Board is aimed at creating **trichinae-safe pork**. The industry has developed two distinctly different methods for safeguarding pork: irradiation and testing. Irradiation of fresh pork at a low level has been approved by USDA and the Food and Drug Administration. While trichinae can be rendered harmless by freezing, curing, or cooking at a high internal temperature, these two new methods of irradiation or testing would provide fresh, trichinae-safe, raw pork. The "other white meat" promotional campaign has also been very successful.

National Dairy Promotion and Research Board. The National Dairy Promotion and Research Board has increased its annual product research budget to $4.4 million to support work at U.S. universities and laboratories. Twenty-five percent of the research budget is spent on new product research.

In the next few years, the board believes that research in biotechnology and genetic engineering can potentially revolutionize the industry.[44]

Presently, the "got milk" promotion campaign, in which many celebrities are pictured with a milk mustache, is very successful. Refer to Figure 13–8 for an address list of several commodity research boards.

MARKET ANALYSIS

Market analysis is collecting information to determine if a product will sell. Careful attention to details is required. In order to get accurate information, research is needed. Research is a methodical study to gain knowledge about the market potential. Conducting research for marketing plans typically involves six steps. A brief discussion of each follows:

1. *Determine the information that is needed.* Do not collect information that will not contribute to the plan. The information needed varies with the product to be marketed and the potential customer.

2. *Develop a means for getting the needed information.* This may include developing written questionnaires, interview schedules, and guidelines for making personal observations.

3. *Word the questions carefully.* Make sure the questions get the information you want. Rating scales can be used to get the opinions of people about products or services. Questions followed by blanks for answers can be used to get more detailed information.

4. *Gather the information.* This involves implementing steps 1 through 3. Step 4 may require more time than thought. It may take six weeks to collect information by mail. Personal interviews require varying amounts of time, but several weeks may be needed to interview the necessary number of people even in a small community.

5. *Tabulate and analyze the information.* First, organize and categorize the information according to the questions asked. Various simple procedures can be used to summarize the information. Once data has been compiled and analyzed, determine what it means.

6. *Draw meaning and interpret the findings.* This involves making judgments that lead to conclusions and recommendations. Try to avoid personal biases and let the information speak for itself. Conclusions must always be based on the findings.[45]

CONCLUSION

Successful agrimarketing management is a matter of perspective. It is like the old story of two shoe salesmen who sail away to a distant land. Upon arrival both note that people in this country do not wear shoes. The first salesman cables home, "Send return ticket immediately; no market here." The second one cables home, "Send warehouse plans; market appears unlimited." Agrimarket opportunities are in the eye of the beholder.

SUMMARY

Little is more important than marketing in the agricultural industry. In 1994, the cost of marketing U.S. domestic food amounted to $401 billion, while the cost of producing farm food products was $110 billion. Agribusiness is the nation's largest industry, and marketing is its largest segment. Over 80 percent of those involved in agribusiness are employed in marketing.

Agribusiness marketing is those processes, functions, and services performed in connection with food and fiber from the farms on which they are produced until their delivery into the hands of the consumer. Selling is taking the farm product to the market and getting whatever the price is for that day. Marketing starts with analyzing the market to see what is needed before beginning production.

In the formative years of this nation, most Americans were farmers producing food and fiber mainly for domestic consumption. As time passed, production agriculturalists found that it was more efficient to specialize. As the country became more industrialized, fewer people produced their own food and depended on farmers to provide food for them. As a result, marketing firms began to develop the services that consumers were demanding.

Supply and demand have long been the factors that determine whether the production of a product can be profitable. Supply is the amount of a product available at a specific time and price. Demand is the amount of a product wanted at a specific time and price.

When several prerequisites are met in the general economic system, the marketing system can lead not only to an efficient meeting of producer and consumer needs but also to an efficient allocation of society's limited resources. Three prerequisites to an efficient economic system are: a free market economy, prices that reflect the full value of market resources, and a high degree of interaction between consumers and production agriculturalists.

Consumers are ultimately the determinant of marketing success. Commodities or services may be on the doorsteps of consumers, but if the consumer chooses not to buy and does not submit to persuasion, then there is no sale. Success in agrimarketing involves understanding the consumer and the role of consumption.

Production agriculturalists can do little to control the profitability of their products unless they approach marketing as a vital component of producing agricultural commodities. There are various types of agricultural markets available to production agriculturalists. The primary types of livestock and dairy markets are terminal markets, auction markets, direct sales, cooperatives, government, vertical integration and futures markets. The primary types of grain markets are forward contracting, futures contracts, harvest or spot pricing, and postharvest pricing. The primary types of fruit and vegetable markets are: roadside markets, farmers' markets, pick-your-own, food brokers, wholesalers, and processors. Speciality markets are also important to those producers that have a special niche. These include fee fishing, organically grown crops, hormone-free beef, on-farm restaurants, and wineries.

Once farm commodities have been marketed, most of the products are processed and marketed again by agribusinesses. Agribusiness-marketing firms and commodity groups spend more than $10 billion annually to advertise and promote food. Four key ingredients of marketing agribusiness products are advertising, price promotion, merchandising promotions, and public relations activities. The agrimarketing mix is composed of four items: product, price, place, and promotion.

Food and fiber commodity research and promotion boards are the result of congressional legislation. This program is referred to as the check-off program for the given commodity. The seller contributes a small amount for each unit sold to a money pool to be used to purchase advertising for the specific commodity. This amounts to about $500 million annually spent for research and promotion. Some examples of these commodity boards include the Potato Board, Cattlemen's Beef Promotion and Research Board, National Honey Board, Cotton Board, American Egg Board, National Pork Board, and National Dairy Promotion and Research Board.

Market analysis helps determine if a product will sell. Careful attention to details is required. In order to get accurate information, research is needed. Conducting research for marketing plans typically involves six steps.

Remember, successful agrimarketing management requires a positive attitude. A positive attitude can help you discover hidden opportunities.

END-OF-CHAPTER ACTIVITIES

Review Questions

1. Define the Terms to Know.
2. Besides marketing, what are three other major decisions that production agriculturalists face?
3. List twenty potential steps involved in agrimarketing.
4. Distinguish between selling and marketing.
5. What development or period in American history caused agrimarketing to develop at a faster rate?
6. Briefly explain how supply and demand offsets prices.
7. What are three prerequisites to an efficient economic system?
8. What are four results of an efficient economic system?

9. List seven marketing strategies that can be used to market farm commodities more profitably.

10. List and briefly describe seven types of livestock and dairy markets.

11. List and briefly describe four types of grain markets.

12. What are four main objectives of forward contracting?

13. What are four potential risks associated with forward contracts?

14. List three advantages forward contracts have over futures contracts.

15. List and briefly discuss six types of fruit and vegetable markets.

16. List and briefly discuss five examples of speciality markets.

17. List and briefly discuss four key ingredients of marketing agribusiness products.

18. Name and define the four items ("P's") that make up the agrimarketing mix.

19. What are five functions that packaging can perform in adding value to an agricultural product?

20. Who monitors the commodity research and promotion boards?

21. How are the commodity research and promotion boards funded?

22. List the names of seven commodity and research boards and give one example of something they have produced or promoted.

23. List the six steps needed to conduct research for marketing plans.

Fill in the Blank

1. Little is more important than _____ in the agricultural industry.

2. Of the $511 billion spent by consumers for food products, less than _____ was returned to producers.

3. Over 80 percent of those involved in agribusiness are employed in _____.

4. _____ are ultimately the determinant of agrimarketing success.

5. _____ is the link between _____ and _____.

6. _____ vote everyday in the marketplace with their dollars and the market hears what they are saying.

7. The world's population increases at the rate of _____ every hour.

8. At the current rate of population growth, the world's population will double every _____ years.

9. Production agriculturalists have historically been _____ _____ rather than price makers.

10. Agrimarketing firms and commodity groups spend more than _____ annually to advertise and promote food.

11. Food and fiber commodity research and promotion boards are the results of _____.

12. The commodity research and promotion boards collect approximately _____ dollars yearly.

Matching

a. 5.6 billion	e. vertical integration	i. 444 million
b. terminal market	f. 260 million	j. 401 billion
c. cooperatives	g. direct sales	k. harvest pricing
d. 10,000	h. auction markets	l. postharvest pricing

_____ 1. cost (in billions of dollars) of marketing U.S. domestic food in 1994

_____ 2. number of new products or product variations that are put on grocery store shelves yearly

_____ 3. earth's population

_____ 4. population of the United States, Canada, and Mexico

_____ 5. population of the United States

_____ 6. this market never actually owns the animals

_____ 7. place where animals are sold by public bidding on individual or groups of animals

_____ 8. producer sells animals (crops) directly to processors

_____ 9. group of producers who join together to market a commodity

_____ 10. process through which several steps in the production, marketing, and processing of animals are joined together

_____ 11. storage of grain in elevators is the important element in this type of grain market

_____ 12. traditional method of pricing used by producers

Activities

1. Give an example of a farm product that you can sell and one that you can market. Write this in report form and present it to the class to see if they agree.

2. Give an example of an agricultural product and how supply and demand could affect the price. (Note: An example would be the mad cow disease scare). Write this in report form and present it to the class.

3. List five examples of the way food is packaged (marketed) today compared to earlier years that are the result of consumer demand and lifestyle. Share these with your class and compile several other examples from your classmates.

4. Suppose you had one hundred beef cows and sold over ninety calves annually at weight of approximately 500 pounds each. Explain your marketing strategy in report form and present it to the class.

5. Suppose you have 200 acres of either corn, wheat, or soybeans. Explain your marketing strategy in report form and present it to the class.

6. Suppose you had an acre of strawberries. Explain your marketing strategy in report form and present it to the class.

7. There are many producers who develop a speciality or niche market. Write a brief report about someone in your community or elsewhere who has developed a speciality or niche market. Present this report to your class.

8. Select a raw agricultural commodity, then, select an end-product of that commodity (for example, wheat and bread). List as many steps or processes that you can determine that add value (value added) to the product from the farm to the consumer.

9. A wheat producer must choose between selling corn at harvest or storing it 6 months. The price at harvest is $3.10 per bushel. The elevator charges .05 per bushel for placing the corn in storage plus .015 per bushel per month storage fee. If the producer sells the corn at harvest, the proceeds will pay off a loan that has a 10 percent interest rate. What price must the corn producer receive to justify the storage costs and break even?

10. Production, labor, capital, and marketing are four decisions that a production agriculturalist faces. Write a brief report stating which you believe is the most important and why. Present this to your class.

11. A wheat producer's cost of production is $3.10 per bushel and the expected selling price is $3.65 per bushel. The producer expects to produce 30,000 bushels. What is the expected net income? Suppose the wheat producer could cut costs by 5 percent and increase selling price by 10 percent if wheat is sold in another market. What is the new expected net profit?

12. A producer sells eighty-two steer calves for 74 cents per pound. The sales contract states the calves are to be weighed at the ranch and the pay weight or actual selling weight will be based on a 4 percent shrink. On the sale date, the eighty-two calves weighed a total of 34,400 pounds.

    a. What was the average pay weight per calf?

    b. How many total dollars did the producer receive from selling the calves?

    c. What did the calves cost in dollars per head?

NOTES

1. United States Department of Agriculture (1997), *Agriculture Fact Book, 1997* (Washington, D.C.: U.S. Government Printing Office, 1997).

2. *Agribusiness Management and Marketing,* #8720B, (College Station, Tex.: Instructional Material Service, 1988), 8714A, p. 1.

3. USDA, *Agriculture Fact Book, 1997.*

4. *Agribusiness Management and Marketing,* 8714A, p. 1.

5. Ewell P. Roy, *Exploring Agribusiness* (Danville, Ill.: Interstate Printers and Publishers, 1980), p. 57.

6. Beierlein and Woolverton, *Agribusiness Marketing,* pp. 119–120.

7. Ibid.

8. Ibid.

9. Elmer L. Cooper, *Agriscience: Fundamentals and Applications,* 2nd ed. (Albany, N.Y.: Delmar Publishers, 1997), pp. 653–654.

10. Ibid.

11. Beierlein and Woolverton, *Agribusiness Marketing,* pp. 26–28.

12. Ibid.

13. Elmer L. Cooper, *Agriscience,* p. 654.
14. Jasper S. Lee, James G. Leising, and David E. Lawver, *Agrimarketing Technology: Selling and Distribution in the Agricultural Industry* (Danville, Ill.: Interstate Publishers, 1994), p. 271.
15. Ewen M. Wilson, "Marketing Challenges in a Dynamic World," in *Marketing U.S. Agriculture 1988 Yearbook of Agriculture* (Washington, D.C.: U.S. Government Printing Office, 1988), pp. 2–4. Deborah Takiff Smith, ed.
16. Ibid.
17. Ibid.
18. Lester Brown, *State of the World* (New York, N.Y.: W. W. Norton & Co., 1998), p. 174.
19. Ibid., p. 173.
20. Cooper, *Agriscience,* pp. 657–658.
21. Ibid., p. 661.
22. Ibid.
23. Ibid., p. 662.
24. *Agribusiness Management and Marketing,* 8714D, p. 6.
25. Ibid., 8714E, pp. 6–7.
26. Ibid.
27. Lee, Leising, and Lawver, *Agrimarketing Technology,* pp. 189–190.
28. Cooper, *Agriscience,* p. 660.
29. Lee, Leising, and Lawver, *Agrimarketing Technology,* p. 189.
30. Ibid.
31. Ibid., p. 191.
32. Ibid.
33. Ibid., p. 192.
34. Wilson, "Marketing Challenges in a Dynamic World," p. 5.
35. Ibid., pp. 4–5.
36. Cooper, *Agriscience,* p. 655.
37. E. Jerome McCarthy and William D. Perreautt, Jr., *Basic Marketing: A Managerial Approach,* 9th ed. (Homewood, Ill.: Richard D. Irwin, 1987).
38. Beierlein and Woolverton, *Agribusiness Marketing,* p. 58.
39. Lee, Leising, and Lawver, *Agrimarketing Techniques,* p. 137.
40. Ibid., pp. 138–139.
41. J. Patrick Boyle, "Commodity Boards Help Develop New Products," in *Marketing U.S. Agriculture* (Washington, D.C.: U.S. Government Printing Office, 1988), p. 160. Deborah Takiff Smith, ed.
42. Cooper, *Agriscience,* pp. 655–656.
43. Boyle, *Marketing U.S. Agriculture,* p. 160.
44. Ibid., pp. 160–163.
45. Lee, Leising, and Lawver, *Agrimarketing Techniques,* pp. 307–309.

CHAPTER 14

Commodity (Futures) Marketing

OBJECTIVES

After completing this chapter, the student should be able to:

- discuss the history of the commodity (futures) market
- describe commodity (futures) exchanges
- analyze the process of buying futures contracts
- describe hedgers and speculators in the futures market
- explain how the price of a commodity is determined
- explain how to make a futures trading transaction
- analyze the process of buying futures options
- discuss the regulations within the commodity (futures) exchanges

TERMS TO KNOW

basis	futures market	open outcry
bear market	futures options	performance bond margin
brokerage office	hedgers	premium
bull market	initial margin	price discovery
call	intrinsic value	put
cash market	long hedge	risk transfer
commodity	maintenance margin	short hedge
commodity (futures)	margin call	speculators
exchanges	margining system	ticks
contract specifications	minimum price fluctuations	time value
forward contracts	offset	trading pit
futures contracts		

INTRODUCTION

Since the early development of agricultural markets in the United States, production agriculturalists have attempted to protect themselves against falling commodity prices at harvest. Many production agriculturalists ignored marketing techniques and sold their commodities at harvest regardless of the price. Today, product agriculturalists (whether livestock or crop producers) realize that a marketing strategy is equal in importance to production, capital, and labor strategies.

The development of **forward contracts** was a major step in allowing producers to reduce marketing risks. There are now many different strategies one can use in forward contracting. In this chapter, we will discuss using the commodity (futures) market to reduce marketing risk.

HISTORY OF COMMODITY (FUTURES) MARKETS

Commodity markets originated in Chicago in the 1800s to help producers reduce price risk. Chicago was the hub that joined all parts of the United States. Many of the country's main railroads, waterways, and roads passed through the city. Production agriculturalists brought their grain to sell by wagons, oxcarts, and river barges. Often the quantity so greatly exceeded the market that many producers ended up dumping the grain out on the streets of Chicago.[1]

No Early Standards

The development of standards, grades, and measures had not yet occurred. Therefore, some buyers and sellers attempted to cheat on their measurements, which caused heated disagreements.[2] A uniform weight per bushel was set for selected grain commodities. For example, fifty-six pounds in a bushel of corn was the standard, where previously some people sold bushels with as little as forty-six pounds.

Few people understood how crucial agricultural futures were to maintaining the stability of markets. Without a price stability mechanism and standards, the price of grains would continue to fluctuate drastically.

Formation of the Chicago Board of Trade

The price fluctuating problems and increasing production created the need for a year-round, centralized marketplace. Eighty-two Chicago businessmen filled this need in 1848 when they founded the Chicago Board of Trade. This allowed buyers and sellers to get together in one place to exchange commodities. Both buyers and sellers began to see the advantages of making contracts to buy or sell commodities in the future. Forward pricing contracts were the result of this need.[3]

As the market grew, the Board of Trade established rules and regulations to help keep a truly competitive market. Members who held seats on the Chicago Board of Trade set up the rules. They had to abide by the rules or be subject to strict disciplinary action.

Forward Contracts Increase in Popularity

What made the Chicago Board of Trade increasingly popular as a centralized marketplace was the growing use of "to arrive" contracts. These contracts allowed buyers and sellers of agricultural commodities to specify delivery of a particular commodity at a predetermined price and date.

These early forward contracts in corn were first used by river merchants who received corn from farmers in late fall and early winter but had to store the corn until it reached a low enough moisture content to ship and until the river and canal were free of ice. Seeking to reduce the price risk of storing corn through the winter, these river merchants would travel to Chicago, where they would enter into contracts with processors for the delivery of grain at an agreed-upon price in the spring. In this way, they assured themselves a buyer and a price for grain. The earliest recorded forward contract in corn was made on March 13, 1851, for 3,000 bushels of corn to be delivered in June of that year.[4]

Change from Forward Contracts to Futures Contracts

Cash forward contracts did have drawbacks. They were not standardized according to quality or delivery time, and merchants and traders did not always fulfill their forward commitments. In 1865, the Chicago Board of Trade took a step to formalize grain trading by developing standardized agreements called **futures contracts**. Futures con-

The Chicago Board of Trade (CBOT): A Chronological History

Year	Historic Event
1848	Chicago's strategic location at the base of the Great Lakes, close to the fertile farmlands of the Midwest, contributed to the city's rapid growth and development as a grain terminal. Problems of supply and demand, transportation, and storage, however, led to a chaotic marketing situation and the logical development of the futures market.
1851	The earliest "forward" contract for 3,000 bushels of corn is recorded. Forward contracts gain popularity among merchants and processors.
1854	CBOT adopts standard weight of a bushel of wheat at 60 pounds, and oats at 32 pounds.
1865	Forward contracts create confusion for users and subsequent defaults. CBOT formalizes grain trading by developing standardized agreements called *futures contracts*. CBOT also begins requiring performance bonds, called *margin,* to be posted by buyers and sellers in its grain markets.
1866	First transatlantic cable completed. The time required to send a message to Europe from Chicago is reduced from three days to three hours.
1877	Futures trading becomes more formalized and speculators enter the picture. Growth in futures trading increases in late nineteenth and early twentieth centuries as new exchanges are formed. Many types of commodities are traded on these exchanges, including cotton, butter, eggs, coffee, and cocoa.
1925	One of the CBOT's biggest years—26.9 billion bushels of grain traded. Western Union installs automatic ticker to replace slower Morse service for improved quotation system.
1967	New electronic price display boards are installed on the walls above the trading floors. Price reporting time is cut to seconds.
1977	CBOT introduces the U.S. Treasury bond futures contract—today the most actively traded contract in the world.
1984	Trading in agricultural futures options begins as CBOT launches options on soybean futures.
1990	CBOT sets new world trading volume at 154 million contracts.
1995	Full membership trades at $710,000, a new record high.

Figure 14–1   Selected significant historic dates of the Chicago Board of Trade.

tracts, in contrast to forward contracts, were standardized as to quality, quantity, and time and location of delivery for the commodity being traded. The only variable was price—which was set through an auction-like process on the trading floor of an organized exchange.[5]

Standardized Agreements in Futures Contracts

Because futures contracts were standardized, sellers and buyers were able to exchange one contract for another and actually offset their obligation to deliver the cash commodity underlying the futures contract. **Offset** in the futures market means taking a futures position opposite, and equal to, one's initial futures transaction, for example, buying a contract if a previous one was sold.

Standardization of contract terms and the ability to offset contracts led to the rapidly increasing use of the futures markets by agricultural firms. Grain merchandisers, processors, and other agricultural companies found that by trading futures contracts, they were able to protect themselves from erratic price movements in commodities they were planning to either sell or purchase.[6]

Initiation of the Margining System

In the same year the Chicago Board of Trade introduced futures contracts, it initiated a **margining system** to eliminate the problems of buyers and sellers not fulfilling their contracts. A margining system requires traders to deposit funds with the exchange or an exchange representative to guarantee contract performance. Although early records were lost in the Great Chicago Fire of 1871, it has been quite accurately established that by 1865, most of the basic principles of futures trading as we know them today were in place.[7] But no one could have guessed how this infant industry would change and develop in the next century and beyond.

Growth in futures trading increased in the late nineteenth and early twentieth centuries as more and more businesses adopted futures trading into their business plans. Still, most of the dramatic growth and successful contracts in the futures industry were yet to come. Refer to Figure 14–1 for historic highlights of the Chicago Board of Trade.

COMMODITY (FUTURES) EXCHANGES

Although most of our references have been made to the Chicago Board of Trade (CBOT), there are many other **commodity (futures) exchanges**. Refer to Figure 14–2. Each exchange tends to specialize in certain commodities although there are overlaps. For example, the Chicago Board of Trade specializes in grain commodities, whereas the Chicago Mercantile Exchange specializes in livestock commodities.

Exchanges Do Not Trade Anything Physical

The hardest concept for a beginner to understand is that exchanges do not really physically trade grain or livestock commodities. It serves as a forum or meeting place for exchange members, buyers, and sellers of commodities. The traders are individual members and member firms who seek to trade agricultural commodities for their customers or themselves.

The questions become, "If grain is traded at an exchange, where is the physical commodity, or, Why are there no piles of grain outside of the Chicago Board of Trade?" Originally, the grain was brought to the regional exchanges. Farmers, who brought grain and livestock to the markets at a certain time each year, often found that immediate quantity far outweighed the market need. The

U.S. and Canadian Futures Exchanges

Chicago Board of Trade
Market Development Department
141 W. Jackson Blvd.
Chicago, IL 60604
800/THE-CBOT

Chicago Mercantile Exchange
30 S. Wacker Drive
Chicago, IL 60606
312/930-1000

Coffee, Sugar & Cocoa Exchange, Inc.
Four World Trade Center
8th Floor
New York, NY 10048
212/938-2800

Commodity Exchange, Inc. (COMEX)
Four World Trade Center
New York, NY 10048
212/938-2900

Kansas City Board of Trade
4800 Main Street
Suite 303
Kansas City, MO 64112
816/753-7500

MidAmerica Commodity Exchange
Market Development Department
141 W. Jackson Blvd.
Chicago, IL 60604
800/THE-CBOT

Minneapolis Grain Exchange
130 Grain Exchange Building
400 South Fourth Street
Minneapolis, MN 55415
612/338-6212

Montreal Exchange
The Stock Exchange Tower
P.O. 61
800 Victoria Square
Montreal, Quebec, Canada H4Z 1A9
514/871-2424

New York Cotton Exchange
Four World Trade Center
New York, NY 10048
212/938-2702

New York Futures Exchange
20 Broad Street
New York, NY 10005
212/656-4949

New York Mercantile Exchange
Four World Trade Center
New York, NY 10048
212/938-2222

Philadelphia Board of Trade
1900 Market Street
Philadelphia, PA 19103
215/496-5000

The Winnipeg Commodity Exchange
Suite 500
360 Main Street
Winnipeg, Manitoba, Canada R3C 3Z4
204/949-0495

Figure 14–2 There are many commodity (futures) exchanges throughout the world. Those listed here are the exchanges in the United States and Canada.

grain buyers, seeing such a large supply, bid the lowest price. The result was very low prices.

Today, the contracts that are traded are for grain to be delivered at some time in the future, say September or December at specified delivery points, if delivery is desired (this rarely happens). This keeps grain producers from getting terrible prices due to oversupply (quantity). In reality, at the time the contracts are traded, the grain is still in the fields, in grain elevators, or perhaps not even planted. This keeps too much grain from being delivered at the same time, thus stabilizing prices.

Commodities in Addition to Agricultural Commodities Are Also Traded

Although this chapter will focus on agricultural commodities, commodity (futures) exchanges trade many other commodities (the futures contract process did start with agriculture). Today, there are futures contracts for interest rates, insurance, stock indexes, manufactured and processed products, crude oil, nonstorable commodities, precious metals, and foreign currencies and foreign bonds.[8] Refer to Figure 14–3 for a list of commodities that are traded.

Purpose of Commodity (Futures) Markets

Futures exchanges, no matter how they are organized and run, exist because they provide two vital economic functions for the marketplace: **risk transfer** and **price discovery**. Futures markets make it possible for those who want to manage price risk—**hedgers**—to transfer some or all of that risk to those who are willing to accept it—**speculators**. These are topics of discussion that follow in this chapter.

BUYING FUTURES CONTRACTS

When discussing contracts, you may think of pieces of paper with lots of fine print. However, there is substantial documentation or paperwork behind the existence of a futures contract, and it is not a formalized contract written by an attorney. Futures contracts are legally binding agreements, made on the trading floor of a futures exchange, to buy or sell something in the future. That "something" could be corn, soybeans, gold, live beef futures, or some other commodity.

In other words, a futures contract establishes a price today for a commodity that will be delivered at a later date. Buyers and sellers in the futures markets look at current economic information (supply and demand) and anticipate how it may impact a commodity's price.[9]

What Does a Futures Contract Look Like?

Although the futures contract is not a written contract, it is a verbal agreement between a buyer and a seller (made on the floor of a futures exchange). Although not written, each futures contract specifies the time of delivery or payment, where the commodity should be delivered, and the quality and quantity of the item. That is what makes futures contracts attractive to those who want to plan ahead and protect themselves from dangerous price swings and to investors wanting to profit from market fluctuations.[10]

The standard features (time of delivery, etc.) are called **contract specifications**. You will find most of them listed in the newspaper along with prices from the previous trading day. The futures exchange where your commodity is traded also can provide contract specifications for the commodity you're studying. The *Wall Street Journal* is one of the best sources for commodity market information. Figure 14–4 illustrates a common example of how commodity prices may appear.

By referring to Figure 14–4, you will notice that several contract specifications are listed for each commodity. Corn will be used as our example:

🌾 name of commodity: corn

🌾 where it is traded: Chicago Board of Trade

🌾 contract months: Mar. '98, May '98, July '98, Sept. '98, Dec. '98, Mar. '99, July '99, Dec. '99

🌾 contract size: 5,000 bushels

🌾 price quote: price per bushel in ¼-bushel increments

Open is the first price anyone paid for corn on January 5, 1998. *High* is the highest price anyone paid for corn on January 5, 1998. *Low* is the lowest price anyone paid for corn on January 5, 1998. *Settle*, or *settlement*, is the last price anyone paid for corn on January 5, 1998. *Change*, or *net change*, is the difference between the settlement price on January 5, 1998, and the previous trading day.

Practice Reading a Price Quote

Just for practice, read one line of price quotes for the July corn futures contract (shown on page 365) and fill in the following blanks, on a separate sheet of

Futures Trading Facts

Commodity	Name of Exchange	Trading Hours N.Y. Time, Mon. thru Fri.	Contract	Minimum Fluctuation Per Lb. etc.	Minimum Fluctuation Per Contract	Daily Trading Limits (From Previous Close)
BROILERS, FRESH	Chicago Mercantile Exchange	10:10 a.m.–2:00 p.m.	30,000 lbs.	2½/100¢	$7.50	2¢
CATTLE (FEEDER)	Chicago Mercantile Exchange	10:05 a.m.–1:45 p.m.	42,000 lbs.	2½/100¢	$10.50	1½¢
CATTLE (LIVE BEEF)	Chicago Mercantile Exchange	10:05 a.m.–1:45 p.m.	40,000 lbs.	2½/100¢	$10.00	1½¢
COCOA	Coffee, Sugar & Cocoa Ex.	9:30 a.m.–3:00 p.m.	10 tonnes	$1.00/tonne	$10.00	$88.00*
COFFEE "C"	Coffee, Sugar & Cocoa Ex.	9:45 a.m.–2:25 p.m.	37,500 lbs.	1/100¢	$3.75	4¢*
COPPER	Commodity Exch., Inc., N.Y.	9:50 a.m.–2:00 p.m.	25,000 lbs.	5/100¢	$12.50	5¢*
COTTON #2	New York Cotton Exchange	10:30 a.m.–3:00 p.m.	50,000 lbs.	1/100¢	$5.00	2¢*
Currencies						
BRITISH POUND	International Monetary	8:30 a.m.–2:24 p.m.	25,000 BP	$.0005	$12.50	$.0500*
CANADIAN DOLLAR	Market of	8:30 a.m.–2:22 p.m.	100,00 CD	$.0001	$10.00	$.0075*
DEUTSCHE MARK	Chic. Merc. Exch.	8:30 a.m.–2:20 p.m.	125,000 DM	$.0001	$12.50	$.0100*
JAPANESE YEN		5:30 a.m.–2:26 p.m.	12.5 Mil. JY	$.000001	$12.50	$.0001*
SWISS FRANC		8:30 a.m.–2:16 p.m.	125,000 SF	$.0001	$12.50	$.0150*
GINNIE MAE MTGES.	Chicago Board of Trade	9:00 a.m.–3:00 p.m.	$100,000 @ 8%	1/32 Pt.	$31.25	64/32*
GOLD	IMM—Chicago Merc. Exch.	9:25 a.m.–2:30 p.m.	100 Troy oz.	$.10/Oz.	$10.00	$50.00
	Commodity Exch. Inc., N.Y.	9:25 a.m.–2:30 p.m.	100 Troy oz.	$.10/Oz.	$10.00	$25.00*
	MidAmerica Com. Ex., Chi.	9:25 a.m.–2:40 p.m.	33.2 Troy oz.	$.025/Oz.	$0.83	$50.00*
	Winnipeg Commodity Exch.	9:25 a.m.–2:30 p.m.	20 Troy oz.	$.10/Oz.	$2.00	$30.00* (U.S.)
Grains—Chicago WHEAT, SOYBEANS, CORN, OATS	Chicago Board of Trade / MidAmerica Com. Ex., Chi.	10:30 a.m.–2:15 p.m. / 10:30 a.m.–2:30 p.m.	5,000 Bus. / 1,000 Bus. (Oats 5,000)	1/4¢ / 1/8¢	$12.50 / $1.25 (Oats $6.25)	Wheat 20¢, Soybeans 30¢ Corn 10¢, Oats 6¢
Grains—Minneapolis WHEAT	Minneapolis Grain Exchange	10:30 a.m.–2:15 p.m.	5,000 Bus.	1/8¢	$6.25	20¢*
Grains—Kansas City WHEAT	Kansas City Board of Trade	10:30 a.m.–2:15 p.m.	5,000 Bus.	1/4¢	$12.50	Wheat—25¢
Grains-Winnipeg BARLEY, OATS, RYE, RAPESEED, FLAXSEED	Winnipeg Commodity Ex.	10:30 a.m.–2:15 p.m.	20 Tonnes	10¢/Tonne	$2.00	Barley, Oats, Rye $5.00/Tonne (Cdn.) Rapeseed & Flaxseed $10.00/Tonne (Cdn.)

(continued on following page)

Figure 14–3 Although the discussion of this chapter focuses on agricultural futures, many other, nonagricultural futures commodities are traded. (Courtesy of Prentice Hall)

Futures Trading Acts (concluded)

Commodity	Name of Exchange	Trading Hours N.Y. Time, Mon. thru Fri.	Contract	Minimum Fluctuation Per Lb. etc.	Minimum Fluctuation Per Contract	Daily Trading Limits (From Previous Close)
HOGS (LIVE)	Chicago Mercantile Exchange	10:10 a.m.–2:00 p.m.	30,000 lbs.	2½/100¢	$7.50	1½¢
LUMBER	Chicago Mercantile Exchange	10:00 a.m.–2:05 p.m.	130,000 Bd. Ft.	10¢/1000 Board Ft.	$13.00	$5.00
OIL, HEATING #2	N.Y. Mercantile Exchange	10:30 a.m.–2:45 p.m.	42,000 gals.	1/100¢	$4.20	2¢*
ORANGE JUICE (frozen concentrate)	New York Cotton Exchange	10:15 a.m.–2:45 p.m.	15,000 lbs.	5/100¢	$7.50	5¢*
PLATINUM	N.Y. Mercantile Exchange	10:15 a.m.–2:30 p.m.	50 Troy oz.	10¢	$5.00	$20.00*
PLYWOOD	Chicago Board of Trade	10:00 a.m.–2:00 p.m.	76,032 Sq. Ft.	10¢/1000 Sq. Ft.	$7.60	$7.00*
PORK BELLIES	Chicago Mercantile Exchange	10:10 a.m.–2:00 p.m.	38,000 lbs.	2½/100¢	$9.50	2¢
POTATOES	N.Y. Mercantile Exchange	10:00 a.m.–2:00 p.m.	50,000 lbs.	1¢/Cwt.	$5.00	50¢*
SILVER	Commodity Exch., Inc., N.Y.	9:40 a.m.–2:15 p.m.	5,000 Tory oz.	10/100¢	$5.00	50¢*
SILVER OLD	Chicago Board of Trade	9:40 a.m.–2:25 p.m.	5,000 Troy oz.	10/100¢	$5.00	40¢–80¢**
SILVER NEW	Chicago Board of Trade	9:40 a.m.–2:25 p.m.	1,000 Troy oz.	10/100¢	$1.00	40¢–80¢**
SILVER	Winnipeg Commodity Exch.	9:40 a.m.–2:25 p.m.	200 Troy oz.	1¢	$2.00	50¢* (U.S.)
SOYBEAN MEAL	Chicago Board of Trade	10:30 a.m.–2:15 p.m.	100 tons	10¢/Ton	$10.00	$10.00*
SOYBEAN OIL	Chicago Board of Trade	10:30 a.m.–2:15 p.m.	60,000 lbs.	1/100¢	$6.00	1¢*
SUGAR world #11 domestic #12	Coffee, Sugar & Cocoa Ex.	10:00 a.m.–1:43 p.m.	112,000 lbs.	1/100¢	$11.20	1/2¢*
SUNFLOWER SEED	Minneapolis Grain Exch.	10:25 a.m.–2:20 p.m.	100,000 lbs.	1¢/Cwt.	$10.00	50¢*
T-BILLS (13 WEEKS)	IMM-Chicago Merc. Exch.	9:00 a.m.–2:40 p.m.	$1,000,000	.01	$25.00	.60*
T-BILLS	N.Y. Futures Exchange	9:00 a.m.–3:00 p.m.	$1,000,000	.01	$25.00	1.00*
T-BONDS (LONG TERM)	Chicago Board of Trade	9:00 a.m.–3:00 p.m.	$100,000 @ 8%	1/32 Pt.	$31.25	64/32*
T-BONDS	N.Y. Futures Exchange	9:00 a.m.–3:00 p.m.	$100,000 @ 9%	1/32 Pt.	$31.25	96/32*

*Expanded limits go into effect under certain conditions.

**Limit based on price range and can be expanded under certain conditions.

Commissions and Margins: Contact your broker for all information.

Monday, January 5, 1998: Open Interest Reflects Previous Trading Day

Grains and Oilseeds

	Open	High	Low	Settle		Change	Lifetime High	Lifetime Low	Open Interest
CORN (CBT) 5,000 bu.; cents per bu.									
Mar	262½	266	261¾	265¾	+	3	305	236	168,498
May	269	273	268¾	272¾	+	3½	310	241¾	51,767
July	275	277¾	274	277½	+	2¼	315½	245	60,236
Sept	275½	277¾	274½	277½	+	2	301	244	7,593
Dec	280	281½	278½	281¼	+	1½	299½	247	34,389
Mr99	285	287½	284½	287¼	+	1½	305	283	1,356
July	293¼	294	293¼	294	+	1¾	312	256¼	401
Dec	274½	275½	273¾	275½	+	1½	291½	265	675
Est vol 40,000; vol Fri 8,115; open int 64,983, + 960.									
OATS(CBT) 5,000 bu.; cents per bu.									
Mar	149¾	153¾	149¾	153¼	+	3	180	148¼	8,269
May	153¼	157½	153¼	157	+	2½	182½	151	2,184
July	157¾	160½	157½	160½	+	2¾	184	153	750
Sept	161	162½	161	162½	+	1¾	177	155	296
Dec	165	165	164	164	+	1¼	177½	162	119
Est vol 1,000; vol Fri 56; open int 2,324, −8									
SOYBEANS (CBT) 5,000 bu.; cents per bu.									
Jan	660½	669	660	667¼	+	6¾	752	583	11,907
Mar	668	676	667	674	+	6½	749½	593	56,345
May	675	682½	675	681½	+	6¾	752	601	27,416
July	679½	687	679½	685½	+	6¾	753	611½	26,720
Aug	682	684	681	683½	+	6½	745	631	4,197
Sept	669	670½	667	669	+	3¼	723	637	270
Nov	660½	663½	659½	662¾	+	4½	717	597	10,711
Est vol 45,000; vol Fri 0-86; open int 27,566, +60									
SOYBEAN MEAL (CBT) 100 tons; $ per ton.									
Jan	200.30	202.50	199.60	200.40	+	.80	239.50	185.50	13,621
Mar	197.40	200.00	197.00	198.20	+	1.50	234.40	184.50	43,296
May	197.00	199.50	196.70	197.90	+	1.50	231.00	185.50	25,507
July	199.50	201.00	198.20	199.80	+	1.60	231.50	188.50	16,864
Aug	200.50	201.50	199.50	200.10	+	1.60	231.50	189.00	5,340
Sept	201.00	201.80	199.50	200.40	+	.70	231.50	192.00	3,820
Oct	200.00	200.80	200.00	200.40	+	1.70	226.00	190.00	903
Dec	202.00	202.00	200.50	201.20	+	1.20	231.00	193.00	3,676
Est vol 16,000; vol Fri 24,832; open int 113,032, +2,042.									
SOYBEAN OIL (CBT) 60,000 lbs.; cents per lb.									
Jan	24.85	24.90	24.70	24.79	−	.05	27.45	21.98	4,534
Mar	25.15	25.25	25.05	25.12	−	.03	27.50	22.20	55,929
May	25.53	25.55	25.33	25.41	−	.02	27.55	22.35	18,641
July	25.70	25.75	25.55	25.58	−	.06	27.40	22.40	12,104
Aug	25.55	25.55	25.50	25.50	−	.04	26.70	22.72	3,716
Sept	25.40	25.45	25.31	25.31	−	.10	26.15	22.90	1,379
Oct	25.15	25.20	25.05	25.05	−	.05	26.20	22.80	679
Dec	24.98	25.10	24.90	24.95	−	.08	26.30	23.00	2,184
Est vol 17,000; vol Fri 16,345; open int 99,183, −115.									
WHEAT (CBT) 5,000 bu.; cents per bu.									
Mar	330¾	334	327	331	+	¼	470	325	58,122
May	337¾	341	335	338¼	+	½	439½	333	13,336
July	344½	347½	342	344¾	+	1½	425	333	20,193
Sept	351¼	352½	350	350½	+	1½	403	346½	1,069
Dec	361	363½	361	361	+	1	417	358	2,907
Est vol 12,000; vol Fri 2,394; open int 19,135, +222.									
WHEAT (KC) 5,000 bu., cents per bu.									
Mar	341½	344	337	338¾	−	2½	491	335	5,449
May	350½	351½	345½	345½	−	1½	450	343	1,477
July	357½	359	354½	354½	−	2¾	408½	333	1,913
Sept	359	359	359	359½	+	2	410	335	124
Dec	369	370	367	367	+	2	418½	363	103
Est vol 3,983; vol Fri 3,743; open int 45,449, +175.									
WHEAT (MPLS) 5,000 bu.; cents per bu.									
Mar	366	370	364½	365½	−	¼	469	361	2,660
May	374½	376	371	372¾	−	½	440	363	788
July	380	382	378½	380¼	+	1½	416	368	373
Sept	385	385	380½	383	+	2	418	360	191
Est vol na; vol Fri 1,807; open int 20,179; +372.									
CANOLA (WPG) 20 metric tons; Can. $ per ton									
Jan	380.70	381.60	380.10	381.60	+	2.30	407.00	337.00	2,837
Mar	385.90	387.70	385.50	387.60	+	2.20	412.30	339.00	27,090
May	390.50	392.50	390.50	392.50	+	2.00	416.40	341.00	3,589
July	395.00	398.30	395.00	398.30	+	2.50	417.40	366.00	975
Sept		....	....	376.10		....	391.00	363.50	220
Nov	376.50	377.80	376.00	377.60	−	.60	384.00	365.00	514
Est vol 2,895; vol Fr 5,778; open int 35,305, +630.									
WHEAT (WPG) 20 METRIC TONS; Can. $ per ton									
Mr98	163.00	164.50	162.60	164.50	+	1.90	183.80	150.00	4,636
May		....	....	165.20	+	1.40	184.40	152.00	1,289
July		....	....	166.20	+	1.90	184.50	164.00	2,217
Oct		....	....	164.20	+	1.70	177.00	164.60	1,161
Est vol 150; vol Fr 282; open int 9,303, −86.									
BARLEY-WESTERN (WPG) 20 metric tons; Can. $ per ton									
Mr98	144.00	144.70	143.10	144.50	+	1.60	161.00	126.30	6,911
May	146.80	147.90	146.70	147.90	+	1.60	162.00	128.70	5,390
July	149.00	149.90	149.00	149.90	+	1.40	162.60	145.00	1,140
Oct	148.00	148.90	148.00	148.90	+	.90	158.00	145.10	956
Est vol 175; vol Fr 330; open int 4,402, −37.									

Livestock and Meat

CATTLE-FEEDER (CME) 50,000 lbs.; cents per lb.									
Jan	74.75	75.27	74.40	74.67	−	.22	84.95	74.40	6,482
Mar	74.05	74.70	73.55	73.90	−	.45	84.50	73.55	5,593
Apr	75.15	75.62	74.65	74.85	−	.47	84.45	74.65	2,515
May	76.25	76.70	75.80	76.07	−	.40	84.40	75.80	1,896
Aug	78.50	78.70	77.90	78.25	−	.25	83.25	77.90	925
Sept	78.75	78.75	78.10	78.25	−	.25	83.05	78.10	200
Oct	78.50	78.75	78.10	78.20	−	.30	83.00	78.10	214
Est vol 4,298; vol Fr 3,477; open int 17,828, +151.									
CATTLE-LIVE (CME) 40,000 lbs.; cents per lb.									
Feb	64.55	64.85	63.62	63.92	−	1.02	73.92	63.62	46,019
Apr	67.20	67.35	66.30	66.85	−	.42	75.55	66.30	28,968
June	67.15	67.30	66.65	66.90	−	.12	72.37	66.65	16,426
Aug	67.80	67.95	67.37	67.62	−	.05	72.15	67.37	6,934
Oct	70.50	70.62	70.00	70.22	−	.27	74.05	70.00	1,909
Dec	71.80	71.80	71.47	71.50	−	.30	74.20	71.47	350
Est vol 22,842; vol Fr 18,360; open int 100,605, −859.									
HOGS-LEAN (CME) 40,000 lbs.; cents per lb.									
Feb	57.60	57.60	57.07	57.20	−	.32	71.90	57.05	21,937
Apr	56.52	56.70	56.22	56.62	+	.20	68.15	55.95	8,602
June	63.50	63.65	62.97	63.32	−	.15	73.70	62.10	6,353
July	62.80	62.95	62.45	62.50	−	.30	71.75	62.45	1,916
Aug	61.00	61.05	60.60	60.80	−	.17	69.85	60.60	429
Oct	57.50	57.60	57.35	57.50		....	64.90	57.00	679
Dec	56.10	56.10	55.70	55.90	+	.15	58.50	55.25	197
Est vol 6,875; vol Fr 4,920; open int 40,169, +1,331.									
PORK BELLIES (CME) 40,000 lbs.; cents per lb.									
Feb	50.45	51.20	49.75	50.17	−	.90	81.02	49.50	5,585
Mar	50.50	51.10	49.35	49.62	−	1.22	79.40	49.35	1,525
May	51.25	51.80	50.60	51.30	−	.47	79.00	50.50	1,167
July	50.25	50.85	49.50	50.20	−	.65	79.00	49.50	678
Aug	46.80	47.50	46.80	47.05	+	.32	75.00	46.05	121
Est vol 2,028; vol Fr 1,373; open int 9,112, +170.									

Food and Fiber

COCOA (CSCE)—10 METRIC TONS; $ PER TON.									
Mar	1,611	1,615	1,592	1,596	−	34	1,803	1,375	40,029
May	1,643	1,645	1,628	1,630	−	31	1,817	1,399	20,952
July	1,666	1,670	1,653	1,655	−	31	1,835	1,485	5,260
Sept	1,695	1,695	1,691	1,680	−	30	1,836	1,456	5,805
Dec	1,714	1,714	1,714	1,706	−	26	1,863	1,510	9,143
Mr99	1,741	1,741	1,738	1,735	−	26	1,901	1,634	8,579
May		....	....	1,753	−	26	1,911	1,688	2,165
July		....	....	1,770	−	26	1,759	1,705	601
Sept		....	....	1,784	−	26	1,778	1,773	416
Est vol 10,844; vol Wd 1,698; open int 92,950, −616.									
COFFEE (CSCE)—37,500 lbs; cents per lb.									
Mar	161.90	171.00	161.90	170.70	+	8.25	203.00	96.25	16,556
May	157.50	165.00	157.50	164.95	+	7.20	195.00	101.00	5,673
July	151.50	159.00	151.50	159.00	+	7.00	191.00	120.00	2,841
Sept	147.50	151.00	147.50	153.00	+	6.75	186.00	122.50	1,354
Dec	142.25	145.00	142.25	148.50	+	6.75	157.50	124.75	1,230
Mr99		....	....	145.50	+	6.25	154.00	124.00	293
Est vol 7,528; vol Wd 4,254; open int 27,977, +76.									
SUGAR-WORLD (CSCE)—112,000 lbs.; cents per lb.									
Mar	12.13	12.22	12.00	12.02	−	.20	12.55	10.13	96,487
May	11.95	12.02	11.86	11.86	−	.13	12.45	10.20	33,056
July	11.61	11.63	11.51	11.52	−	.08	12.11	10.30	30,751
Oct	11.49	11.55	11.47	11.47	−	.08	11.97	10.45	26,528
Mr99	11.38	11.43	11.38	11.38	−	.06	11.87	10.41	7,362
May	11.33	11.33	11.33	11.33	−	.07	11.68	10.75	1,220
July	11.30	11.30	11.28	11.28	−	.06	11.68	11.28	1,399
EST VOL 26,357; VOL WD 9,808; OPEN INT 196,819, −457.									
SUGAR-DOMESTIC (CSCE)—112,000 lbs.; cents per lb.									
Mar	21.89	21.89	21.89	21.89	−	.03	22.51	21.81	2,585
May	22.10	22.10	22.10	22.10	−	.03	22.56	22.06	3,038
July	22.40	22.40	22.40	22.40		....	22.67	22.33	2,713
Sept	22.50	22.50	22.50	22.50	−	.03	22.68	22.35	2,555
Nov	22.43	22.43	22.43	22.43		....	22.51	22.00	1,048
Ja99		....	....	23.34	+	.02	22.44	22.24	301
Mar		....	....	22.37	+	.02	22.42	22.30	223
Est vol 722; vol Wd 448; open int 12,473, +121.									
COTTON (CTN)—50,000 lbs.; cents per lb.									
Mar	66.98	67.85	66.90	67.54	+	.43	81.00	65.75	41,852
May	68.32	69.00	68.25	68.80	+	.38	81.00	67.40	14,985
July	69.70	70.25	69.55	69.94	+	.24	79.25	68.80	14,971
Oct	72.15	72.35	72.15	72.15	+	.30	78.00	70.80	1,444
Dec	72.50	73.35	72.50	73.14	+	.46	76.60	71.75	13,638
Mr99		....	....	74.14	+	.51	77.25	72.95	497
May		....	....	74.65	+	.40	75.30	73.40	313
July	74.95	75.00	74.95	75.15	+	.40	77.31	74.00	122
Est vol 15,000; vol Fr 4,147; open int 87,853, +213.									
ORANGE JUICE (CTN)—15,000 lbs.; cents per lb.									
Jan	78.80	79.50	78.10	78.40	−	1.60	119.75	69.10	1,971
Mar	82.00	82.50	80.00	81.25	−	2.45	100.25	75.20	25,616
May	85.10	85.30	83.80	84.70	−	2.10	97.75	75.30	6,232
July	88.00	88.40	87.50	87.40	−	2.70	105.00	78.25	4,002
Sept	91.00	91.00	90.25	90.40	−	2.80	99.00	80.95	1,680
Nov	91.00	91.00	91.00	92.65	−	3.05	105.00	83.40	1,564
Ja99	97.00	97.00	97.00	94.90	−	3.30	107.50	86.05	444
Mar	95.70	95.70	95.70	97.15	−	3.55	109.50	90.80	495
Est vol 10,000; vol Wd 26; open int 42,075, −2,715.									

Figure 14–4 Commodity contract specifications are listed in the format above. The *Wall Street Journal* is one of the best sources for commodity market information. (Courtesy of the *Wall Street Journal*)

paper. July corn opened at (1) _____ and closed or settled at (2) _____ on April 16. The lowest price anyone paid for July corn on April 16, was (3) _____, and the highest price anyone paid for July corn on that day was (4) _____. The contract size is (5) _____, prices are quoted in (6) _____, and it is traded at the (7) _____ (answers follow).

Friday, April 16

CORN (CBOT) 5,000 bu.; cents per bu.

	Open	High	Low	Settle	Change
May	231¼	231¾	227	227¾	−4
July	236½	237¼	232¼	233	−4¼
Sept.	241¼	241¾	237½	237¾	−3¾
Dec.	246	246¾	242¾	243½	−3½

(1) $2.36½, (2) $2.33, (3) $2.32¼, (4) $2.37¼, (5) 5,000 bushels, (6) cents per bushel, (7) Chicago Board of Trade.

This is just one example of how commodity futures prices can appear in the newspaper. Other papers may carry the same information or more, and they may publish it in a different format. Prices are listed in the paper according to the contract specifications. Thus if corn is quoted in cents per bushel, then a price written as *232* means corn is trading for *232* cents, or $2.32, per bushel.

Determining the Value of One Futures Contract

To figure out the value of a commodity, multiply the settlement price by the contract size: settlement price times contract size equals contract value. We can use an agricultural example (corn) to demonstrate how to determine the value of a futures contract.

Suppose the settlement price for December corn futures is 200 cents a bushel, that is, $2.00 a bushel. To calculate the dollar value of one corn contract, multiply the $2.00 settlement price by the contract size. In the case of Chicago Board of Trade (CBOT) corn futures, each contract equals 5,000 bushels of corn, so if 1 bushel of corn is worth $2.00, then a 5,000-bushel contract is worth $10,000: $2.00 per bushel times 5,000 bushels equals $10,000.

Futures contracts do not always trade in even numbers; sometimes they move in fractions. These fractions are the smallest price unit at which a futures contract trades and are called **minimum price fluctuations**. In futures lingo, minimum price fluctuations also are referred to as **ticks**. The tick size of a futures contract varies according to the commodity.

The minimum price fluctuation for a CBOT corn futures contract, for instance, is ¼-cent per bushel, or $12.50 a contract. Using fractions, you can figure out how we came up with $12.50. Because a $.01-change in a 5,000-bushel contract equals $50.00 (5,000 × $.01), a ¼-cent change equals $12.50 ($50 × ¼). A shorter calculation is: 5,000 bushels times $.0025 equals $12.50.

Here is a table to help in calculating the value of CBOT grain (corn) futures:

(Cents/bu.; 5,000-bu./contract)

¼ cent per bu. = $12.50 per contract

½ cent per bu. = $25 per contract

¾ cent per bu. = $37.50 per contract

1 cent per bu. = $50 per contract

Keeping in mind that the minimum price fluctuation for CBOT corn futures is ¼-cent per bushel, the next few higher prices following corn trading at 200 cents per bushel ($2.00/bu.) would be 200¼ cents, 200½ cents, 200¾, and 201 cents.

To calculate the value of a futures contract when fractions are involved, just use the same equation of settlement price times contract size. For example, if December corn futures are trading at 200¼ cents/bu. ($2.00¼), then the contract value is: $2.0025/bu. times 5,000 bushels equals $10,012.50.

Determining Profit or Loss on a Futures Contract

Suppose you read in the paper that soil moisture in the Midwest was below normal for the month of June and the forecast does not look promising for rain. Limited rainfall during the growing season could cause the production of corn to decrease, increasing the price. Anticipating higher corn prices, you buy one December corn futures contract at 250 cents/bushel ($2.50/bu.) On July 1, if you were right and corn prices rise, you will make a profit. Remember, you want to satisfy the golden rule of trading: *Buy low, sell high.*

Throughout the month of July, there is no rain in the Corn Belt. The end result is higher corn prices, so you decide to offset your position on July 30 by selling one December futures contract at $2.55/bu.

Did you make a profit or a loss, and how much? First of all, ask yourself if you satisfied the golden rule of Buy low, sell high? You did: you bought a corn futures contract at $2.50/bu. and sold it 5 cents higher at $2.55/bu. To figure out the actual dollar amount of your profits, use either one of the following calculations. (You know that the contract size is 5,000 bushels.)

Calculation #1

Jul. 1	BUY 1 Dec. corn futures at	$2.50/bu.
Jul. 30	SELL 1 Dec. corn futures at	$2.55/bu.
Profit		$.05/bu.

To calculate the profit per contract, multiply your profits by the contract size: $.05/bu. profit times 5,000 bushels equals $250 per contract. If you traded more than one contract, just multiply the number of contracts by $250 minus the profit per contract. Remember that brokerage fees will always be subtracted from this.

Another way to figure out your profits is by calculating the values of the December corn futures contract on July 1 and July 30 and then subtracting the difference between the two.

Calculation #2

Jul. 1	BUY 1 Dec. corn futures valued at $12,500 ($2.50 × 5,000)
Jul. 30	SELL 1 Dec. corn futures valued at $12,750 ($2.55 × 5,000)
Profit	$250/contract
Total Profit	$250 × 1 contract traded = $250

As you can see, the results from both calculations are the same. Pick the method you feel most comfortable using to figure out trading profits and losses.

HEDGERS AND SPECULATORS

There are two main categories of futures traders that utilize futures contracts. These are the hedger and the speculator. Hedgers either now own, or will at some time own, the commodity they are trading. These can be either production agriculturalists, elevator owners, or any others in the agribusiness input or output sectors.

Speculators, on the other hand, will likely never own or even see the physical commodity. In other words, they are not production agriculturalists or agribusiness owners. They just buy and sell futures contracts and hope to make a profit on their expectations of future price movements. Speculators always offset their position by buying (selling) futures contracts they originally sold (bought).

At planting time, for example, corn producers are uncertain what commodity prices will be at harvest. Hedging is a tool production agriculturalists and agribusinesses can use to shift some of the risk resulting from price uncertainty to speculators trading in the futures market. Hedging involves taking an equal but opposite position in the futures market to what one has in the cash market. Should prices fall, a producer who placed a hedge will be protected.[11] Therefore, hedgers willingly give up the opportunity to benefit from favorable price changes in order to achieve protection against unfavorable changes.

Long (Buying) and Short (Selling) Hedgers

Two terms used to explain buying and selling are *long* and *short*. If you first buy a futures contract, this is called going long, or being **long hedge**. If you first sell a futures contract, this is called going short, or going **short hedge**.

Hedging in the futures market is a two-step process. Depending upon your cash market situation, you will either buy or sell futures as your first position. For instance, if you are going to buy a commodity in the cash market at a later time, your first step is to buy a futures contract, whereas if you are going to sell a cash commodity at a later time, your first step in the hedging process is to sell futures contracts.

The second step in the process occurs when the cash market transaction takes place. At this time, the futures position is no longer needed for price protection and should therefore be offset (closed out). If your hedge was initially long, you would offset your position by selling the contract back. If your hedge was initially short, you would buy back the futures contract. Both the opening and closing positions must be for the same commodity, number of contracts, and delivery month.[12]

Example of a Long Hedge. A food processor is planning to buy 240,000 pounds of soybean oil over the next few months to cover production needs. Currently, soybean oil is quoted at 26 cents a pound, but the company management is concerned that prices will rise by the time they are ready to purchase and take delivery of the oil. To take advantage of current prices, the company decides to buy four Chicago Board of Trade September soybean oil futures contracts at 26 cents a pound. The standard contract size for CBOT soybean oil is 60,000 pounds.

The management's greatest fear comes true. The price of soybean oil jumps to 31 cents a pound by the time the food processor is ready to purchase it. The company offsets its futures position by selling four Chicago Board of Trade September soybean oil contracts for 31 cents a pound. The company's hedging activities result in the following:

Cash Market	Futures market
June	**June**
Plans to purchase 240,000 lbs. soybean oil in the cash market at $.26 lb.	Buys 4 CBOT Sept. soybean oil futures contract at $.26/lb.
August	**August**
Purchase 240,000 lbs. soybean oil in the cash market at $.31/lb.	Sells 4 CBOT Sept. soybean oil futures contract at $.31/lb.
Purchase price of cash soybean oil	$.31/lb.
Less futures gain ($.31 − $.26)	$.05/lb.
Net purchase price	$.26/lb.

By using Chicago Board of Trade soybean oil futures, the food processor lowered its purchase price from 31 cents to 26 cents a pound. That was exactly what the company expected to pay.

This type of hedge is a long hedge because the food processor initially purchased futures. If futures are initially sold, the hedge is a short hedge.[13]

Example of a Short Hedge.   Rather than selling for whatever price the local elevator is willing to pay come harvest, a producer decides to explore a variety of marketing alternatives, including futures. Her ultimate goal is to improve her bottom line. Suppose she figures that it costs her $2 to produce one bushel of corn. When corn prices enter a range where the producer can make a profit, she decides to hedge a portion of her crop by selling one CBOT December corn futures contract. The standard contract size for one CBOT corn futures contract is 5,000 bushels.

By early May, CBOT December corn futures hit $2.60 a bushel. To "lock in" a selling price of $2.60, the grain producer sells one CBOT December corn futures contract. As it turns out, the Midwest experiences near-perfect growing conditions that year and corn yields are above normal, causing prices to drop. By harvest, corn has fallen to $1.90 a bushel. The producer therefore offsets her futures position by purchasing one CBOT December corn futures contract. The corn producer's hedging activities result in the following:

Cash market	Futures market
May	**May**
Plans to sell 5,000 bu. corn in the cash market at $2.60/bu.	Sells one CBOT Dec. corn futures contract at $2.60/bu.
October	**October**
Sells 5,000 bu. corn in the cash market at $1.90/bu.	Buys one CBOT Dec. corn futures contract at $1.90/bu.
Sales price of cash corn	$1.90/bu.
Plus futures gain ($2.60 − $1.90)	$.70/bu.
Net sale price	$2.60/bu.

By using Chicago Board of Trade corn futures, the producer increased her final sale price from $1.90 to $2.60 a bushel. That was exactly what she wanted to receive. Better yet, the final sale price was $.60 a bushel higher than her production expenses.

You should keep in mind that these examples are simplified to give you an overview of hedging. In both examples, the hedgers were holding two positions: a cash position and a futures position. By definition, a hedge consists of both a cash and a futures position, even when the cash position is anticipated, as shown in the first example.[14] As you will recall, the firm was planning to purchase 240,000 pounds of soybean oil in the cash market.

Transactions Can Be Unpredictable.   Be aware that the market doesn't always move as you might anticipate, but a hedger's primary objective is to achieve price protection. This means there could be times that you may not be able to take advantage of a price move after a futures hedge is initiated. However, an experienced hedger is willing to forfeit this opportunity so he or she can have price protection.

Relationships between the Cash and Futures Market.   Understanding relationships between the cash and futures markets is crucial to understanding the practice of hedging. A **cash market** is a market where actual commodities are bought and sold. A **futures market** is a market where traders buy and sell futures contracts. Many times the price of the underlying commodity and the related futures contract will move up and down together, but the price moves are not always equal.

Basis.   In both examples, the cash price of the commodity equaled the futures price, but in real life, this does not happen often. Typically, either the price one pays or receives from corn at a county elevator (cash market), for example, is either higher or lower than the futures price (Chicago Board of Trade). (There are many economic reasons why this happens, but this is beyond the scope of our discussion.) Anyway, the relationship between cash prices and futures prices is referred to as **basis**. The cash price minus the futures price of a commodity equals the basis. Successful hedgers understand how basis ultimately offsets their final results.[15]

Speculators

Agricultural producers, commodity processors, exporters, food manufacturers, and others use the futures market to shift market risk (the risk of adverse price movements) to someone else. The party who assumes the risk is the speculator.

Gambler or Businessperson.

A commodity speculator is a rational businessperson who is willing to take a price risk in hopes of making a profit. Such speculation is not just gambling such as that found in Las Vegas. The difference is in the nature of the risk. In the casino the risk is deliberately created. For example, in the games of blackjack or roulette, the risk of loss is created. People are willing to take the risk in the name of entertainment. They are hoping to profit even though they know the odds are against them. When people assume this artificially created, but real, risk of loss, there is no positive benefit to society.[16]

Speculators Are a Positive Influence in the Futures Market.

The commodity speculator, on the other hand, assumes a risk that occurs naturally. In agribusiness, price risk is a natural part of holding inventories of agricultural commodities. If agribusiness firms were not able to shift price risk to speculators, their costs would increase because of the losses incurred from adverse prices for food and fiber products. Because the commodity speculator assumes price risk, agribusiness firms can produce products at lower cost, and society will benefit.[17]

Some people are concerned that commodity speculators exert undue influence on futures market commodity prices. At times in the distant past this may have been true, but it is unlikely to occur today. First of all, competition in the marketplace quickly punishes those who try to raise or depress prices. Second, there are two layers of supervision that would soon detect the efforts of a large-scale buyer or seller trying to corner the market or a group of speculators conspiring to manipulate prices.[18]

Speculators Are Not Interested in the Physical Commodity.

In contrast to hedgers, speculators generally have no interest in buying or selling the actual commodity, whether corn, soybeans, or live cattle futures. Speculators buy and sell futures contracts expecting to make a profit based on their interpretations of future price changes. The presence of speculators in the futures market increases the effectiveness of the market and provides hedgers an opportunity to reduce their price risks.[19]

The profit potential is proportional to the amount of risk that is assumed and the speculator's skill in forecasting price movement. Potential gains and losses are as great for the selling (short), speculator as for the buying (long) speculator. Whether long or short, speculators can offset their positions and never have to make or take delivery of the actual commodity.[20]

Trading Philosophy of Speculators.

The golden rule of speculators is, buy low, sell high. If a trader expects prices to rise, he or she will buy a futures contract. Then, at a later time, if prices rise, the trader will make an offsetting sale at a profit. The opposite also holds true: if a trader expects prices to fall, he or she will sell a futures contract. Again, at a later time, if prices fall, the trader will make an offsetting purchase at a profit. Experienced speculators can become very sophisticated in their trading techniques.

Types of Speculators.

Speculators can be classified by their trading methods: position trader, day trader, scalper, and spreaders. A brief discussion of each follows:

🌾 A *position trader* is a public or professional trader who initiates a futures or options position and holds it over a period of days, weeks, or months.

❧ A *day trader* holds market positions only during the course of a trading session and rarely carries a position overnight. Most day traders are exchange members who execute their transactions in the trading pits.

❧ A *scalper* trades only for him- or herself in the pits. He or she trades in minimum fluctuations, taking small profits and losses on a heavy volume of trades. Like a day trader, a scalper rarely holds positions overnight.

❧ *Spreaders* trade on the shifting price relationships between two or more different futures contracts. Examples include: different delivery months of the same commodity, the prices of the same commodity on different exchanges, products and their by-products, and different but related commodities.[21]

Speculation Example. Let us say that due to increasing world demand for corn, an investor anticipates corn prices rising and so buys one Chicago Board of Trade December corn futures contract at $2.75 a bushel. Within two weeks, the price rises to $3.00 a bushel. The investor offsets his or her position by selling one Chicago Board of Trade December corn futures contract at a profit of 25 cents a bushel ($3.00 selling minus $2.75 buying price). This equals a total profit of $1,250 ($.25 profit times 5,000 bushel contract size). Of course, if the price of the corn falls 25 cents per bushel, the investor would lose $1,250.

DETERMINING THE PRICE OF A COMMODITY

Determining the price of a commodity is called *price discovery*, which also refers to the price of a futures contract agreed to by a buyer and a seller through a futures exchange. Futures exchanges do not set price. They are free markets where the forces that influence prices are brought together in open auction. As this marketplace assimilates new information throughout the trading day, it translates the information into a single benchmark figure—a fair market price that is agreed on by both buyer and seller.[22]

Price Discovery Trading on the Trading Floor

The heart of a futures exchange is the **trading pit** where buyers and sellers meet each day to conduct business. Each trader who enters these trading pits brings along specific market information, such as supply and demand figures, currency exchange rates, inflation rates, and weather forecasts, which contributes to the ongoing price discovery function of the futures markets.[23]

For example, a grain exporter in the southern United States will be kept constantly informed of grain prices through the futures price quotes disseminated by the Chicago Board of Trade. This information will assist the export company in offering a purchase price for the grain it buys from American farmers and allow it to set an appropriate selling price for the grain it exports to other countries.

As trades between buyers and sellers are executed, the fair market value (price) of a given commodity is discovered and these prices are disseminated instantaneously by state-of-the-art telecommunications throughout the world.[24] These prices are carried by daily newspapers and various on-line quotation services. In looking at the newspaper listing of futures prices in Figure 14–5, you'll notice the several contract specifications.

Forecasting Market Direction through Government Reports

Among the different factors market participants use to forecast market direction are USDA reports. Released throughout the year, the market antici-

Example Price Listing September 12

CORN (CBOT) 5,000 bu; cents per bu

	Open	High	Low	Settle	Change	Open Interest
Sep	293½	294½	290¼	292	+1¾	4,370
Dec	295	298¾	293¾	298½	+5¾	282,106
Mr96	303	306¼	301	306	+5¾	78,564
May	305¼	308¾	304¼	308½	+5½	11,796
Jul	304¾	308¾	304	308½	+5½	30,727
Sep	284	287½	283¾	287¼	+4	2,292
Dec	269	273½	268½	272¼	+4	14,140
Dec97	263	264	263	264	+2	201

Est vol 57,000; vol Mon 35,804; open int 424,916

Commodity: Chicago Board of Trade corn

Contract months: Sep '95, Dec '95, Mar '96, May '96, July '96, Sep '96, Dec '96, Dec '97

Contract size: 5,000 bu

How the contract is quoted: cents per bu

Figure 14–5 The price per bushel of corn shown here is not set by the futures exchanges. Instead, the price is discovered by a buyer and seller agreeing on a price.

pates their release and uses the reported information in developing market opinion. Not surprisingly, attention to USDA reports is heightened during the growing season and climaxes at harvest. Some of the more closely watched reports are:

Crop Progress (issued weekly during growing season) lists overall condition of selected crops in major producing states.

Crop Production (coincides with the growing season, issued around the tenth of each month) reports production estimates based on actual field surveys.

Export Sales (released weekly) summarizes U.S. export sales by importing country.

Grain Stocks (released quarterly) reports the number of stocks held by state.

Supply and Demand (issued around tenth of each month) gives an overall look of supply versus demand.[25]

MAKING THE FUTURES TRADING TRANSACTION

Opening an Account

Before any futures trading can be done, you must open an account with a **brokerage office** (commission house, wire houses, futures commission merchants). Local brokerage offices are usually the most convenient outlet for futures transactions. These offices are located in most towns and cities throughout the United States. To be sure that you use a reputable broker, you should contact current and past customers and then check to see if the broker is registered with the National Futures Association (NFA) and that no complaints have been registered.[26] You can contact the NFA by calling 1-800-621-3570 (outside of Illinois) or 1-800-572-9400 (within Illinois).

When you open an account, you should also inform the broker whether you will be using it for hedging or speculating purposes. Brokers charge customers a commission fee to buy or sell futures contracts. Commission rates will vary depending on the kind and amount of service offered, but rates are generally competitive. The customer also signs several legal documents concerning both parties' responsibilities regarding the account.[27]

Types of Accounts.   There are various types of futures accounts. The two most used for beginners are the individual and joint accounts. *Individual accounts* are used when trading decisions are made by the customers. *Joint accounts* are set up when both the broker and customer are to have input into trading decisions. When an account is set up, money is deposited in order to get ready to trade.

Margins

When a customer places an order, they are required to post a **performance bond margin**, which is a financial guarantee required by both buyers and sellers to ensure they fulfill the obligation of the futures contract. Minimum margin requirements for futures contracts usually range between 5 and 18 percent of a contract's face value and are set by futures exchanges, but brokerage firms can require a larger margin than the exchange minimum.[28]

The initial amount a market participant must deposit into his or her account when he or she places the order is called **initial margin**. On a daily basis, the margin is debited or credited based on the close of that day's trading session (referred to as *marking-to-the-market*) with respect to the customer's open position. A customer must maintain a set minimum margin known as **maintenance margin** in his or her account. If the debits from a market loss reduce the funds in the customer's account below the maintenance level, the broker will call the customer for additional funds to restore the account to the initial margin level. This request for additional money is called a **margin call**. Margin in excess of the required amount is available for withdrawal by the customer.[29]

Placing Orders

A vital function of the commission house is to properly relay customer orders. Orders must be written and entered without vagueness to ensure that they are properly handled on the trading floor. A mistake in this process can be very costly; therefore, both the customer and his commodity representative must communicate orders clearly. There are three major types of orders: market order, limit order, and stop order. Keep in mind that it is up to the individual brokerage firm and exchange which kinds of orders are accepted. Be sure to talk to your broker before making a decision on the orders you want to use. A brief discussion of each follows.

Market Orders.   The market order is the most common type of order. In a market order, the customer states the number of contracts of a given delivery month that he or she wishes to buy or sell. The customer does not specify the price but simply wants it executed as soon as possible at the best possible price.

When a market order is filled, it is usually close to the price that was trading at the time the order was placed. However, in a fast market, the price could be different from what it was when the order was entered.

Limit Orders.   Limit orders have a price limit at which they must be executed. They can be executed only at that price or better. The advantage of a limit order is that the customer knows the worst price that he or she will receive if the order is executed. The disadvantage is that the order might not be filled.

Stop Orders. Stop orders are normally used to liquidate earlier transactions. Let us assume that a customer bought a Chicago Board of Trade corn futures contract for $3.00 per bushel. To prevent a large loss in a **bear market** in the event corn prices fall, the customer could place a stop order at $2.75 per bushel. The customer's position is liquidated if and when the corn futures drop to $2.75. The stop order is executed when the stop price is reached, but not necessarily at the stop price.

Stop orders also can be used to enter the market. Suppose a trader expected a **bull market** only if it surpassed a specific price level. In that case, he could use a *buy-stop order* when and if the market reached this point.

One variation of a stop order is a *stop-limit order*. With a stop-limit order, the trade must be executed at the exact price (or better) or held until the stated price is reached again. If the market fails to return to the stop-limit level, the order is not executed.[30]

Order Route

Orders received from customers are sent as soon as possible by a commission house to the exchange on which the order is to be executed. Telephone stations are located on the outskirts of the trading floor to handle the many orders coming to the exchange from companies and individuals. All orders are time-stamped at various stages along the order route as a check to ensure that the order is being expedited in the best possible fashion. After the order is received by the exchange floor phone clerk, it is time-stamped and passed to a broker by a runner (messenger).

A pit broker may receive an order to "buy five [meaning 5,000 bushels] December wheat" [at the market price, which is understood]." The pit broker looks for another trader who is selling. By voice and hand signals, the pit broker flashes the bid to buy, receives a sold signal from the seller, and the order is completed. The broker writes the price and the seller's identity on the order blank and endorses it, and a runner returns it to the floor phone clerk.

When the phone clerk receives the endorsed order indicating that the trade has been executed, he or she time-stamps the order and phones the report to the firm's wire room where it is relayed to the initiating office. The commission house confirms the transaction with the customer.[31] Refer to Figure 14–6 for detailed step-by-step procedures of the order flow.

Action on the Trading Floor

All market participants have indirect access to the trading floor through their brokers, but only exchange members have the privilege of trading on the floor. To the average person, the system appears to be totally chaotic, but in fact it is very well organized. At the Chicago Board of Trade, trading is done in pits (raised octagonal platforms), which permits all buyers and sellers to see each other. On each business day, trading is officially opened and closed by the clanging of a large gong. No trades can be made before or after this official signal. The rules and regulations of the Chicago Board of Trade require pit traders to use an **open outcry** in buying and selling. In addition, they use standard hand signals to clarify their verbal bids and offers. Refer to Figure 14–7.

Because all trades must be recorded, traders wear colored badges, which indicates their trading privilege. Exchange rules dictate that personnel (both male and female) on the trading floor must wear jackets and ties. Everyone on the floor performs very specific and important job functions and the colorful jackets help distinguish their jobs. In order for brokerage firms to identify their staff on the trading floor easily, the staff wear trading jackets of a certain color.

Order Flow

1. After you place an order with your broker, he or she will call in the order to the floor of the Chicago Board of Trade.
2. The order is time-stamped.
3. A member firm employee, known as a runner, rushes the order to a floor broker in the trading pit. (Some orders are flashed by hand signals to brokers in the trading pit, and some fills are flashed back to the order station.)
4. The floor broker fills the order using open outcry and hand signals to communicate price and quantity.
5. Once the order is filled, the floor broker reports the price to the appropriate Chicago Board of Trade price reporter overlooking the trading pit.
6. The price reporter enters the price change into a laptop quotation computer.
7. The computer transmits the price information to electronic price boards facing the trading floor and around the world via information services.
8. The completed order is picked up by a runner and returned to the telephone desk.
9. The order is time-stamped again.
10. The customer is notified that his or her order has been filled.

Figure 14–6 Step-by-step procedure for the order flow of a commodity (futures) contract.

Figure 14–7 The action on the trading floor of an exchange appears to be chaotic, but it is very well organized.

CAREER OPTION

Career Area: Commodity Trader/Broker

Many of the agricultural crops grown in the United States are commodities valued around the world. Corn, cotton, wheat, sugar cane, sugar beets, live cattle, hogs, pork bellies, and orange juice are some examples of commodities. A commodity broker buys and sells the predictions for crop success or failure on an open market. The Chicago Board of Trade is one of the best known commodity markets where trade is conducted Monday through Friday every week of the year. Speculators in the commodity market are located all over the world. They buy and sell portions of these and other crops, or futures, through a broker or trader.

Experience and education are needed to be a commodity broker or trader. An understanding of agricultural production, crop trends over time, grains, animals, oils, and specialty crops is essential for success in this field. A bachelor's degree in agribusiness, agricultural economics, finance, or marketing is beneficial. A commodity broker or trader is licensed by the Board of Trade to conduct business and must adhere to a strict code of ethics.

Imports and exports of commodities are somewhat controlled by the futures market and help the United States to maintain a balance of payments with other countries. Commodity brokers generally work out of an office taking buy and sell orders from speculators. They relay the orders to the commodity trader, who is active on the buying and selling floor of the market itself.

Related jobs that are important to the commodity market are farmer, hog or cattle producer, grain producer, orange grove owner, grain elevator operator, trucker, sugar processor, crop forecaster, and market reporter.

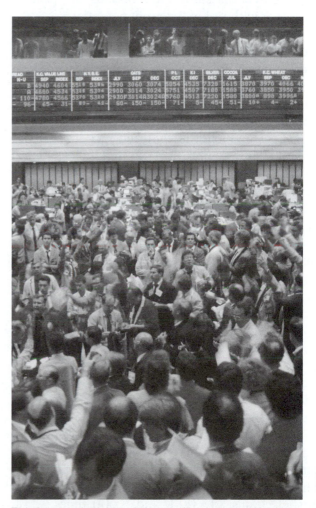

This is an action shot of commodity traders at the Chicago Board of Trade. (Reprinted with permission of the Chicago Board of Trade. From *Action in the Marketplace.* © 1994)

FUTURES OPTIONS

In 1982, another market innovation, *options on futures*, was instituted. In contrast to futures, options on futures allow investors and risk managers to define risk and limit it to the cost of a premium paid for the right to buy or sell a futures contract. At the same time, options can provide the buyer with unlimited profit potential.

Futures options are much more attractive for many hedgers and speculators. Futures options are similar to making a down payment on a house. Let us say you paid $1,500 to a real estate agent as a down payment on a $125,000 house that you intended to buy. While you were making arrangements to complete the contract, you discovered a $95,000 house with just as much floor space, which you liked as well as the $125,000 house. Therefore, you lose your $1,500 down payment that was used to hold the house for you, but you avoid losing $30,000 by selecting a better deal.

With options, once you make a transaction, you can predict your maximum losses. For example, let us say you bought a futures options contract for December corn at $3.00 per bushel, anticipating it would go to $3.50 per bushel. The premium cost $35.00 for the futures options contract. Unfortunately, due to overproduction, imports, embargoes and other unexpected happenings, corn drops to $2.00 per bushel. Since the market goes against you, you can simply let the option expire. Your loss will be $35.00 plus a commission.

What if this transaction had been a futures contract? Remember, one Chicago Board of Trade futures contract of corn includes 5,000 bushels. Therefore, 5,000 bushels of corn times $3.00 per bushel equals $15,000, and 5,000 times $2.00 per bushel equals $10,000. ($15,000 minus $10,000 equals $5,000.) As you can see, with the same drop in prices, a futures contract loses $5,000 plus commission, but a futures option contract loses only $35 plus commission. What if you had bought 10 futures contracts of corn? Options allow the hedger and speculator to have their cake and eat it too. As you can see from the previous examples, an option gives the buyer the right, but not the obligation, to buy or sell an agricultural futures contract at some specified time in the future for a price set at the time the option is purchased.

Calls (Purchases) and Puts (Sells)

A **call** is an option that gives the option buyer the right (without obligation) to purchase a futures contract at a certain price on or before the expiration date of the option for a price called the premium, determined in open "outcry" trading in pits on the trading floor. A **put** is an option that gives the option buyer the right (without obligation) to sell a futures contract at a certain price on or before the expiration date of the option.

Premium Cost

The cost of the futures options is called the **premium**. In agricultural options traded at the Chicago Board of Trade, the premium is the only variable in the contract. The premium depends on market conditions such as volatility, time until an option expires, and other economic variables. The price of an option is discovered through a trading process regulated by the exchange and the futures industry.

Different call and put options trade simultaneously. Trading months for options are the same as those of the underlying futures contract, while strike prices are listed in predetermined multiples for each commodity. Initially, the strike prices are listed in a range around the current underlying futures price. As futures prices increase or decrease, additional higher and lower strike prices are listed. For each agricultural option traded, you will find several option prices listed. Each price listed represents the price for a given contract month and strike price.[32] You can find option prices in daily newspapers, through on-line quotation services, or at local grain elevators. Refer to Figure 14–8.

The premium is the only element of an option contract that is negotiated in the trading pit; all other elements of an option contract—such as the strike price and expiration date—are predetermined or standardized by the Chicago Board of Trade or other mercantile exchanges.[33]

The Most You Can Lose Is the Premium

Regardless of how much the market fluctuates, the most an option buyer can lose is the premium. You deposit the premium with your broker, and the money goes to the option seller. Because of this limited and known risk, buyers are not required to maintain margin accounts, as they must when entering into futures contracts. (Margin is a cash deposit used to secure a contract.) Option sellers, on the other hand, face the same types of risks as participants in the futures market and must post margin with their brokers. The amount of margin required for option sellers depends upon their overall position risk. At the Chicago Board of Trade, the exchange uses a margining system referred to as SPAN®, which evaluates a customer's overall risk and determines the amount of required margin based on that risk. In addition to the premium, the broker you use to buy and sell an option will charge a commission fee.

Intrinsic Value

The option premium—the price of an option—equals the option's intrinsic value plus its time

---

Example of Newspaper Listing

CORN (CBOT) 5,000 bu; cents per bu

Strike	Calls-Settle			Puts-Settle		
Price	Dec	Mar	May	Dec	Mar	May
280	20¾	29¼	31½	2½	3⅛	4½
290	13¾	21¼	24½	5⅜	6½	8
300	8⅝	16	19	10⅝	10½	.....
310	5½	11½	14	17¾	16	.....
320	3⅜	8	11	25½	22	.....
330	1⅞	5¾	8	34	29½	.....

Est vol 20,000; vol Mon 9,120 calls; 4,021 puts

Open int Mon 209,380 calls; 120,729 puts

1. The first column lists several strike prices for Chicago Board of Trade corn options. Typically, the exchange lists corn options in 10-cent increments.

2–4. Columns 2–4 list option premiums for Dec, Mar, and May CBOT corn call options. CBOT corn options trade in ⅛ cent intervals and premiums are quoted in cents and eighths of a cent per bushel.

5–7. Columns 5–7 list option premiums for Dec, Mar, and May CBOT corn put options. Below the columns appear estimated trading volume figures for the day reported followed by actual trading volume and open interest for the preceding day divided between calls and puts.

Figure 14–8 Once you decide to purchase a futures option, locate a newspaper or other service to find the listing of the option's price. Notice that there are many choices of option, as illustrated here.

value. An option has **intrinsic value** if it would be profitable to *exercise* the option. Call options, for example, have intrinsic value when the strike price is below the futures price (meaning you can purchase the underlying futures contract at a price below the current market price). For example, when the December corn futures price is $2.50, a December corn call with a strike price of $2.20 has an intrinsic value of 30 cents a bushel. If the futures price increases to $2.60, the option's intrinsic value increases to 40 cents a bushel. If the futures price declines to $2.40, the intrinsic value declines to 20 cents a bushel.

Put options, on the other hand, have intrinsic value when the strike price is above the futures price (meaning you can sell the underlying futures contract at a price above the current market price). For example, when the July corn futures price is $2.50, a July corn put with a strike price of $2.70 has an intrinsic value of 20 cents a bushel. If the futures price increases to $2.60, the option's intrinsic value declines to 10 cents a bushel.

Another way to say an option has intrinsic value is to say the option is *in the money*. Options also can be *at the money* (the option has no intrinsic value—strike price equals futures price), or *out of the money* (the option has no intrinsic value and would not be profitable to exercise). If an option has no intrinsic value, then the premium is equal to its *time value*.[34] Refer to Figure 14–9.

Time Value

Time value—sometimes called *extrinsic value*—reflects the amount of money that buyers are willing to pay hoping that an option will be worth exercising at or before expiration. For example, if July corn futures are at $2.16 and a July corn call with a strike price of $2 is selling for 18 cents, then the intrinsic value equals 16 cents (the difference between the strike price and futures price) and the time value equals 2 cents (difference between the total premium and the intrinsic value). Note that an option's time value declines as it approaches expiration. It will have time value at expiration, and any remaining premium will consist entirely of intrinsic value. Major factors affecting time value include the following.

Relationship between the Underlying Futures Price and the Option Strike Price. Time value is typically greatest when an option is at the money. This is because at-the-money options have the greatest likelihood of moving into the money before expiration. In contrast, most of the time value in a deep in-the-money option is eliminated because there is a high level of certainty the option will not move out of the money. Similarly, most of the time value of a deep out-of-the-money option is eliminated because it is unlikely to move in the money.

Time Remaining until Option Expiration. The greater the number of days remaining until

Determining Option Classifications

	Call Option	Put Option
In-the-money	Futures price > Strike price	Futures price < Strike price
At-the-money	Futures price = Strike price	Futures price = Strike price
Out-of-the-money	Futures price < Strike price	Futures price > Strike price

Figure 14–9 Deciding whether you want to exercise, offset, or expire your futures options is based on the price of the commodity being in-the-money, at-the-money, or out-of-the-money.

expiration, the greater the time value will be. This occurs because option sellers will demand a higher price since it is more likely that the option will eventually be worth exercising (in the money).

Market Volatility. Time value also increases as market volatility increases. Again, option sellers will demand a higher premium since the more volatile or variable a market is, the more likely the option will be worth exercising (in the money).

Interest rates. Although the effect is minimal, interest rates affect the time value of an option: as interest rates increase, time value decreases.[35]

Using Futures Options to Benefit Agribusiness

Food manufacturers, processors, millers, and other agribusinesses can use options to better manage their raw material positions so they increase their profitability. Options are also a good way to protect the value of their on-hand inventories.

Elevator operators can use agricultural options to offer their customers a wider variety of marketing alternatives giving them a competitive edge. One popular contract now possible with agricultural future options is a minimum price contract, where producers establish a minimum price for their crop but leave open the opportunity to increase their final sale price.

If you're a production agriculturalist you can lock in a desirable selling price of your crops or obtain "price insurance" when you store your crop.

Feedlot managers can use options to protect themselves against rising feed costs by using agricultural futures options to set a maximum purchase price.

Investors: There are a number of option strategies, at a variety of risk levels, that can provide profit opportunities. For example, purchasing agricultural futures options to profit from an expected price change can provide unlimited profit potential and limited risk.[36]

Ways to Exit a Futures Option Position

Once an option has been traded, there are three ways you can get out of a position: exercise the option, offset the option, or let the option expire.

Exercise. Only the option buyer can decide whether to exercise an option. When an option position is exercised, both the buyer and the seller of the option will be assigned a futures position. The option buyer first notifies his or her broker, who then submits an exercise notice to the Board of Trade Clearing Corporation. The exercise is carried out that night. The Clearing Corporation creates a new futures position at the strike price for the option buyer. At the same time, it assigns an opposite futures position at the strike price to a randomly selected customer who sold the same option. The entire procedure is completed before trading opens the following business day.

Offsetting. Offsetting is the most common method of closing out an option position. You do this by purchasing a put or call identical to the put or call you originally sold; or by selling a put or call identical to the one you originally bought.

Offsetting an option before expiration is the only way you can recover any remaining time value. Offsetting also precludes the risks of being assigned a futures position if you originally sold an option and want to avoid the possibility of being exercised against. Your net profit or loss, after a commission is deducted, is the difference between the premium you paid to buy (or received to sell) the option and the premium you receive (or pay) when you offset the option. Market participants face the risk that there may not be an active market for their particular options at the time they choose to offset, especially if the option is out of the money or the expiration date is near.

Expiration. The other choice you have is to let the option expire by simply doing nothing. In fact,

Example #1

Situation: A farmer wants protection in case soybean prices fall by harvest

Strategy: Farmer buys a CBOT put option

If falling commodity prices are a threat to your profitability, buying put options allows you to establish a minimum selling price without giving up the opportunity to profit from higher prices.

Assume you're a soybean producer and you have just finished planting your crop. You are worried that prices may fall before harvest and would like to establish a floor price for a portion of your expected soybean crop. You go ahead and purchase a November soybean put with a strike price of $6.50—this gives you a minimum selling price of $6.50 a bushel (excluding basis, commissions, and the cost of the option).

If, just before harvest, the November futures price declines to $5.75 (your put option is in the money), you can exercise your option and receive a short futures contract at the $6.50 strike price. Buying back the futures contract at the lower market price of $5.75 gives you a 75-cent per bushel profit (the difference between the strike price and the futures price), which should roughly offset the decline in the price of soybeans. Or, rather than exercising the option, you might be able to earn more by selling back the option to someone else. This would allow you to profit from any remaining time value as well as the 75 cents' intrinsic value, both of which would be reflected in the option premium.

If November soybean futures were trading at $5.75, the $6.50 soybean put would be worth at least 75 cents plus any remaining time value. Generally, when an option is in the money and has time value, it's common for someone to offset—sell back the option rather than exercising it. That's because exercising the option will yield only its intrinsic value. Any time value that remains will be forgone unless it is offset. Also, an extra brokerage commission may be incurred when exercising an option.

If the price of soybeans increases just before option expiration and is above the option strike price (your put option is out the money), you can simply let the option expire or sell it back prior to expiration to capture any remaining time value. In either case, your loss on the option position can be no larger than the premium paid, and you still will be able to sell your crop at the higher market price.

Figure 14–10   A grain producer who is worried about falling prices, can buy a *put option,* as illustrated here.

the right to hold the option up until the final day for exercising is one of the features that makes options attractive to investors. Therefore, if the change in price you have anticipated does not occur or if the price initially moves in the opposite direction, you have the assurance that the most you can lose is the premium you paid for the option. On the other hand, option sellers have the advantage of keeping the entire premium they earned, provided the option does not move in the money by expiration.

Many options expire worthless on their last trading day. It is important to recognize that options on futures frequently follow a different trading schedule than the underlying futures contract, typically

Example #2

Situation: A food processor wants protection against rising raw material costs

Strategy: Food processor buys a CBOT soybean oil call option

If rising commodity prices are a threat to your profitability, buying call options allows you to establish a maximum purchase price without giving up the opportunity to profit from falling prices.

For example, suppose you're a food processor and you expect to buy soybean oil during the spring. You are worried that prices may rise over the next few months and would like to establish a ceiling price for your eventual purchase. So you buy a May soyoil call with a 25-cent strike price—this gives you a maximum purchase price of 25 cents a pound (excluding basis commissions and the cost of the option).

If, in April, the May futures price increases to 30 cents (your call option is in the money), you can exercise your option and receive a long futures contract at the 25-cent strike price. Selling back the futures contract at the higher market price (30 cents per pound) gives you a 5-cent a pound profit, which should roughly offset the increase in the cost of soyoil. Or, rather than exercising the option you might be able to earn an even larger profit by selling the option to someone else (offsetting the option position) at a higher premium. This would allow you to profit from any remaining time value as well as the increase in intrinsic value, both of which would be reflected in the option premium.

The 25-cent soyoil call would be worth at least its intrinsic value of 5 cents (the difference between the strike price and the futures price) plus any remaining time value. In most cases, when an option is in the money and has time value it's common for someone to offset, in this case sell back the option rather than exercising it. That's because exercising the option will yield only its intrinsic value. Any time value that remains will be forgone unless it is offset. Also, an extra brokerage commission may be incurred when exercising an option.

If the price of soyoil in April declines and is below the option strike price (your call option is out of the money), you can simply let the option expire or sell it back prior to expiration. In either case, your loss on the option position can be no greater than the premium paid, and you still will be able to purchase your soyoil at the lower market price.

Figure 14–11   A food processor who is worried about rising prices, can buy a *call option,* as illustrated here.

expiring a month before the underlying futures contract does. For example, options on December CBOT corn futures stop trading the third week of November, whereas the December CBOT corn futures contract does not stop trading until mid-December.

Options that have value are usually offset, rather than exercised, before expiration. However, if you are holding an in-the-money option at expiration, the Board of Trade Clearing Corporation will automatically exercise the option unless you give notice to the Clearing Corporation before expiration.[37] Refer to Figures 14–10 and 14–11 for two examples of futures options.

REGULATIONS WITHIN THE COMMODITY (FUTURES) EXCHANGES

Is Your Money Safe?

If you are seriously considering changing your current business practices to include futures in order to have better protection against changing prices, one of your concerns is surely whether your money is safe. It is really to everyone's best interest that the futures industry provide a safe, well-capitalized arena for trading futures contracts. Exchanges, market participants, and federal regulators work together to protect the integrity of the marketplace.

Exchanges are responsible for ensuring that their daily market operations are handled in an efficient manner in keeping with the highest standards of ethical business conduct. From an exchange standpoint, one of the foundations for maintaining safe markets is its margining system.[38]

Protecting the Marketplace

A regular part of the exchange's financial surveillance activities is monitoring the positions of all traders with large orders within a firm. This continuous review of traders holding large positions gives the exchange the ability to anticipate potential liquidity concerns created by open positions held by a member firm.

Protecting Investors in U.S. Markets

The Commodity Futures Trading Commission (CFTC) was created by the U.S. Congress in 1974 to oversee the trading of contracts on U.S. futures exchanges. Empowered with jurisdiction over futures trading, the responsibilities of this commission include regulating exchange trading of futures contracts; approving all futures exchanges' rules and regulations, as well as new contracts to be traded; and enforcing rules to protect customers' funds.

In addition, the National Futures Association (NFA) was established in 1982 as a self-regulatory organization to ensure the highest standards of professional conduct among individuals, firms, and organizations that make up the futures industry. Membership in the NFA is mandatory for futures commission merchants, commodity pool operators, introducing brokers, and commodity trading advisors. The overall objective of the NFA is to protect the futures-trading public. It registers companies and individuals dealing with public customers; enforces compliance with its regulations and those of the CFTC; and provides a forum for resolving futures-related disputes.[39]

CONCLUSION

Rules and regulations of the exchanges are extensive and are designed to support competitive, efficient, liquid markets. These rules and regulations are scrutinized continuously and are periodically amended to reflect the needs of market users. The success of the system is obvious. Trading futures and futures options can be a very exciting profitable learning experience. Before jumping into the trading market, do your homework, consider the risk, and evaluate your financial situation. The goal for trading futures is to improve your financial situation.

SUMMARY

Since the early development of agricultural markets in the United States, production agriculturalists have attempted to protect themselves against falling commodity prices at harvest. Production agriculturalists soon realized that a marketing strategy was equal in importance to production, capital, and labor strategies. Thus, forward contracting led to commodity (futures) markets to reduce marketing risk.

Commodity (futures) markets originated in Chicago in the 1800s. The price fluctuating and increasing production created the need for a year-round, centralized marketplace. Eighty-two Chicago businessmen filled this need when they founded the Chicago Board of Trade. In 1865, the Chicago Board of Trade developed standardized agreements called futures contracts.

Besides the Chicago Board of Trade, there are many other commodity (futures) exchanges. The hardest concept for a beginner to understand is that exchanges do not really physically trade grain, livestock or other commodities. Trades are made for contracts of commodities, but delivery rarely occurs. Besides agricultural commodities, there are future contracts for interest rates, insurance, stock indexes, manufactured and processed products, nonstorable commodities, precious metals, and foreign currencies and bonds.

Futures contracts are legally binding agreements, made on the trading floor of a futures exchange, to buy or sell something in the future. Actually, there is no such thing as a written futures contract. It is a verbal agreement between a buyer and a seller made on the floor of a futures exchange.

There are two main categories of futures traders that utilize futures contracts. These are the hedger and the speculator. Hedgers either now own, or will at some time own, the commodity they are trading. Speculators, on the other hand, will likely never own or even see the physical commodity. They just buy and sell futures contracts and hope to make a profit on their expectations of future price movements.

Determining the price of a commodity is called price discovery. Futures exchanges do not set prices. They are free markets where the forces that influence prices are brought together in open action.

Before any futures trading can be done, you must open an account with a brokerage office. When you open an account, you should inform the broker whether you're going to be using your account for hedging or speculating purposes. When customers place an order, they are required to post a performance bond margin, which is a financial guarantee required by both buyers and sellers to ensure they fulfill the obligation of the futures contract.

A vital function of the commission house is to properly relay customer orders. There are three major types of orders: market orders, limit order, and stop order. Orders received from customers are sent as soon as possible by a commission house to the exchange on which the order is to be executed. After several stops on the trading floor are completed, the commission house confirms the transaction with the customer.

In 1982, futures options were instituted. In contrast to futures, options on futures allow investors and risk managers to define risk and limit it to the cost of a premium paid for the right to buy or sell a futures contract. At the same time, options can provide the buyer with unlimited profit potential.

It is really to everyone's best interest that the futures industry provide a safe, well-capitalized arena for trading futures contracts and options. Exchanges, market participants, and federal regulators work together to protect the integrity of the marketplace.

Trading futures and futures options can be a very exciting and profitable learning experience. Before jumping into the trading market, do your homework, consider the risk, and evaluate your financial situation. Improving your financial situation is the goal of futures trading.

END-OF-CHAPTER ACTIVITIES

Review Questions

1. Why was the Chicago Board of Trade started?
2. When was the Chicago Board of Trade started?
3. List the names of twelve futures exchanges other than the Chicago Board of Trade.
4. What is the hardest concept for a beginner to understand about futures exchanges?
5. Other than grain and livestock, list eight other futures contracts that are sold.
6. Explain the difference between hedgers and speculators.
7. Explain how a production agriculturalist uses hedging to protect or enhance his or her financial situation.
8. Explain how speculators have a positive influence in the futures market.
9. How are the prices of a commodity determined?
10. List five USDA reports that help to forecast market direction.
11. List ten steps in the procedure for filling a futures contract order.
12. Compare futures contracts and futures options.
13. What are four major factors affecting the time value of a futures option?
14. Give five ways (examples) of how using futures options can benefit agribusiness.
15. Name and briefly explain three ways to exit a futures option position.
16. List two organizations whose objectives are to protect commodity (futures) traders and investors.

Fill in the Blank

1. A _____ strategy is equal in importance to production, capital, and labor strategies.
2. Standardization of contract terms and the ability to _____ contracts led to the rapidly increasing use of the futures markets by agricultural firms.
3. By the year _____, most of the basic principles of futures trading as we know them today were in place.
4. Futures exchanges exist because they provide two vital economic functions for the marketplace: _____ and _____.
5. The primary objective of a hedger is to achieve _____.
6. _____ are not interested in the physical commodity.
7. Only _____ members have the privilege of trading on the floor.
8. The rules and regulations of the Chicago Board of Trade require pit trades to use _____ _____ _____ in buying and selling.

9. On the trading floor, since all the trades must be recorded, traders wear _____ _____, indicating their trading privilege.

10. _____ _____ are similiar to making a down payment on a house.

11. A _____ is an option that gives the option buyer the right to purchase a futures contract.

12. A _____ is an option that gives the option buyer the right to sell a futures contract.

13. Another way to say an option has _____ _____ is to say the option is in the money.

14. _____ _____ reflects the amount of money that buyers are willing to pay in hopes that an option will be worth exercising at or before expiration.

Matching

a. spreader

b. premium

c. position trader

d. day trader

e. joint account

f. National Futures Association

g. individual account

h. brokerage office

i. stop order

j. margin call

k. initial margin

l. limit order

m. market order

n. margin call

o. bank

p. scalper

_____ 1. a public or professional trader who initiates a futures or options position and holds it over a period of days, weeks, or months

_____ 2. holds market only during the course of a trading session and rarely carries a position overnight

_____ 3. trades in the trading pit only for him- or herself, taking minimum fluctuations to make a profit or loss

_____ 4. trades the shifting price relationship between two or more different futures contracts

_____ 5. place you must go in order to open a futures account

_____ 6. place or organization to call to see if the brokerage office you selected is a quality business

_____ 7. used when trading decisions are made by the customers

_____ 8. set up when both the broker and customer have input into trading decisions

_____ 9. amount of money a market participant must deposit into the futures account when he or she places the order

_____ 10. request for additional money to be added to the futures account

_____ 11. customer states the number of contracts of a given delivery month without specifying the price at which to buy or sell

_____ 12. a trade can only be executed at the desired price or better

_____ 13. a customer's order is liquidated if a contract falls to a certain price

_____ 14. the most futures option buyer can lose

Activities

1. Write a short essay explaining how grain prices were stabilized with the formation of futures exchanges.

2. Refer to the following line of price quotes on soybeans, by referring to Figure 14–4 and complete the blanks: September soybeans opened at _____ and closed or settled at _____ on January 5. The lowest price anyone paid for September soybeans on January 5 was _____ and the highest price anyone paid for September soybeans on that day was _____. The contract is _____, prices are quoted in _____, and it is traded at the _____.

3. Suppose you buy two CBOT July wheat futures contracts on January 15 for $3.30/bu. Then, on March 2, you offset the position by selling two CBOT July wheat futures contracts for $3.50¼/bu. What is your profit or loss?

4. Give an example of a long hedge similar to the example in this chapter. Use a commodity in which you are the most familar or would use if you ever traded futures.

5. Give an example of a short hedge similar to the example in this chapter. Again, use a familar commodity.

6. Some say speculating is the same as gambling. Others say speculating is a calculated business decision. Write a short essay defending your position. Present this to classes along with other students in the form of a discussion or debate.

7. Give a speculating example similar to the one in this chapter.

8. If you decided to trade futures commodities either as a hedge or speculator, would you trade futures contracts or futures options? Defend your choice.

9. Suppose you're raising a crop or livestock commodity which is traded at the Chicago Board of Trade and Chicago Mercantile Exchange. Give an example similar to the two examples in the chapter on the strategy of how you would use futures options.

10. In order to gain experience as a commodity futures trader, complete the following activities. The assignment is designed for five weeks, but you can adjust it as your instructor directs. Samples are given to assist you with each activity. The four activities are tracking prices, charting prices, economic factors, and trading positions.[40]

    Pick one contract month to trade and trade the same contract month throughout the trading exercise. Be sure to choose a contract month that's at least two months away from the time you begin your trading exercise. For example, if you begin your trading exercise on October 10, trade the contract month of December or later. Or if you begin the trading exercise on January 15, set the contract month of March or later.

    a. *Tracking Prices:* This activity is designed to get you used to monitoring the markets. Later you'll use this information to create price charts.

        (1) **Every business day** during the five-week trading period, record the daily opening, high, low, settlement prices, and net change for the contract month of the commodity you are trading. This information is published in the business sections of several newspapers, or you can contact a local commodity broker.

(2) **Follow the same contract month** for the entire period. (You should be tracking prices of the same contract month and commodity you are trading.)

(3) **Create a tracking table** like the one below, filling in the prices you collected during the five-week trading period. You should have twenty-five days of price history when you are done.

SAMPLE: Tracking Prices
(Prices obtained from quotes listed by CBOT below)

Name	J. Trader		Commodity	Wheat
Trading Firm	Wheat Commodities		**Contract Month**	July

Date	Open	High	Low	Settle	Change
5/10	292½	294½	291	293¾	+¾
5/11	293¾	295	293¼	293½	−¼
5/12	295	298¼	295	297½	+4

Monday, May 10

WHEAT (CBOT) 5,000 bu.; cents per bu.

	Open	High	Low	Settle	Change
May	346	350	346	348	+1
Jul	292½	294½	291	293¾	+¾
Sep	295½	297¼	294¼	297	+1¼
Dec	307	308	305½	307¾	+1
Mar	311	313	310¾	313	+1½
May	311½	312	311½	312	+¾

Tuesday, May 11

WHEAT (CBOT) 5,000 bu.; cents per bu.

	Open	High	Low	Settle	Change
May	349	354	348	353¾	+5¾
Jul	293¾	295	293¼	293½	−¼
Sep	297	298¼	296¼	296½	−½
Dec	308	209	307	307¼	−½
Mar	313½	314¼	313	313	. . . .
May	313½	313½	313	313	+1

Wednesday, May 12

WHEAT (CBOT) 5,000 bu.; cents per bu.

	Open	High	Low	Settle	Change
May	357	363½	356¾	363	+9¼
Jul	295	298¼	295	297½	+4
Sep	298½	300½	298	299¾	+3¼
Dec	309	311½	309	310¼	+3
Mar	314¾	316	314¾	315¾	+2¾
May	315	316	315	315½	+2½

b.   *Charting Prices:* Although economic factors affect price, not every price change can be attributed to economic factors. In fact, some traders make buying and selling decisions based on what a particular chart looks like. Using charts to make trading decisions is known as *technical analysis.* One type of chart used by traders is a bar chart that shows the high, low, and settlement price for each trading day of a given time period.

(1)   Create a bar graph, using the prices you collected in the previous activity, "Tracking Prices." Your graph can be hand-drawn, computer generated, black-and-white, color—whatever you like. Don't be afraid to be creative. Your bar chart should illustrate twenty-five consecutive business days of prices—high, low, and settlement—that you collected in "Tracking Prices."

(2)   On the vertical axis, list prices. Make sure to use price intervals appropriate to your commodity's current market prices. For example, if wheat prices ranged from $3.00 to $3.50, you would use 5-cent intervals.

(3)   On the horizontal axis of your chart, list the twenty-five consecutive business dates of your five-week trading period.

(4)   For every day of the trading period:

🌾 Mark the day's high with a dot.

🌾 Mark the day's low with a dot.

🌾 Draw a vertical line connecting the two points. This line is referred to as the day's trading range.

🌾 Draw a horizontal dash indicating the settlement price that intersects the vertical line.

(5)   Draw all twenty-five daily bar charts on the same graph.

A sample bar chart is given below. The numbers used to create the chart come from the sample price table in section (a). Since there only three days of prices reported in the previous activity, the sample below includes charts for just those days. However, by the time J. Trader from Wheat Commodities is done, there will be twenty-five bar charts recorded.

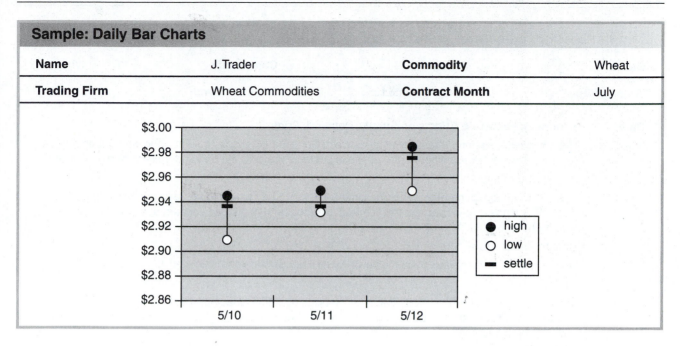

Sample: Daily Bar Charts

Name	J. Trader	Commodity	Wheat
Trading Firm	Wheat Commodities	Contract Month	July

c. *Economic Factors:* Using economic information to make trading decisions is known as fundamental analysis. You make trading decisions based on your predictions of how economic factors and events may change the supply and/or demand of your commodity and how prices may be affected. This combines your understanding of economic theory with real-world events.

(1) For the entire five-week trading period (twenty-five business days), keep a daily diary of the economic factors/events that may have caused a change in the supply of and/or demand for your commodity and how price may have been affected. Daily newspapers, weekly magazines, radio/television broadcasts, and local market experts, such as commodity brokers, are excellent sources of information for your diary. Your diary should include:

❧ date

❧ detailed summary of economic event (can include local, domestic, and/or international news)

❧ description of how an event may affect supply of and/or demand for your commodity

❧ explanation of how an event may impact the price of your commodity—higher, lower, no change—and note of your trades

(2) In your diary, write down whether or not you made a trade for each of the twenty-five business days. Note that not every move in the market can be attributed to economic factors, but do include as many descriptions as possible.

Sample: Economic Factors			
Name	J. Trader	**Commodity**	Wheat
Trading Firm	Wheat Commodities	**Contract Month**	July

Date:	**Economic Factors & Effects on Supply, Demand, Price**
May 5	Dry conditions continue throughout the Midwest. Continued dryness could hurt the wheat crop. With a possible drop in the supply of wheat and no expected change in the demand, prices should increase. Bought wheat futures.
May 6	There's an announcement that Russia is in the market to buy wheat. Continued dryness in the Midwest. With a possible increase in the demand for wheat and continued concern over lower supply, wheat prices should increase. Bought wheat futures.
May 7	No important economic factors in the news. Market prices should remain stable. Did not trade.
May 8	Much needed rainstorms soak the Midwest. Reports of a big Canadian wheat harvest expected. Both of these factors would increase the supply of wheat. Without any expected change in demand, wheat prices should decrease. Sold wheat futures.

d. *Trading Positions:* Test your skills as a commodity trader using current economic news to make trading decisions. To begin trading in the futures market, you can either buy or sell one or more futures contracts. This is your "opening position." At a later time, you can "offset the position" (or close out the position). To offset the position, you do just the opposite of your opening position. Suppose you bought one wheat futures contract on Monday. Then, on Friday, you decide to offset your position by selling one wheat contract. Likewise, if you sold one wheat futures contract on Monday, then on Tuesday, you decide to offset the position by buying one wheat contract.

Here is another trading tip: Each futures contract has specific trading months. If you buy a December corn futures contract, the only way you can offset it is if you sell a December corn futures contract. Contract months are included in your commodity's contract specifications.

(1) Trade the same contract month and commodity throughout the five-week trading period.

(2) Make at least two trades a week. The combination of an opening position and an offsetting position counts as one trade. Once you have completed the five-week trading exercise, you should have made at least ten trades. Note that you can make more than two trades per week. Decide if you want to open and offset them the same week or carry them over to another week. Example: Suppose you have made two trades in one week and then decide to open another position. You can either offset the third trade the same week you opened it or offset it a week or two later.

(3) All open positions must be offset by the last day of the five-week trading period.

(4)   Each trade should include one to ten contracts. Example: If you buy three December wheat futures contracts on Tuesday, then, on Thursday, you sell three December wheat futures contracts.

(5)   Calculate your profits and losses after you offset a position.

(6)   Create a chart similar to the one below to record your trading positions, profits, and losses.

Sample: Trading Positions

Name		J. Trader				Commodity		Wheat
Trading Firm		Wheat Commodities				**Contract Month**		July

Trade	Date	Position	#Contracts Buy	Sell	Price	Unit Profit/Loss	Total Profit/Loss (contract size x unit profit/loss x # of contracts)
1	5/5	Open	1		$3.50		
	5/6	Offset		1	$3.55	+.05	+$250
2	5/7	Open		1	$3.60		
	5/8	Offset	1		$3.70	-.10	-$500
3	5/9	Open	2		$3.68		
	5/13	Offset		2	$3.80	+.12	+$1,200
4	5/12	Open	1		$3.75		
	5/13	Offset		1	$3.80	+.05	+250

Explanation of the Trading Positions

Trade 1:   You start trading. You open a position on Monday, May 5, buying one July wheat futures contract at $3.50. On Tuesday, May 6, you offset the position by selling one July wheat futures contract at $3.55.

Trade 2:   On Wednesday, May 7, you open another position—selling one July wheat futures contract at $3.60. On Thursday, May 8, you offset the position by buying one July wheat futures contract at $3.70.

Trade 3:   On Friday, May 9, you buy two July wheat futures contracts at $3.68. You offset the position by selling two July wheat futures contracts at $3.80 on Tuesday, May 13.

Trade 4:   On Monday, May 12, you buy another July wheat futures contract at $3.75. You offset that position on Tuesday, May 13, by selling one July wheat futures contract at $3.80. Note that since you bought two July wheat contracts on Friday, May 9, and one July wheat contract on Monday, May 12, you offset these positions by selling three July wheat futures contracts on Tuesday, May 13, at $3.80. In total, these positions account for Trade 3 and Trade 4.

NOTES

1. *Agribusiness Management and Marketing,* #8720B (College Station, Tex.: Instructional Materials Service, 1988), 8714F, pp. 2–3.
2. Ibid.
3. Ibid.
4. Chicago Board of Trade, *Action in the Marketplace* (Chicago: Chicago Board of Trade, Publications Department, 1994), p. 3.
5. Ibid.
6. Ibid.
7. Ibid.
8. Ibid., p. 6.
9. Patrick Catania, ed., *Commodity Challenge* (Chicago: Chicago Board of Trade, Market Development Department, 1993), p. 19.
10. Chicago Board of Trade, *Agricultural Futures for the Beginner* (Chicago: Chicago Board of Trade, Market and Product Development Department, 1996), p. 4.
11. Randall D. Little, *Economics: Applications to Agriculture and Agribusiness* (Danville, Ill.: Interstate Publishers, 1997), p. 312.
12. Chicago Board of Trade, *Action in the Marketplace,* p. 9.
13. Chicago Board of Trade, *Agricultural Futures for the Beginner,* pp. 6–10.
14. Ibid.
15. Ibid.
16. James G. Beierlein and Michael W. Woolverton, *Agribusiness Marketing: The Management Perspective* (Englewood Cliffs, N.J.: Prentice-Hall, 1991), pp. 296–297.
17. Ibid.
18. Ibid.
19. Catania, *Commodity Challenge,* p. 22.
20. Chicago Board of Trade, *Action in the Marketplace,* p. 20.
21. Ibid.
22. Chicago Board of Trade, *Agricultural Futures for the Beginner,* p. 12.
23. Ibid., pp. 10–11.
24. Ibid.
25. Ibid., p. 10.
26. Ibid., p. 22.
27. Chicago Board of Trade, *Action in the Marketplace,* p. 12.
28. Ibid.
29. Ibid.
30. Ibid., p. 13.
31. Ibid., p. 14.
32. Chicago Board of Trade, *Agricultural Options for the Beginner* (Chicago: Chicago Board of Trade, Market and Product Development Department, 1996), p. 5.
33. Ibid.
34. Ibid., p. 7.
35. Ibid., pp. 8–9.
36. Ibid., p. 10.
37. Ibid., pp. 11–13.
38. Chicago Board of Trade, *Agricultural Futures for the Beginner,* p. 13.
39. Ibid., pp. 15–18.
40. These activities are adapted from the Commodity Challenge Program of the Chicago Board of Trade. This program was sponsored in conjunction with the National FFA Organization. Copies of the *Commodity Challenge* workbook can be purchased from Beth Hill, Chicago Board of Trade, 141 W. Jackson Blvd., Suite 2250, Chicago, IL 60604-2994.

CHAPTER 15

International Agriculture Marketing

OBJECTIVES

After completing this chapter, the student should be able to:

🌾 explain the importance of trade

🌾 examine the basis for trade

🌾 discuss exporting agricultural products

🌾 discuss importing agricultural products

🌾 describe the agricultural trade balance

🌾 differentiate between the general economic importance of free trade and protectionism

🌾 explain export and import trade barriers

🌾 describe international and regional trade agreements

🌾 examine careers related to international agriculture

TERMS TO KNOW

absurdities	economy of scale	imports
advocate	embargoes	law of comparative advantage
commodity	European Union (EU)	monetary policy
common market	exchange rates	opportunity cost
currencies	exports	protectionism
customs union	free trade	sanctions
deficits	free trade area	subsidies
domestic	humanitarian	tariffs
economic isolation	import quotas	trade barriers

INTRODUCTION

International trade has been a part of the agricultural industry since colonial times. Tobacco was the first of many agricultural **exports**. Early Americans found quickly that the exchange of goods and services between other countries was best for both countries if the product could be produced more cheaply abroad than at home. The colonists also learned quickly that international trade could increase the new country's standard of living.

On the surface it would appear that international trade would be no different than trading within the nation. For example, trade with California or Florida for citrus products, Michigan for automobiles, Georgia for peaches, Maine for fish, or Texas for beef cattle is easily done. Factors of production (land, labor, capital, and management) differ substantially in different regions of the country. Since every region of the country has varying amounts of these production factors, each state has different advantages in producing products.

International trade is based upon the same production factors. However, there are some differences that create challenges. Some of these challenges include: **currencies** and **exchange rates**; **monetary policy**, or systems; language barriers; technological advancement; national policies; national security; and politics. For example, former President Jimmy Carter canceled grain shipments to the USSR as punishment for its invasion of Afghanistan. Each country has its own **domestic** issues and problems and its own approach to addressing them. These challenges influence how each country views international trade.

IMPORTANCE OF TRADE

Economic Benefits

Gains from international trade occur because of low foreign prices of selected products. With each country having varying levels of resources of land, labor, capital, and management, international trade allows a country to specialize in the production of those products it best produces.[1]

A country is better off using its resources to produce those goods for which it is particularly suited and to buy from other countries the goods they more efficiently produce. Trading and production specialization among nations permit higher living standards and real income than would otherwise be possible and permit more goods to be produced from a given set of resources. Both allow a more efficient allocation of resources.

Quality of Life

We import many things that increase our quality of living or eating preferences. For example, every time American consumers have a cup of coffee, a glass of iced tea, a banana split, or a chocolate candy bar or add vanilla, cinnamon, or pepper to their food, they are benefiting from agricultural trade. The same is true when they wear silk shirts or buy rubber tires. These and many other agricultural products consumed or used in the United States are produced elsewhere. Without **imports**, these products would be unavailable. Refer to Figure 15–1.

Jobs

In a recent USDA study, it was estimated that more than 1 million full-time jobs were related to agricultural exports. It was also estimated that for every dollar of agricultural exports sold by the U.S., $2.05 was actually added to the U.S. economy. Additionally, it has been observed that for

Figure 15–1 Besides common imported items such as coffee, tea, bananas, cocoa, vanilla, cinnamon, and pepper, foods (like those shown here) from around the world are finding a place on American tables. Can you identify the gooseberries, tomatillo, kiwi fruit, mountain soursop, papaya, and pomelo? (Courtesy of USDA)

every dollar generated by food and fiber exports, $.75 goes to the nonfarm business sectors of the economy. Thus, agricultural exports have resulted in such positive impacts on the U.S. economy as increased income, more employment, and the strengthening of the trade balance.[2]

U.S. exports provide increased income to production agriculturalists, as well as to those employed in agribusiness that supply purchased imports to production agriculturalists. U.S. exports also provide income to those that assemble, process, and transport agricultural exports to foreign countries. Although it varies from year to year, approximately 23 percent of farm income, or $1.00 out of every $4.26 received from the sale of farm products, is derived from exports.[3]

Impact on States. Because such a great variety of commodities is produced by production agriculturalists throughout the United States, all regions of the country benefit from agricultural exports. Agricultural exports account for a third to a half of total farm income in 16 states: Arkansas, Illinois, Indiana, Iowa, Kansas, Louisiana, Minnesota, Mississippi, Missouri, Montana, Nebraska, North Carolina, North Dakota, Ohio, Oklahoma, and South Carolina.[4]

The consequence of abandoning the export market would be a reduction in production of about 50 percent in wheat, rice, and soybeans, and 25 percent in feed grains. Refer to Figure 15–2. A loss of this market would be a disaster for U.S. production agriculturalists as well as the related agribusiness input and output sectors. The agribusiness input sector could lose as many as 400,000 jobs, and the agribusiness output sector (export-related transportation, storage, merchandising, and port operations) could lose as many as 1 million jobs.[5]

Consumer Benefit from International Trade

Without international trade, production agriculturalists would be burdened with large surpluses.

Figure 15–2 If there were no exports, there would be a reduction in production of about 50 percent in wheat, rice, and soybeans, and of 25 percent in food grains. (Courtesy of USDA)

These excess surpluses would be dumped into domestic markets or forced into government stocks. Prices would be severely depressed and farm income drastically lowered. More production agriculturalists would lose money and be forced out of business.

Initially, consumers would benefit from depressed food prices, but as production agriculturalists would be forced to sell out due to low prices and excess stocks would become depleted, food supplies would dwindle and prices would rise again, and to even higher levels than before. Without international trade, fewer production agriculturalists would be producing for a smaller domestic market. Benefits or profit would be substantially reduced due to **economy of scale** and prices would be significantly higher than they were when international trade was allowed. Therefore, in the long run, even consumers would be hurt by lack of international trade.[6]

BASIS FOR TRADE

At the time of the Declaration of Independence, Adam Smith stressed the **absurdities** of **economic isolation** and advocated free trade as a means for each nation to increase its wealth. Adam Smith made the following quote from *The Wealth of Nations*, "It is the maxim of every prudent master of a family, never to attempt to make at home, what it will cost him more to make than buy."[7]

Therefore, the key concept in understanding the basis for international trade is the following: countries export those commodities they are relatively most efficient in producing (those requiring least resources) and import those commodities they are relatively least efficient at producing.[8]

It is important to remember that gains from international trade do not depend on a nation being absolutely (absolute advantage) more efficient than other countries in producing some commodities and absolutely less efficient in producing other commodities. Rather, gains from trade occur when a country is *relatively* more efficient in the production of some commodities than others. This concept is called the **law of comparative advantage**. Consider the following examples.

Absolute Advantage

An absolute advantage exists if one country can produce a **commodity** less expensively than another country. Suppose that one resource unit (worker-day) in Mexico is being used to produce 3 tons of soybeans and another worker-day is being used to produce 12 tons of pepper. For the United States, let us assume that one worker-day is being used to produce 15 tons of soybeans and another worker-day is being used to produce 5 tons of pepper. Figure 15–3 shows that total production in the two countries together equals 18 tons of soy-

Absolute Advantage Example

	Soybeans (tons) Produced by 1 Man Day	Pepper (tons) Produced by 1 Man Day	Total
Before trade:			
Mexico	3	12	15
United States	15	5	20
	18	17	35

	Soybeans (tons) Produced by 1 Man Day	Pepper (tons) Produced by 1 Man Day	Total
After trade when countries specialize:			
Mexico	0	24	24
United States	30	0	30
	30	24	54

Figure 15–3 Mexico has an absolute advantage in pepper production, and the United States has an absolute advantage in the production of soybeans.

beans and 17 tons of pepper, or a total of 35 tons of raw food and fiber products when they do not trade. However, Mexico is more efficient in using a resource unit to produce pepper (one worker-day produces 12 tons of pepper compared to 5 in the United States) and less efficient in soybean production (3 tons compared to 15 in the United States per worker-day). Therefore, Mexico has an *absolute advantage* in pepper production and the United States has an absolute advantage in the production of soybeans.

Comparative Advantage

The principle of comparative advantage explains how gains from trade are possible even when one nation can produce all products more efficiently than another because of differences in **opportunity cost** among countries. Opportunity cost, defined in this context, is the output of one commodity sacrificed to produce more of another.

Let's assume that the efficiency of pepper production in the United States suddenly increases such that a one-man day can now produce either 15 tons of soybeans or 15 tons of pepper. We shall continue to assume that one worker-day in Mexico can produce either 3 tons of soybeans or 12 tons of pepper. Refer to Figure 15–4. Thus, with two resource units available in each country and no specialization, world output would be 18 tons of soybeans and 27 tons of pepper. Notice that the United States has a greater productive capacity than Mexico in both commodities. The United States is 5.00 times (15 to 3) more efficient in producing soybeans and 1.25 times (15 to 12) more efficient in producing pepper.

Because of the law of comparative advantage, it would be beneficial for each of these nations to specialize and trade (assuming no transportation costs). The United State's *greatest comparative advantage* would be in soybean and Mexico's *smallest comparative disadvantage* would be in pepper.[9]

Importance of Specialization

Again, refer to Figure 15–3. If both countries specialized in producing what they have an absolute advantage in and traded with the other, the United States could produce 30 tons of soybeans and Mexico could produce 24 tons of pepper. Thus, world output would jump from 35 tons to 54 tons, a gain of 19 tons of output.

Now, refer to Figure 15–4. By specializing on this basis, world output in this two-country example would still increase by 9 tons (i.e., 54 tons minus 45 tons). *"The key point to remember is that even if a nation has an absolute advantage in both products, total world output can be increased if the opportunity costs differ among countries."*[10]

Comparative Advantage Example

	Soybeans (tons) Produced by 1 Man Day	Pepper (tons) Produced by 1 Man Day	Total
Before trade:			
Mexico	3	12	15
United States	15	15	30
	18	27	45
	Soybeans (tons) Produced by 1 Man Day	**Pepper (tons) Produced by 1 Man Day**	**Total**
After trade when countries specialize:			
Mexico	0	24	24
United States	30	0	30
	30	24	54

Figure 15–4 The greatest comparative advantage of the United States would be in soybeans, and Mexico's smallest comparative disadvantage would be in pepper.

U.S. Agricultural Exports: Value by Commodity, Selected Calendar Years

Commodity	1970	1980	1982	1986	1992	1995
			Million Dollars			
Animals and animal products						
Dairy products	127	175	347	438	726	711
Fats, oils, and greases	247	769	663	411	525	827
Hides and skins, excluding fur skins	187	694	1022	1521	1346	1621
Meats and meat products	132	890	978	1113	3339	4522
Poultry products	56	603	515	496	1211	2345
Other	101	660	410	530	778	907
Total animals and animal products	850	3791	3935	4509	7925	10,933
Cotton, excluding linters	372	2864	1955	786	1999	3681
Fruits and preparations	334	1335	1376	2040	2732	3240
Grains and preparations						
Feed grains, excluding products	1064	9759	6444	4330	5737	8153
Rice, milled	314	1289	997	621	725	996
Wheat and flour	1111	6586	6927	3279	4675	5734
Other	107	357	273	398	3035	3654
Total grains and preparations	2596	17,991	14,641	8629	14,172	18,537
Oilseeds and products						
Cottonseed and soybean oil	244	915	692	343	432	786
Soybeans	1228	5880	6218	4321	4380	5400
Protein meal	358	1654	1447	1302	1398	1701
Other	91	944	784	493	980	1036
Total oilseeds and products	1921	9393	9141	6459	7190	8923
Tobacco, unmanufactured	517	1334	1547	1209	1651	1400
Vegetables and preparations	206	1188	1174	1024	2871	3889
Other	463	3360	2853	1349	4389	5211
Total exports	7259	41,256	36,622	26,064	42,929	55,814

Figure 15–5 Feed grains, soybeans, wheat and flour, and meats and meat products are among the largest U.S. agricultural exports. (Source: USDA, Economic Research Service, *Foreign Agricultural Trades of the United States,* January/February/March 1998.)

EXPORTING AGRICULTURAL PRODUCTS

Because of excess production of many agricultural commodities in the United States, the export market has been extremely important for the agribusiness output sector. With the recent changes in the demand for U.S. agricultural commodities, it appears that those in the agribusiness output sector will have an opportunity to sell more of their commodities in foreign markets. Agricultural exports amount to approximately 10 percent of all merchandise exports of the United States.[11] In the last several years, the United States has made great strides in developing international markets for feed grains, soybeans, and wheat.

Major Commodities That Are Exported

Figure 15–5 shows U.S. agricultural exports from 1970 to 1995, by individual products. Feed grains, wheat, and soybeans are the largest components of exports. Wheat and flour are next followed by meats and meat products. Cotton and tobacco are still significant, but grains and preparations, oilseeds, fruits and vegetables, animals, and animal products are all important exports.

At present, 25 percent of U.S. farm cash receipts are derived from the export market. The United States exports approximately 54 percent of the wheat crop, 43 percent of the soybean crop, 21 percent of the corn crop, 38 percent of the grain sorghum crop, and about 49 percent of the rice crop. Refer to Figure 15–6.

Major Buyers of U.S. Exports

Asia and Western Europe are the largest markets for U.S. agricultural exports (Figure 15–7). Asia has been increasing its imports of rice, wheat, soybeans, and feed grains. Recently, Japan became

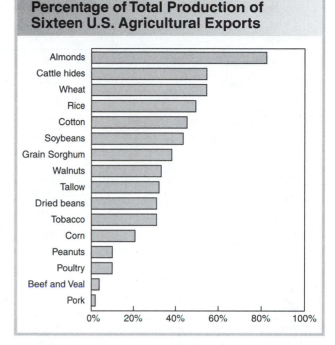

Figure 15–6   The percentages shown here represent the amount of total production of the commodity product in the United States that is exported. (Source: USDA)

our first $10 billion agricultural customer, the largest single importer of U.S. farm products.

Humanitarian Reasons for Exporting

Agricultural exports are also desirable for **humanitarian** reasons. Many parts of the world are incapable of producing enough food to feed their people. Others can produce the food, but only at a prohibitively high cost. Since the United States is a more efficient producer of food, it can help. Countries with limited resources can buy food from the United States at prices substantially below the cost of production. The United States operates the world's largest food aid program to needy countries shipping to about seventy countries under the Food for Peace Program.[12]

U.S. Agricultural Exports by Regions in Selected Years

Region	1973	1980	1982	1986	1992	1995
				Million Dollars		
Western Europe	5605	12,351	11414	7026	7804	9003
Enlarged EU-15	4526	9256	8398	6604	7290	8537
Other Western Europe	1079	3095	3016	423	514	466
Eastern Europe and FSU	1498	3285	2718	1091	2663	1639
FSU	920	1130	1871	658	2346	1346
Eastern Europe	577	2155	847	433	317	293
Asia	6509	15,037	13,675	10,537	17,923	27,939
West Asia	521	1418	1428	1334	1782	2478
South Asia	1564	734	800	450	629	1019
SE Asia, excluding Japan and PRC	1851	4281	4387	3586	1545	3012
Japan	2998	6331	5555	5106	8437	10,957
People's Republic of China (PRC)	575	2273	1505	61	545	2633
Latin America	1692	6176	4438	3641	6669	7926
Canada, excluding trans-shipments	1034	1905	1805	1533	4902	5738
Canada trans-shipments	677	12	15	14	0	0
Africa	583	2303	2287	1997	2570	3070
North Africa	307	1244	1223	1340	1461	2144
Other Africa	276	1059	1064	657	1109	926
Oceania	83	188	270	224	398	499
Total	17,680	41,256	36,622	26,064	42,929	55,814

Figure 15–7 Asia and Western Europe are the largest markets for U.S. agricultural exports. (Source: USDA, Economic Research Service, *Foreign Agricultural Trades of the United States,* January/February/March 1996.)

Beneficial Side Effects of Exporting

Providing adequate diets improves a country's ability to develop and grow. As economic growth improves in developing countries and people's incomes rise, they are more able to purchase the food they need and also are more likely to want more and higher-quality food.

Since the United States is a major producer of agricultural products, it is in our best interest to promote world economic growth. Economic growth increases the size of our future markets for agri-cultural commodities. This growth in international demand is one of the driving factors underlying U.S. economic growth.[13]

IMPORTING AGRICULTURAL PRODUCTS

The United States is one of the largest importers of agricultural products. The United States imports, annually, agricultural products worth over $26

Value of Agricultural Imports, by Selected Major Commodity Group

Commodity Group	Value $1,000	Leading Countries of Origin
Competitive products		
Meat and products	2,657,548	Canada, Australia, New Zealand
Dairy products	963,376	New Zealand, Italy, Ireland
Grains and feeds	2,339,092	Canada, Italy, Thailand
Fruits, nuts, and vegetables	4,725,954	Mexico, Chile, Spain
Sugar and related products	1,128,712	Canada, Dominican Republic, Guatemala
Wine and malt beverages	2,084,650	France, The Netherlands, Italy
Oilseeds and products	1,562,955	Canada, Italy, Philippines
Tobacco, unmanufactured	613,182	Turkey, Brazil, Greece
Noncompetitive products		
Bananas and plantains	1,071,834	Costa Rica, Colombia, Ecuador
Coffee and products	2,485,433	Brazil, Colombia, Mexico
Cocoa and products	1,033,906	Canada, Ivory Coast, Indonesia
Rubber and gums	965,390	Indonesia, Thailand, Malaysia
Tea	185,515	China, Argentina, Germany
Spices	326,890	Indonesia, India, Madagascar
Total imports	26,818,015	Canada, Mexico, Brazil

Figure 15–8   Over $26 billion of agricultural exports are purchased yearly. Fruits, nuts, and vegetables are the largest imports. (Source: USDA)

billion. Refer to Figure 15–8. Agricultural imports make up under 5 percent of total U.S. imports.

Imports provide consumers with commodities that are either not produced in the U.S. or not available in sufficient quantities. Major imports not produced in the United States include bananas, some spices, coffee, tea, cocoa, and rubber. Domestic production of other goods, such as certain cheeses and tobaccos, is insufficient to satisfy domestic demand, so imports meet the excess demand.

Canada, Mexico, and Brazil are the top sources of agricultural products imported into the United States. However, if the (**European Union**) (**EU**) is considered as a collective, then it is the top exporter of agricultural goods into the United States.

AGRICULTURAL TRADE BALANCE

When the outflow of dollars by purchasing imports exceeds the inflow of dollars by selling exports we have a deficit in our international trade balance. International trade balance **deficits** can be corrected in the long run by reducing imports (purchases) and by increasing exports (sales). The main purpose of tracking the trade balance is to inform monetary authorities in each nation and to aid banks, firms, and individuals engaged in international trade in their business decisions.[14]

Although the trade balance is presented by the media as having a major impact on our economy, it is not as bad as it seems. For centuries, countries thought it was desirable to have an export trade surplus in order to accumulate gold and other precious metals that they thought were measures of wealth. It was Adam Smith who pointed out (in his *Wealth of Nations*) that goods rather than gold are the true wealth of a nation.[15] However, the issue remains as to how much of a trade imbalance is too much.

In most years throughout its history, our nation has maintained a surplus of exports over imports. Following World War II, the balance of trade averaged almost $5 billion per year. A negative trade balance began to occur in the early 1970s, averaged about − $30 billion from 1977 through 1980, then peaked near − $160 billion in 1987. The trade balance has improved somewhat since then, with a balance of − $140 in 1995.[16]

Agricultural exports have done their part to lower the trade deficit. Through 1995, there have been thirty-six consecutive years of agricultural trade surpluses, in contrast to the persistent trade deficits in nonagricultural trade. In 1995 the U.S. agricultural trade surplus was approximately $25 billion, the second largest trade surplus ever.[17]

FREE TRADE OR PROTECTIONISM

Some people favor **free trade** and some favor **protectionism**. It depends upon how you are personally affected as to which side you take. For example, let us say you are a beef producer and you **advocate** free trade. However, you learn that three of the major processing companies signed contracts for shipments of beef from neighboring countries. In the long run, the action taken by the beef processing companies may help everyone except the U.S. beef producers and ranchers. Their prices could drop to the levels below production costs. Now, are the beef producers and ranchers for or against free trade? This incident could happen with any commodity.

Agricultural Trade Policy

Whether there should be free trade or protectionism is not all a black and white issue. There are many gray areas. Because of these gray areas, trade policies are formulated in the "national interest." Effective trade policy takes into account the following objectives:

- 🌿 to maintain certain industries, such as steel and aviation, in the interest of national security
- 🌿 to seek to stabilize economic activity
- 🌿 to have a positive impact on the balance of payments
- 🌿 to safeguard consumer costs
- 🌿 to increase or maintain employment
- 🌿 to promote equal income distribution[18]

General Economic Arguments for Free Trade

The leading arguments in favor of free trade center around the law of comparative advantage. Embedded in the law is the concept of specialization and the idea that society gains from specialization. It is argued that if markets are not obstructed by protectionistic measures and private entrepreneurs are allowed to pursue their interests, resource allocation would be optimized and overall welfare would be maximized. In essence, the proponents of free trade say that resources will adjust across national boundaries as well as within.

General Economic Arguments for Protectionism

There are several economic arguments for those that favor protectionism. These will be briefly discussed.

Infant Industry. This argument contends that when an industry is in its infancy stage its per unit costs are greater because of insufficient volume produced because of lack of demand or of investment capital. The argument states that infant industries should be protected for a period of time so that they are not eliminated before they become operationally efficient.[19]

Balance of Payments Argument. This argument against free trade suggests that it can lead to excessive imbalance of payments and resources imbalances during adjustment periods. The need for this kind of protection is closely related to the development of national economic policies which emphasize goals such as growth and employment rather than efficiency, which is the fundamental goal of international specialization (free trade).[20]

Unfair Competition Argument. This argument proposes that foreign countries implement unfair trade practices. Currently, this is the argument against the "invasion" of Japanese automobiles and European dairy products in the United States. The argument suggests that coordination between governments, firms, and banks to offer excessive **subsidies** is unfair competition for the "unprotected" U.S. industries.[21]

Self-Sufficiency for National Security. This argument suggests that in the event of war, either of a military or economic nature, each country would like to be self-sufficient, accordingly, trade restrictions are needed to strengthen industries producing essentials for survival. This would include food as well as defense-related material.[22]

Diversification for Stability. Some countries, due to size, geographic location, and resource endowments, are essentially economically dependent on exports of only a few commodities. Barriers against imports of other goods, it is alleged, would induce domestic producers to diversify and produce those items formerly imported. The benefits would be realized in the form of greater economic independence and domestic stability.[23]

Protection of Wages and the Standard of Living. This argument by "protectionists" is that developed-country industries cannot compete against low-wage countries, particularly in the production of labor intensive products. This is an argument currently being used by the U.S. textile and footwear industries against low-cost Asian and Latin American producers.[24]

EXPORT AND IMPORT TRADE BARRIERS

It stands to reason that if we are not going to allow free trade, there must be ways to protect or restrict trade. These ways are called **trade barriers**. There are many forms of protection that governments use with respect to international trade. The two major methods are those that influence imports and those that influence exports.

Import Controls

Tariffs are government-imposed taxes on goods when they cross national boundaries. Tariffs are the most commonly used trade restriction tool. Import tariffs make goods more expensive than they would otherwise be. As a result, consumption of domestically produced goods is encouraged.

Import quotas impose a limit on the amount of a good that may be imported in a given period. Once the quota is filled, no more of that good may be imported, regardless of the price. Quotas and tariffs have similar effects on restricting trade, but they differ in who receives the revenue. With a tariff, the government gets the revenue; with a quota, it accrues to firms with the right to import the protected good.[25]

Regulatory trade restrictions are government-imposed rules and regulations that limit imports. For example, to protect its citizens, a country may have restrictions in place that forbid importing fruits and vegetables sprayed with certain insecticides.

Export Controls

Export embargoes prohibit shipments of commodities to certain other countries. Nations use these embargoes or **sanctions** to maintain domestic supplies and to support foreign policy. For example, in 1973, the United States placed an embargo on soybeans. This lowered soybean prices and increased American supplies of soybeans. An example of a foreign policy embargo occurred in 1980. As mentioned earlier, a grain embargo was enacted against the USSR. This embargo was put into effect because of the USSR's foreign policy actions in Afghanistan.[26] However, instituting export embargoes can result in loss of confidence in a nation as a dependable supplier and result in the loss of trade.

Voluntary export restraints are an agreement in which foreign governments are asked to limit their exports of specific commodities to a given quantity. This controls the importation of those commodities, protecting domestic producers.[27]

INTERNATIONAL AND REGIONAL TRADE AGREEMENTS

Somewhere between the two extremes of free trade and trade barriers are trade agreements. Reducing trade barriers is an obvious way to increase trade. Most governments are not willing to reduce these barriers without receiving something in return. Therefore, many governments enter into formal trade agreements with other nations. These agreements specify the terms for international trade.

Two main approaches have been used in the last half of the twentieth century to liberalize trade. One is the international approach under the General Agreement on Tariffs and Trade (GATT), and the World Trade Organization (WTO), and the second is a regional approach.[28] This latter approach is illustrated by organizations such as the EU and NAFTA, where a number of countries agree to liberalize trade among themselves, but maintain common and uniform tariffs against others.

International Approach to Trade Agreements

The General Agreement on Tariffs and Trade (GATT) was created in 1947 to secure a substantially free and competitive market as the best means of conducting international trade and serving the best interests of all nations. Since its creation, eight major negotiating conferences have been held by GATT. The last round, the Uruguay Round, was conducted between 1986 and 1994 in which 120 countries participated. It resulted in agreement to reform all the negotiating areas and to establish a new intergovernmental organization, the World Trade Organization (WTO). The five basic principles of GATT are:

- 🌾 Trade must be nondiscriminatory.
- 🌾 Domestic industries receive protection mainly from tariffs.
- 🌾 Agreed-upon tariff levels bind each country.
- 🌾 Consultations are provided to settle disputes.
- 🌾 GATT procedures may be waived if other members agree and compensation is made to them.[29]

CAREER OPTION

Career Area: International Agriculture Trader/Marketer

All nations strive to attain a higher standard of living for their people. This could not be maintained if the United States chose to isolate itself and not engage in a wide range of international trade activities. With each country having varying levels of resources of land, labor, capital, and management, international trade allows a country to specialize in the production of those products it best produces. As transportation and communication improve, we have become familiar with more places and people around the world, so more business is being conducted between and among people and nations of different continents. Without international trade, production agriculturalists would be burdened with large surpluses and forced to sell out due to low prices.

Careers in international trade are challenging and exciting. Some of the benefits include travel, working with people in different countries, and learning about the broad scope of world agriculture. Importers and exporters will rely on professionals who can see the broad picture of production, processing, marketing, and trade. They employ those who can interpret complex policies and trade regulations, while also exhibiting competence in working with people of different cultures and speaking other languages. Production agriculturalists and agribusinesses in America will continue to depend on exports as a market for U.S. products, while consumers will continue to depend on

commodities and products from other countries to be available. Therefore, people who can sell U.S. products, help expand markets, and develop new markets for them will constantly be in demand.

Being an international trader requires the most up-to-date technological assistance, including, among other devices, fax machines, computers, and access to the Internet.

The World Trade Organization (WTO), which officially began functioning January 1, 1995, essentially succeeds GATT. Nations that ratified GATT are members of the WTO. The structure of the WTO is intended to provide for more effective decision making and greater involvement in trade relations. The WTO has been given authority to oversee adherence to trade agreements on goods and services, as well as intellectual property. Five functions of the WTO are as follows:

❦ facilitate the implementation, administration, and operation of the GATT terms and further the established objectives

❦ provide a forum for negotiation among member countries concerning their multilateral trade relations, and then assist in implementing the results of those negotiations

❦ oversee dispute settlement between member countries

❦ through the Trade Policy Review Mechanism, improve adherence by member countries to rules, disciplines, and commitments made under the multilateral trade agreements

❦ cooperate, as appropriate, with the International Monetary Fund and the International Bank for Reconstruction and Development (the World Bank) to achieve greater coherence in global economic policy making.[30]

Regional Approach to Trade Agreements

Customs unions, common markets, and free trade areas are types of regional integration. A **customs union** is an economic and political organization between two or more countries abolishing trade restrictions among themselves and establishing a common and uniform tariff to outsiders. A **free trade area** is the same as a customs union except there is no common and uniform tariff imposed on those excluded from the union. Likewise, a **common market** is the same as a customs union, except that it includes the free mobility of factors of production.[31] The following are two examples of regional trade agreements.

The *European Union,* a custom union, is an economic federation that has evolved from the European Community (EC) and the European Economic Community (EEC). The EEC was organized in 1959 by the Treaty of Rome to form an economic integration of six western Europe countries. Presently, fifteen countries are in the EU: France, West Germany, Italy, Belgium, the Netherlands, Luxembourg, United Kingdom, Ireland, Denmark, Greece, Portugal, Spain, Sweden, Austria, and Finland. EU is intended to be a unified area in which commerce is carried on freely, much as it is among the states of the United States. The area has common policies with respect to agriculture, transportation, taxes, and international trade.[32]

The *North American Free Trade Agreement (NAFTA)* was completed in 1994. This agreement was created to establish a free trade area between the United States, Canada, and Mexico. Canada and Mexico are the second and third largest export markets for U.S. agricultural products (Japan is the largest). Under NAFTA, tariffs and other trade barriers among these countries will be reduced through the 1990s. The outlook for increasing exports of corn, wheat, and oilseeds to Mexico is very good, while meat exports have already expanded.

NAFTA does not include any agreement to form common foreign policies, stabilize exchange rates, or coordinate welfare or immigration policies. The trade preferences only apply to goods made in NAFTA countries, defined to be at least 62.5 percent domestic parts used to manufacture product.

CAREERS RELATED TO INTERNATIONAL AGRICULTURE

Export-Import Services

The great volume and the technical nature of the export-import business requires specialists who know the fundamentals of international trade and who know the commodities of agribusiness. It needs efficient technical and commercial services: transportation, grading, storage, inspection, packaging, finance, marketing, and insurance. Export companies purchase commodities from U.S. production agriculturalists, store these products, arrange for their sale to foreign countries, and provide for their shipment.

Export companies need contracts and good relations with all foreign countries. They need people who can sell U.S. products and help expand markets and develop new markets for them. They need people who know how to listen as well as how to talk; people who can converse with citizens of foreign governments. Learning more than one language would also be an asset.[33]

Foreign Agricultural Services

The Foreign Agricultural Service (FAS) is an agency of the U.S. Department of Agriculture (USDA). The agency represents U.S. agricultural interests throughout the world. There are more than 70 different offices that provide assistance and expertise to more than 100 different countries. Through its staff of agricultural professionals and officers, the FAS helps promote trade of U.S. agricultural commodities and products, works to improve trade conditions throughout the world, and monitors agricultural production in other countries. It is also involved in providing information to the USDA as it develops agricultural policy and programs. Professionals in these positions are employed in Washington, D.C., and throughout the world.[34]

International Agricultural Research

Agricultural research is vital. During the past thirty years the amount of food produced has increased at a higher rate than the world population, supporting the importance of research. Most agricultural research is conducted through public agencies, universities, or agribusiness companies. The results of research are shared worldwide

through professional associations, journals, visiting scholars, or international agribusiness companies. Agricultural scientists or producers can apply this new technology to their own setting.[35]

An important network for developing countries is a worldwide group of research centers that support agriculture. The centers are all connected by an umbrella organization called the Consultative Group on International Agricultural Research (CGIAR). The list includes:

- 🌿 International Center for Tropical Agriculture (CIAT), Cali, Colombia

- 🌿 International Center for the Improvement of Maize and Wheat (CIMMYT), El Batan, Mexico

- 🌿 International Potato Center (CIP), Lima, Peru

- 🌿 International Board for Planned Genetic Resources (BPGR), Rome, Italy

- 🌿 International Center for Agricultural Research in the Dry Areas (ICARDA), Aleppo, Syria

- 🌿 International Crops Research Center for the Semi-Arid Tropics (ICRISAT), Hyberabad, India

- 🌿 The International Food Policy Research Institute (IFPRI), Washington, D.C.

- 🌿 The International Institute of Tropical Agriculture (IIITA), Ibadan, Nigeria

- 🌿 The International Livestock Center for Africa (ILCA), Addis Ababa, Ethiopia

- 🌿 International Laboratory for Research on Animal Diseases (ILRAD), Nairobi, Kenya

- 🌿 International Rice Research Institute (IRRI), Los Banos, Philippines

- 🌿 International Service for National Agricultural Research (ISNAR), The Hague, Netherlands

❦ West Africa Rice Development Association (WARDA), Bouake, Cote d'Ivoire[36]

The Peace Corps

The Peace Corps was started in 1961. It was an idea proposed by President John F. Kennedy. It was established by Congress to promote world peace and friendship. The Peace Corps Act contained three specific goals. They were:

❦ to help the people of interested countries and areas in meeting their needs for trained people

❦ to help promote a better understanding of Americans on the part of the peoples served

❦ to help promote a better understanding of other peoples on the part of Americans[37]

Peace Corps volunteers spend two years working with people in other countries. The volunteers work to help them meet their own needs in areas such as education, food production, health care, water sanitation, and housing. Over its history, more than 125,000 volunteers and staff have served with the Peace Corps in ninety-four countries. In the mid-1980s, the corps was active in sixty-three different countries, including countries in Latin America, South America, the Caribbean, Africa, Southeast Asia, and the Pacific region.[38]

CONCLUSION

The world of international trade is complex and ever-changing. International trade can be profitable, but it can also increase risk. In the future, there is every prospect that a higher and higher percentage of world output will be traded internationally. The question for managers will no longer be, "Should we trade internationally?" but instead, "How can we trade more effectively?" Expertise in international trade will become an increasingly important skill for managers to obtain.

Careers in international trade are challenging and exciting. Travel, working with people in different countries, and learning the broad scope of world agriculture are all benefits of this exciting career. Production agriculturalists in America and agribusinesses will continue to depend on exports as a market for U.S. products. Consumers will continue to depend on commodities and products from other countries to be available, to be safe and wholesome and fairly priced.

To accomplish these tasks importers and exporters will rely on professionals who understand international agriculture trade. They will need people who can see the broad picture of production, processing, marketing, and trade. They will need employees who can interpret complex policies and trade regulations. International companies with offices or production facilities in other countries seek and promote to international positions employees who are skilled in management and marketing. They also want these people to be competent and comfortable working with people who are of different cultures and who speak different languages.

SUMMARY

International trade has been a part of the agricultural industry since colonial times. Early Americans found quickly that the exchange of goods and services between other countries was best for both countries if the product could be produced more cheaply abroad than at home. Gains from international trade occur because of low foreign prices of selected products. With each country having varying levels of resources of land, labor, capital, and management, international trade allows a country to specialize in the production of those products it best produces.

It is estimated that more than one million full-time jobs are related to agricultural exports. It is also estimated that for every dollar of agricultural exports sold by the United States, $2.05 is actually added to the U.S. economy. Additionally, it has been observed that for every dollar generated by food and fiber exports, $.75 goes to the nonfarm business sectors of the U.S. economy.

The key concept in understanding the basis for international trade is that countries export those commodities they are relatively most efficient in producing and import those commodities they are relatively least efficient at producing. Gains from international trade do not depend on a nation being absolutely more efficient than other countries in producing some commodity; rather, gains from trade occur when a country is relatively more efficient in the production of some commodities than others.

Agricultural exports amount to approximately 10 percent of all merchandise exports of the United States. At present, 25 percent of U.S. farm cash receipts are derived from the export market. The United States exports approximately 54 percent of the wheat crop, 43 percent of the soybean crop, 21 percent of the corn crop, 38 percent of the grain sorghum crop, and 49 percent of the rice crop. The United States is one of the largest importers of agricultural products. Annually, the United States imports agricultural products worth over $26 billion. However, agricultural imports make up under 5 percent of total U.S. imports.

When the outflow of dollars from purchasing of imports exceeds the inflow of dollars from selling of exports, the country has a deficit in the international trade balance. International trade balance deficits can be corrected in the long run by reducing imports (purchases) and by increasing exports (sales).

Some people favor free trade and some favor protectionism. The leading arguments in favor of free trade center around the law of comparative advantage. It is agreed that if markets are not obstructed by protectionistic measures and private entrepreneurs are allowed to pursue their interests, resource allocation will be optimized and overall welfare will be maximized. There are several economic arguments for those that favor protectionism. They are as follows: infant industry argument, balance of payments argument, unfair competition argument, self-sufficiency of national security, diversification for stability, and protection of wages and standard of living argument.

It stands to reason that if we are going to allow free trade, there must be ways to protect or restrict trade. These ways are called trade barriers. Import trade barriers include tariffs, import quotas, and regulatory trade restrictions. Export trade barriers include export embargoes and voluntary export restraints. Somewhere between the two extremes of free trade and trade barriers lie trade agreements. Two major international trade agreements are GATT and WTO. Two major regional trade agreements are EU and NAFTA.

There are many careers related to international agriculture. These include careers in export-import services, the Foreign Agricultural Service (an agency of the USDA), international agricultural research, and the Peace Corps.

END-OF-CHAPTER ACTIVITIES

Review Questions

1. What are the four major factors of production?

2. Although production factors are the same internationally, what are eight challenges of international trade?

3. Name nine imports that increase our quality of living or eating preferences.

4. List the sixteen states in which a third to a half of total farm income comes from agricultural exports.

5. Explain how consumers benefit from international trade.

6. What is the key concept in understanding the basis for international trade?

7. Differentiate between absolute advantage and comparative advantage in relationship to international trade.

8. Explain why it is important for a country to specialize even when it can raise an agricultural commodity.

9. List ten agricultural commodities that are exported.

10. What are the humanitarian reasons for exporting?

11. Briefly explain the beneficial side effects of exporting.

12. Name six major imports not produced in the United States.

13. What is the main purpose of tracking the trade balance?

14. List six objectives effective trade policy takes into account.

15. Explain the general economic arguments for free trade.

16. List and briefly explain the six general economic arguments for protectionism.

17. Briefly explain three trade barriers to control imports.

18. Briefly explain two trade barriers used to control exports.

19. What are five basic principles of GATT?

20. What are five functions of the WTO?

21. List the fifteen countries that participate in European Union (EU).

21. List three general characteristics needed for people seeking careers in export-import services.

22. What are three goals of the Peace Corps?

Fill in the Blank

1. _____ was the first exported agricultural product.

2. International trade allows a country to _____ in the production of those products it best produces.

3. _____ and _____ _____ among nations permit higher living standards and real income than would otherwise be possible.

4. More than _____ _____ full-time jobs are related to agricultural exports.

5. For every dollar of agricultural exports sold by the United States, _____ dollars was actually added to the U.S. economy.

6. The consequence of abandoning the export market would be a reduction in production of about _____ percent of wheat, rice, and soybeans and _____ percent in feed grains.

7. The United States exports approximately _____ percent of the wheat crop, _____ percent of the soybean crop, percent of the corn crop, and about _____ percent of rice crop.

8. _____, _____, and _____ are the top countries (sources) of agricultural products imported into the United States.

9. The international trade balance deficits can be corrected in the long run by _____ imports and _____ exports.

10. Through 1995, there have been _____ consecutive years of agricultural trade surpluses.

11. _____ helps promote trade of U.S. agricultural products, works to improve trade conditions through the world, and monitors agricultural production in other countries.

12. The Peace Corps was an idea proposed by _____ .

Matching

a. 25 million

b. Peace Corps

c. Germany

d. 10 percent

e. 23 percent

f. 15 percent

g. 26 billion

h. 600,000

i. USSR

j. 5 percent

k. 1 billion

l. free trade area

m. customs union

n. common market

o. NAFTA

p. GATT

q. CGIAR

r. Japan

_____ 1. percent of farm income derived from exports

_____ 2. without exports, number of jobs lost in agribusiness input sector

_____ 3. without exports, number of jobs lost in agribusiness output sector

_____ 4. percent of total exports which were agricultural

_____ 5. became U.S.'s first $10 billion agricultural customer

_____ 6. amount of agricultural products imported annually to the United States

_____ 7. percentage of total imports that are agricultural products

_____ 8. approximate amount of annual U.S. agricultural trade surplus (in dollars)

_____ 9. agreement created to establish free trade between the United States, Canada, and Mexico

_____ 10. an economic and political organization between two or more countries abolishing trade restrictions among themselves and establishing a common and uniform tariff to outsiders

_____ 11. worldwide group of research centers that support agriculture

_____ 12. volunteers spend two years working with people in other countries

Activities

1. Prepare a report (research paper) on a major international trade agreement that influences the marketing of U.S. agricultural products.

2. Read and study both free trade and protectionism. Select which you believe and defend it by presenting the agreement(s) to your classes.

3. Write a one-page essay supporting or not supporting the opinion as to whether U.S. agriculture should feed a hungry world. Present this paper to your class.

4. Is a negative U.S. trade balance detrimental to the country? Explain.

5. Prepare a paper as to whether you are for or against NAFTA. Read articles from various agriculture magazines and other sources. Present your paper to the class or debate the topic.

6. Go to your local grocery and identify as many products as you can that are imported. If possible, list the country of origin.

7. Explain comparative advantage. In what commodities does the United States have a comparative advantage? Can America have a comparative advantage in the production of every commodity? Explain.

8. Prepare a one page essay on how the arguments for and against free trade relate to the law of comparative advantage.

Notes

1. Larry B. Martin, ed., *U.S. Agriculture in a Global Economy* (Washington, D.C.: U.S. Government Printing Office, 1985), p. 21.

2. Gail L. Cramer, Clarence W. Jensen, and Douglas D. Southgate, Jr., *Agricultural Economics and Agribusiness,* 7th ed. (New York: John Wiley & Sons, 1997), pp. 37–38.

3. Ibid.

4. Ibid., pp. 465–467.

5. Ibid.

6. Martin, *U.S. Agriculture in a Global Economy*, p. 25.

7. Adam Smith, *The Wealth of Nations*, quoted in Randall D. Little, *Agriculture and Agribusiness* (Danville, Ill.: Interstate Publishers, 1998), pp. 497–498.

8. John B. Penson, Jr., Rulon D. Pope, and Michael L. Cook, *Introduction to Agricultural Economics* (Englewood Cliffs, N.J.: Prentice-Hall, 1986), p. 475.

9. Ibid., pp. 476–478.

10. Ibid, p. 478.

11. Cramer, Jensen, and Southgate, Jr., *Agricultural Economics and Agribusiness*, p. 464.

12. Martin, *U.S. Agriculture in a Global Economy*, p. 27.

13. Ibid., p. 28.

14. Dominick Salvatore, *International Economics*, 5th ed. (New York: MacMillan Publishing Co., 1995), p. 362.

15. Cramer, Jensen, and Southgate, *Agricultural Economics and Agribusiness*, p. 470.

16. Ibid.

17. Randall D. Little, *Economics: Applications to Agriculture and Agribusiness* (Danville, Ill.: Interstate Publishers, 1997), p. 504.

18. *Advanced Agribusiness Management and Marketing*, 8735B (College Station, Tex.: Instructional Materials Service, 1990), p. 235.

19. Penson, Pope, and Cook, *Introduction to Agricultural Economics*, p. 499.

20. Ibid.

21. Ibid.

22. Little, *Economics*, p. 504.

23. Ibid., p. 505.

24. Penson, Pope, and Cook, *Introduction to Agricultural Economics*, p. 24.

25. Ibid.

26. *Introduction to World Agricultural Science and Technology*, 8379B-2 (College Station, Tex.: Instructional Materials Service, 1991), p. 22.

27. Little, *Economics*, p. 503.

28. Mordechai E. Kreinin, *International Economics: A Policy Approach*, 2nd ed. (New York: Harcourt Brace Jovanovich, 1975), p. 307.

29. *Advanced Agribusiness Management and Marketing*, pp. 236–237.

30. Philip Raworth and Linda C. Reif, *The Practitioner's Deskbook Series: The Law of the WTO* (New York: Oceania Publications, 1995).

31. Cramer, Jensen, and Southgate, *Agricultural Economics and Agribusiness*, p. 462.

32. Little, *Economics*, p. 512.

33. Marcella Smith, Jean M. Underwood, and Mark Bultmann, *Careers in Agribusiness and Industry* (Danville, Ill.: Interstate Publishers, 1991), p. 340.

34. Ibid., p. 339.

35. Ibid., p. 340.

36. Ibid., pp. 341–342.

37. Gerald T. Rice, *Peace Corps in the 80's* (Washington, D.C.: Office of the Peace Corps, Public Affairs, 1985), p. 6.

38. Smith, Underwood, and Bultmann, *Careers in Agribusiness and Industry*, p. 343.

Agrimarketing Channels

OBJECTIVES

After completing this chapter, the student should be able to:

- describe the historical evolvement of agrimarketing channels
- discuss assemblers of agricultural products
- discuss agricultural commodity processors and manufacturers
- explain agribusiness wholesaling
- explain agribusiness retailing
- analyze the food service industry
- explain the related and contributory services to the marketing channels
- analyze the careers in agrimarketing

TERMS TO KNOW

agricultural commodities	convenience stores	specialized food stores
agricultural products	extended care	standardization
agrimarketing channels	grading	superettes
assemblers	kilns	supermarket
blanching	merchandising	superstores
broilers	millinery	vending
chain store	noncommercial food	warehouse/limited-assortment
commercial food	establishment	supermarket
establishment	perishable	wholesalers

INTRODUCTION

Former commissioner of agriculture for the State of Tennessee William Wayne Walker III claimed that while he was in office, 50 percent of the workforce was employed because of the agricultural industry. Although the statistical reporting service does not substantiate this claim, former Commissioner Walker based his claims on the amount of people employed in the agribusiness input sector, production agriculture, and especially the agribusiness output sector.

The agribusiness output sector, which includes the marketing channels and those jobs and careers that provide support to the **agrimarketing channels**, employ millions of workers. Few outside the agricultural industry realize just how many jobs exist because of agriculture. The primary objective of this chapter is to explore the jobs and career opportunities that are available because of the marketing channels of **agricultural commodities** and **agricultural products**.

Agrimarketing channels are the path that an agricultural commodity follows from the "farmer's gate to the consumer's plate."[1] The length of this path depends on the product. For example, pick-your-own strawberries have a very short path: producer to consumer. However, the market channel for most food products is as follows: producer—assembler—food processor—food wholesaler—food retailer—consumer.[2] Jobs and careers that provide support to these marketing channels include storing, warehousing, financing, insuring, risk taking, transporting, promoting, advertising, communicating, inventorying, packaging, labeling, merchandising, selling, distribution, inspecting, and regulating. As you can see, Commissioner Walker could substantiate his claims if all the facts were known. Refer to Figure 16–1.

HISTORICAL EVOLVEMENT OF AGRIMARKETING CHANNELS

In the beginning years of this nation, most Americans were farmers producing food and fiber mainly for home consumption. They did more than just produce it. They grew or fabricated most of the inputs needed to produce these products. They also processed farm products into a form that could be consumed: grain into flour, flour into bread, fibers into cloth, hides into leather. They packaged their produce to meet consumers' needs. They cured the meat and processed the milk so it could be stored for future consumption. They took their produce to the village or to loading docks for export. They sought out buyers and saw to all of the financial matters involved in transferring ownership of their produce. All of these activities—beyond growing the basic raw materials—added value to the farm commodities.[3]

Evolvement of Specialists

As farmers and nonfarmers alike developed and applied new technologies, and as people carved out areas of specialization, it became physically impossible for farmers to perform some of these functions, and economically infeasible for them to perform others.

Specialists evolved to provide building supplies, machinery, and tools; to process and package food; to transport the goods; and to buy and sell farm produce. They were not only more efficient at the tasks, they could also capitalize on the economies of large-scale operations. Farmers also became better at producing the raw material, since they no longer had to divide their skills between farming and a host of other activities.[4]

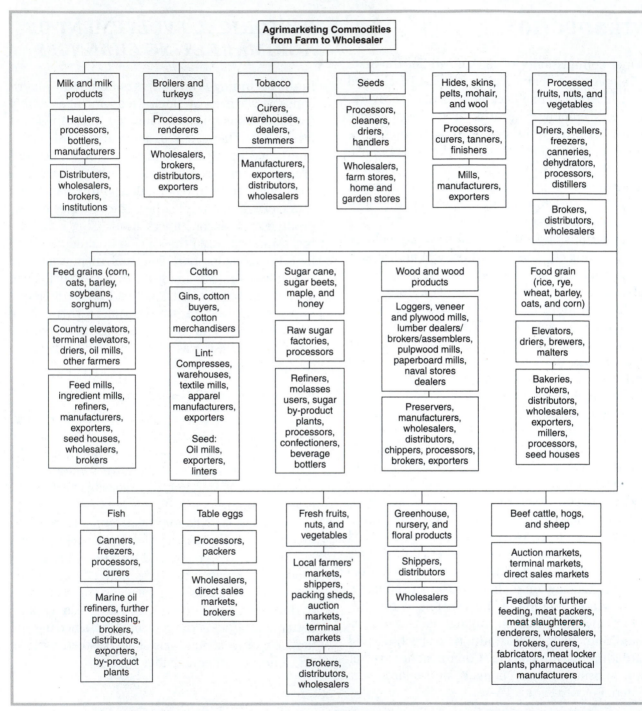

Figure 16–1 Many jobs and processes make up the agrimarketing channels as foods travel from the producer to the consumer. This chart stops at the wholesaler. Can you complete the chart as the food continues to the consumer?

The Farmers' Share of the Consumer's Dollar Falls

In the early days, farmers captured almost all of the consumer's food dollar, but the food was expensive because producers were not efficient at providing all of the marketing functions in addition to producing the raw material. Today, production agriculturalists capture less than 30 cents of the consumer's food dollar, but collectively the price of food and all of the services that the marketing sector adds is less than it otherwise would be.[5]

Both Producer and Consumer Gain

In the end, everyone gains. Consumers have access to a greater quantity and variety of products at a lower unit cost; production agriculturalists can sell more produce at a higher unit price; and the agrimarketing sector can employ more people, because the increased demand for food and fiber (and the related services) requires more workers.

Assemblers of Agricultural Products

Over 10,000 **assemblers** can be found between the production agriculturalist and the processor.[6] Assemblers do not usually change the form of a product. They put small lots of a commodity together to provide a larger, more economical unit. This permits the processors to capture more of the economies of size possible in shipping and plant operation. For grain, this function is performed by assemblers such as the local elevators. Refer to Figure 16–2. For livestock, this function is performed by a commission buyer who, for a fee, buys animals from individual producers on behalf of a processor. Other examples include: vegetable packing sheds, livestock sale barns, transporters that go from one farm to another, such as bulk milk trucks that pick up raw milk on the farm and egg-sorting plants. A further explanation of assemblers follows.

Resident buyers are individuals who either own or operate local facilities through which they buy products. These buyers may work as representatives of other firms or may be representing themselves.

Order buyers serve as buyers for chain stores or large wholesale firms. They pay the producer for the kind and quality of produce that customers demand. The buyers are usually interested in livestock. The buyer, himself, may be employed by a firm or paid on a commission basis.

Traveling buyers travel from region to region as crops mature. They buy the commodity directly from the production agriculturalist in truck lots or boxcar lots. They may be representatives of terminal buyers for local shipping points or may operate independently.

Auction markets are locations farmers transport a commodity to, and have buyers bid for the commodity. Livestock, fruits and vegetables, and tobacco are examples of types of auction markets available to production agriculturalists. Over 90 percent of the domestically grown tobacco is marketed through auctions, and the amount of other products sold is substantial.

Having consumers as assemblers places the production agriculturalist next to the consumer. The consumer buys the commodity directly from the producer. Consumers see this as a way to purchase fresher, higher quality foods at a more economical price and in a convenient and friendly atmosphere. Many production agriculturalists see consumers as assemblers (direct marketing) as a means of increasing income. It is estimated that there are over 13,000 consumers as assemblers or direct marketing outlets including roadside stands, pick-your-own, and farmers' markets.[7]

As you can see, the role of the assembler is to increase the efficiency of the marketing system by getting "the right quality of products, at the right time, at the right price, at the right place, and in the right quantity to meet the needs of the processor."[8]

Figure 16–2 Grain elevators are an example of food assemblers in the agrimarketing channel. (Courtesy of USDA)

AGRICULTURAL COMMODITY PROCESSORS AND FOOD MANUFACTURERS

Although we have used the words commodity and products interchangeably through this book, food manufacturers differentiate the terms. A commodity is an agricultural crop, livestock, milk, etc. which comes off the farms in its raw state. Once the raw commodity has had its initial form altered by a processor and further changes its form it becomes a food product.[9] For example, milk is a commodity, whereas cheese is a product.

Technology in Commodity Processing and Food Manufacturing

The commodity-processing and food-manufacturing industries owe their existence to advancements in technology. New technologies permit better utilization of existing food supplies by extending the useful life of **perishable** food items. Many of the advances in technology came as a result of efforts to develop low-cost, high-volume commodity pro-

cessing and food-manufacturing procedures. The result was the development of a separate processing-manufacturing sector in agribusiness.[10]

Size, Volume, and Structure

Size. There are approximately 22,000 firms in the commodity processing and food manufacturing sector. They add over $202 billion of value to the items they handle. About half of the sector's output comes from the 100 largest firms.[11] They include such well-known companies as General Foods, General Mills, Kelloggs, Pillsbury, Sunkist Growers, Land O'Lakes, and a host of others.

Volume. The food processing sector in the United States was very profitable in the early 1990s. Processed food shipments, which total about $395 billion annually, account for about 13 percent of all U.S. manufacturing activity and represent the largest sector in the economy.[12]

American consumers have over 230,000 packaged food products from which to choose. Including new size introduction, over 16,000 new grocery products have been introduced in some years. However, industry estimates put the failure rate of new food products at 90 to 99 percent.[13]

Structure. The structure of the food processing industries has changed considerably. Many smaller processors have sold out because they were unable to capitalize and manage needed changes. There were over 1,250 sales of businesses between 1982 and 1992. Others have merged with stronger companies to survive. Almost 2,900 mergers took place during this time. The 50 largest food processors control 47 percent of sales.

Competition. Despite the increased size of the larger firms, competition among the 16,000 firms in the 49 food processing industries is intense. Even in periods of bad economic times, food processors use price and nonprice competition to gain both consumer acceptance and retail shelf space in the

$160 billion name-brand retail food market and in the $240 billion food-processing market.[14]

Processors

The processor is normally the first one in the marketing system to alter the form of a raw agricultural commodity. Processing involves many different activities. It includes all of the changes that occur as the product is prepared for consumption. There are eleven principal types of food processors: canned food, frozen food, sugar, beverage, meat, dairy, bakery, grain mill, confectionery, paper, and cotton. These account for over 20,000 plants. A brief discussion of some of these areas follows:

Beef cattle processing involves slaughtering the animal and preparing it into the desired cuts or food items. In some cases, the carcasses or parts of the carcasses may be made into canned or frozen food products. Hamburger, steak, wieners, and ribs all require different processing procedures. Refer to Figure 16–3.

Potato processing varies widely. Some potatoes are bagged for sale through supermarkets as whole potatoes. Other potatoes undergo considerable processing, such as potato chips and instant mashed potatoes.

Cotton processing involves separating the lint (fiber) and seed, weaving the fiber into cloth, and constructing garments or other products from the cloth.

Wheat processing includes milling into flour and baking the desired products. There are many steps along the way to the consumer that ensure a quality, wholesome product.

Milk is *pasteurized* (heated to a specified temperature to kill bacteria) and *homogenized* (treated to disperse fat particles throughout the milk to keep the cream from rising to the top). It is carefully packaged and stored to maintain its quality. It is placed in cartons of convenient sizes for the customer's use. Milk must be stored in refrigerated areas.

Poultry undergoes considerable preparation for the consumer. Processing involves several steps in slaughtering the animal, including removing the feathers and internal organs, preparing the desired cuts, and packaging in a variety of ways for the consumer. Fresh poultry must always be stored under refrigeration or in frozen form.

Canning and freezing fruits and vegetables involves cleaning, peeling, cutting, shelling, breaking (beans), **blanching**, cooking, canning, freezing and other processing depending upon the commodity. Warehousing is a critical factor for canning and freezing plants since they cannot immediately sell all of what they have canned.

Paper is processed from wood. Principal paper products include paper for printing, fine and coarse paper, industrial and tissue paper, container boards, cardboard boxes, and construction paper. About 500 pounds of paper per person is used each year, requiring the net annual wood growth from about 3/4 of an acre of commercial forest. A large New York newspaper uses the equivalent of the yearly growth from 6,000 acres of commercial forest land for its Sunday edition alone.[15]

Figure 16–3 Processing is part of the agrimarketing channel. The workers shown here are producing beef into hamburger patties. (Courtesy of Oklahoma Department of Agriculture)

Some of the large and better-known processors include Cargill (grain and meat), IBP (meat), Central Soya (oilseeds), AMPI (dairy), and Monfort (beef).

Food Manufacturers

Food manufacturers continue the job started by the processor by adding value to crops and livestock by increasing the level of preservation, convenience, and quality. Food manufacturing includes firms that bake bread, cookies, and crackers; firms that manufacture prepared breakfast cereals; firms that manufacture ice cream; and so on. In each case the firm takes a raw agricultural commodity (wheat, corn, milk) that has had its initial form altered by a processor (turned into flour, processed corn, pasteurized milk) and further changes its form into a manufactured food product (bread, corn flakes, ice cream) that is desired by consumers.

In this process the raw agricultural commodities (wheat, corn, milk) completely lose their identity and emerge from the food manufacturing process as part of a food product. The commodities no longer exist but are now part of the ingredients found in a food product such as corn flakes—a highly differentiated, branded, manufactured food product. This transformation from raw commodity to food product is a major breaking point in the agribusiness marketing system.[16] Some of the largest and best-known food manufacturers are Pillsbury, General Mills, Coca-Cola, and H.J. Heinz.

AGRIBUSINESS WHOLESALING

The transfer of food and fiber from processing and manufacturing plants to **wholesalers** and then to retailers and other establishments is an important link in the agrimarketing channels. Wholesalers are operators of agribusinesses engaged in the purchase, assembly, transportation, storage, and distribution of groceries and food products for sale to retailers, institutions, and business, industrial, and commercial users.

Food wholesalers buy railroad car or truckload quantities of food products for delivery to their warehouses where they break these larger-sized lots into smaller units of pallet or case size for shipment to individual food retailers and food service establishments. Refer to Figure 16–4.

Size and Volume of Sales

There are over 476,000 wholesale business establishments with annual sales totaling over $2,525 trillion. Employment exceeds 6.3 million persons, with an annual payroll of over $181 billion. Wholesaling grocery and farm-related raw materials involves over 880,000 employees in some 54,700 firms with $500 billion of sales annually.[17]

Figure 16–4 Wholesaling is part of the agrimarketing channel. The grain is being stored in the unit shown here. It can then be divided into smaller units for shipment to individual food retailers and food service establishments. (Courtesy of Oklahoma Department of Agriculture)

Trend toward Wholesale-Retail Affiliation

At one time the wholesaler maintained the major economic position in the food marketing channel. The large number of small independent retailers were dependent on wholesalers to stock the shelves, provide credit for retail inventories and numerous other functions. On the other hand, processors looked to wholesalers as the major outlet for their products. However, the economic power began to decrease with the development of the food retail chain stores.

Chain stores began to bypass independent wholesalers and deal directly with processors. The large volume handled allowed retailers to provide their own warehousing facilities. Processors, on the other hand, began to increase in size and provide their own wholesaling. To survive, independent wholesalers aligned with small independent retailers who were also being squeezed by the retail chain store development. Some marketing specialists contend that the trend toward wholesale-retail affiliation is the most significant development in food wholesaling in recent years.

Reasons Why Wholesalers Exist

Food wholesalers exist in the agrimarketing food channel because they provide food retailers with a range of services involving credit, savings, and variety.

Credit. Wholesalers give retailers a wide assortment of products and in some cases provide short-term credit.

Savings. By purchasing large quantities of single products, wholesalers obtain price savings in the form of volume discounts that retailers would be unable to obtain for themselves.

Variety. Wholesalers maintain a variety of products and inventory levels that the retailer could not sustain.[18]

Types of Wholesalers

Wholesalers are classified in many ways, depending on the interest of the classifier. The most common methods of classification are based on economic structure, the number of functions performed, the variety of goods they handle, and their affiliation with retailers. It should be obvious that any given wholesaler may fit several of these classifications at one time. Based on economic structure, wholesalers are normally classified as (1) merchant wholesalers, (2) manufacturer's sales branches, and (3) agents and brokers.

Merchant Wholesalers. Merchant wholesalers take title to the goods that they handle. This is the largest group of wholesalers, numbering over 50,000 in the food industry with total sales over $100 billion annually. These agribusiness wholesalers account for about 30 percent of all merchant wholesalers in the United States and handle about 50 percent of all sales volume. These statistics suggest that food wholesalers are larger than wholesalers of most other products.

Merchant wholesalers can be divided into several types.

General-line wholesaler merchants handle a broad line of dry groceries, health and beauty aids, and household products.

Limited-line wholesaler merchants handle a narrow range of dry groceries dominated by canned foods, coffee, spices, bread, and soft drinks.

Specialty wholesale merchants handle perishables, such as frozen foods, dairy products, poultry, meat, fish, fruit, and vegetables.

Wholesale clubs are hybrid wholesale-retail establishments selling food, appliances, hardware, office supplies, and similar products to their individual and small-business members at prices slightly above wholesale.

There was a rapid increase in the number of wholesale clubs in the late 1980s and early 1990s.

This growth intensified competition between wholesale clubs and between wholesale clubs and traditional supermarkets. Existing supermarkets responded to the increase in wholesale clubs by offering their own club packs offering unadvertised specials, featuring attractive bakery and restaurant (deli) items, and putting special emphasis on the various services supermarkets provide that wholesale clubs do not.[19]

Manufacturers' Sales Branches.   Manufacturers' sales branches are extensions of food processing firms' marketing activity at the wholesale level. They are owned and operated by food manufacturers, and they provide a wide variety of services. They are the second largest group based on number, and they usually deal in larger volume than merchants or brokers. A major portion of all food manufacturers' sales branches are controlled by companies that have more than twenty-five units in different locations.

Agents and Brokers.   Agents and brokers are wholesalers who do not take title to the goods they handle. The ownership of the goods is retained by the clients. Therefore, agents and brokers have no price risks in their business. This type wholesaler tends to be smaller than other types, and many are highly specialized, although some do handle a wide variety of goods. Also, agents and brokers deal more with other wholesalers than retailers. Refer to Figure 16–5.

AGRIBUSINESS RETAILING

Agribusiness retailers include those businesses selling groceries, prepared foods, soft drinks, floral products, clothing, shoes, furniture, home furnishings (from agriculturally derived products), and other products. There are over 588,000 retail food outlets in the United States.[20]

The history of food retailing in this country is marketed by three major developments. The first was the development of the retail food chain store. The second was the development of the supermarket concept.[21] A third development is the superstore and **warehouse/limited-assortment supermarket**. Chain stores will be discussed next. Superstores and warehouse limited-assortment will be discussed within the types of grocery stores.

Chain Stores

A **chain store** is defined as the operation of eleven or more stores under a single owner. Grocery chain stores now account for about 65 percent of food store sales, up from 37 percent in 1948 and 25 percent in the 1920s. The chain store's share of total grocery store sales was consistent through the mid-1980s.[22]

The chain stores revolutionized the way food was sold at retail. Their lower prices and wider selection led to a decline in the family-run, single-outlet food stores, which are sometimes called "ma and pa" stores. Some of the larger retail food chains include Safeway, Kroger, and A&P. The largest chains operate as many as 5,000 stores. In addition to these, there are many other chains that operate fewer stores on a local or regional basis.

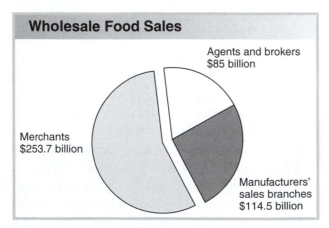

Wholesale Food Sales

Agents and brokers $85 billion

Merchants $253.7 billion

Manufacturers' sales branches $114.5 billion

Figure 16–5   Food wholesalers are usually classified as either merchant wholesalers, manufacturer's sales brokers, or agents and brokers.

Types of Grocery Stores

The retail grocery store is near the end of the marketing channel where agricultural products, as well as other products, are purchased by the consumer. There are about 165,000 grocery stores with annual sales of about $360 billion and employment of about 3.2 million people.[23]

Retail grocery stores are usually classified into three types: supermarkets, convenience stores, and superettes. Refer to Figure 16–6.

Supermarkets. Supermarkets offer the customer a full line of 10,000 to 15,000 items, plus many nonfood items such as household cleaners, health and beauty aids, and so on.

Interestingly, it was not the chain food stores that first adopted the supermarket concept. Supermarkets were started by several independent "ma and pa stores" as a way to compete with the national chains. They were so successful that it became hard for the chains to ignore them. The national chains finally accepted the concept and by the mid-1950s the **supermarket** method of food retailing was firmly established.[24]

Supermarkets represent just under 10 percent of all retail food stores but account for almost 72 percent of total grocery sales.[25] Two key trends affecting supermarkets are superstores and warehouse/limited-assortment supermarkets.

Superstores offer a greater variety of products than conventional supermarkets. Superstores account for 25 percent of all supermarkets resulting in 33 percent of supermarket sales. *Warehouse/limited-assortment supermarkets* offer larger sizes of fewer items at lower prices than the typical supermarket. These nonconventional formats (superstores and warehouse/limited-assortment supermarkets) presently account for more than 50 percent of all supermarkets resulting in approximately 66 percent of supermarket sales.[26]

Convenience Stores. **Convenience stores** were developed in the late 1950s starting in the South and West. To some extent, they filled the role in the expanding suburbs of the "ma and pa" grocery stores in older communities. A number of them started as dairy stores, with milk accounting for as much as 40 to 50 percent of sales. These helped to fill the niche of home delivery of milk, which was declining. With skyrocketing gasoline prices in the 1970s and 1980s, many convenience stores added self-service gasoline pumps. More

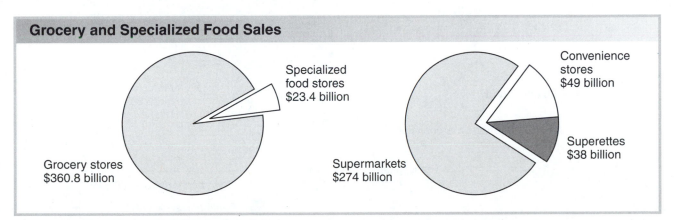

Grocery and Specialized Food Sales

Specialized food stores $23.4 billion

Grocery stores $360.8 billion

Convenience stores $49 billion

Superettes $38 billion

Supermarkets $274 billion

Figure 16–6 Grocery stores are usually classified as either supermarkets, convenience stores, or superettes. Specialized food stores, which make up about 6.5 percent of food sales, includes candy and nut stores, dairy stores, produce markets, meat and fish markets, retail bakeries, and miscellaneous food stores.

recently, carry-out foods, including hot sandwiches, and in-store eating have become staples of convenience stores.[27]

Few people buy all their weekly groceries at convenience stores, but just about everyone stops at them occasionally for one or two items. Many convenience stores are part of a chain. The largest chain is the 7-Eleven, Inc., which operates 7-Eleven stores.[28]

Convenience stores make up about 20 percent of retail stores, accounting for 12.6 percent of grocery

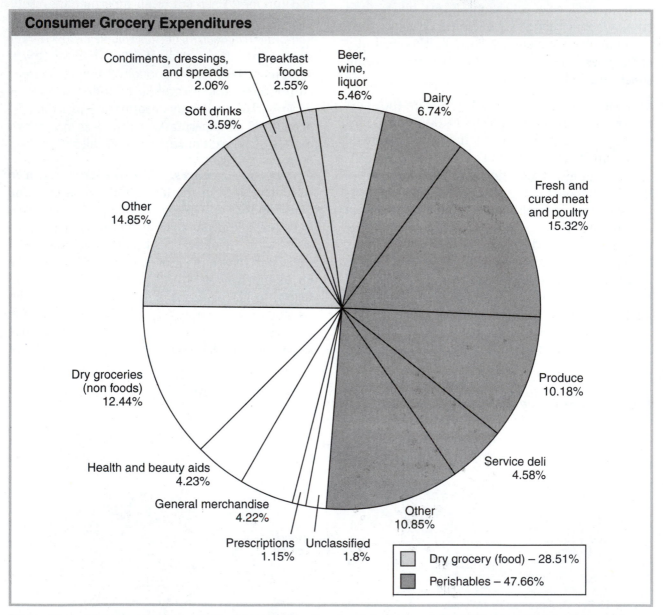

Consumer Grocery Expenditures

Condiments, dressings, and spreads 2.06%
Breakfast foods 2.55%
Beer, wine, liquor 5.46%
Dairy 6.74%
Soft drinks 3.59%
Fresh and cured meat and poultry 15.32%
Other 14.85%
Dry groceries (non foods) 12.44%
Produce 10.18%
Health and beauty aids 4.23%
Service deli 4.58%
General merchandise 4.22%
Other 10.85%
Prescriptions 1.15%
Unclassified 1.8%

Dry grocery (food) – 28.51%
Perishables – 47.66%

Figure 16–7 Do you ever wonder what percentage of sales of various products are sold in grocery stores? If so, this graph should help.

sales. Lately, traditional convenience stores have gotten increased competition from gasoline stations that have added convenience stores. These gasoline stations are not counted as food stores because food product sales make up less than half of total sales.[29]

Superettes. **Superettes** and "ma and pa" stores represent about 37 percent of all retail food stores accounting for 10 percent of sales. These smaller grocery stores, which offer a variety of food and nonfood grocery products, are usually located in areas not adequately served by supermarkets, such as in densely populated urban areas or sparsely populated rural areas.[30]

Specialized Food Stores. **Specialized food stores** make up 6 percent of retail grocery sales.[31] They usually sell a single food category. Examples include retail bakeries; meat and fish markets; candy and nut stores; natural and health food stores; coffee, tea, and spice stores; and ice cream stores. Figure 16–7 shows the percentage of sales of various products sold in grocery stores.

Other Agribusiness Retailing

Besides food sales, agribusiness retailing also involves the sales of alcoholic beverages, tobacco products, clothing, shoes, **millinery**, furniture and home furnishings, and floral products. These products are mainly derived from grain, tobacco, cotton, wool, mohair, hides, skins, feathers, wood, nursery, flower and ornamental plants.[32]

FOOD SERVICE INDUSTRY

One of the fastest-growing segments of agribusiness is the food service industry: hotels, restaurants, and institutions. It includes all those people and firms involved in serving food at hotels, restaurants, fast-food outlets, schools, hospitals, prisons, military installations, and other away-from-home eating locations. Sales at these outlets represent approxi-

mately 33 percent of all the money spent on food at the retail level, and the percentage is rising.[33]

Growth and Volume of the Food Service Industry

In 1954, before the federal interstate highway system paved the way to suburbia, and before Ray Kroc opened his first McDonald's, the U.S. food service industry was a $15–billion-a-year business. Except for the rich, Americans generally ate out only when they were truly away from home, on business trips or on vacation, or for a very special occasion. Only 195,000 restaurants were in the country at that time, and food service accounted for only 25 percent of total consumer spending for food.

Today, eating out is part and parcel of our lives, more a necessity than a luxury, and food service accounts for more than 50 percent of total food expenditures. The food service industry has increased its share of market relative to its traditional rival; the supermarket or retail grocery. It has expanded to become over a $260 billion industry, with 600,000 outlets all across America. It employs approximately 8 million people.[34]

Percentage of Selected Foods Consumed in the Food Service Industry

Everyday, approximately 100 million Americans, including 42 percent of the total U.S. adult population, eat out at least once. Over the course of a year, more than 78 billion meals and snacks are served in the food service industry. This industry uses more than 40 percent of all the meat, 55 percent of all the lettuce, 60 percent of all the butter, 65 percent of all the potatoes, and 70 percent of all the fish produced in the United States.[35]

Reasons for Growth in the Food Service Industry

This growth reflects a change in American lifestyles. Increasing numbers of households include a husband and wife who both work, or a single head of household who must work. There is simply less time for these people to spend on food shopping, preparation, and cleanup. People are willing to substitute money for time by eating away from home. The population is also more mobile. This leads to more people being away from home at mealtime; and even if they are at home, many like to drive to a restaurant to eat. These factors have combined to transform eating out from a luxury to a necessity for many Americans.[36]

Types of Food Service Establishments

Food service establishments are divided into two major categories: commercial and noncommercial.

Commercial.   **A commercial food establishment** is open daily to the general public. They are stand-alone buildings or part of a host establishment such as a hotel or shopping mall. They prepare meals and snacks, serve and sell them for profit to the general public. Commercial establishments include: fast food, restaurant, lodging places, retail host, recreation, cafeterias, drinking places, and social caterers. Commercial food service sales of meals and snacks account for nearly $191 billion which represents approximately 76 percent of food service sales.[37] Refer to Figure 16–8.

Fast foods are familiar to everyone. Three of the larger fast-food establishments are Pizza Hut, Kentucky Fried Chicken (KFC) and McDonalds. Agribusiness corporations are owners of many fast-food establishments. Examples are Ralston Purina's operation of the Jack-in-the-Box fast-food chain for a number of years; Pillsbury's ownership of Burger King and Steak-n Ale; and Pepsico's operation of Pizza Hut, Taco Bell, and Kentucky Fried Chicken.[38]

Full-service restaurants, where waiters or waitresses serve full meals of breakfast, lunch, and dinner to their customers, have long been a staple of the away-from-home food industry. The secrets to a restaurant's success have always been good food served in a pleasant environment, with good service and reasonable prices.

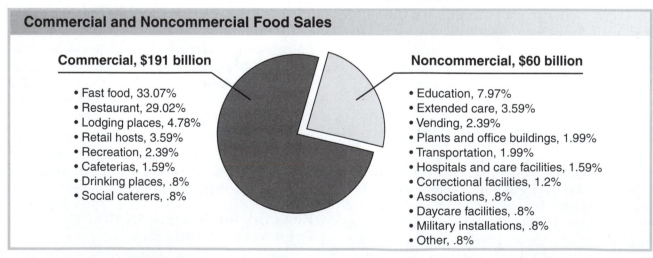

Commercial and Noncommercial Food Sales

Commercial, $191 billion
- Fast food, 33.07%
- Restaurant, 29.02%
- Lodging places, 4.78%
- Retail hosts, 3.59%
- Recreation, 2.39%
- Cafeterias, 1.59%
- Drinking places, .8%
- Social caterers, .8%

Noncommercial, $60 billion
- Education, 7.97%
- Extended care, 3.59%
- Vending, 2.39%
- Plants and office buildings, 1.99%
- Transportation, 1.99%
- Hospitals and care facilities, 1.59%
- Correctional facilities, 1.2%
- Associations, .8%
- Daycare facilities, .8%
- Military installations, .8%
- Other, .8%

Figure 16–8   Commercial food sales account for approximately 76 percent of all food sales, and noncommercial food sales account for the remainder.

Lodging places are establishments that provide both lodging and food service to the general public. These include hotels, motels, and tourist courts.

Retail hosts are food service operations that operate in conjunction with, or as part of, retail establishments, such as department stores, drugstores, and other miscellaneous retailers.

Recreation places include food service operation in theaters, in bowling alleys or billiard halls, in commercial sports establishments, in country clubs, at public golf courses, and in miscellaneous amusement and recreational establishments.[39]

Cafeterias are similar to restaurants except most include a walk through line to select salads, bread, vegetables, meats, deserts, and beverages. Cafeterias as a rule tend to be less formal than restaurants.

Drinking places are those establishments that primarily service alcoholic beverages for on-premise or immediate consumption but that also provide food service. These include bars, taverns, nightclubs, and saloons. At 57.5 percent, beer sales are the primary component of alcoholic beverage sales, followed by distilled spirits at 30.9 percent, and wine at 11.6 percent.[40]

Social caterers bring the food on site for special occasions such as weddings, graduation parties, company picnics, and so forth.

Noncommercial Establishments. Noncommercial food establishments are those in which meals and snacks are served as a supportive service rather than as the primary service. These include education, extended care, vending, plants and office buildings, transportation, hospitals and care facilities, associations, day care facilities, military installations, and others. Noncommercial food service accounts for approximately $60 billion, or 24 percent of total food service sales.[41]

Taste Trends and New Operating Styles

It is hard to say whether the food service industry really adapts to meet existing consumer needs, or if it often anticipates the needs and wants of the consumer and creates new markets where none existed before.

Salad bars, the away-from-home breakfast market, the frozen pizza market, popular Mexican foods, and more recently, gourmet cookies, Cajun cooking, and pasta—in all these cases, was the industry truly responding to consumer demand, or was it, by its own marketing savvy and skill, devising new ways to expand the market and make eating away-from-home ever more inviting and enjoyable to consumers?[42]

Future of the Food Service Industry

Food service is an industry which, like the food industry overall, is relatively recession proof. It is driven more by employment, consumer confidence, and disposable income than by deficits, interest rates, and gross domestic product.

Despite all the economic uncertainties, the food service industry should be able to adapt and respond to this changing environment. Staying in tune with consumers always has been, and will continue to be, the strength of this industry.[43]

RELATED AND CONTRIBUTORY SERVICES

There are many functions that relate or contribute to the agrimarketing channels. These services are essentials in the movement of food and fiber from the farm to the consumer. Some of these related and contributory services include: standardizing and grading, packaging, order processing, storing, inventory control, transportation, market communications, financing, product development, market research, and regulatory services. A brief discussion of each follows.

Standardization and Grading

Standardization means the establishment of standards for matters such as quality, size, weight, and color. Once the standards are set, **grading** is possible using the criteria set forth. In food markets, for example, meat can be graded prime, choice, and good. The use of standardization and grading make for easier contractual and exchange agreements between marketers.

USDA is involved extensively in grading agricultural products. For example, in a recent year, the USDA graded approximately 37 percent of eggs, 95 percent of butter, 55 percent of frozen fruits and vegetables, 81 percent of beef, 80 percent of turkeys, 56 percent of **broilers** and other poultry, 97 percent of tobacco, and 97 percent of cotton.[44]

Packaging

Packaging serves two primary purposes: to preserve contents and to merchandise and advertise the product. With the availability of self service retailing, the packaging of a product does the majority of the selling. The color, shape, label and other characteristics entice consumers into purchasing a product many times. Packaging is a multimillion-dollar industry.

Storing

Timing is essential in the marketing process. Making goods available at the time consumers need them has a beneficial effect on prices of the commodity as well, since it allows a commodity to be marketed throughout the year rather than flooding the market at harvest time.

Unprocessed, as well as processed, agricultural products may be stored. Warehouses, bins, sheds, coolers, elevators, tanks, freezer lockers, and **kilns** are but a few of the types of storage available.

Order Processing

Order processing refers to the procedures followed when an order is received. The cost of the materials must be calculated accurately and the payment collected. Sales tickets are prepared to document a sale. Order processing also involves picking what the customer wants from the inventory and preparing it for shipment.[45]

Inventory Control

Inventory control is maintaining an adequate stock of products. Records of receipts and shipments must be kept. Inventory control also involves producing what is needed. Computer systems are used by most agribusinesses to keep inventory records.[46]

Transportation

Products have to move from where they are produced to where they are needed. This involves several loading and unloading steps including: farm to local market, local market to processor, processor to retailer. The four most common methods of transporting commodities are by trucks, railroad, boats, and air transport.

Truck Transportation. Truck transportation offers rapid movement of perishable foods, flexible schedules, lower costs, dependability, speed, and less loss in transit as its main benefits. Truck transport has increased more when compared to the other forms of hauling raw and processed foods and fibers. Grain, livestock, milk, vegetables, and fruits account for about 85 percent of the truck haulings of farm produce in the United States.[47]

Railroads. Railroads provide a means of moving large volumes of bulk commodities safely and economically over long distances. Rail service is the oldest form of general transportation. Approximately one-fifth of the railroads' profits come from transporting farm commodities. The primary commodities carried by rail are grains, animal feeds, canned goods, lumber, and wood products.

Boat Transport. Boat transport has increased on both rivers and oceans. Boat transporting is best adapted to transport of grains or the large volume bulk commodity shipping where speed is not critical.

Air Transport. Commodities that have a high value per unit are well suited for air transport. The speed and dependability of air transport make it a consideration in the shipping of certain commodities such as flowers, nursery items, seafood, and specialty crops.

Market Communications

In distributing farm supplies, agribusinesses use newspaper advertising, articles, telephone, radio, television, direct mail, fax machines, the Internet, and satellites. Production agriculturalists utilize newspapers to obtain farm stories, advertising, and general farm news. Use of the telephone and fax machines is varied and increasingly more important. Radio is relied upon for market, advertising, and farm news reports.

The U.S. Department of Agriculture maintains more than 170 year-round market news offices to furnish daily and weekly reports on market prices, supplies, and other market conditions related to all major farm commodities.[48] Also, trade associations represent and educate their members as well as inform the general public concerning trade activities, problems, and responsibilities. Refer to Figure 16–9. The communication industry is massive and hires thousands of people.

Financing

Lags between the time of production and of sale and between the time of sale and receipt of payment are reasons for financing. Financing must be provided to operate plants and perform the functions necessary to market a commodity.

Product Development

This involves research and development involving new food products and the modification of existing products. These efforts are conducted to gain the market share from competitors, develop new products or new forms of existing products, and innovating and creating new technology.

Market Research

Attempting to determine the needs and wants of consumers has become an important marketing function. If producers and marketers are to produce and market effectively, they must have accurate knowledge about their customers. Marketing research attempts to provide such information through consumer interviews, taste panels, store experiments, control displays, and other methods.[49]

Merchandising

Merchandising concerns all phases of marketing. That is, it refers to all the characteristics associated with the distribution, display, and sale of a product. The goal of merchandizing is the marketing of a commodity that is appealing and satisfies a desire or need to the consumer.

Advertising

Agrimarketing firms spend about $11 million annually on advertising. Consumers are the primary target of advertisers. Newspapers have about 24 percent of all advertising expenditures. Television, radio, and magazine advertising have 22 percent, 7 percent, and 5 percent, respectively, of all advertising expenditures. Advertising for food and food products makes up about 16 percent of all television and 7 percent of all magazine advertising expenditures.[50]

Regulatory

The USDA provides many regulatory functions as inspectors and graders as mentioned earlier. The quality of agricultural commodities is a very important consideration in agrimarketing channels. The Grade Standards Program of the USDA's Agricultural Marketing Service helps to insure that pur-

American Association of Nurserymen
1250 Eye Street, NW, Suite 500
Washington, DC 20005

American Bakers Association
1111 14th Street, NW, Suite 300
Washington, DC 20005

American Cotton Shippers Association
1725 K Street, NW, Suite 1210
Washington, DC 20006

American Dairy Products Institute
130 North Franklin
Chicago, IL 60606

American Farm Bureau Federation
600 Maryland Avenue, SW, Suite 800
Washington, DC 20024

American Feed Industries Association
1701 North Fort Myer Drive
Arlington, VA 22209

American Frozen Food Institute
1764 Old Meadow Road, Suite 350
McLean, VA 22102

American Goat Society
Rt. 2, Box 112
De Leon, TX 76444

American Meat Institute
1700 North Moore Street
Arlington, VA 22209

American Plywood Association
7011 South 19th, P.O. Box 11700
Tacoma, WA 98411

American Quarter Horse Association
Amarillo, TX 79160

American Rabbit Breeders Association
1925 South Main St., Box 426
Bloomington, IL 61701

American Seed Trade Association
1030 15th St., NW
Washington, DC 20005

American Sheep Producers Council
200 Clayton Street
Denver, CO 80206

American Soybean Association
600 Maryland Avenue, SW
Washington, DC 20024

Burley and Dark Leaf Tobacco Export Association
1100 17th St., NW, Suite 902
Washington, DC 20036

California Avocado Commission
17620 Fitch, 2nd Floor
Irvine, CA 92714

California Cling Peach Advisory Board
P.O. Box 7111
San Francisco, CA 94120

California Pistachio Commission
5114 East Clinton Way
Suite 113
Fresno, CA 93727

California Raisin Advisory Board
P.O. Box 5335
Fresno, CA 93755

California Table Grape Commission
P.O. Box 5498
Fresno, CA 93755

Corn Refiners Association
1001 Connecticut Avenue
Washington, DC 20036

Cotton Council International
1030 15th Street, NW
Suite 700, Executive Building
Washington, DC 20005

Farm and Industrial Equipment Institute
410 N. Michigan Ave., Suite 680
Chicago, IL 60611

Florida Department of Citrus
1115 East Memorial Boulevard
P.O. Box 148
Lakeland, FL

Flue-Cured Tobacco Cooperative Stabilization Corporation
Box 12600
Raleigh, NC 27605

Food Marketing Institute
1750 K St., NW, Suite 700
Washington, DC 20006

Grocery Manufacturers of America
1010 Wisconsin Avenue, NW
Washington, DC 20007

Independent Bakers Association
Box 3731
Washington, DC 20007

International Apple Institute
6707 Dominion Drive, Box 1137
McLean, VA 22101

International Ice Cream Association
888 16th St., NW
Washington, DC 20006

Millers National Federation
600 Maryland Ave., SW, Suite 305
Washington, DC 20024

Minnesota Grain Exchange
Minneapolis, MN

National Ag. Chemicals Association
1155 15th St., NW, Suite 900
Washington, DC 20005

National Association of Conservation Districts
1025 Vermont Ave., NW, Suite 730
Washington, DC 20005

National Association of Meat Purveyors
8365-B Greensboro Drive
McLean, VA 22101

National Association of Wheat Growers
415 Second St., NE, Suite 300
Washington, DC 20002

National Broiler council
1155 15th St., NW, Suite 614
Washington, DC 20004

National Cheese Institute
699 Prince Street, Box 20047
Alexandria, VA 22320

National Corn Growers Association
201 Massachusetts Ave., NE, Suite C4
Washington, DC 20002

National Cotton Council of America
1030 15th St., NW, Suite 700
Washington, DC 20005

National Council of Commercial Plant Breeders
1030 15th St., NE, Suite 964
Washington, DC 20005

National Council of Farmer Cooperatives
50 F St., NW, Suite 900
Washington, DC 20001

National Dairy Promotion Board
2111 Wilson Blvd., Suite 600
Arlington, VA 22201

National Food Processors
1401 New York Ave., NW
Washington, DC 20005

National Forest Products Association
1250 Connecticut, NW, Suite 70
Washington, DC 20005

National Hay Association, Inc.
P.O. Box 99
Ellensburg, WA 98926

National Milk Producers Federation
1840 Wilson Blvd.
Arlington, VA 22201

National Pecan Marketing Council
741 Piedmont Ave., NE
Atlanta, GA 30308

National Potato Promotion Board
1385 South Colorado Blvd., #512
Denver, CO 80222

National Pork Producers Council
1015 15th St., NW, Suite 200
Washington, DC 20005

National Turkey Federation
11319 Sunset Hills Road
Reston, VA 22090

North American Blueberry Council
P.O. Box 166
Mamora, NJ 08223

North American Export Grain Association
1747 Pennsylvania Ave., NW
Suite 1175
Washington, DC 20006

Northern Hardwood and Pine Manufacturers Association
P.O. Box 1124
Green Bay, WI 54305

Northwest Cherry Growers
1005 Tieton Drive
Yakima, WA 98902

Northwest Horticultural Council
P.O. Box 570
Yakima, WA 98907

Nursery Marketing Council
1250 Eye St., NW, Suite 500
Washington, DC 20005

Oregon-Washington-California Pear Bureau
Woodlark Building
Portland, OR 97205

Papaya Administrative Committee
First Insurance Building
1100 Ward Avenue, Rm. 860
Honolulu, HI 96814

Protein Grain Products International
6707 Old Dominion Drive
Suite 240
McLean, VA 22101

Rice Council for Market Development
P.O. Box 740123
Houston, TX 77274

Rice Millers Association
1235 Jefferson Davis Hwy., Suite 302
Arlington, VA 22202

Southern Forest Products Association
P.O. Box 52468
New Orleans, LA 70152

Sugar Association
1511 K St., NW
Washington, DC 20005

The Fertilizer Institute
1015 18th St., NW, Suite 11
Washington, DC 20036

Tobacco Associates
1101 17th St., NW
Washington, DC 20036

United Egg Producers
3951 Snapfinger Parkway, Suite 580
Decatur, GA 30035

United Fresh Fruit & Vegetable Association
727 North Washington St.
Alexandria, VA 22314

United States Beet Sugar Association
1156 15th St., NW
Washington, DC 20005

U.S. Brewers Association
1750 K St., NW
Washington, DC 20006

U.S. Chamber of Commerce
1615 11 St., NW
Washington, DC 20062

U.S. Feed Grains Council
1400 K St., NW, Suite 1200
Washington, DC 20005

U.S. Meat Export Federation
3333 Quebec St., Suite 7200
Stapleton Plaza
Denver, CO 80207-2391

U.S. Wheat Associates, Inc.
1620 I St., NW, Suite 801
Washington, DC 20006

USA Dry Pea and Lentil Council, Inc.
Stateline Office
P.O. Box 8566
Moscow, ID 83843

Washington State Apple Commission
P.O. Box 18
Wenatchee, WA 98801

Western Wood Products Association
1500 Yeon Building
Portland, OR 97204

Wine Institute
165 Post Street
San Francisco, CA 94109

Figure 16–9 Trade associates communicate to consumers about their products. Many jobs and careers are created by trade associations.

chasers of agricultural commodities receive the quality they want and for which they pay.

CAREERS IN AGRIMARKETING

The number of careers in the field of agrimarketing is massive. Many jobs have been mentioned throughout this chapter including assemblers, food processors, food wholesalers, food retailers and the food service industry. Careers are also abundant in the many related areas that contribute to the agrimarketing channels. These areas include careers in standardizing and grading, packaging, storing, order processing, inventory control, transportation (trucking, rail, boat and air transport), market communication, financing, product development, market research, merchandising, advertising, and regulation.

Space does not permit a discussion of the volume and size of each of these areas. However, there will be a brief discussion of the opportunities in the meat and livestock, feed, fruit and vegetable industries, ornamental horticulture industry, cotton industry, and dairy industry. The Career Option on pages 432–433 illustrates the magnitude of the career areas in agrimarketing.

Meat and Livestock Industry

There are over 1,400 federally inspected slaughter facilities in the United States, plus smaller plants that provide local custom slaughter and meat processing that are under state and local inspection. There are several thousands of wholesalers of meat and meat products and about 261,000 retail markets that sell meat directly to consumers. Over 15,000 of these stores specialize in meat and seafood. There are some 350,000 restaurants that market over 40 percent of the meat products sold in the United States.[51]

Feed Industry

The feed industry processes and sells to farmers and feeders some $18 billion worth of feed each year. Nationally, 14 percent of all farm costs are for formula and concentrated feeds. Much of the feed grain leaves the farm to return later as processed commercial feeds. Corn that leaves the farm could come back in the form of antibiotics, enzymes, pharmaceuticals, amino acids, molasses, or corn gluten feed or meal.

Today some 7,000 animal feed–processing plants produce over 110 million tons of feed each year, making this one of the largest manufacturing industries in the United States. This manufacturing industry employs some 100,000 workers, including local feed mill workers, custom grinders and mixers, at least 2,000 wholesale feed dealers, over 20,000 retailers, and several thousand hatcheries that sell feed.[52]

Fruits and Vegetables

The fruit and vegetable processing industries provide approximately 220,000 jobs annually. Approximately 22,000 processing establishments turned raw products into canned, frozen, dried, or specialty products valuing approximately $375 billion.[53]

The per capita consumption of fruit is approximately 217 pounds annually. The consumption of fresh fruit increased by 17 pounds in a 17 year period to a total of 98 pounds. The consumption of fruit juices has also increased. The farm value of all of the fruit and tree nuts marketed was over $8 billion and represented 11 percent of all of the agricultural crops grown in the United States.

Vegetables are also a large part of the consumer's diet. Americans consume an average of 325 pounds of vegetables annually. Of that total, consumers ate approximately 130 pounds of potatoes, sweet potatoes, and dry peas; 70 pounds of fresh vegetables; 100 pounds of canned vegetables; and 50 pounds of frozen vegetables. The value of vegetable sales to producers was over $13 billion.[54]

CAREER OPTION

Career Areas: Inspector, Quality Controller, Seller, Processor, Trucker, Wholesaler, Distributor, Produce Manager, Meat Cutter, Packer, Grader, Auctioneer, Co-op Manager

The food industry is massive and includes both plant and animal products. It includes the producers, processors, distributors, wholesalers, retailers, fast-food establishments, and restaurants. Careers in food science; store management; produce management; meat cutting; laboratory testing; field supervision; research; diet and nutrition; health and fitness; and promotion are all possibilities in the food industry.

The processing-and-packaging industry is enormous and employs a large number of individuals in the United States. Field supervisors and coordinators direct the work of crews to harvest crops at the peak of their quality and transport them to processing or packing plants. In many cases, huge packing and/or processing machines are used in the field or orchard or on boats. Quality-control personnel collect food specimens, label them, test them, and maintain records to ensure quality control on each batch of food coming out of the plant.

Excellent jobs are available in supermarkets for produce stockers and foremen. Products include fruits, vegetables, and, possibly, ornamental plants. Tasks may include inventorying, ordering, handling, stocking, and displaying produce to keep it fresh and attractive. In large cities as well as small towns, street vendors are frequently seen selling premium fruit and vegetables. Roadside stands provide opportunities for younger members of families to develop business skills while earning money for present and future needs.

Corn, wheat and other small grains, sugar cane, soybeans, sugar beets, and other specialty crops are grown on large acreages in the United States, Canada, and many nations of the world. In the United States, grain, oil, and specialty crops account for large amounts of exports and do much to help maintain our balance of payments in foreign trade. Grain brokers, futures brokers, market reporters, grain elevator operators, and others owe their jobs to these crop enterprises.

After agribusiness (food and fiber) products go through the agrimarketing channels, they end up at the consumer. Satisfaction, quality, and food safety are all goals of the individuals in agrimarketing. (Courtesy of USDA)

CAREER OPTION (concluded)

Along with crop enterprises are jobs in building and storage construction; systems engineering; machinery sales and service; welding; irrigation; custom spraying; hardware sales; agricultural finance; chemical sales; and seed distribution.

The marketing of agricultural products occurs in all segments of our society on a continuing basis. This is evidenced by the year-round continuance of bulging displays of produce, dairy, meat, seafood, bread, frozen foods, canned goods, deli foods, and ready-to-eat foods in the supermarkets.

Marketing careers may be launched through programs in the technical schools, colleges, or universities in the plant and animal sciences, food science, agricultural economics, marketing, or communications. Additionally, studies in the biological sciences are appropriate for laboratory work associated with agribusiness products and marketing.

Ornamental Horticulture

Approximately, 23,500 ornamental horticulture establishments produce cut flowers, flowering and foliage plants, bedding plants, and cultivated forest greens. The value of the products they sell is more than $22 billion annually. About 16,000 establishments sell lawn and garden supplies and nursery stock. The value of their sales exceeds $3.4 billion each year.[55]

Cotton Industry

The cotton industry takes cotton from production agriculturalists and, by many activities and services, transfers it to users in this country and throughout the world. It employs thousands of workers and affords many opportunities. There are 729,000 persons employed in textile mills and over 1,000,000 employed in the apparel industry.[56]

Dairy Industry

The dairy industry buys over 154 billion pounds of milk from producers each year. This represents over 17.9 billion gallons of milk. Each year, Americans consume an average of 224 pounds of milk and dairy products.

The industry employs over 145,000 workers along with the necessary supervisors, managers, administrators, and specialists. They work in thousands of plants, distributing centers, retail stores and offices. The gross farm income from dairy products is approximately $22 billion annually. Consumer expenditures on dairy products are approximately $66 billion, which represents 12 percent of all consumer food expenditures.[57]

CONCLUSION

Agribusiness is big business, especially in the agribusiness output sector or agrimarketing. Marketing channels define the path that a food product follows from the "farmer's gate to the consumer's plate." The future belongs to those who prepare for it. Many careers are available in the agrimarketing sector for those who are willing to apply themselves. Refer to Figure 16–10 for a list of careers and employment projections.

Distribution of Employment Opportunities for Graduates

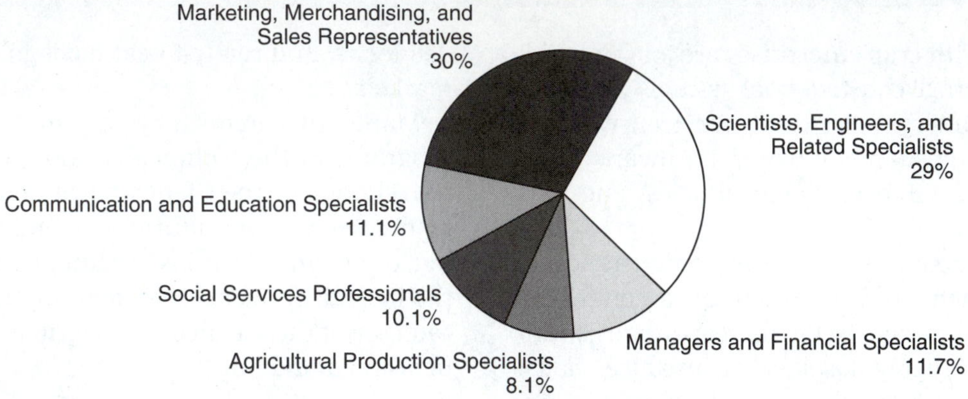

Marketing, Merchandising, and
Sales Representatives
30%

Scientists, Engineers, and
Related Specialists
29%

Communication and Education Specialists
11.1%

Social Services Professionals
10.1%

Agricultural Production Specialists
8.1%

Managers and Financial Specialists
11.7%

**Marketing, Merchandising, and
Sales Representatives**
Account Executive
Advertising Manager
Commodity Broker
Consumer Information Manager
Export Sales Manager
Food Broker
Forest Products Merchandiser
Grain Merchandiser
Insurance Agent
Landscape Contractor
Market Analyst
Marketing Manager
Purchasing Manager
Real Estate Broker
Sales Representative
Technical Service Representative

Social Services Professionals
Career Counselor
Caseworker
Community Development
Specialist
Conservation Officer
Consumer Counselor
Dietitian
Food Inspector
Labor Relations Specialist
Naturalist
Nutrition Counselor
Oudoor Recreation Specialist
Park Manager
Peace Corps Representative
Regional Planner
Regulatory Agent
Rural Sociologist
Youth Program Director

**Managers and Financial
Specialists**
Accountant
Appraiser
Auditor
Banker
Business Manager
Consultant
Contract Manager
Credit Analyst
Customer Service Manager
Economist
Financial Analyst
Food Service Manager
Government Program Manager
Grants Manager
Human Resource Development
Manager
Insurance Agency Manager
Insurance Risk Manager
Landscape Manager
Policy Analyst
Research and Development
Manager
Retail Manager
Wholesale Manager

**Communication and Education
Specialists**
College Teacher
Computer Software Designer
Computer Systems Analyst
Conference Manager
Cooperative Extension Agent
Editor
Educational Specialist
High School Teacher
Illustrator
Information Specialist
Information Systems Analyst
Journalist
Personnel Development Specialist
Public Relations Representative
Radio/Television Broadcaster
Training Manager

**Agricultural Production
Specialists**
Aquaculturalist
Farmer
Feedlot Manager
Forest Resources Manager
Fruit and Vegetable Grower
Greenhouse Manager
Nursery Products Grower
Farm Manager
Rancher
Turf Producer
Viticulturist
Wildlife Manager

**Scientists, Engineers, and
Related Specialists**
Agricultural Engineer
Animal Scientist
Biochemist
Cell Biologist
Entomologist
Environmental Scientist
Food Engineer
Forest Scientist
Geneticist
Landscape Architect
Microbiologist
Natural Resources Scientist
Nutritionist
Pathologist
Physiologist
Plant Scientist
Quality Assurance Specialist
Rangeland Scientist
Research Technician
Resource Economist
Soil Scientist
Statistician
Toxicologist
Veteranarian
Waste Management Specialist
Water Quality Specialist
Weed Scientist

Figure 16–10 Many careers are available in the agrimarketing channels and related services. (Courtesy of USDA)

SUMMARY

Marketing channels are the paths that an agricultural product follows from the "farmer's gate to the consumer's plate." The length of the path depends on the product. Pick-your-own-strawberries has a short path: producer to consumer. However, the market channel for most food products would be producer to assembler to food processor to food wholesaler to food retailer to consumer.

Early farmers took the food from production to retailing. As farmers and nonfarmers alike developed and applied new technologies, specialization became more popular. Farmers became better at producing the raw material, since they no longer had to divide their skills between farming and a host of other activities. In the early days, farmers captured almost all of the consumer's food dollar. Today, production agriculturalists capture less than 30 cents of the consumer's food dollar due to specialization and agrimarketing.

Over 10,000 assemblers can be found between the production agriculturalist and the processor. Assemblers do not usually change the form of a product. They put small lots of a commodity together to provide a larger, more economical volume.

There are approximately 22,000 firms in the commodity processing and food manufacturing sector. They add over $215 billion of value to the items they handle. The processor is normally the first party in the marketing system to alter the form of a raw agricultural commodity. Food manufacturers continue the job started by the processor by adding value to crops and livestock by increasing the level of preservation, convenience, and quality.

The transfer of food and fiber from processing and manufacturing plants to wholesalers and then to retailers and other establishments is an important link in agrimarketing channels. There are

over 476,000 wholesale business establishments with annual sales totaling over $2,525 trillion. Wholesaling grocery and related products and farm-related raw materials involves over 880,000 employees in some 54,700 firms with $500 billion of sales annually.

Agribusiness retailers include those businesses selling groceries, prepared foods, soft drinks, floral products, clothing, shoes, furniture, home furnishings (from agriculturally derived products) and other products. There are over 588,000 retail outlets in the United States. The history of food retailing in this country is marked by three major developments: retail food chain stores, supermarkets, superstores and warehouse/limited-assortment supermarkets.

One of the fastest growing segments of agribusiness is the food service industry: hotels, restaurants, and institutions. It includes all those people and firms involved in serving food at hotels, restaurants, fast-food outlets, schools, hospitals, prisons, military installations, and other away-from-home eating locations. The food service industry has increased its share of the market relative to retail groceries and supermarkets. It has expanded to over a $260 billion industry with 600,000 outlets across the country employing over 10 million people.

Besides the obvious agrimarketing channels, there are many functions which relate or contribute to the agrimarketing channels. Some of these related and contributory services include: standardizing and grading, packaging, order processing, storing, inventory control, transportation, market communications, financing, product development, market research, and regulatory services. Thousands of jobs are available in these areas.

There are millions of careers in the field of agrimarketing. Many jobs and the number of people employed in various segments of the agrimarketing channels has already been presented. Other jobs in various areas include: *meat and livestock industry*—261,000 retail markets sell meat, 15,000 stores specialize in meat and seafood, and 350,000 restaurants sell meat products; *feed industry*—employs over 100,000 workers and sells over $18 billion worth of feed each year; *fruit- and vegetable-processing industries*—provide over 219,000 jobs in 2,000 processing establishments producing $30 billion worth of products; *ornamental horticul-*ture—includes 23,500 establishments producing $1 billion annually and 16,000 establishments selling lawn and supplies and nursery stock totaling $3.1 billion; *cotton industry*—employs 729,000 persons employed in textile mills and 1,000,000 employed in the apparel industry; *dairy industry*—employs over 145,000 and produces $54 billion worth of dairy products.

END-OF-CHAPTER ACTIVITIES

Review Questions

1. Define the Terms to Know.
2. Briefly explain the historical evolvement of agrimarketing channels.
3. Name and briefly describe five examples of assemblers.
4. Give three examples of consumers as assemblers (direct-marketing outlets).
5. Differentiate between an agricultural commodity and an agricultural product.
6. List and briefly explain eight examples of processors.
7. Explain the role of food manufacturers.
8. List five functions that wholesalers perform.
9. Briefly discuss the trend toward wholesale-retail affiliation.
10. Name and briefly discuss three reasons wholesalers exist.
11. Name and briefly discuss three types of wholesalers.
12. Who started supermarkets? Why?
13. Explain the difference between traditional supermarkets, superstores, and warehouse/limited-assortment supermarkets.
14. List seven other agribusiness retail items besides food.
15. List seven segments of the food service industry.
16. What percent of each of the following food items are served in the food service industry (a) meat, (b) lettuce, (c) butter, (d) potatoes, (e) fish?
17. List and briefly discuss eight types of commercial food establishments.
18. List ten noncommercial food establishments.
19. Production agriculturalists sell corn, but name seven products that production agriculturalists buy that include corn.

20. How many pounds of each of the following products do Americans consume annually (a) potatoes, sweet potatoes, and dry peas; (b) fresh vegetables; (c) canned vegetables; (d) frozen vegetables; (e) milk and dairy products?

Fill in the Blank

1. Production agriculturalists capture less than _____ cents of the consumer's food dollar.

2. Over _____ assemblers can be found between the product agriculturalist and processor.

3. The commodity processing and food manufacturing industries owe their existence to advancements in _____ .

4. There are approximately _____ firms in the commodity processing and food manufacturing sector.

5. American consumers have over _____ package food items from which to choose.

6. Industry estimates put the failure rate of new food products at _____ to _____ percent.

7. There are over _____ wholesale businesses with annual sales totaling over _____ trillion with an employment exceeding _____ .

8. Wholesaling grocery and related products and farm related materials involves over _____ employees in some 54,7000 firms with $500 billion of sales annually.

9. There are over _____ retail food outlets in the United States.

10. Chain stores account for _____ percent of food store sales.

11. There are about 165,000 grocery stores with annual sales of about $360 billion and employment of about _____ million people.

12. Superstore and warehouse supermarkets presently account for more than 50 percent of all supermarkets resulting in approximately _____ percent of supermarket sales.

13. Convenience stores make up about 20 percent of retail food stores accounting for _____ percent of grocery sales.

14. Superettes represent about 37 percent of all retail food stores accounting for _____ percent of sales.

15. Specialized food stores make up _____ percent of retail food sales.

16. Eating out accounts for _____ percent of total food expenditures.

17. The food service industry has become a _____ _____ dollar industry employing _____ _____ people.

18. Each day approximately _____ percent of Americans "eat out" at least one meal.

19. Commercial food service sales account for _____ percent while noncommercial accounts for _____ percent of sales.

20. Newspapers have about _____ percent of all advertising expenditures.

Matching

a. order processing   e. transportation   i. financing

b. inventory control   f. product development   j. advertising

c. merchandising   g. storing   k. market communication

d. packaging   h. regulatory agency   l. standardizing and grading

_____ 1. warehouses, bins, sheds, elevators, tanks, freezers, lockers, and kilns

_____ 2. needed often due to the time of sale and receipt of payment

_____ 3. insures that purchasers of agricultural commodities receive the quality they want and for which they pay

_____ 4. consumers are the primary targets

_____ 5. makes for easier contractual and exchange agreements between marketers

_____ 6. newspaper advertising, articles, telephone, radio, television, fax machine, the Internet, and satellites

_____ 7. maintaining an adequate stock of products

_____ 8. involves picking what the customer wants from inventory and preparing it for shipment

_____ 9. involves researching new food products

_____ 10. distribution, display, and sale of a product

_____ 11. moving products from where they are produced to where they are needed

_____ 12. does the majority of the selling

Activities

1. Identify the job of your six closest relatives or guardians. Determine if their jobs are a part of the agrimarketing channel and/or related services (example: accountant at grocery store). Along with the others in your class, compile a list of the total workers and determine the percentage of those employed because of agrimarketing channels.

2. Give five examples of assemblers of farm commodities in your community or adjoining communities.

3. List the agricultural commodity processors and/or food manufacturers in your community and adjoining communities.

4. List the agribusiness wholesalers in your community that sell or deliver products to restaurants, grocery stores, and/or institutions.

5. Compile a list of ten of your friends or schoolmates. Determine if and where they work part-time. What percent of them are employed in phases of the agricultural industry?

6. List the commercial and noncommercial food service establishments in your community and surrounding areas.

7. Have the teacher divide the class into two groups and debate the following topic: Does the food service industry adapt to meet consumer needs or does it anticipate needs and wants for the consumer where none existed before?

8. List as many related and contributory services to the agrimarketing channels as you can that exist in your community and/or surrounding areas. Compare your list with others in class. How many jobs could you identify?

9. Several thousands and, in some cases, millions of job numbers were given for several different areas of agrimarketing channels and their related and contributory services. List the job areas and the job numbers for each area. What is your total? Remember, this is only an estimate.

10. Beyond this chapter, look at our areas in the agribusiness input sector and list job areas and numbers. This information can be found in this book and other areas.

11. Refer to Figure 16–1. Complete the chart, being sure to include food retailers, the food service industry, and the related and contributory services.

NOTES

1. Kevin J. Bacon and Robert Birkenholz, *Careers I Unit for Agricultural Sciences I Core Curriculum* (Columbia, Miss.: University of Missouri, Instructional Materials Laboratory, 1988), p. 1.
2. James G. Beierlein and Michael W. Woolverton, *Agribusiness Marketing: The Management Perspective* (Englewood Cliffs, N.J.: Prentice-Hall, 1991), pp. 129–130.
3. Milton C. Hallberg, "U.S. Food Marketing—A Specialized System," in *Marketing U.S. Agriculture: 1988 Yearbook of Agriculture* (Washington, DC: U.S. Government Printing Office, 1988), pp. 12–13. Deborah Takiff Smith, ed.
4. Ibid.
5. Ibid.
6. Ewell Paul Roy, *Exploring Agribusiness* (Danville, Ill.: Interstate Printers and Publishers, 1980), p. 69.
7. Ibid., p. 73.
8. Beierlein and Woolverton, *Agribusiness Marketing*, p. 130.
9. Ibid., p. 131.
10. Ibid., p. 128.
11. United States Department of Agriculture, *Agricultural Statistics, 1998* (Washington, D.C.: U.S. Government Printing Office, 1998).
12. Randall D. Little, *Economics* (Danville, Ill.: Interstate Publishers, 1997), p. 325.
13. Ibid., p. 326.
14. Ibid., p. 325.
15. Roy, *Exploring Agribusiness*, p. 78.
16. Beierlein and Woolverton, *Agribusiness Marketing*, p. 131.
17. Little, *Economics*, p. 328.
18. Beierlein and Woolverton, *Agribusiness Marketing*, p. 144.
19. Little, *Economics*, p. 329.
20. Ibid., p. 330.
21. Beierlein and Woolverton, *Agribusiness Marketing*, p. 141.
22. Little, *Economics*, p. 333.
23. Ibid., p. 331.
24. Beierlein and Woolverton, *Agribusiness Marketing*, p. 143.
25. Little, *Economics*, p. 331.

26. Ibid., p. 333.

27. Alden C. Manchester, "Food Marketing Industry Responds to Social Forces," in *Marketing U.S. Agriculture: 1988 Yearbook of Agriculture* (Washington, D.C.: U.S. Government Printing Office, 1988), p. 10.

28. 7-Eleven, Inc., *A Company Profile* (May 1, 1999) Available online: http://www.7-11.com/investor/profile.html

29. Little, *Economics*, p. 332.

30. Ibid.

31. Ibid.

32. Roy, *Exploring Agribusiness*, p. 113.

33. George J. Seperich, Michael W. Woolverton, and James G. Beierlein, *Introduction to Agribusiness Marketing* (Englewood Cliffs, N.J.: Prentice-Hall, 1994), p. 140.

34. N. Omri Rawlins, *Introduction to Agribusiness* (Murfreesboro, Tenn.: Middle Tennessee State University, 1999), p. 189.

35. Ibid, p. 190.

36. Beierlein and Woolverton, *Agribusiness Marketing*, p. 150.

37. Little, *Economics*, p. 335.

38. Beierlein and Woolverton, *Agribusiness Marketing*, p. 151.

39. Little, *Economics*, p. 336.

40. Ibid.

41. Ibid.

42. Mayer, "U.S. Food Service Industry," p. 88.

43. Ibid., p. 89.

44. Little, *Economics*, p. 303.

45. Jasper S. Lee, James G. Leising, and David E. Lawver, *Agrimarketing Technology: Selling and Distribution in the Agricultural Industry* (Danville, Ill.: Interstate Publishers, 1994), p. 390.

46. Ibid., p. 392.

47. Little, *Economics*, p. 307.

48. Ibid., p. 309.

49. Ibid., p. 315.

50. Ibid., pp. 309–310.

51. Marcella Smith, Jean M. Underwood, and Mark Bultmann, *Careers in Agribusiness and Industry* (Danville, Ill.: Interstate Publishers, 1991), pp. 137–138.

52. Ibid., p. 119.

53. Seperich, Woolverton, and Beierlein, *Introduction to Agribusiness Marketing*, p. 121.

54. United States Department of Agriculture, *Agricultural Statistics, 1998* (Washington, D.C.: U.S. Government Printing Office, 1998).

55. Ibid.

56. Ibid.

57. Ibid.

UNIT

5

Agricultural Economics

CHAPTER 17

Agricultural Economics and the American Economy

OBJECTIVES

After completing this chapter, the student should be able to:

- define the Terms to Know
- define economics
- explain three major components of economics
- discuss three basic economic questions
- explain six types of economic systems
- discuss economics from a historical perspective
- discuss the role of government versus individuals in the economic system
- describe the characteristics of the American economy
- differentiate between macroeconomics and microeconomics
- explain agricultural economics

TERMS TO KNOW

aggregate
agricultural economics
allocation
applied science
basic science
capital
capitalism
classical economic theory
communism
cost-price squeeze
economics

externalities
fascism
free markets
goods
initiative
Keynesian
laissez faire
macroeconomics
microeconomics
needs
Payment In Kind (PIK)

profit
scarce
scarcity
services
socialism
subsectors
subsidies
synthesis
wants

INTRODUCTION

Today's world is complex and continually changing. The successful agribusiness manager must possess a basic understanding of economic principles in order to react to these changes. To understand **agricultural economics**, the agribusiness manager must first understand basic economic principles.

DEFINITIONS OF ECONOMICS

There are many definitions of **economics**. Consider each of the following definitions, look for key words and phrases and then form your own definition:

- 🌾 Economics is the study of allocation of scarce resources among competing alternatives.[1]

- 🌾 Economics is the study of how individuals and nations make choices about how to use scarce resources to fulfill their wants.[2]

- 🌾 Economics is the study of how society allocates scarce resources and goods.[3]

- 🌾 Economics is defined as the science of allocating scarce resources (land, labor, capital, and managements) among different and competing choices and utilizing them to best satisfy human wants.[4]

- 🌾 Economics is a study of how to get the most satisfaction for a given amount of money or to spend the least money for a given need or want.[5]

- 🌾 Economics is the study of how we work together to transform scarce resources into goods and services to satisfy the most pressing of our infinite wants, and how we distribute these goods and services among ourselves.[6]

- 🌾 Economics is the study of the decisions involved in producing, distribution, and consuming goods and services.[7]

- 🌾 Economics is a social science that studies how consumers, producers, and societies choose among the alternative uses of scarce resources in the process of producing, exchanging, and consuming goods and services.[8]

- 🌾 Economics is being concerned with overcoming the effects of scarcity by improving the efficiency with which scarce resources are allocated among their many competing uses, so as to best satisfy human wants.[9]

THREE MAJOR COMPONENTS OF ECONOMICS

Three key words or phrases can be gotten from each of these definitions: **scarcity**, types of resources, and wants and needs.

Scarcity

Scarcity is the economic term that describes a situation where there are not enough resources available to satisfy people's needs or wants. Economics is the study of society's **allocation** of **scarce** resources. These resources are considered scarce because of a society's tendency to demand more resources than are available.

While most resources are scarce, some are not, such as the air that we breathe except in places where smog or other air pollutants are severe. A resource that is not scarce is called a free resource or good. However, economics is mainly concerned with scarce resources and goods. It is the presence of scarcity that motivates the study of how society allocates resources.[10]

Shortage versus Scarcity. It is important not to confuse shortage and scarcity. Scarcity always exists because it is an unlimited or unsatisfied want, whereas shortages are always temporary. Shortages often exist after hurricanes or floods destroy goods and property.[11] Temporary shortages of products such as gasoline may be caused when imports are dramatically decreased for any reason.

Types of Resources

Resources are the inputs that society uses to produce outputs. Traditionally, economists have classified resources as: natural resources (land), human resources (labor), and manufactured resources (capital), and entrepreneurship (management).

Natural Resources (Land). Land and the mineral deposits it contains constitute a major natural resource in the agricultural industry. The population growth in this country is resulting in an increasing demand for land for commercial and residential uses. This has resulted in the permanent loss of fertile farmland for agricultural use. Water is another natural resource that is limited in many of the western states.[12] Refer to Figure 17–1.

Human Resources (Labor). The services provided by laborers and management to the production of goods and services are human resources and are also considered scarce. Agribusinesses may not be able to hire all the labor services they desire at the wage they wish to pay.[13]

Manufactured Resources (Capital). All the property people use to make other goods and services is **capital**. These resources take the form of machines, equipment, and structures.

Entrepreneurship (Management). Entrepreneurship refers to the ability of individuals to start new businesses and to introduce new products and techniques. It involves **initiative** and individual willingness to take risks and make a profit. Without proper management or entrepreneurship, agribusinesses would cease operating efficiently. Refer to Figure 17–2.

Today the four resources just discussed are called the factors of production. They are used to produce **goods** and **services**. Goods are the items that people buy. Services are the activities done for others for a fee.

Figure 17–1 Land is a valuable resource. Cities, manufacturing, highways, and other factors associated with population growth have resulted in the permanent loss of much fertile farmland ideal for agricultural use. (Courtesy of USDA)

Figure 17–2 Without proper management, agribusinesses will fail. It is not how much you make, but how you spend. Excellent management can lead to a profit. (Courtesy of Farm Credit Services)

Wants and Needs

Needs include things that are really crucial to daily living. These basic needs include enough food, clothing, and shelter to survive. Most of us would also consider a good education and adequate health care as needs. Of course, there are other needs depending on your situation.

Wants are things which are not crucial to daily living. A basic tractor is a need to a farmer, but a tractor with a cab, air conditioner, and radio may be a want. The difference between needs and wants is not a clear one. The air-conditioned cab tractor may be a need if the operator has severe allergies. Refer to Figure 17–3.

Economists use a term called *insatiable* (unlimited, unsatisfied) wants. This means that human wants cannot be satisfied no matter how much or how many goods we have. You have probably heard individuals say, "If I had a million dollars I could buy everything I want." The fact is that most people are not satisfied with whatever level of consumption they have. The more they get, the more they want. The people who have thousands of dollars want millions, and the people who have millions want billions. People who have small houses want large houses, and those who have small cars want large cars. As soon as we get what we thought would satisfy us, we suddenly discover a need (really a want) for something else. This human trait in conjunction with scarcity of resources provides economic problems. The efforts to solve these problems are the basis for the discipline of economics.

THREE BASIC ECONOMIC QUESTIONS

Because of the relationship between scarce resources and unlimited/unsatisfied wants, all societies have to answer some basic economic questions. These questions entail trying to decide what to sell, how to sell it, and who should receive the benefits. Therefore, all economic systems must solve the following three questions:

- 🌾 What goods should be produced, and how much of each?
- 🌾 How should these goods be produced?
- 🌾 Who should get what and how much?

What Goods, and How Much, to Produce

This question is answered every time people buy goods. In a market economy consumers determine what products will be made when they buy things. For example, when sales are made, agribusiness retailers usually order additional products to replace those that were sold. Manufacturers receive these orders and set their production schedules.[14] The consumer also tells the manufacturer how much to produce.

How to Produce Goods

This question is answered by agribusiness producers or manufacturers according to what will yield the greatest profits. Agribusiness firms are in business to make money. One way to improve profits is to find the least expensive means to produce. Therefore, the least expensive way should be the

Figure 17–3 Whether this big tractor is a need or a want depends on the situation. (Courtesy of Keith Mason)

most efficient way. Consumers tell the producer or manufacturer what to produce, but the producer or manufacturer determines how to produce it.[15]

Who Should Get What?

This question refers to who will receive the benefits of the goods. The question is answered by determining who has the greatest needs, wants, and ability to pay.

Combining the Economic Ingredients

To simplify the economic questions, assume that all the goods and services in our society are represented by a pie.

What Type of Pie to Produce? We have to decide what type of pie to produce, a chocolate, strawberry, or lemon pie, or if more than one and how much of each. The type and amount of resources as well as the current level of technology limits the size and the variety of pies available to us.

What Combination of Ingredients to Use? Once we decide the type and amount of pie or pies to produce, we have to decide what combination of ingredients to use; how much labor versus capital, and so on. The ideal combination depends on the availability of each and the management skills of the pie makers. Some countries emphasize more labor, while others emphasize more capital.

How to Divide the Pie? The most controversial economic question that societies have to deal with is how to divide the pie. Should everyone get an equal share? Should those who are more productive get a larger share? How much pie, if any, should people get, who do not, for reasons beyond their control, make any contribution to producing the pie? Once a decision is made about how much goes to each individual or group, we must determine which economic system will accomplish this

goal best, and what role should governments play in the economy.

ECONOMIC SYSTEMS

Each society answers the three basic questions (what, how, and for whom) according to its view of how best to satisfy the needs and wants of its people. The values and goals that a society sets for itself determine the kind of economic system it will have. Economists have identified six types of economic systems: traditional, **capitalism**, **fascism**, **socialism**, command (**communism**), and mixed economic system. These terms are often used to designate political systems as well as economic systems. The major distinction between these classifications is the degree of control by private individuals versus the group which is represented by government.

Traditional System

A pure traditional economic system answers the three basic questions according to tradition. In a traditional system, things are done "the way they have always been done." Economic decisions are based on customs, religious beliefs, and ways of doing things that have been handed down from generation to generation. Traditional economic systems exist today in very limited parts of Asia, the Middle East, Africa, and Latin America.[16]

Capitalism

Capitalism is an economic system where individuals own resources and have the right to use their time and resources however they choose. There are minimum legal controls from government. The economic system labeled capitalism is a self-regulating system, free from government involvement in economic decisions. Capitalism is dependent on market forces to determine prices, to assign resources, and to distribute income. Market prices signal the value of resources and economic goods.[17]

Private ownership and control of production factors allow individuals to make decisions. Each owner makes production decisions motivated by a desire to realize a **profit**. Production begins or ends by individual choice. The profits earned or losses incurred are a direct result of right or wrong business decisions.[18]

Free markets are the sole guides to individuals and firms in making production, exchange, and compensation decisions. Competition is a driving force in all types of economic activities and decisions. Refer to Figure 17–4.

Socialism

The basis of socialism as an economic system is public ownership of all productive resources. The government or "the state" directs all decision making. A process of central planning determines utilization of human and nonhuman resources across various sectors of the economy. Society, through the state, jointly owns all industries and holds control of all property for the mutual benefit of the people. This calls for centralized decision making by government planners and limits individual economic incentives. There is only one major employer and one owner—the government.[19]

State socialism involves the complete planning of all economic effort, with resources assigned according to these plans. The state also establishes and administers all prices for consumers and producers. The lack of free prices and **free markets** eliminates any direct economic competition. The state alone initiates new business activity.[20] Refer to Figure 17–5.

Fascism

Fascism is an economic system in which productive property, although owned by individuals, is used to produce goods that reflect government or "the state's" preferences. Private individuals have a high degree of economic power if they support the government in power. If they do not, they have little or no control.

Major Components of Capitalist Economic System

- There is private ownership of property and resources.
- The free market determines prices.
- There is freedom of occupational choice.
- There is minimal involvement from government.
- There is freedom for economic gain through profit or wages.
- Competition is a driving force.
- Efficiency is rewarded and inefficiency is penalized.

Figure 17–4 The United States tends to be a capitalistic economic system, even though government guides production in many industries.

Components of a Socialist Economic System

- The state owns all public utilities and large-scale industries.
- Property is owned by society for the benefit of society.
- Prices are determined by the government.
- Occupations are assigned by the government.
- The government makes all economic decisions.
- There are no economic incentives.
- There is no economic competition.
- There is inefficient production.

Figure 17–5 The basis of socialism as an economic system is that the government or "the state," directs all decision making.

Fascism suppresses opposition, censors criticism, and thus denies some freedoms to individuals. Although property is privately owned and business firms control production, the government maintains firm control over labor, employers, and consumers.[21]

Communism

Communism is a totalitarian system of government in which a single, authoritarian party controls government-owned means of production with the intended aim of establishing a "puppet" society. In other words, the government has total control of economic matters and private individuals have none. Production is organized according to an economic plan. Everyone contributes according to ability and distribution is made according to need.[22]

Mixed Economic Systems

With the exception of the traditional economic system, it is doubtful whether any country fits either of these classifications exactly. The real question is which classification most closely approaches the actual system of each country. No society has been willing to grant all economic decision making to individuals as characterized in capitalism. Conversely, no society has been willing to give all power to the government as characterized by *communism*. Each country has to decide how much power to give to each. Obviously, there is not one ideal ratio for all countries or even for any one country because all societies are constantly changing the ratio of each. During some periods, more government control is emphasized while in other periods less government and more individualism is emphasized.

American Economy Is Mixed. Most people commonly refer to the United States as capitalistic. However, the U.S. government guides production in many industries. Government regulations exist in nearly all sectors of the U.S. economy. Government **subsidies** and grants provide incentives to increase or reduce production of goods and services.[23]

American Agriculture and the Mixed Economic System. Government's intervention in agriculture with supply control programs and demand expansion programs influences the supply and demand of specific agricultural products. The **Payment In Kind (PIK)** program was an example of government involvement in agriculture. Agriculture is not the only economic sector in which the U.S. government intervenes. Loan guarantees to large firms such as Lockheed and Chrysler are forms of government intervention in the private sector.[24]

ECONOMICS—A HISTORICAL PERSPECTIVE

Normally, an academic discipline evolves over a long period of time and it is difficult to pinpoint a place or time when one begins. This is not the case with economics. Of course, economic problems have been around since resources became limited relative to the number of people. However, economic historians generally agree that modern economics as a discipline began in 1776 with the publication of a book by Adam Smith entitled, *An Inquiry into the Causes of The Wealth of Nations.* This book is commonly referred to as *The Wealth of Nations* and is generally considered to be one of the literary classics of that period.

The Father of Economics

England was the major center of intellectual activity during the 18th century and a major portion of current economic policy is based on many economic ideas presented during that period. Economic historians contend that all the ideas presented by Adam Smith had been introduced before by other economists. However, his contribution was one of synthesis. He organized the ideas of predecessors into a major body of thought relative to economic problems of his time. As a result, Adam Smith is commonly referred to as the Father or Founder of Economics.

In *The Wealth of Nations,* Adam Smith was responding to the economic environment of his period. In the latter part of the eighteenth century the government in England was highly involved in economic matters. With that system in control, a small percentage of the population was wealthy, but the masses were extremely poor with no hope for improvement. Most economists of that period were discussing economic problems and proposing solutions.

Adam Smith's Proposed Economic System

Adam Smith, like most economists of that time, was looking for new directions in economic policy to deal with the severe economic conditions he saw around him. He proposed a system completely opposite to the system operating in England and many other countries at that time. Several names have been coined to explain the system he proposed. The major ones include pure capitalism, free enterprise, and **laissez faire**. All of the terms are used interchangeably, and they mean basically the same thing. The basic meaning is that individuals can control the economy without any government interference.

The Influence of David Ricardo and Thomas Malthus

David Ricardo and Thomas Malthus were also prominent English economists during the eighteenth century, and each made a major contribution to economic thought in a special area. Ricardo was concerned primarily in the role of land as a resource and its effect on the total economic picture. Malthus was interested primarily in the effect of increasing population relative to limited food production.

THE ROLE OF GOVERNMENT VERSUS THE ROLE OF INDIVIDUALS

A major question in all economic systems is "What is the role of government versus the role of individuals?" Adam Smith said, "The government should provide a police force for internal and external protection and nothing else. Economic matters should be the sole responsibility of private individuals." He said "Let each individual seek his or her own personal gain," and that society would be guided by an "invisible hand" in such a way that as a whole it would be better off than it would with government intervention. In other words, let each individual operate independently rather than let some government agency decide what should and should not be done.

Economic Classification According to the Role of Government in Economic Decision Making

Economic theory is often classified according to the role of government in economic decision making. These two classifications are **classical economic theory** and **Keynesian** economics.

Classical Economic Theory. Classical economic theory contends that an economic system is self-sufficient in itself and any outside interference by government does more harm than good. Adam Smith and his followers are called classical economists. They advocated little or no government intervention in economic matters. This type of economic thought, in various degrees, was dominant in the United States until the 1930s. However, the Depression raised serious questions concerning the effectiveness of this system, and numerous economists, as well as political leaders, began to look for something different.

Keynesian Economics. The demise of the concept of classical economics began with the publication of *The General Theory of Employment Interest and Money* by John Maynard Keynes in 1936. Keynes, an Englishman, said that economic systems are not always self-sufficient and sometimes need outside help. Of course, the only help available outside the private sector is the public sector or government. He did maintain that public intervention should only be temporary. Once the system was working properly the private sector should take over again.

Keynesian economics, which requires the government to play whatever role necessary to accomplish economic goals, began to play a major role in the United States with President Franklin D. Roosevelt's New Deal program. Ironically, the temporary role of government in economics evolved into permanent government intervention and the major question today is not whether the government should be involved in economic matters, but rather, how much should it be involved. The economic policy in the United States has definitely shifted from classical economics to *Keynesian economics*.

Competition

One might ask, "What is the major characteristic of the capitalist economic system that would balance it out and keep a small group from getting control of a majority of the wealth?" The answer is a high degree of competition. Adam Smith envisioned a society of large numbers of small businesses. If there was excess profits in one industry, more firms would be developed, increasing the output of that product and the price would be reduced to a level that would provide normal or acceptable profits. If an industry did not provide an acceptable profit, the least efficient firms would shift to other industries that provided potential for profit, thus keeping the economy balanced.

Individual Economic Freedom

The major characteristic of *pure capitalism* is individual economic freedom. However, a high degree of individual economic freedom requires a high degree of responsibility from each individual with pure capitalism; if an individual cannot make a decent living, it is his or her fault, not that of some government agency. Those individuals who cannot get a job or cannot feed and clothe their family have no government agency to blame. The only source of economic help for individuals under capitalism is the extended family, civic or religious groups. Pure capitalism is a very impersonal system, which provides a high degree of freedom but turns a deaf ear to individuals who cannot seem to get their fair share of the pie.

Political Party Philosophy

Adam Smith's concepts of pure capitalism had a major impact on many countries around the world, especially the United States. The economic system in the United States is generally considered as the major example of capitalism, but it is very mixed with small portions of other economic systems as discussed earlier. Americans are also mixed on their philosophy of the role of government versus individuals. The Republican Party philosophy has a tendency toward *individual* economic control, whereas the Democratic Party philosophy has a tendency toward *economic assistance* from the government.

CHARACTERISTICS OF THE AMERICAN ECONOMY

A common term for the U.S. economic system is the free enterprise system. It is the freedom of private businesses to organize and operate for profit in a competitive environment. Government interference is necessary only for regulation to protect public interest and to keep the national economy in balance.[25]

The American economy has six major characteristics. They are as follows:

- 🌾 little or no government control
- 🌾 freedom of enterprise
- 🌾 freedom of choice
- 🌾 the right to own private property
- 🌾 profit incentive
- 🌾 competition[26]

These characteristics are interrelated and to varying degrees, all are present in the American economy. The six characteristics could be referred to as *free enterprise with some regulations*. Notice the influence of Adam Smith in these six characteristics. A brief discussion of each of the six characteristics follows.

The Role of Government

Although the role of government was discussed earlier, a few more comments will be made to put it in perspective. Capitalism, as practiced in the United States today, would be best defined as an economic system in which private individuals own the factors of production and decide how to use them within the limits of the law.[27]

Originally, the Founders of the United States limited the role of government mainly to national defense and keeping peace. Lately, the role of government has increased significantly in the United States. This is especially true in the areas of regulating business and providing public services. Today, the work of federal agencies includes regulating the quality of various foods and drugs, watching over the nation's money and banking system, inspecting workplaces for hazardous conditions, and guarding against damage to the environment. Refer to Figure 17–6.

Freedom of Enterprise

Our economic system is also called the free enterprise system. This term emphasizes that individuals are free to own and control the factors of production.[28] Yet there are limits on free enterprise. In most states, teenagers must be sixteen before they can work, and then laws are set to limit how many hours they can work. Minimum wages are also set by law to protect workers. Refer to Figure 17–7.

Freedom of Choice

Part of freedom of choice is the freedom to fail. Freedom of choice means that buyers make the decisions about what should be produced. The success or failure of a good or service in the marketplace depends on individuals freely choosing what they want. Nevertheless, the government sets laws to protect buyers. Laws set safety standards for such things as labels or tags, appliances, and automobiles. Laws also regulate public utilities.

Figure 17–6 Although many individuals are opposed to government intervention, we take for granted a safe food supply. Most food products are inspected by the USDA. (Courtesy of USDA)

Figure 17–7   The United States is not a pure capitalist economy because we have laws to control factors of production. Can you imagine working these children eight hours daily? Some countries continue to operate sweatshops. (Courtesy of USDA)

Only in the success of our abilities to work together, coupled with our skill in assessing the land, will we realize our public as well as individual conservation objectives.

Figure 17–8   The freedom to own our land, manage it, and conserve it is a great characteristic of the American economy. (Courtesy of USDA)

Private Property

Private property is simply what is owned by individuals or groups rather than by the federal, state, or local government. You are free to buy whatever you can afford, whether it is land, an agribusiness, a home, or a car. You can also control how, when, and by whom your property is used. If you own an agribusiness, you can keep any profit you make.[29] Refer to Figure 17–8.

Profit Incentive

Whenever a person spends time, knowledge, money, and other capital resources in an agribusiness, that investment is made with the idea of making a profit. The desire to make a profit is called the profit incentive. The goal of profit is what moves people to produce things that others want to buy.[30]

Competition

Competition is the rivalry among producers or sellers of similar goods to win more business by offering the lowest prices or better quality. Businesses have to keep prices low enough to attract buyers yet high enough to make a profit. This forces business to keep the cost of production as low as possible. Competitors succeed because they are able to produce products at a price that makes people want to buy.[31]

MACROECONOMICS VERSUS MICROECONOMICS

As disciplines evolve over time, the body of knowledge accumulates and eventually the discipline is broken into divisions and specialties. Economics is no exception. The total body of economic knowledge today is too extensive for a person to be a general economist. The most common division of economics is into **macroeconomics** and **microeconomics**.

Macroeconomics

The prefix *macro,* means "large," indicating that macroeconomics is concerned with the study of the economy on a large scale or nationally. Macroeconomics looks at the **aggregate** (total) performances of all the markets in the national economy and is concerned with the choices made by large **subsectors** of the economy. These subsectors include: the household sector, which includes all consumers; the business sector, which includes all firms; and the government sector, which includes all government agencies.[32] Seven key economic concepts are related to macroeconomics: gross domestic product, aggregate (total) supply, aggregate (total) demand, unemployment, inflation and deflation, monetary policy, and fiscal policy. A short explanation of each follows.

Gross Domestic Product (GDP). GDP is defined as the market value of the total output of all final goods and services produced within a country's boundaries during one year.

Aggregate Supply. Aggregate supply is the total amount of goods and services produced by the economy in a period of time.

Aggregate Demand. Aggregate demand is the total amount of spending on goods and services in the economy during a period of time.

Unemployment. Unemployment is defined as the number of people without jobs who are actively seeking work.

Inflation and Deflation. Inflation is a sustained increase in the average price level of the entire economy. Deflation is a sustained decrease in the average price level of an entire economy.

Monetary Policy. Monetary policy seeks to affect the amount of money available in the economy and its costs.

Fiscal Policy. Fiscal policy consists of changes in taxes, in government expenditures on goods and services, and in transfer payments that are designed to affect the level of aggregate (total) demand in the economy.[33]

Microeconomics

The prefix *micro* means "small," indicating that microeconomics is concerned with the study of the economy on a small scale. Microeconomics considers the individual markets that make up the economy and is concerned with the decisions made by small economic units such as individual consumers, individual firms, and individual government agencies such as the USDA and state governments.

Six key economic concepts are related to microeconomics: markets and prices, supply and demand, competition and market structure, income distribution, market failures, and the role of government. A short explanation of each follows.

Markets and Prices. Markets are institutional arrangements that enable buyers and sellers to exchange goods and services. Prices are the amounts of money that people pay in exchange for a unit of a particular good or service.

Supply and Demand. *Supply* is defined as the different quantities of a resource, good, or service that will be offered for sale at various possible prices during a specific time period. *Demand* is defined as the different quantities of a resource, good, or service that will be purchased at various possible prices during a specific time period. Refer to the Appendix for a comprehensive explanation of the laws of supply and demand.

Competition and Market Structure. *Competition* is determined by the number of buyers and sellers in particular markets. *Market structures* refer to the extent to which competition prevails in particular markets.

Income Distribution. Income distribution may be classified into functional distribution—the division of an economy's total income into wages and salaries, rent, interest, and profit; or income distribution may be classified into personal distribution of income, classifying the different population groups by the number of them receiving different amounts of income.

Business Failures. Business failures occur when there is inadequate competition, lack of access to reliable information, resource immobility, externalties, and not enough demand for public goods.

Role of Government. The role of government includes establishing a framework of law and order in which a market economy functions.[34]

AGRICULTURAL ECONOMICS

There are several ways that agricultural economics may be defined, but basically it is *the application of economic concepts to agricultural problems*. Therefore, as stated earlier, to be a productive agricultural economist, one must first understand basic economic principles. Economics is sometimes called a **basic science**, whereas agricultural economics is called an **applied science**. The follow-

CAREER OPTION

Career Area: Agricultural Economist

Agricultural economics could be described as the application of economic concepts to agriculture problems. Obviously, the job of economists is to alert us when the factors within the economy are beginning to get unbalanced. Therefore, to be a productive agricultural economist, one must first understand basic economic principles, including its relationship with federal, state, and local governments. Agricultural economics deals with how humans choose to use their knowledge and scarce productive resources such as land, labor, capital, and management to produce and market food and fiber.

The field of agricultural economics includes a wide array of content knowledge. It is such a broad area that it is very difficult for an individual to absorb all the economic information, so one usually specializes in one or more phases of the discipline. Some of the specialties include production economics (farm management), agriculture finance, rural development, price analysis, agriculture policy, international development, and marketing. If one chooses the marketing phase, he or she usually concentrates on a specific prod-

uct or group of products such as grain marketing, livestock marketing, cotton marketing, or fruit and vegetable marketing. Agricultural economists tend to have become economists working with agriculturalists rather than agriculturalists working with economists.

Beyond production, the key to making a profit in the agricultural industry is having a keen knowledge of agricultural economics. These agricultural economists are evaluating an array of options for their client's cattle-breeding program. (Courtesy of USDA)

ing definition is a good example of merging general economics with agriculture. Agricultural economics is an applied social science dealing with how humans choose to use technical knowledge and scarce productive resources such as land, labor, capital, and management to product food and fiber and to distribute it for consumption to various members of society over time. Refer to the Career Option on page 454 for a further explanation of a career as an agricultural economist.

Agricultural economics, like most disciplines, did not begin in a given year with a specific publication like economics did. Instead, it evolved from specific farm problems in the United States around the beginning of the twentieth century. It evolved as a special study of agricultural problems, not as a speciality within the discipline of economics. Refer to Figure 17–9 for an example of a key agricultural economic concept.

Example of Key Agricultural Economic Concept

Utilization of Existing Resources: A Key Concept in Agricultural Economics

A very simple economic principle is that of the utilization of existing resources. Whatever business venture you plan, look around and determine what resources are already available for your use. Remember, the four factors of production discussed in an earlier chapter (land, labor, capital, and management). These resources should be considered in any business decision.

Let us suppose that you are trying to decide whether to go into the dairy farming business or the flower and plant production business. One of your uncles has a 200-acre farm that he wants to keep in the family, and another uncle has just retired and has three 30- by 80-foot greenhouses equipped with heat, water, and tables. Both relatives have agreed that your only cost for using the facilities would be to keep them maintained, neat, and orderly, plus pay the property taxes. The taxes are approximately $2,000 yearly.

Which would you choose? Both are rare opportunities. Considering that dairying and flower and plant production are equally attractive, the greenhouse facilities would be the best utilization of available resources because most of the resources are in place. Except for consumable supplies, with the greenhouse selection, you are ready for production. With the dairy, 200 acres in land is available, but over $200,000 in capital would still be needed to build the dairy barn and silos and to purchase tractors, equipment, and cows. Assuming an interest rate of 8 percent on $200,000 of borrowed capital, $16,000 interest would have to be paid the first year alone. These costs can be avoided by selecting the greenhouse option.

Although the utilization of resources is a simple concept, it is often overlooked. Even for those who have already started a business, it is wise to make production decisions based on resources that already exist whether it be family labor, existing land, or other utilized tractors and equipment. Dr. Warren Gill, University of Tennessee Beef Specialist, jokingly presents the "UDE" ("use daddy's equipment") principle of agricultural economics to explain the concept of utilization of resources. You should utilize whatever resources are available and affordable to maximize profit.

Figure 17–9 Many agribusinesses fail because they spend unnecessary money by failing to utilize the resources already at their disposal.

Early Agricultural Economics Centered around Farm Production

Agricultural economics began as farm economics with an emphasis on farm production and management problems. During the early 1900s farmers were being pressured by the increasing cost of resources, on one hand, and lower prices received from marketing firms, on the other, which is commonly known as the **cost-price squeeze**. In an effort to solve these problems, specialists from a wide variety of established disciplines were recruited and as a result, a new discipline evolved. The major emphasis of agricultural economics is still farm production economics. However, the discipline has expanded considerably, and today it encompasses the total agricultural industry.

Speciality Areas in Agricultural Economics

The study of agricultural economics for almost a century has accumulated a vast body of knowledge and new knowledge is being added constantly. As a result, it is very difficult for an individual to absorb the total amount of economic information so we have to specialize in one or more phases of the discipline.

Some of the specialties include production economics (farm management), agricultural finance, rural development, price analysis, agricultural policy, international development, and marketing. Those students interested in marketing usually concentrate on a specific product or group of products such as grain marketing, livestock marketing, cotton marketing, or fruit and vegetable marketing.

CONCLUSION

Agricultural economics has many subspecialities, including crop, livestock, poultry, land, resource, forestry, marine, and water economics. Also, among the many concerns of agricultural economists are marketing, finance, business organization, policy, statistics, and resource development.[35]

Agricultural economics is also related to consumer economics. The home of the production agriculturalist and his or her family are integral parts of the business. Agricultural economists cannot ignore the economics of the production agriculturalists' home, since the family competes for funds with the agribusiness itself and is an integral component of its processes.[36]

Agricultural economists are involved increasingly as advisors or consultants to other social sciences and disciplines because of their knowledge of, and concern for, the efficient use of human, natural, and capital resources. Agricultural economists have become economists working with agriculture rather than agriculturalists working with economists.[37]

SUMMARY

Today's world is complex and continually changing. The successful agribusiness manager must possess a basic understanding of economic principles in order to react to these changes. To understand agricultural economics, the agribusiness manager must first understand basic economic principles.

There are many definitions of economics. Three key words or phrases can be gotten from each of these definitions: scarcity, types of resources, and wants and needs.

Because of the relationship between scarce resources and unlimited/unsatisfied wants, all societies have to answer some basic economic questions. These questions entail trying to decide what to sell, how to sell it, and who should receive the benefits.

Each society answers these basic questions according to its view of how best to satisfy the needs and wants of its people. The values and goals that a society sets for itself determine the kind of economic system it will have. Economists have identified six types of economic systems: traditional, capitalism, fascism, socialism, command (communism), and mixed economic systems.

Economic historians generally agree that modern economics as a discipline began in 1776 with the publication of a book by Adam Smith entitled, *An Inquiry into the Causes of the Wealth of Nations*, which is considered to be one of the literary classics of that period. Because of the influence of this book, Adam Smith is commonly referred to as the Father or Founder of Economics.

A major question in all economic systems is, "What is the role of government versus individuals?" Therefore, economics is often classified as to the role of government in economic decision making. These two classifications are classical economic theory and Keynesian economics. Classical economic theory contends that an economic system is self-sufficient in itself and any outside interference by government does more harm than good. Keynesian economics contends that economic systems are not always self-sufficient and sometimes they need outside help.

A common term for the U.S. economic system is *free enterprise system.* This guarantees the freedom of private businesses to organize and operate for profit in a competitive environment. Government interference is necessary only for regulation to protect public interest and to keep the national economy in balance. The American economy has six major characteristics: little or no government control, freedom of enterprise, freedom of choice, private property, profit incentive, and competition.

The two most common classifications of economics is macroeconomics and microeconomics. The prefix, *macro* means "large," indicating that macroeconomics is concerned with the study of the economy on a large scale or nationally. The prefix, *micro* means "small," indicating that microeconomics is concerned with the study of the economy on a small scale. Small scale economic units include individual consumers, firms, farms, or agribusinesses.

Agricultural economics is the application of economic concepts to agricultural problems. To be a productive economist, you must first understand basic economic principles. Agricultural economists have become economists working with agriculture rather than agriculturalists working with economists.

END-OF-CHAPTER ACTIVITIES

Review Questions

1. Define the Terms to Know.
2. Explain the difference between shortage and scarcity.
3. Name and briefly discuss four types of resources that are classified by economics and that affect the economy.
4. Explain the difference between a need and a want.
5. Name and briefly explain the three basic economic questions.
6. Compare or match the three basic economic questions with three analogies of goods and services in our society as represented by making a pie.

7. List seven major components of a capitalistic economic system.

8. List eight components of a socialistic economic system.

9. Briefly describe Adam Smith's economic beliefs.

10. List the three major characteristics of an American economy.

11. Name three large subsectors of the economy when macroeconomics is studied.

12. Name and briefly discuss seven economic concepts related to macroeconomics.

13. Name and briefly discuss six economic concepts related to microeconomics.

14. Briefly describe how agricultural economics began.

15. Name seven speciality areas of agricultural economics.

Fill in the Blank

1. Economics is being concerned with overcoming the effects of _____ by improving the efficiency with which scare _____ are allocated among their many competing uses, so as to best satisfy human _____.

2. _____ is the study of how society allocates scare resources.

3. A resource that is not scarce is called a _____ resource or good.

4. The book _____ was written by Adam Smith and is considered the beginning of economics as we know it today.

5. A common term for the U.S. economic system is _____ enterprise system.

6. Part of freedom of choice is the freedom to _____.

7. The goal of _____ is what moves people to produce things that others want to buy.

8. _____ is the rivalry among producers or sellers of similar goods to win more business by offering the lowest price or better quality.

9. The prefix *macro* means "large," indicating that _____ is concerned with the study of the economy on a large scale or nationally.

10. The prefix *micro* means "small," indicating that _____ is concerned with the study of the economy on a small scale.

11. Economics is sometimes called a basic science whereas agricultural economics is called an _____ science.

12. Agricultural economics have become economists working with agriculture rather than agriculturalists working with _____.

Matching

a. fascism   e. Thomas Malthus   i. Republican Party

b. Adam Smith   f. traditional economic system   j. John Maynard Keynes

c. David Ricardo   g. Democratic Party   k. communism

d. capitalism   h. mixed economic system   l. socialism

_____ 1. "the way things have always been done"

_____ 2. the basis of this economic system is public ownership of all productive resources

_____ 3. this economist was interested primarily in the effect of increasing population relative to limited food production

_____ 4. the government has total control of economic matters and private individuals have none

_____ 5. Father of Economics

_____ 6. an economic system that possesses both the characteristics of capitalism and socialism

_____ 7. an economic system where individuals own the land but produce goods that reflect government preferences

_____ 8. economist who believed that economic systems are not always self-sufficient and sometimes need outside help

_____ 9. political party that tends to favor individual economic control

_____ 10. political party that tends to favor government assistance

_____ 11. This economist was concerned primarily with the role of land as a resource and its effect on the total economic picture.

_____ 12. economic system where individuals own resources and have the right to use their time and resources however they choose

Activities

1. Read the nine definitions of economics. Select the definition which makes the most sense to you and write a paragraph explaining why.

2. List five wants and five needs that you have. Share these with the class. Are they in agreement with you as to those that are wants and those that are needs.

3. Write a 100-word essay on why entrepreneurship or management is considered a factor of production.

4. Pretend that you wish to start an agribusiness. Select the type of agribusiness and answer the three basic economic questions.

5. Select and defend the economic system that you favor. Write a short essay and present it to class. Give an explanation of features that appeal to you along with any weaknesses.

6. Select an economist such as Adam Smith, David Ricardo, Thomas Malthus, John Maynard Keynes, or anyone else. Using the library, Internet, or other sources, write a one- to two-page report on

who they were and their economic philosophy or the contribution they made to the field of economics.

7. Select and defend your position as to whether you favor individual economic control or governmental assistance with the economy. Your teacher may have you debate this or participate in a panel discussion. Either way, write a one-page essay defending your position.

8. Study the six major characteristics of the American economy covered in this chapter. Select the one that you could most easily give up and the one that you would best want to hold onto. Write a one-page essay defending your choice and share this with the class.

9. There are seven key economic concepts related to macroeconomics. Refer to these in this chapter. Select one concept and prepare a one- to three-page report further elaborating on your selection.

10. Do the exercise in question 9 with one of the key microeconomic concepts.

11. Write a paragraph (or more if needed) explaining why economics is a basic science whereas agricultural economics is an applied science.

12. Select a product that you use every day. Identify as many factors as you can that were used in making the product. Classify each factor as either a natural resource, human resource (labor), manufactured resource (capital), or management (entrepreneurship).

NOTES

1. *Advanced Agribusiness Management and Marketing* (College Station, Tex.: Instructional Materials Service, 1990), 8735B, p. 33.

2. Roger LeRoy Miller, *Economics Today and Tomorrow* (New York: Glencoe McGraw-Hill, 1995), p. 8.

3. John Duffy, *Economics* (Lincoln, Neb.: Cliffs Notes, 1993), p. 1.

4. Randall D. Little, *Economics* (Danville, Ill.: Interstate Publishers, 1997), p. 3.

5. Ibid.

6. Fred M. Gottheil, *Principles of Economics* (Cincinnati, Ohio: South-Western Publishing Company, 1996), p. 6.

7. Stanford D. Gordon and Alan D. Stafford, *Applying Economic Principles* (New York: Glencoe McGraw-Hill, 1994), p. 5.

8. John Penson, Rulon Pope, and Michael Cook, *Introduction to Agricultural Economics* (Englewood Cliffs, N.J.: Prentice-Hall, 1986), p. 7.

9. Gail L. Crammer, Clarence W. Jensen, and Douglas D. Southgate, Jr., *Agricultural Economics and*

Agribusiness (New York: John Wiley and Sons, 1997), p. 5.

10. Duffy, *Economics*, p. 1.

11. Miller, *Economics Today and Tomorrow*, p. 9.

12. Penson, Pope, and Cook, *Introduction to Agricultural Economics*, pp. 8–9.

13. Ibid.

14. Gordon and Stafford, *Applying Economic Principles*, p. 41.

15. Ibid.

16. Miller, *Economics Today and Tomorrow*, p. 32.

17. *Advanced Agribusiness Management and Marketing*, p. 45.

18. Ibid.

19. Ibid., p. 46.

20. Ibid.

21. Little, *Economics*, p. 27.

22. Gottheil, *Principles of Economics*, p. 863.

23. *Advanced Agribusiness Management and Marketing*, p. 47.

24. Ibid.

25. Ibid.
26. Miller, *Economics Today and Tomorrow*, p. 37.
27. Ibid.
28. Ibid.
29. Ibid.
30. Ibid., p. 39.
31. Ibid., p. 40.
32. Duffy, *Economics*, p. 1.
33. Miller, *Economics Today and Tomorrow*, p. 3.
34. Ibid., pp. 2–3.
35. Little, *Economics*, p. 6.
36. Ibid.
37. Ibid., p. 7.

18 *Economic Activity and Analysis*

OBJECTIVES

After completing this chapter, the student should be able to:

- discuss economic activity in America
- measure economic activity through economic indicators
- explain the function of money and banks
- explain the role the federal government has in promoting economic stability
- explain the basis of economic policy
- discuss the nation's economic goals
- describe various types of economic analysis
- explain ways to attain economic efficiency

TERMS TO KNOW

anthropology
artificial insemination
astute
bank money
barter
base year
business cycles
cloning
commodity money
consumer price index
contraction
deflation
depression
discouraged workers
economic fluctuations
economic indicator

economic policy
economic recovery
equal product curve
Federal Reserve System
fiat money
fiscal policy
gross domestic product (GDP)
hypothesis
implicit GDP price deflator
inflation
law of increasing cost
marginal analysis
marginal benefits
marginal cost
market basket
monetary policy

opportunity cost
peak
political science
producer price index
production possibilities
prosperity
psychology
recession
scenario
sociology
standard of living
stock prices
trough
value judgments
variables

INTRODUCTION

It is sometimes difficult to separate the study of agricultural economics from the study of other social sciences, such as **sociology**, **anthropology**, **political science**, and **psychology**. Agricultural economics and these other social sciences study individual and social behavior. Agricultural economics concentrates on those aspects of behavior that affect the way we, as individuals and a society, produce and consume goods and services.[1] A part of understanding agricultural economics is understanding economic activity and its relationship with federal, state, and local governments. For example, it is difficult to appreciate what federal, state, and local governments do without understanding the *economic* circumstances underlying their actions. In reality, government budgets are economic documents. Taxes and government spending are economic tools used by the political system to meet economic as well as political and social objectives. The national debt, budget deficits, and the welfare system require an understanding of economics.[2]

The role of the family in society and how the family behaves as an economic unit is tied to agricultural economics. To some extent, even when and whom we marry, the number of children we have, and interpersonal relationships within the family are determined by economics. Therefore, an understanding of economic activity, economic goods of the nation, and how to analyze these goods and activities are crucial to understanding the economic system of the United States.

ECONOMIC ACTIVITY IN THE UNITED STATES

Many factors are involved in maintaining a strong economy. These factors include efficient production of agricultural products freeing workers from industrial work; efficient production of goods and services; and good wages so consumers can purchase goods, services, and entertainment. The more money spent, the stronger the economy. The cycle continues.

Purpose of the Economist

Obviously, the job of economist is to alert us when the factors within the economy are beginning to get unbalanced. This unbalanced economy can cause a snowball effect leading to unemployment which leads to people buying less goods and services. If less goods and services are being purchased, the industrial segment of the economy produces less, causing higher unemployment, resulting in meaning even less money available to purchase goods and services. Again, the snowball or cycle continues.

Every day, economist and business leaders provide forecasts, reports, and predictions about the state of the economy. These reports appear as often as reports about the weather.

The Great Depression

Why does the state of the economy require daily attention? What is so important about economic information? The answer is the fear of another Great Depression. History has taught the world a great deal about the effects of changes in the state of the economy. The Great Depression, which occurred in 1929, is an event that illustrates the results of a severe economic decline.

In summer 1929, there was a peak of economic prosperity. A few months later, on a day that has become known as Black Thursday (October 24, 1929) **stock prices** collapsed. From Black Thursday in 1929 to about 1939, the United States was in a **depression**. Investors lost millions of dollars and the downward economic snowball began. Great declines in production and in employment occurred. Factories shut down. Businesses and banks failed. As businesses failed, more and more people lost their jobs. By 1932, 12 to 15 million people were unemployed, or 25 to 30 percent of the workforce.[3]

Business Cycles and Patterns

If you study the economic changes that have taken place in the past, you will notice an irregular cycle of ups and downs. The ups and downs in economic activity are called **economic fluctuations**. The repeated rise and fall of economic activity over time is called a **business cycle**.[4]

Economic Recovery. Figure 18–1 illustrates a business cycle. A rise in business activity is **economic recovery**. The rise indicates an expansion of the economy or economic growth. A recovery is a period of economic growth or expansion following a **recession** or depression.

Peak. A **peak** is the highest level of economic activity in a business cycle. A peak indicates **prosperity** and means the economy is rapidly expanding. A moderate rate of inflation accompanies a peak in the business cycle. At the peak of the business cycle, new businesses open, production of goods and services is high, many jobs are available, and employment is high.[5]

Contraction. A contraction is a noticeable drop in the level of business activity, which indicates a slowdown in the growth of the economy. A **contraction** may mean stable or falling prices for goods and services or it may mean unemployment is increasing.[6]

Trough. A valley or **trough** is the lowest level of business activity in a particular business cycle. The terms *recession* and *depression* are often used to describe long periods of time in the valley or trough of a business cycle. A recession is a period of severe economic decline in which spending decreases, so fewer goods and services are demanded, and unemployment rises. A depression is a severe recession that lasts for several years.

Circular Flow of the Economy

Figure 18–2 shows the circular flow of economic activity. This diagram illustrates how the economy's resources, money, goods, and services flow between households and firms through resource and product markets.

In Figure 18–2, people are both consumers and producers. They live in households, where they consume the goods and services they buy on the product market, and supply their resources (land, labor, capital, and management) on the resource market to firms who use the resources to produce the goods and services that appear on the product market. Entrepreneurs provide the vision and drive associated with the firm's production of goods and services, anticipating a profit in return.[7]

Specific Explanation of Circular Flow. Figure 18–2 illustrates the relationship of the different parts of our economic system. Individuals sell the factors of production (land, labor, capital, and management) to businesses and, in return, gain income such as wages, rents, and profits. Businesses

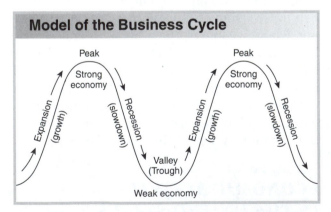

Figure 18–1   A business cycle is the repeated rise and fall of economic activity over time. A business cycle is made up of four phases: expansion, peak, recession, and valley.

buy the factors of production and use them to produce goods and services. Businesses sell the finished products to individuals, who buy the goods and services with their income. Businesses pay taxes on their income and property to the government, which provides public services and payments to individuals. Individuals, in turn, pay taxes to the government and provide the government with land and labor, thus completing the circle.[8]

Your Effect on the Circular Flow. Suppose you get a job selling flowers and shrubs at a lawn and garden center. You are a factor of production

to the garden center because you sell your labor to it. You receive a paycheck from the garden center and the money flows back to you. Some of your paycheck flows to the government in taxes to provide roads and other services. With some of your weekly paychecks you buy a lawnmower for your lawn care business. Money flows from you to the agribusiness selling it. Taxes on the sale of the lawnmower and the business's profits flow to the government, completing the circle.[9]

Interdependence. *High degrees of dependence among the different parts of an economic systems are*

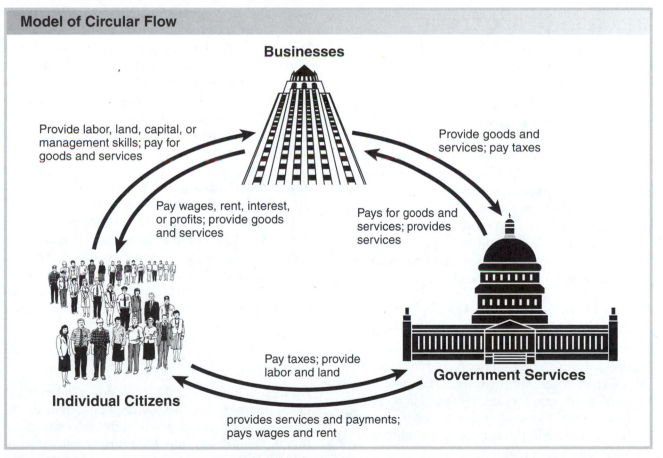

Model of Circular Flow

Businesses

Provide labor, land, capital, or management skills; pay for goods and services

Provide goods and services; pay taxes

Pay wages, rent, interest, or profits; provide goods and services

Pays for goods and services; provides services

Individual Citizens

Pay taxes; provide labor and land

Government Services

provides services and payments; pays wages and rent

Figure 18–2   The circular flow model illustrates the high degree of dependence among the different parts of our economic system.

illustrated by the circular flow. Consider the following explanation:

Suppose that people begin to worry about the economy. They may decrease their spending, which in turn means fewer goods and services need to be produced, which means fewer resources will be needed, which means less income is earned and distributed. On the other hand, suppose that people have a positive outlook about the economy. They may be apt to increase their spending, which could lead to increased total income put into the economy in a reversal of the sequence.[10]

MEASURING ECONOMIC ACTIVITY VIA ECONOMIC INDICATORS

The circular flow of the economy illustration showed us how economists study and define economic activity. However, it does not determine the future state of the economy. This is done by **economic indicators**. Economic indicators are important data or statistics that measure economic activity and business cycles. These indicators normally change before the rest of the economy does. There are many economic indicators, but three major economic indicators will be discussed: gross domestic product, inflation, and unemployment.

Gross Domestic Product

One indicator of how well our economy is performing is to determine how many goods and services it produces during a certain period of time, usually a year. **Gross domestic product (GDP)** is the total value of goods and services produced in a country in a given year. To calculate the *gross domestic product*, economists add the sum of goods and services which includes four main areas:

❦ consumer goods and services

❦ business goods and services

❦ government goods and services

❦ goods and services sold to other countries[11]

Determining Economic Growth. A dollar value is assigned to the goods and services to find the sum. The gross domestic product in 1970 was over $1 trillion; in 1980 it was just over 2.7 trillion. In 1997, the gross domestic product was approximately $7 trillion. Evaluating the change in gross domestic product from one year to the next is one way to determine whether or not there has been economic growth.[12]

By measuring the changes in the growth of the gross domestic product (GDP) annually, economists are able to track the health of the economy. When the GDP fails to grow significantly, it is a sign that there may be underlying problems in the nation's economy. As a result, government officials may offer suggestions and take steps to influence the direction of the economy. If the overall GDP grows substantially, it indicates that economic policies do not need fixing.[13]

Standard of Living. The gross domestic product is one way to measure how well people are living in a particular nation. The **standard of living** depends on the amount and kinds of goods and services that people of a nation enjoy. An economic system that can provide for its citizens' basic needs and can product the goods and services its citizens want and can afford, that nation will have a high standard of living.[14]

Inflation

Rising or falling prices affect the dollar value of the gross domestic product. A prolonged rise in the general price level of goods and services is called inflation. Business and government leaders consider the **inflation** rate to be an important general economic indicator. For example, last year a gallon of milk may have cost $2.00. This year, it may cost

$2.25. The physical output (the gallon of milk) has not changed, only the money value.

Inflation and the Purchasing Power of Money.
Inflation reduces the "purchasing power" of money. Your money buys less. Sometimes people describe inflation as a time when "a dollar is not worth a dollar anymore." A dollar's purchasing power is the real goods and services that it can buy. If your income stays the same, but the price of one good that you are purchasing goes up, your effective purchasing power falls. **Deflation**, a prolonged decline in the general price level, also affects the dollar value of the gross domestic product. However, deflation rarely occurs.[15]

Who Loses from Inflation?
Perhaps more than any other single group, people on fixed incomes, such as retirees, have reason to worry about inflation. Retired people cannot count on an increase in income as prices rise. Anyone on a fixed income cannot stay up monetarily (due to declining purchasing power) as fast as prices rise. Many retired people have fewer goods and services to enjoy because of the rising prices. Refer to Figure 18–3 for a **scenario** on the plight of retirees with inflation.

Landlords worry about inflation when the rental income is tied to long-term rental leases. Workers who accept union-negotiated long-term fixed wages can lose due to inflation. For example, a contract negotiated during the 1980s when the inflation rate was 4.6 percent allowed you to buy, perhaps, $100 worth of groceries. By 1990, you could buy only $63.70 worth of groceries (assuming the price of groceries went up at the same rate as the price level).[16] Savers can lose if the inflation rate exceeds the interest rate. However, if compound interest is taken into account over several years, as discussed in an earlier chapter, savers probably will never lose.

Who Gains from Inflation?
Not everyone loses from inflation. Homeowners, landowners, and investors gain from inflation. Brick homes with three bedrooms, kitchen, den, and one bathroom were purchased for about $15,000 in the early 1960s. The same house today is worth approximately $75,000 or more. Farms purchased in certain regions of the Southeast for $200 per acre in the early 1960s are now selling for $3,000 or more per acre. Typically, the rise in inflation

Fixed Income and Inflation

Perhaps more than any other single group, people living on fixed incomes, such as retirees, have to worry about inflation. Back in 1958 Maria and her husband bought a deferred annuity that cost them $100 monthly. In 1998, it started to pay them $700 a month in retirement benefits. They were excited about the prospect of living comfortably in retirement on the savings they had accumulated. After all, back in 1958, when they put their retirement plans together, their rent was $135 per month; a new automobile was $1,500; milk was $.25 a quart; and a first-run movie was $.35. What they failed to factor into their plans, however, was inflation.

When Maria and her husband retired in 1998, they began to receive their $700 each month, yet their retirement dreams were shattered. The $700 does not come close to covering their apartment rent, and a new car—now at $17,000—is simply out of the question. They, along with millions of other people who live on fixed incomes, are big losers under inflation.

Figure 18–3. More than for any other group, inflation is especially difficult for retirees living on a fixed income.

causes investments to increase in value. Obviously, every investment is not a winner but an **astute** businessperson will constantly monitor investments, buy and sell, to get the most profit possible.

Measuring Inflation. Prices in different sectors of the economy rise at different rates. As a result, the government uses several measures of inflation. The three most commonly used are the **consumer price index**, the **producer price index**, and the **implicit GDP price deflator**. Only the consumer price index will be discussed here in measuring inflation.

The consumer price index is found through a survey of several hundred goods sold in about 21,000 outlets.[17] These goods called **market basket** includes about 400 goods and services such as food, housing, transportation, clothing, entertainment, medical care, and personal care.[18] Refer to Figure 18–4 for the consumer price index for several years.

The Federal Bureau of Labor Statistics compiles the consumer price index. The prices for the goods and services are set from a **base year**. The average prices for 1982, 1983, and 1984 are the base years used by many economists and are given a value of 100. Base years are chosen because they are considered typical of most years. Change in price levels cause the consumer price index to either rise or fall.

Calculating Inflation from the Consumer Price Index. Let us consider two examples. The *consumer price index* (CPI) for later years indicates the percentage that the market basket price has risen since the base year (100).

Example 1: If the index for the current year is 140, it means that prices have gone up 40 percent since the base year (140 − 100 for the base year = 40 percent increase).

Example 2: The consumer price index (CPI) also can be used to calculate inflation from year to year. Refer to Figure 18–4. At the end of 1993, the CPI is 144.4. At the end of 1992, it was 140.3, which is a difference of 4.1 (144.4 − 140.3 = 4.1). If we use 1992 as the base year, we can find out what percentage consumer prices on average rose from 1992 to 1993. We do this by dividing 4.1 by 140.3, which gives us 0.0292 (4.1 ÷ 140.3 = .0292 × 100 = 2.92). The .0292 is multiplied by 100 to give the result or a percentage (2.92%).

Unemployment

Another important economic indicator is the number of people who are unemployed during a given period of time. The U.S. Bureau of Labor Statistics gathers data related to employment and unemployment in our economy each month. Roughly 50,000 households are questioned each month.[19] These homes are chosen from all geographic, ethnic, social, and occupational groups to

Consumer Price Index	
1985	107.6
1986	109.6
1987	113.6
1988	118.3
1989	124.0
1990	130.7
1991	136.2
1992	140.3
1993	144.4
1994	148.4
1995	152.4
1996	156.9
1997	160.5

Figure 18–4 The consumer price index is a major indicator for measuring inflation. (Source: Bureau of Labor Statistics)

represent a cross-section of the American population. The unemployment rate in March 1998 was the lowest in years (4.6 percent). In 1982 and 1983, the United States had an unemployment rate near 10 percent, which is considered high.[20]

Calculating the Unemployment Rate. The unemployment rate measures the percentage of the total civilian labor force that are currently unemployed. The formula is as follows:

$$\text{unemployment rate} = \frac{\text{number of people unemployed}}{\text{number of people in the civilian labor force}}$$

Types of Unemployment. There are different kinds of unemployment. Some types are a greater problem than others. Four specific types have been identified by economists. The types are structural, frictional, seasonal, and cyclical unemployment.

Structural unemployment—the most serious—is made up of workers whose skills are no longer in demand due to technological advances or discoveries of natural resources. *Frictional unemployment* is made up of workers who are between jobs, and who will quickly find new ones. Unemployment is due to firings, layoffs, voluntary searches for new jobs or training. *Cyclical unemployment* is made up of workers who have lost their jobs because of a recession or a downturn in the economy. *Seasonal unemployment* occurs where demand for labor varies over the year, as with the harvesting of crops.

Discouraged workers are workers who are not actively searching for work. They are not considered a part of the civilian labor force and therefore are not counted among the unemployed. These are workers who have given up looking for a job.

Why Unemployment Matters as an Economic Indicator. High unemployment is usually a sign that all is not well with the economy. Also, the waste of human resources that unemployment causes is an extremely serious problem. Unemployment can reduce living standards, disrupt families, and reduce an individual's self-concept. As a result, maintaining a low unemployment rate is one of the major goals in stabilizing the economy.[21]

Value of Economic Indicators

Although the economic indicators are not always exact, they are exact enough to show what is generally happening in a segment of the economy. Various groups and agencies use and study the economic indicators. The government uses the indicators to control the rate of inflation. Business people check the economic indicators and adjust their production of goods and services. Investors watch the economic indicators to determine when to buy or sell stocks. If inflation is moderate and the unemployment rate low, investors might want to buy stocks to make a profit due to future economic prosperity.

MONEY AND BANKS

Economists view money as a claim on something that has value. Money usually has little value of its own, but it represents value. People must agree that something is money.[22] In some places, beads, shells, special rocks, cows, or corn are money because the people there accept them as representing value.

Different Types of Money

Economists differentiate three different types of money: fiat money, commodity money, and bank money. A brief discussion of each follows.

Fiat Money. *Fiat* means "order." **Fiat money** has value because a government fiat, or order, has established it as acceptable for payment of debts. Dollar bills are an example of fiat money because their value as slips of printed paper is less than their value as money. Look at any dollar bill. To the left of George Washington's picture you will find a statement that the bill is "legal tender." This means that it must be accepted to satisfy a debt.[23]

Commodity Money. **Commodity money** is a *good* whose value serves as the value of money. Gold coins are an example of commodity money. Mediums of exchange such as cattle and gems are considered commodity money. They have a value as a good aside from their value as money. For example, cattle are used for food and, in some countries, for food, power, and transportation, whereas gems are used for jewelry.[24]

Bank Money. **Bank money** consists of the bank credit that banks extend to their depositors. Transactions made using checks drawn on deposits held at banks involve the use of bank money. While fiat money or currency makes up only 20 to 25 percent of the total money supply in the U.S., bank money or checkbook dollars make up the remaining 75 to 80 percent of the money supply.

Functions of Money

Money is often defined in terms of the three functions or services that it provides. Money serves as a medium of exchange, storage value, and standard of value.

Medium of Exchange. Money's most important function is as a medium of exchange to facilitate transactions. Buyers and sellers use money as a means of payment for goods and services. Money is the "go between" that facilitates each exchange. Without money, all transactions would have to be conducted by barter, which involves the exchange of one good or service for another. Barter is a clumsy method of exchange.

Storage Value. Money has the ability, called store of value, to hold its value over time—it doesn't spoil, rot, or disintegrate.[25] Money can be accumulated and saved, hopefully without losing value. If there were no inflation, the value of a dollar today would be the same as its value was three years ago.

Standard of Value. Money also functions as a *unit of account,* providing a common measure of the value (yardstick) of goods and services being exchanged. It helps us compare the value of different goods and services. In this country, we measure economic value in terms of our basic unit of money, the dollar. In Japan, it is the yen; in France, the franc.

How Banks Create Money

Consider what happens when the bank receives a $100,000 deposit for one of its depositors. The bank is required to set aside 10 percent of this deposit, or $10,000 as reserves. Refer to Figure 18–5. It then lends out its excess reserves, in this case, the remaining $90,000 of the initial deposit.

For the sake of better understanding, suppose that all borrowers redeposit their loans in the same bank. The bank receives another $90,000 in new deposits of which the bank sets $9,000 aside as required and lends out all of its excess reserves. Let us pretend that all borrowers redeposit their loans in the same bank, and the bank again sets aside 10 percent of these deposits. The bank again lends out the remainder which is redeposited in the bank and the cycle continues.[26]

If you were to follow this multiple deposit expansion to its completion, the end result would be that the bank's deposits would increase by $1 million, its loans would increase by $900,000, and its reserves would increase by $100,000, all due to the initial deposit of $100,000. Again, refer to Figure 18–5.

In reality, loan recipients do not deposit all of their loans into a bank; they hold a fraction of their loan funds as currency. If some loan funds are held as currency, then there is a leakage of money out of the bank system, and the 10 percent of new deposits from the loans would not hold true.[27] However, we learned earlier that 75 to 80 percent of the money flow in the United States is deposited in banks and checking accounts.

Multiple Expansion of Deposits			
Round	New Deposits	New Reserves	New Loans
1	$100,000	$10,000	$90,000
2	90,000	9,000	81,000
3	81,000	8,100	72,900
4	72,900	7,290	65,610
5	65,610	6,561	59,049
6	59,049	5,904	53,145
7	53,145	5,314	47,831
8	47,831	4,783	43,048
9	43,048	4,304	38,744
10	38,744	3,874	34,870
11	34,870	3,487	31,383
12	31,383	3,138	28,245
13	28,245	2,824	25,421
14	25,421	2,542	23,879
15	23,879	2,387	21,492
16	21,492	2,149	19,343
17	19,343	1,934	17,409
18	17,409	1,740	15,669
19	15,669	1,566	13,103
20	13,103	1,310	11,793
21	11,793	....	....

Figure 18–5 With an original deposit of $100,000, a bank could create $1 million if borrowers redeposited all of their borrowed money there.

THE FEDERAL GOVERNMENT AND ECONOMIC STABILITY

Economic stability is one of our economic goals. However, some changes in our business and economic world are expected. Earlier we discussed the Great Depression. A fear of another Great Depression keeps the federal government involved in managing the economy. When the economy is in a recession, the government tries to prevent a depression and bring about a recovery. When economic times are good, the government attempts to maintain it. The government also tries to control the rate of inflation.

Government economic policies designed to influence economic activity are **fiscal policy** and **monetary policy**. Fiscal policy involves using government spending and taxation to influence the economy. Monetary policy involves controlling the supply of money and credit to influence the economy. However, the goals of both the fiscal and monetary policies are the same. The goals are to promote price level stability, full employment, and the achievement of a strong gross domestic product.

Fiscal Policy

Fiscal policy is carried out by the legislative and/or the executive branches of government. The two main instruments of fiscal policy are government expenditures and taxes.[28]

Government Expenditures.   Sometimes the government acts to slow down economic activity. If inflation is too high, the government may cut its own spending. Since the government is a big consumer of goods and services, less government spending will slow economic activity.[29]

Taxes.   Another way the federal government tries to either stimulate or dampen the economy is through taxes. Cutting taxes indirectly increases consumer spending. When the government cuts taxes, workers have more take-home pay. By increasing their spending on goods and services, the consumer helps spend the economy out of a recession.

Monetary Policy

Monetary policy is the management of the money supply and interest rates. Monetary policy is conducted by a nation's central bank. In the United States, monetary policy is carried out by the **Federal Reserve System**.

Slowing Down the Economy.   The Federal Reserve is one of the sources of money in the economy; it can add or subtract money from the economy as it sees fit. One way the Federal Reserve can subtract money is by increasing the interest rate on money it lends to banks. Banks pass the increase on to businesses causing the economy to slow as the businesses borrow less.[30]

Speeding Up the Economy.   The Federal Reserve can also "speed up" the economy by lowering the interest rates it charges banks. For example, when unemployment gets too high, the Federal Reserve may put more money into the economy by lowering the interest rates. This stimulates spending and encourages business growth, which leads to hiring of more people.[31]

Independence of the Federal Reserve System. The Federal Reserve System operates independently of the President and Congress. It has the goal of keeping the economy growing without causing inflation. As shown above, it does this by managing the money supply and interest rates.

BASIS OF ECONOMIC POLICY

An **economic policy** is a course of action that is intended to influence or control the behavior of the economy. Economic policies are typically implemented and administered by the government. Examples of economic policies include decisions made about government spending and taxation, about the redistribution of income from rich to poor, about the supply of money, and about what to include in the farm bill. The effectiveness of economic policies can be accessed in one of two ways: positive economics (truth) and normative economics (**value judgments**).

Positive Economics

Positive economics focuses on "what is" and "what would happen if" questions and policy issues. No value judgments are made. Instead, the economic behavior of producers and consumers is explained or predicted.[32] For example, the **hypothesis** that "an increase in the supply of money leads to an increase in prices" belongs to the realm of positive economics because it can be tested by examining the data on the supply of money and the level of prices.[33] In other words, economists call these things *positive statements* because they can be proven.

Normative Economics

Normative economics involve the use of value judgments to assess the performance of the economy and economic policies. The focus of normative economics is on determining "what should be" or "what ought to be." Therefore, normative economic *hypotheses* cannot be tested. For example, the hypothesis that "inflation rate is too high" is an example of normative economics because it is based on a value judgment and therefore cannot be tested or confirmed.[34]

THE NATION'S ECONOMIC GOALS

Nations have values and set goals based on these values. These goals are evident in government policies and in the actions of people. The values and goals of a nation determine the kind of economic system they choose (if they have a choice). As discussed earlier, we learned that the United States tends toward a market or capitalist economic system.

Before we take a trip, we usually decide where we want to go and plan the best route to get there. Otherwise, how can we know when we have arrived if we do not know where we are going? As a nation we also need to know where we are going economically so we can plan the best route. We do this by setting some specific goals and evaluate our progress periodically relative to where we are.

Over the years several general economic goals have been established for the nation as a whole. Some of the major ones include economic growth, economic stability, economic justice, economic security, and economic freedom. A brief discussion of each follows.

Economic Growth

Economic growth means an expansion of the economy to produce more goods, services, jobs, and wealth. Economic growth represents an effort to raise our standard of living. It also relates to levels of employment or unemployment and numerous other economic concepts that affect society directly. High rates of growth may cause some problems such as environmental pollution, but low growth tends to cause more problems.

Before the 1930s we assumed that the private sector of the economy would accomplish this goal without intervention of government. Today, we hold the federal government responsible for providing appropriate monetary and/or fiscal policies to attain the acceptable level of economic growth.

Economic Stability

The goal of economic stability is to reduce extreme ups (inflation) and extreme downs (deflation) in the economy. The most preferred goal is a low rate of inflation; usually 2 to 4 percent is considered acceptable. Some inflation is desirable because it tends to stimulate economic growth. However, when it gets to above 10 percent, economists and others become concerned.

Production Agriculturalists and Economic Stability. Economic instability is a major problem for production agriculturalists, much more so than for other sectors of the economy. Due to the large number of U.S. producers and the impact of

world production on American agriculture markets, agricultural product prices are constantly changing. For most nonfarm business, if prices change, they usually go up. For production agriculturalists, prices go down about as much as they go up. This results in "boom and bust" production agriculture. It is very difficult to plan future production when you have very little control over the price you receive. During the first half of the 1980s, most production agriculturalists received no returns for their labor and very little return on their investment. To reduce the impact of market price variation, the government has provided minimum price guarantees on a wide variety of agricultural products to help with economic stability.

Economic Justice

One of the most difficult economic goals to accomplish is how to divide the economic pie so that everyone gets a fair share, yet it is probably the most important goal. When a society consists of millions of people and each has a different perspective of what is fair and just, it is almost impossible to accomplish this goal. However, the members of all societies must feel that the system is at least acceptable or a new system will be developed. Many revolutions have been fought for just this reason.

Ways to Attain Economic Justice. In an effort to provide equal opportunity for everyone, the following programs have been developed: a free public school system, special programs for the handicapped, minimum wage laws, price support programs, and laws prohibiting discrimination against individuals based on sex, race, religion, and age.

Progressive Income Tax. In addition, the United States has a progressive income tax so that those who have more income are required to pay a larger share for public services. We have an inheritance tax so increasing amounts of wealth will not accumulate over time in the hands of a few families. As our concept of what is fair changes, new programs will be developed to accomplish the continuous goal of economic justice. However, we are not likely to find a system that will provide complete economic justice for everyone.

Economic Security

Security means protecting people against poverty and supplying them with the means to provide for a medical emergency through an increasing number of government social programs directed, among others, to the elderly. One of the best examples of government efforts to provide economic security for American citizens is the social security program. This program is designed primarily to provide financial assistance to citizens over 65 or 67, depending on the year of birth, and to assure financial support to those who cannot work due to some medical problem.

Economic Freedom

The goal of economic freedom allows each member of society to enjoy the freedoms of entrepreneurship, choice, and private property. Economic freedom also allows individuals to make their own decisions in the marketplace.

Economic freedom in America means that individuals can select a career of their choice. They can establish a business of their own or work for someone else. However, all economic activity has restrictions. There are zoning rules that restrict where certain types of businesses can be located. There are health restrictions to protect workers and consumers. In fact, there are thousands of economic rules established by government that specify what individuals can and cannot do.

Production Agriculturalists and Economic Freedom. Even production agriculturalists, who are usually associated with a high degree of economic freedom, do not have complete freedom to do as they please. For example, they are often told what

Figure 18–6 Even though production agriculturalists have economic freedom, there are many government restrictions on who can produce tobacco and how much they can produce. (Courtesy of Jamie Mundy)

they can produce and how much. Only those producers who have grown tobacco in the past can produce and sell tobacco today, or they can buy an "allotment" from someone else who has raised tobacco. Those who now produce tobacco are restricted by government programs relative to how much they can sell. Those that are not restricted by government are restricted by the marketplace. Refer to Figure 18–6. Any production agriculturalist who expects to make a profit must produce the type and amount of products that buyers are willing to purchase. However, even with all these restrictions, Americans still have more economic freedom than any society in the world.

TYPES OF ECONOMIC ANALYSIS

There are many ways to analyze economic situations. Whole textbooks have been written on economic analysis. However, our purpose is to briefly introduce six economic analysis concepts which have been developed to help us understand more clearly what economics is and how it is applied.

The six economic concepts are: **marginal analysis**, ceteris paribus (all else held constant), **production possibilities**, **opportunity cost**, **law of increasing cost**, and **equal product curve**.

Marginal Analysis

In marginal analysis, one examines the consequences of adding to or subtracting from the current state of affairs. For example, suppose you are a small agribusiness owner which operates a lawn and garden center. You are considering hiring a new worker. You determine the *marginal benefit* of hiring the additional worker as well as the **marginal cost**.

Marginal Benefit.   The **marginal benefit** of hiring a new worker is the value of the additional goods or services that could be produced by hiring, for example, a new lawn and garden center employee. In other words, can more product such as supplies, shrubs, and flowers be produced by hiring the new employee?

Marginal Cost.   The marginal cost is the additional wages the employer will have to pay the new lawn and garden center employee.

Economic Decisions.   An economic analysis of the decision to hire the lawn and garden center employee involves weighing the marginal benefits against the marginal cost. If the marginal benefits are greater than the marginal costs, then it makes sense for the owner of the lawn and garden center to hire a new employee. If not, then the new worker should not be hired. In other words, if the cost of the new worker was $8,000 per year, net profits from the sale of trees, flowers, shrubs, and so on from the lawn and garden center would have to exceed more than $8,000 per year.

Ceteris Paribus ("All Else Held Constant")

In performing economic analysis, it is sometimes difficult to separate out the effects of different factors, decisions, or outcomes. For example, let us say sales of the lawn and garden center doubled after the new worker was hired. However, the same week the new worker was hired, television advertising began, an ad was placed in the newspaper, and a new sign was erected in front of the lawn and garden center.

Determining the Cause of an Increase in Sales.   Four things or variables happened which could have caused the sales to double: new worker, television advertising, newspaper advertising, and a new sign. If we really wanted to determine the increase in sales, we could hold constant three of the **variables**. Since the new worker has two young children, let us stop the television and newspaper advertising and remove the sign. If sales remain doubled after holding three variables constant, our economic analysis will have "separated out" the new worker as the reason for doubling sales. The variable that was allowed to vary was the new worker. Of course, we could have held constant any of the four variables, but in the final analysis, we would have found that the new worker is what made the difference in sales doubling.

Production Possibilities

Production possibilities is another type of economic analysis. The total amount of products produced by a society or any business producing a product is limited by the type and amount of resources available in combination with a specific level of technology. To simplify this concept, let us assume that there are resources available for the lawn and garden center owner to produce only two products, poinsettias and mums. The owner can produce all poinsettias, all mums, or several combinations of the two. A hypothetical example of several production possibilities is shown in Figure 18–7.

Figure 18–7 is used to make a specific economic point. In modern terms it shows that you can't

Selected Production Options for Poinsettias and Mums

Production Possibilities	Poinsettias (6" pots)	Mums (6" pots)
A	0	270
B	25	247
C	50	215
D	75	170
E	100	105
F	125	0

Figure 18–7 Only a certain quantity can be produced by any one individual or firm. To produce more of one item, you must produce less of one or more other items.

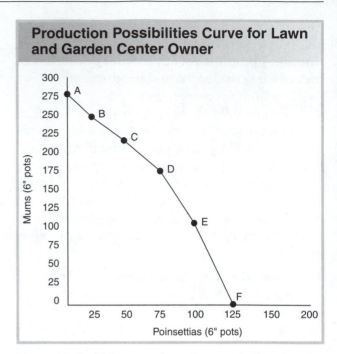

Figure 18–8 This curve shows how much of one product can be produced relative to production of another product.

have your cake and eat it too. Specifically, in order to produce poinsettias, the lawn and garden center owner must give up some mums. To gain the first 25 six-inch pots of poinsettias, 23 six-inch pots of mums must be sacrificed. This concept continues throughout the example, and if the owner wants to produce 125 six-inch pots of poinsettias, the owner cannot produce any mums. Conversely, if the owner wants to produce more mums, he or she will have to give up some poinsettias. This concept not only applies to the lawn and garden center owner but to society as a whole. All societies are limited by what they can produce. Therefore, to produce more of one item we must give up the production of one or more others.

This concept is often illustrated with a graph as well as a table. The graphic illustration is called a production possibilities curve. This curve shows all production possibilities between two products. This type curve is sometimes called a production possibilities frontier as well as a transformation curve. Refer to Figure 18–8.

Opportunity Cost

One of the most important economic analysis concepts is called opportunity cost. Economists define opportunity cost as the value of a service or product that must be given up in order to obtain another good or service. To illustrate this concept refer again to Figure 18–8, the production possibilities curve. The opportunity cost of producing poinsettias is the value of the mums the owner has to give up. Specifically, the opportunity cost for producing the first 25 six-inch pots of poinsettias is the value of the 23 six-inch pots of mums the owner has to give up. The opportunity cost of producing 125 six-inch pots of poinsettias is the value of 270 six-inch pots of mums.

The concept of opportunity cost applies to all aspects of our lives. There is an opportunity cost to every decision we make. For example, the total cost of a college education is not just the direct cost of tuition, books, and room and board. It also includes an *opportunity cost*—the amount you could have made if you worked full-time instead of going to college. From an economic analysis point of view, the best decision is one in which the alternative you select (college) provides a greater return than your opportunity cost (money loss in wages while obtaining college).

Law of Increasing Cost

We have established the economic relationship between the production of two products using limited resources that as you produce more of one you have to produce less of another. However, a closer look at the illustration shows an even more significant relationship—the rate of substitution between two products is not equal. The law of increasing cost states that to produce equal extra amounts of one product or service, an increasing amount of another product or service must be given up. Figure 18–9 shows that the first 25 six-inch pots of poinsettias require a sacrifice of only 23 six-inch pots of mums. However, the second 25 six-inch pots of poinsettias require a sacrifice of 32 six-inch pots of mums. To add 25 more six-inch pots of poinsettias, the owner of the lawn and garden center would have to give up 105 six-inch pots of mums.

You may be asking at this point why the sacrifice in mum production increases rather than remains constant or decreases. The answer is because there is not perfect flexibility or interchange ability among resources used to produce both crops. For example, poinsettias can be inserted into a rotation, making use of surplus labor, spreading out the growing seasons, and using greenhouse space that is more efficient for poinsettia than mums. However, these advantages decline progressively as more poinsettias are produced. This also explains why a production possibilities curve is bowed rather than a straight line. Again, refer to Figure 18–8.

Equal Product Curve

Figure 18–8 illustrated a production possibilities curve showing that one product can be substituted

Law of Increasing Cost

Production Possibilities	Poinsettias (6" pots)	Mums (6" pots)	Mums given up for Poinsettias
A	0	270	—
B	25	247	23
C	50	215	32
D	75	170	45
E	100	105	65
F	125	0	105

Figure 18–9 The law of increasing cost states that to produce an extra amount of one product (poinsettias), an increasing amount of another product (mums) must be given up.

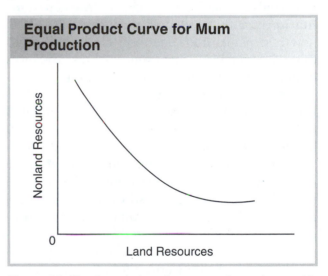

Equal Product Curve for Mum Production

Figure 18–10 An equal product curve shows the combinations of two sets of resources that will produce the same amount of total output.

for another although not at a constant rate. However, different *resources* can also be substituted for one another in the production of a given product. There are numerous combinations of land, labor, and capital that will produce the same amount of product. This concept is often expressed by an equal product curve. Refer to Figure 18–10. An equal product curve shows the combinations of two sets of resources that will produce the same amount of total output. It should be noted that the curve is bowed in rather than bowed out. This is because we have resources instead of *products* on the axes and the fact that there is not perfect substitutability between resources.

Productive Power of Advanced Technology and Innovations

Ideas, more so than any factor of production, are the most revolutionizing force able to shift the production possibilities of any economy. Ideas can shift the curve out beyond imagination.[35] Who would have thought just a century ago that we could breed thousands of cows from one bull (**artificial insemination**), or that we could produce thousands of new rose plants from one rose (**cloning**). Economists describe ideas that eventually take the form of new applied technology as innovations.

Earlier we emphasized the fact that a production possibilities curve is fixed due to limited resources and the producer or society in general cannot produce beyond their production possibilities curve. However, this limitation is not absolute due to technologies and innovations. We do have some flexibility. We can substitute one product for another and, best of all, we can even shift the production possibilities curve outward. *An outward shift of the production possibilities curve is called economic growth.* An outward shift in the equal product curve implies the same thing. How is this possible? We can improve the productivity of our resources through new technologies and by changing the combination of resources used.

For example, through the use of hybrid seed corn we have increased the average yield of corn from 74 bushels per acre in 1965 to over 120 bushels per acre. We have similar results in crossbreeding, artificial insemination and embryo transfer, in cattle. The use of computers has improved the productivity of resources in all phases of our economy. The substi-

tution of capital for labor also shifts our production possibilities curve outward.

Another example is that economic levels of countries who apply a large amount of capital relative to labor tend to be much higher. We also classify countries as developed or underdeveloped relative to their level of economic growth. Figure 18–11 shows a production possibilities curve using food and nonfood products for a developed country (B) versus an underdeveloped country (A). Note that more of both food and nonfood items can be produced.

Common Pitfalls in Economic Analysis

There are two "pitfalls" that should be avoided when conducting economic analysis: the fallacy of composition and the false-cause fallacy.

Fallacy of Composition. Fallacy of composition is the belief that if one individual or firm ben-

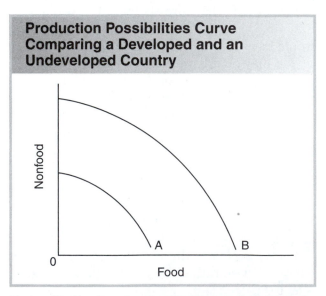

Figure 18–11 An outward shift of the production possibilities curve is called economic growth. Due to new technologies and greater amounts of capital relative to labor, developed countries have a greater outward shift of the production possibilities curve.

efits from some action, all individuals or all firms will benefit from the same action. While this may in fact be the case, it is not necessarily so.

For example, suppose that a purebred Angus breeder decided to start embryo transfer as a management technique to increase profits. As a result, it worked, and the news spread throughout the country. However, when other cattle breeders implemented the new technology, they did not experience the same profits because the market for their calves produced from embryo transfer was not as good as the purebred Angus breeder's. The reason was that they did not have as good a quality of cattle from the start. Hence, the profits were not realized.

False-Cause Fallacy. A similar pitfall in economic analysis is the false-cause fallacy. The false-cause fallacy often arises in economic analysis of two correlated actions or events. When one observes that two actions or events seem to be correlated, it is often tempting to conclude that one event has caused the other. However, by doing so, one may be committing the false-cause fallacy, which is ignoring the simple fact that *correlation does not imply causation.*

For example, let us pretend that new tractor prices have steadily increased over some period of time and that new tractor sales have also increased over this same period. On the surface, it would seem that an increase in the price of new tractors causes an increase in new tractor sales. This false conclusion is an example of the false-cause fallacy; new tractor prices and new tractor sales may be positively correlated, but that correlation does not imply that there is any causation between the two. In order to explain why both events are taking place simultaneously, one may have to look at the other factors or variables such as rising consumer incomes, inflation, or rising producer cost.[36]

ATTAINING ECONOMIC EFFICIENCY

Economic growth is used to describe the results of improving output relative to the limited resources provided by nature. Specifically, how can we change an input-output ratio in order to improve economic efficiency? Five specific steps follow which will improve the input-output ratio at any point in time. Let us assume the following input-output ratio. It takes 500 units of resources to produce 100 units of outputs. How can we improve this input-output ratio of five to one? There are at least five ways to improve:

Strategy 1: increase output with the same input

Result: more milk from the same number of cows

Strategy 2: same output with less input

Result: same amount of milk with fewer cows

Strategy 3: increase output with less input

Result: more milk with fewer cows

Strategy 4: increase inputs and increase output but increase output more than input

Result: 5 percent more cows results in 10 percent more milk

Strategy 5: decrease inputs and decrease output but decrease inputs more than outputs

Result: using 10 percent less feed results in 5 percent less milk

All of these possibilities are difficult, some more than others, but they are all realistic alternatives. As one or more of these alternatives are accomplished by agribusinesses or society in general, our economic standard of living is improved. Some societies have been very successful with this process while others can't seem to make any progress. Some industries have been more successful than others. For example, output per hour of labor for farm workers has increased approximately 30 percent since 1979 while output per hour of labor for nonfarm workers has increased only 10 percent during the same period.

Several careers are related to the economic activity and analysis. The Career Option below explains some of these in the area of agribusiness.

CONCLUSION

Economics is a complex process involving land, labor, capital and management. Much of the research, policies, and laws that are in place today are the result of the Great Depression. The Great Depression was so devastating to the American people both economically and psychologically that efforts are continuously being made to avoid another catastrophe. Besides the impact of the Great Depression, the desire of the American people to have a comfortable standard of living is the other reason we study economic activity and make economic analysis.

SUMMARY

A part of understanding agricultural economics is understanding economic activity and its relationship with federal, state, and local governments.

CAREER OPTION

Career Areas: Ag Banker, Financial Manager, Agriscience Loan Officer, Financial Analyst, Marketing Specialist

The management of finances is the most important single function of any career option associated with agribusiness. Possessing a great deal of technical knowledge and know-how is very important. However, this know-how will not make you a success in your career field unless you are a competent money manager.

As a financial manager, you might work for a bank or a credit agency as an agriscience loan officer, or for a company that sells supplies to producers and growers. As credit manager, loan officer, financial analyst, or marketing specialist you could utilize your knowledge of business and finance as well as production, processing, or distribution. Careful planning of finances is probably more important than planning any other aspect of the agribusinesses. Some very good agribusinesses have failed because the manager was not a good money manager.

Many jobs are available in banks in the position of agricultural loan officer. An overall knowledge of business and finance is necessary, as well as knowledge about the agribusiness input, production, and output sectors. (Courtesy of Cliff Ricketts)

Also, the role of the family in society and how the family behaves as an economic unit is tied to agricultural economics. Therefore, an understanding of economic activity, economic goals of a nation, and how to analyze these goals and activities are crucial to understanding the economic system of the United States.

Many factors are involved in maintaining a strong economy. These factors include efficient production of agricultural products freeing workers for industrial work; efficient production of goods and services; and good wages so consumers can purchase goods, services, and entertainment. The job of economist is to alert us when the factors within the economy are beginning to get unbalanced. Two things that economists study are business cycles and patterns, and the circular flow of the economy.

Economic indicators are important data or statistics that measure economic activity and business cycles. These indicators normally change before the rest of the economy does. There are many economic indicators, but the three major ones are: gross domestic product, inflation, and unemployment.

Economists view money as a claim on something that has value. Money usually has little value of its own, but it represents value. Economists differentiate three different types of money: fiat money, commodity, and bank money. Banks create money by making loans, keeping 10 percent in reserves and reloaning money from deposits of the loaned money.

Economic stability is one of our economic goals. When the economy is in a recession, the government tries to prevent a depression and bring about a recovery. When economic times are good, the government attempts to maintain it. Government economic policies designed to influence economic activity are fiscal policy and monetary policy. Fiscal policy involves using government spending and taxation to influence the economy. Monetary policy involves controlling the supply of money and credit to influence the economy.

An economic policy is a course of action that is intended to influence or control the behavior of the economy. The effectiveness of economic policies can be assessed in one of two ways: positive economics (truth) and normative economics (value judgments). Positive economics focuses on "what is" and "what would happen if" questions and policy issues. Normative economics involve the use of value judgments to assess the performance of the economy and economic policies.

Nations have values and set goals based on these values. These goals are evident in government policies and in the actions of people. Over the years several general economic goals have been established for the nation as a whole. Some of the major ones include economic growth, economic stability, economic justice, economic security, and economic freedom.

There are many ways to analyze economic situations. Six economic analysis concepts have been developed to help us understand more clearly what economics is and how it is applied. These six economic analysis concepts are marginal analysis, ceteris paribus (all else held constant), production possibilities, opportunity cost, law of increasing cost, and equal product curve. However, there are two pitfalls that should be avoided when conducting economic analysis: the fallacy of composition and the false-cause fallacy.

Economic efficiency is getting the most profit possible from a product or economy by having the most optimum input-output ratio. Economic growth is used to describe the results of improving output relative to the limited resources provided by nature.

Economics is a complex process involving land, labor, capital, and management. Besides the impact of the Great Depression, the desire of the American people to have a comfortable standard of living, is the other reason we study economic activity and conduct economic analysis.

END-OF-CHAPTER ACTIVITIES

Review Questions

1. Define the Terms to Know.
2. List four family matters that are affected by economics.
3. List three things that are crucial to understanding the economic system of the United States.
4. What are three factors involved in maintaining a strong economy?
5. What is the purpose of economists?
6. Why does the state of the economy require daily attention?
7. Name the four parts of a business cycle and give two characteristics of each.
8. Differentiate between a recession and a depression.
9. After examining Figure 18–2, explain the circular flow of the economy.
10. What are three economic indicators?
11. In calculating the gross domestic product, economists add the sum of what four main areas of goods and services?
12. What four groups can lose from inflation?
13. Who are three groups that gain from inflation?
14. What are three ways to measure inflation?
15. Name and define the four types of unemployment.
16. Why does unemployment matter as an economic indicator?
17. What are three values of economic indicators?
18. Name and describe the three functions or services of money.
19. Explain how banks create money.
20. What are two economic policies that the government uses to stabilize the economy?
21. Name and explain two fiscal policies the federal government can enact to influence the economy.
22. Name and explain two ways that the Federal Reserve System can influence the economy through monetary policies.
23. State four examples of economic policies.
24. Name and briefly describe five of the nation's economic goals.
25. List six economic analysis concepts.
26. Briefly explain the economic analysis concept of marginal analysis.
27. Briefly explain the economic analysis concept of ceteris paribus.
28. Briefly explain the economic analysis concept of production possibilities.
29. Briefly explain the economic analysis concept of opportunity cost.

30. Briefly explain the economic analysis concept of law of increasing cost.
31. Briefly explain the economic analysis concept of equal product curve.
32. Explain how the production possibilities curve can shift outward due to technology and innovations.

Fill in the Blank

1. In reality, government budgets are _____ documents.
2. During the Great Depression, in 1932, the unemployment rate reached _____ to _____ percent of the workforce.
3. The gross domestic product in 1970 was over $1 trillion, in 1980 it was over $2.7 trillion, in 1997 it was _____.
4. By measuring the changes in the growth of the gross domestic product annually, economists are able to track the health of the _____.
5. Inflation reduces the _____ of money.
6. The _____ _____ _____ _____ _____ compiles the consumer price index.
7. Workers who have given up looking for a job are referred to as _____ _____ and are not included in the unemployment rate.
8. _____ usually has little value of its own, but it represents value.
9. Bank money or checkbook dollars make up _____ to _____ percent of the money supply.
10. In the United States, monetary policy is carried out by the _____ _____ _____.
11. An outward shift of the production possibilities curve is called _____ _____.
12. _____ _____ _____, more so than any factor of production, are able to shift the production possibilities of any economy.

Matching

a. Federal Reserve System
b. dollar bills
c. normative economics
d. checks
e. ceteris paribus
f. legal tender
g. positive economics
h. fallacy of composition
i. monetary policy
j. fiscal policy
k. false cause fallacy
l. gold coins

_____ 1. must be accepted to satisfy a debt
_____ 2. example of fiat money
_____ 3. example of commodity money
_____ 4. example of bank money
_____ 5. involves using government spending and taxation to influence the economy
_____ 6. involves controlling the supply of money and credit to influence the economy

_____ 7. independent of the president and Congress

_____ 8. economic truths

_____ 9. economic value judgments

_____ 10. all else held constant

_____ 11. correlation does not imply causation

_____ 12. belief that if one individual or firm benefits from some action, all will benefit from the same action

Activities

1. Prepare a 100-word essay giving a set of events which could happen to cause another Great Depression. Extra reading and library research may be needed but are not necessary.

2. Write a paragraph examining your effect on the circular flow of the economy.

3. Refer to Figure 18–4. Calculate the rate of inflation from 1985 to 1990.

4. Refer to formula for calculating unemployment in this chapter. Calculate the unemployment rate if there are 100,000,000 people in the work force and 5,000,000 are unemployed.

5. Suppose you make a $10,000 loan from the bank. Using the example in Figure 18–5 show graphically how much money the bank can create from your loan.

6. Opportunity costs face us daily. Give an example of an opportunity cost that you have to decide upon. Remember, it does not have to involve money. Share this with the class in a large discussion.

7. Suppose you went to the grocery store to buy milk, blank video tapes, and paper towels. You also consider buying a 2-liter soda but you notice the price is 30 cents higher than last week. The prices of the other items seem to be the same as always. Is the soda price an example of inflation? Why or why not?

8. Work in a group of three to five people to find graphs, tables, and charts showing the gross domestic product (GDP) and other economic indicators over the past few years. Use the library and any computer resources you can. Create a poster of your findings. What phase of the economic cycle are we in now? What groups of people need to know such information? Share the findings of your group with the class.

9. Consider the two statements: "Fifteen percent of our people live below the poverty line," and "Too many people live below the poverty line." Distinguish the positive economic statement from the normative economic statement.

10. Read an editorial from your local newspaper. Cite at least one *positive economic statement* from the editorial and at least one *normative economic statement*. Explain how you know which is positive and which is normative.

11. What does ceteris paribus mean? Cite an example why the concept is useful to economists.

12. Identify one of the most widely debated economic policies or issues and discuss the various viewpoints with your classmates and teacher. Choose a position and be ready to defend it by researching facts and figures to support it.

NOTES

1. Fred M. Gottheil, *Principles of Economics* (Cincinnati, Ohio: South-Western College Publishing, 1996), pp. 6–7.
2. Ibid.
3. Betty J. Brown and John E. Clow, *Introduction to Business: Our Business and Economic World* (New York: Glencoe/McGraw-Hill, 1997), pp. 38–41.
4. Ibid.
5. Ibid.
6. Ibid.
7. Gottheil, *Principles of Economics*, p. 9.
8. Brown and Clow, *Introduction to Business*, p. 31.
9. Ibid., p. 32.
10. Ibid.
11. Ibid., p. 43.
12. Ibid.
13. Sanford D. Gordon and Alan D. Stafford, *Applying Economic Principles* (New York: Glencoe/Macmillan/McGraw-Hill, 1994), p. 220.
14. Brown and Clow, *Introduction to Business*, p. 43.
15. Roger L. Miller, *Economics: Today and Tomorrow* (New York: Glencoe/McGraw-Hill, 1995), p. 339.
16. Gottheil, *Principles of Economics*, p. 584.
17. Gordon and Stafford, *Applying Economic Principles*, p. 272.
18. Miller, *Economics*, p. 339.
19. Gordon and Stafford, *Applying Economic Principles*, p. 269.
20. Brown and Clow, *Introduction to Business*, p. 46.
21. Miller, *Economics*, p. 438.
22. Gordon and Stafford, *Applying Economic Principles*, p. 292.
23. Ibid.
24. Miller, *Economics*, p. 366.
25. Brow and Clow, *Introduction to Business*, p. 313.
26. John Duffy, *Economics* (Lincoln, Neb.: Cliffs Notes, 1993), p. 68.
27. Ibid., p. 69.
28. Ibid., p. 73.
29. Brown and Clow, *Introduction to Business*, p. 48.
30. William G. Nickels, James M. McHugh, and Susan M. McHugh, *Understanding Business* (Chicago, Ill.: Richard D. Irwin, 1996), p. 77.
31. Ibid.
32. John Penson, Rulon Pope, and Michael Cook, *Introduction to Agricultural Economics* (Englewood Cliffs, N.J.: Prentice-Hall, 1986), p. 8.
33. Ibid.
34. Duffy, *Economics*, pp. 2–3.
35. Gottheil, *Principles of Economics*, p. 36.
36. Duffy, *Economics*, p. 9.

Appendix

LAWS OF SUPPLY AND DEMAND

Once you have decided to take advantage of the opportunity to operate an agribusiness, the next decision is what to grow. Many considerations will be analyzed in the decision making process. As you decide which crop(s) to raise and market, another economic principle to consider is the law of supply and demand. Selection of a crop that has already saturated the market would not be economically sound. There would be little *demand* for your product. On the other hand, selection of a crop that would not have much competition with a strong market would be an economically sound decision. The low supply of a crop with a high demand should lead to strong revenue.

Factors Affecting Price

Three factors determine the price of almost everything bought and sold: supply, demand, and the general price level. A single factor cannot explain price change. For example, a rose grower sold roses for $12.00 a dozen. A year later, she sold a dozen roses for $8.00 per dozen. There are several reasons for the difference in price between the two dozen roses. Perhaps the number (supply) of roses available increased. Maybe consumer preference (demand) for roses decreased. Last, the general price level could have caused the change in price. Forces that may have little to do with rose production directly, can change the general price level. Wars, depressions, recessions, periods of recovery, and other things affect the general price level.

Law of Supply

Supply is the amount of a product available at a given time and place at a specific price. It is a direct price and quantity relationship that states what producers are willing to supply at a given price. The law of supply is as follows: as the price of a product increases, producers will soon supply more of that product. As the price falls, the quantity supplied also falls.

Factors That Can Increase or Decrease the Supply of Products

Some of the important factors which can increase or decrease the supply of agricultural products are as follows:

Total Units in Production. Reporting services provide statistics such as numbers (supply) of cattle on feed or total acres of corn planted. This type of data is a good indicator of future supply.

Weather. A sudden freeze or continued drought limits production and reduces supply in an area, causing the supply curve to shift to the left. A freeze or drought must be very severe and widespread to have a major influence on total supply. Yet the price of crops grown in limited areas, such as Georgia peaches, will increase substantially in response to a spring freeze after the trees have blossomed.

Land. In the past, production agriculturalists increased agricultural production by cultivating more land. From 1950 to 1980, 30 percent of the increase in world agricultural production resulted from cultivating new land. In the future, less new land will be available for cultivation.

Water. The amount of water (rainfall or irrigation water) can exert a major influence on the supply of agricultural products. Irrigation can greatly increase the productivity of land.

Changes in Technology. Since 1950, technology has accounted for two-thirds of the increase in world agricultural production. Tractors, fertilizers, pesticides, biotechnology, and hybrid seed are examples of recent agricultural technologies.

Amount in Storage. Wholesalers, producers, and government agencies store many commodities. Storage is a normal part of distributing products to consumers. However, when storage facilities fill to capacity, average production levels may result in average supply.

Government Action. Federal policy often affects supply. When a price support program requires producers to limit production, supply decreases. When a government places sanctions on the export of a product, supply increases.

Time Lags. When a price is unusually high, producers typically decide to increase production. The period of time required for such decisions to affect supply is called a time lag. It takes time to bring idle resources into full production. Time lags usually correlate to a growing season or the animals' reproductive cycle (for beef, 27 to 30 months).

Credit. Abundant, low-interest credit permits faster expansion of production and increases supply. "Tight" money and high interest rates limit expansion, and thus, supply.

Input Prices Changes. Changes in the prices of inputs used in the production process also affect the supply.

Prices of Other Outputs. The prices of other products that can be produced using the same resources also influence the supply of a given product.

Anticipated Prices. If a production agriculturalist expects the price of a commodity—corn, for example—to increase, then he or she will plant more corn, therefore, increasing the supply.

Number of Sellers. If existing producers are generating profits, new producers have the incentive to produce to capture the profits, which results in increased supplies. On the other hand, producers suffering continued losses will exit the industry, which results in decreasing supplies.

Law of Demand

Similar to supply, the economic concept of demand is based on a price and quantity relationship. How much of a product is demanded depends on the amount that consumers are willing and able to purchase at a particular price. The law of demand states that as the price of a product rises, consumers will demand less of that product.

Factors That Can Increase or Decrease the Demand for Products

As with the supply curve, there are important factors relating to demand. These factors include the following:

Income Level. When people have more money, they buy more goods and services, thus demand for certain commodities increases.

Population. As the number of people trying to buy the same item increases, the demand for that item will also increase.

Individual Taste.   Some consumers will buy a product because it is needed or because it is a "slight" luxury. Others will buy a product because it fits into their lifestyle. It may be perceived as the "in" thing to have or to serve. This is harder to measure exactly and may change relatively fast. Still, it can be a useful tool for predicting the success of a new product or determining the value of a proven product.

Competing Products.   If the price of a competing product is much lower than what is usually paid, eventually someone will try it. If consumers stay with the new product, it will lower the demand for the old product. Manufacturers use this as a tool to build demand for their products. They will offer coupons, taste improvements, and better results to gain customers.

Weather.   Extremes in temperature can affect the demand for some commodities. For example, people eat more pork in the winter and drink more lemonade in the summer.

Seasonal Demand.   Consumers demand more turkey during Thanksgiving and ham at Christmas. Saint Valentine's Day and Mother's Day are times of an increase in demand for flowers.

Advertising and Promotions.   Advertising and promotions are ways of trying to increase consumers' awareness and desire to have a certain product. Some methods used are television and radio commercials, magazine ads, and billboards. Promotions include such things as free trial samples and giveaways.

Glossary

absurdities – Things that are meaningless, lacking order or value.

accounting cycle – The sequences of accounting procedures used to record, classify, and summarize accounting information.

accounts payable – The liability arising from the purchase of goods and services on credit.

accounts receivable – A balance due from a debtor on a current account.

accrual method – Method for reporting income that states that the owner reports the income when the services are delivered.

accrued – Accumulated.

actuarial interest rate – Interest rate equal to the actual rate charged on a loan.

A.D. – *Anno Domini* ("in the year of the Lord"), indicating dates since the birth of Christ.

add-on interest – Interest that is added to the loan by taking the total interest paid on a loan and adding it to the loan amount. This then allows the lender to divide by the number of periods to get the installment payment amount.

advocate – To plead in favor of.

aerobic – Pertaining to organisms that grow only in the presence of oxygen, as do some bacteria in a properly prepared compost.

aggregate – Total.

agribusiness – The manufacture and distribution of farm supplies to the production agriculturalist, and the storage, processing, marketing, transporting, and distribution of agricultural materials and consumer products that were produced by production agriculturalists.

agribusiness input sector – All resources that go into production agriculture to produce farm commodities.

agribusiness output sector – Any phase of the agribusiness sector after the commodity leaves the farm, such as sales or marketing.

agricultural commodities – Agricultural animals, crops, and vegetables in their original, unaltered state.

agricultural economics – An applied social science dealing with how humans choose to use technical knowledge and scarce productive resources such as land, labor, capital, and management to produce food and fiber and to distribute it for consumption to various members of society over time.

agricultural products – Agricultural commodities that have been altered from their original state; for example, butter from milk.

agricultural statistics – Facts and figures relating to the agricultural industry.

agrimarketing – Those processes, functions, and services performed in connection with food and fiber from the farms on which they are produced until their delivery into the hands of the consumer.

agrimarketing channels – The paths that an agricultural product follows from "the farmer's gate to the consumer's plate."

agriscience – All jobs relating in some way to plants, animals, and renewable natural resources. It is also the application of scientific principles and new technologies to agriculture.

agriservices – Services offered to production agriculturalists by private firms, such as agricultural research, consulting, artificial insemination, veterinary services, communications, and supplying.

agriservices sector – The portion of the agriculture industry concerned with researching new and better ways to produce and market food, to protect food producers and consumers, and to provide special custom-type services to all of the other phases of agriculture.

agronomic – Any part of agriculture having to deal with field crop production and soil management.

allocation – The distribution or apportioning of something for a specific purpose.

allocation of resources – Land, labor, and equipment that are equally distributed to all segments of the economy.

amortize – Loan that is set up on equal installment payments.

annual percentage rate (APR) – Common name for actuarial interest rate; the rate charged for interest on a loan.

annuities – Sum of money payable yearly or at other regular intervals.

anthropology – The science of human beings in relation to distribution, origin, classification, and relationship of races, physical character, environmental and social relations, and culture.

applied science – The application of basic scientific theories to specific problems.

appreciate – To increase in value.

artificial insemination – The deposition of spermatozoa in the female genitalia by artificial rather than by natural means.

assemblers – Person who puts small lots of a commodity together to provide a larger, more economical unit.

assets – Economic resources which are owned by a business and are expected to benefit future operations. Some examples include: cash, receivables, inventory, investments, equipment, buildings, and prepaid accounts.

astute – Crafty, shrewd, keen in judgment.

attaché – A person on the official staff of an ambassador or minister to a foreign country.

attaches – To bind or to take by legal authority.

auctioneer – The person who is conducting the sale at an auction market.

augmentation – The act or process of making greater, more numerous, larger, or more intense.

average fixed costs – Total fixed costs divided by the quantity.

average product – Amount of output produced divided by the number of units of input; total product expressed relative to some level of input.

average revenue – The amount of revenue expressed relative to each unit of output. It equals total revenue divided by quantity.

average total cost – Total fixed costs plus the total variable costs divided by the quantity.

average variable cost – Total variable costs divided by the quantity.

balance sheet – Reports the business's financial condition on a specific date.

balancing a checkbook – Making sure your checkbook register records are the same as the bank's.

bank money – Money consisting of the bank credit that banks extend to their depositors.

barter – The exchange of goods or services for other goods and services, with no money changing hands.

base mixes – A feed mixture produced by a feed manufacturer to which home-grown feed, such as corn, is added.

base year – Typical years of consumer prices used to help compile the consumer price index.

basic science – Research to discuss general relationships between two or more variables.

basis – The difference between the price of a commodity and the price of a related futures contract; for example, cash price minus futures price equals basis.

B.C. – Before Christ.

bear market – A period of declining market prices.

biopesticides – Effective pesticides that will not be harmful to water, air, soil, wildlife, humans, or the food we consume.

biotechnology – Technology concerning the application of biological and engineering techniques to microorganisms, plants, and animals; for example, gene splicing, cloning, and DNA mapping.

blanching – Scalding in water or steam in order to remove the skin, whiten, or stop enzymatic action (as food for freezing).

blocked currency – The idea that money could exchange only within the nation making the purchase to boost the economies of underdeveloped nations, when surpluses were sold to developing nations.

board of directors – A group of individuals chosen to make decisions for the company.

bovine somatotropin (BST) – A growth hormone that is found in cattle. Injections of extra BST are used to increase the production of meat and milk.

broaches – A cutting tool for removing material from metal or plastic to shape an outside surface or a hole.

broilers – Chickens from 8 to 12 weeks old, weighing 2½ or more pounds, sufficiently tender to be broiled.

brokerage – Firm that buys and sells stock or other products for a commission.

brokerage office – Most convenient outlet for futures transactions, such as commission house, wire houses, and futures commissions merchants. These offices are located in most towns and cities throughout the United States.

budget – Financial plan for an agribusiness.

budgeting – The orderly fashion in which owners of an agribusiness put their financial plans on paper.

bulk feed – Feed sold in large quantities.

bull market – A period of rising market prices.

business cycle – The repeated rise and fall of economic activity over time.

business plan – A written description of a new business venture that describes all aspects of the proposed agribusiness. It helps you focus on exactly what you want to do, how you will do it, and what you expect to accomplish.

business survey – A survey that a business conducts to determine the market potential for a selected business.

buyer's fever – When a buyer acts on quick, thoughtless decisions; buying without thinking.

call – An option that gives the option buyer the right (without obligation) to purchase futures contract at a certain price on or before the expiration date of the option for a price called the premium; determined in open outcry trading in pits on the trading floor.

callus – Plant tissue that is not differentiated into leaf, root, stem, or other specialized tissue.

capital – Goods used to produce other goods and services; the investment that the owner has put in the business.

capital expenditures – These are expenses that are depreciated over a number of years instead of in the year purchased, except for land and homes, which are items for which the cost cannot be recovered until they are sold.

capital expenditures budget – This budget shows how money projected for capital expenditures is to be allocated among various divisions or activities within the agribusiness. This budget is a list of projects (equipment, etc.) that management believes to be worthwhile, together with the essential cost of each.

capital intensive – Having a high capital cost per unit of output; requiring greater expenditure in the form of capital than of labor.

capitalism – An economic system where individuals own resources and have the right to use their time and resources however they choose.

cash flow budget – This budget summarizes the amount and timing of income that will flow in and out of the business during the year.

cash market – Where actual commodities are bought and sold.

castings – An impression taken from an object with a liquid or plastic substance; to make into a mold.

centralization – The placing of a complete line of farm production supplies and services in a central location.

chain store – The operation of eleven or more stores under a single owner.

check-off program – A program in which the seller contributes a small amount for each unit sold to a money pool to be used to purchase advertising for the specific commodity.

chronological – Arranged in, or according to, the order of time.

classical economic theory – Classification of economics that contends that an economic system is self-sufficient in itself and any outside interference by government does more harm than good.

cloning – A process through which genetically identical organisms are produced.

collateral – Something of value deposited with a lender as a pledge to secure repayment of a loan.

collective bargaining – Negotiation between an employer and a labor union usually on wages, hours, and working conditions.

commercial food establishment – Category of food service establishment that is open daily to the general public. These include fast food, restaurants, and lodging places.

commission – A fee the selling agent receives for work in selling animals.

commodity – A transportable resource product with commercial value.

commodity (futures) exchanges – Central marketplaces with established rules and regulations where buyers and sellers meet to trade futures and options on futures contracts.

commodity money – A good whose value serves as the value of money.

common market – Same as a customs union, except that it includes the free mobility of factors of production.

common stock – Gives the owner the right to vote on business matters.

communism – Economic system in which the government has total control of economic matters and private individuals have none.

compatible – Agreeing with or capable of existing together in harmony.

competitive enterprises – Those enterprises in which you must decrease the production of one to increase the production of another.

complete feed – Feed fed to animals without any additional preparation, which accounts for 80 percent of feed produced.

complete fertilizer – One that contains nitrogen, phosphoric acid, and potash.

complimentary enterprises – An increase in one enterprise automatically increases the output of another enterprise.

compound interest – Interest computed on the accumulated unpaid interest as well as the original's principal.

compression – The process of compressing the fuel mixture in a cylinder of an internal combustion engine.

concentrates – Any feed high in energy (usually grain); stock feed low in fiber content and high in digestible nutrients.

consolidation – The process of bringing a complete line of farm production supplies and services together in a location.

consortium – An agreement, combination, or group (as of companies) formed to undertake an enterprise beyond the resources of any one member.

constant substitution – When one resource can be substituted for another at the same rate for each additional unit of input.

consumer price index – General measure of retail prices for goods and services usually bought by urban wage earners and clerical workers. It includes prices of about 400 items, such as food, clothing, housing, medical care, and transportation.

contraction – A noticeable drop in the level of business activity, which indicates a slowdown in the growth of the economy.

contract specifications – The standard features, such as time of delivery.

contractual interest rate – Interest rate not always equal to the actual rate charged on the loan.

convenience stores – Type of retail grocery store which was intended to replace the "ma and pa" grocery stores in older communities. These stores usually sell the major necessities: bread, milk, eggs, etc.

cooperative – An organization in which the profits and losses are shared by all members.

Cooperative Extension Service – The Cooperative Extension Service is responsible for programs in four major areas: agricultural and natural resources, home economics, community development, and 4-H youth development. The Cooperative Extension Service of the land-grant colleges and universities was created under federal legislation with the Smith-Lever Act of 1914. Cooperative Extension Service philosophy is to help people identify their own problems and opportunities, and then to provide practical research-oriented information that will help them solve the problems and take advantage of the opportunities.

corporate charter – A license to operate in a specified state granted when the articles are in agreement with state law.

corporation – A legal classification with authority to act and have liability separate from its owners.

cost – Amount paid or charged for something.

cost-effective – Economical in terms of tangible benefits produced by money spent.

cost of sales – The expense and labor that are directly associated with the production or acquisition of products held for sale.

cost-price squeeze – Pressured by increasing cost of resources on one hand and lower prices received from marketing firms on the other.

cotton gin – A machine used to separate the cotton seed from the lint; invented by Eli Whitney in 1793.

crawler-type – A diesel-powered farm tractor developed by Caterpillar Tractor Company in 1931.

credit – An entry on the right-hand side of a ledger constituting an addition to a revenue, net worth, or liability account.

creditor – The person or company to whom the account payable is owed.

crop rotation – Growing annual plants in a different location in a systematic sequence. This helps control insects and diseases, improves the soil structure and productivity, and decreases erosion.

cultivate – Loosening of the soil and removal of weeds from among desirable plants; the planting, tending, harvesting, and improving of plants.

currencies – Something (coins, treasury notes, bank notes, etc.) that is in circulation as a medium of exchange

custom grinding – Grinding and mixing of feed for a certain person using a specified feed content mixture. It is mixed according to the farmer's needs.

customs union – An economic and political organization between two or more countries abolishing trade restrictions among themselves and establishing a common uniform tariff to outsiders.

cylinder – The piston chamber in an engine.

debit – An entry on the left-hand side of a ledger constituting an addition to an expense or asset account or a deduction from a revenue, net worth, or liability account.

Dairy Herd Improvement Association (DHIA) – Group that, among other things, weighs the milk of cows periodically to provide official milk records for dairy cattle.

debt-equity ratio – The ratio of debt to equity allowed in the business. For example, if the ratio is 2, one can have twice as much debt as equity.

deductible expenses – Expenses that the agribusiness purchases on a day-to-day basis for its operation, which are consumed in a short period of time.

default – Failure to pay financial debts.

deficits – An excess of expenditure over revenue.

deflation – A prolonged decline in the general price level.

defoliants – A type of chemical that, when applied to a plant, causes the foliage to drop off.

delinquencies – A debt on which payment is overdue.

demand – The amount of a product wanted at a specific time and price.

deoxyribonucleic acid (DNA) – The molecule that carries the genetic information for most living systems. The DNA molecule consists of four bases (adenine, cytosine, guanine, and thymine) and a sugar-phosphate backbone, arranged in two connected strands to form a double helix. DNA contains the genetic material for every cell and is found in every cell.

depreciation – The systematic write-off of the value of a tangible asset over its estimated useful life.

depression – A severe drop in income and prices for an area or for a nation.

desiccants – A type of pesticide which dries out or reduces water or moisture content in attempt to destroy pests.

diligence – Steady application; constancy in an effort to accomplish something.

discount (prime) rates – Rates set by the Federal Reserve System. These rates are influenced by the supply and demand factors for money.

discouraged workers – Workers who are not actively searching for work.

distance learning – Communication technology in which services are delivered to people via satellite from another location.

distillation – The process of separating the components of a mixture by differences in boiling point; the evaporation and subsequent condensation of a liquid, as when water is boiled in a retort and the steam is condensed in a cool receiver.

diversification – Not specializing in just one or two crops: a wide variety of crops.

dividends – Money paid to a shareholder.

domestic – Of, relating to, or originating within a country.

domesticate – To bring wild animals under the control of humans over a long period of time for the purpose of providing useful products and services. This process involves careful handling, breeding, and care.

double entry bookkeeping – Practice of writing every transaction in two places.

double taxation – This is where regular corporations must first pay taxes on the profits of the firm and then the income is taxed again as individual income when stockholders receive their dividends.

draft animals – Animal used for work stock and, in some countries, still used for plowing and pulling heavy loads.

drought – A period of insufficient rainfall for normal plant growth which begins when soil moisture is so diminished that vegetation roots cannot absorb enough water to replace that lost by transpiration.

ecologists – One who is concerned with the interrelationship of organisms and their environment.

economic fluctuations – The ups and downs in economic activity.

economic indicators – Important data or statistics that measure economic activity and business cycles.

economic isolation – Countries that do not trade with other countries.

economic policy – A course of action that is intended to influence or control the behavior of the economy.

economic recovery – A rise in business activity.

economics – There are many definitions of economics, but it is most often defined as the science of allocating scarce resources (land, labor, capital, and management) among different and competing choices and utilizing them to best satisfy human wants. Key words include: scarcity, types of resources, and wants and needs.

economists – People who specialize in economics and concern themselves chiefly with description and analysis of the production, distribution, and consumption of goods and services.

economy of scale – The most efficient production size.

efficient economic system – Where there will be: decision making by the consumers as to the who, what, and where of the economic system; a pricing system that brings about an efficient allocation of resources; the maximization of consumers' satisfaction; and the maximization of producers' long-run profits.

embargoes – Prohibit shipments of commodities to certain other countries. They are also called sanctions.

embryo splitting – A form of cloning that is accomplished by dividing a growing embryo into equal parts using a surgical procedure performed with the aid of a microscope.

embryo transfer – Procedure for placing living embryos obtained from a donor animal in the reproductive tract of a recipient female animal.

emulsifier – Substances that aid in the uniform dispersion of oil in water.

enterprises – Projects on a farm, such as the production of any crop or livestock.

entrepreneur – Person who accepts all the risks pertaining to forming and operating a small business. This also entails performing all business functions associated with a product or service and includes social responsibility and legal requirements.

entrepreneurship – The art or process of organizing, managing, and perhaps even owning a business.

environmentalists – Any person who advocates or works to protect the air, water, and other natural resources from pollution.

equal product curve – Concept whereby different resources can be substituted for one another in the production of a given product. There are numerous combinations of land, labor, and capital that will produce the same amount of a product.

equilibrium price – The price in a market at which quantity supplied and quantity demanded are equal.

equities – Money value of a property or of an interest in a property in excess of claims or liens against it.

equity capital – Venture capital; capital invested or available for investment in the ownership element of a new or fresh enterprise.

eradicate – To do away with completely.

erroneous – Mistaken; containing or characterized by error.

ethanol – The alcohol product of fermentation that is used in alcoholic beverages and for industrial purposes.

European Union (EU) – Top exporter of agricultural goods into the United States.

exchange rates – The ratio at which the principal unit of two currencies may be traded.

exports – Shipment of agricultural commodities to foreign countries.

extended care – Care for people who need assistance in living, such as nursing and retirement homes.

externalities – Factors beyond the control of an individual, group, or business.

extrapolate – To infer from known data.

factor-factor decisions – Decisions of which specific combination of resources (land, labor, capital) or factors of production to use.

factor-product decisions – Decisions of how much fertilizer, if any, should the producer add to their product (corn, etc.).

fallow – Cropland left idle in order to restore productivity through accumulation of moisture.

farm assets – Something that is of value or a complement to the farm, whether real estate or non–real estate.

farm contracting – Services performed by individuals with specialized equipment, including clearing land, installing drainage systems, and aerial spraying.

farm commodity – Agricultural products such as wheat, corn, and cattle that are raised on the farm.

fascism – Economic system in which productive property, although owned by individuals, is used to produce goods that reflect government or the state's preferences.

Federal Land Banks – Several banks that are a part of the Farm Credit Administration of the United States; the Federal Land Banks through local national farm loan associations make long-term (5–40 years), amortized, first mortgage, farm real estate loans.

Federal Reserve System – A central banking system in the United States, created by the Federal Reserve Act of 1913 and designed to assist the nation in attaining its economic and financial goals.

fermentation – The processing of food by means of yeast, molds, or bacteria.

fiat money – Money that has value because a government fiat, or order, has established it as acceptable for payment of debts.

fibrous products – Products containing fibers.

fidelity shares – Low-cost shares of money invested.

financial assets – Funds or capital which the person, corporation, or other party has in their possession.

financial institutions – Any business that lends money.

financial resources – Any place where a consumer or business person can acquire money.

financial security – Having no financial strain or monetary problems; being able to pay monthly bills without stress.

fiscal policy – Economic policy designed to influence economic activity which involves using government spending and taxation to influence the economy.

fixed expenses – Items that can be used over and over for a long period of time which results in the same price (expense) each year; bills that are due every month such as house payments.

flaming – The process of setting a field on fire to destroy all organic matter.

flex time – A system that allows employees to choose their own times for starting and finishing work within a broad range of available hours.

foreclosure – A procedure whereby a lender takes steps toward attaining ownership of property given as security for a loan.

forge shop – Place where heavy hammers forge the red-hot steel billets into shape.

formulation – The product of a mixture, such as feed.

forward contracts – A cash contract in which a seller agrees to deliver a specific cash commodity to a buyer sometime in the future. These are privately negotiated and are not standardized.

foundry – Place where molten iron is poured into molds to make castings.

franchise – A contract in which a franchisor sells to another business the right to use its name and sell its products.

franchisee – Person purchasing a franchise.

franchisor – Person who grants a franchise.

fraternal – Of, relating to, or involving brothers.

free enterprise – Economic system that allows individuals to organize and conduct business with a minimum of government control; individuals privately own what they produce.

free market economy – Exists where the consumers provide the answers to the questions of what to produce, how much to produce, when to produce, who should produce, and for whom goods should be produced.

free markets – The movement of goods and services among nations without political or economic obstruction.

free trade – Trade between businesses in different countries without restrictions from any government.

free trade area – Group of nations that abolishes trade restrictions among themselves without also imposing common tariffs on other nations.

full-line companies – Produce and sell tractors as well as a wide variety of equipment,

fumigants – A type of pesticide that produces gas, vapor, fume, or smoke intended to destroy insects and other pests.

fungicide – A type of pesticide used to control fungi that attack plants and animals.

futures – Legally binding agreements made on the trading floor of a futures exchange; trading by means of buying or selling agricultural commodities that are to be produced or are in the process of being produced.

futures contract – A legally binding agreement, made on the trading floor of a futures exchange, to buy or sell something in the future. A futures contract specifies the quantity, quality, time and form of delivery, and a negotiated price.

futures market – Where traders buy and sell futures contracts.

futures options – Type of insurance to protect against price decline.

futures trading – A means of buying or selling agricultural commodities that are to be produced or are in the process of being produced.

gasohol – Registered trade names for a blend of 90 percent unleaded gasoline with 10 percent fermentation ethanol.

gender selection – Managing the reproductive process to produce animals of the desired sex.

gene mapping – The process of finding and recording the locations of genes on a chromosome.

general partnership – Association of two or more people who, as owners, manage a business together.

gene splicing – The process of removing a gene from its location on a chromosome and replacing it with another gene.

genetic engineering – The practice of modifying the heredity of an organism by inserting new genes from other organisms into the chromosome structure.

germ plasm – The heredity material of the germ cells. It is the chromosomes in germ cells that transmit heredity characteristics to the offspring.

glycerol – A substance obtained from fats used as a solvent and plasticizer; used as a semen extender.

goods – Things that have economic utility or satisfy an economic want.

grading – The arranging of something, such as food, according to standards and criteria set forth for quality.

grants – An award given for something deserved or merited after careful weighing of pertinent factors; usually public funds given to a school or individuals for a project.

gross domestic product (GDP) – Total value of goods and services produced in a country in a given year.

gross national product (GNP) – The total monetary value of all final goods and services produced in a country during one year.

gross pay – Total amount earned before deductions are taken from a paycheck.

hatcheries – Companies where eggs are incubated; usually a commercial establishment where newly hatched young chicks are sold.

hedgers – Those who own or will own the actual cash commodity—they use the futures market to establish either a buying or selling price for a commodity they own or are expecting to own; those who manage price risk.

hedging – Buying or selling futures contracts to protect one's profit margin against a possible price change of a cash commodity that he or she plans to buy or sell.

herbicide – A type of pesticide used to control undesirable plants (weeds).

hog cholera – An acute, contagious, viral disease of swine characterized by sudden onset, fever, high morbidity, and mortality.

holding action – A strategy used to try to improve farm prices. Prices usually increased for a short period of time after each action, but few long-term gains were realized.

homogenous – Of the same kind or a similar kind or nature.

hormone – A substance produced in the body and carried by body fluids to tissues where it causes specific body functions to occur.

host – A living plant or animal in which a parasite lives.

humanitarian – Human attributes or qualities.

humus – Organic matter in the soil that has reached an advanced stage of decomposition and has become colloidal in nature. It is usually characterized by a dark color, a considerable nitrogen content, and chemical properties such as a high cation-exchange capacity.

husbandry – The care of a household or scientific control and management of a branch of farming.

hybrid – The result from a cross between parents that are genetically unlike; hybrid plant seed was developed for better-quality, higher-producing crops.

hydraulic lifts – Feature that provides a means of lifting by pressure transmitted when a quantity of liquid is forced through a small orifice, such as a tube or hose.

hydroponics – A plant production system in which plant nutrients are provided in a water solution, and plants are grown without soil.

hydrostatic transmission – A transmission in which fluids tend to be at rest.

hypothesis – A tentative assumption made in order to draw out and test its logical consequences.

implant – A device that is placed under the skin or in the female reproductive tract to slowly release hormones into the bloodstream of the animal.

implements – Tools that aid a person to make work and effort more productive and effective.

implicit GDP price deflator – Price index that removes the effect of inflation from GDP so that the overall economy in one year can be compared to another year.

import quotas – Impose a limit on the amount of a good that may be imported in a given period.

imports – Shipment of agricultural commodities from foreign countries for sale, use, etc.

incidental expenses – Any expenses that occur seldomly but are business related, including unexpected emergencies due to accidents or weather damage.

income stabilization – Steps taken by the Farm Service Agency to stabilize the nation's agricultural economy through price support loans and purchases.

income statement – Reports revenues, costs, expenses, and profits (or losses) for a specific period of time showing the results of business operations during that period.

incomplete fertilizer – Fertilizers containing only two of the three materials—nitrogen, phosphoric acid, and potash.

indigo – A leguminous plant which produces a blue dye used for various purposes.

Individual Retirement Account (IRA) – Account allowed by the government to invest $2,000 yearly, with taxes deferred until retirement.

industrialization – The process of becoming industrial.

inflation – An increase in the volume of money and credit relative to available goods and services resulting in a continuing rise in the general price level.

initial margin – The initial amount a market participant must deposit into his or her account when he or she places the order.

initiatives – Individual willingness to take risks and make a profit.

input – Supplies and services farmers use to produce crops, livestock, and other items; any resource used in production.

insecticide – Type of pesticide used to control various insects.

inseminate – To place semen in the vagina of a female animal.

institutional advertising – Designed to create a favorable image of the firm or institution offering the products or service.

Integrated Pest Management (IPM) – The use of natural insect enemies and limited chemical applications to control harmful insects while providing protection for useful insects.

interdisciplinary – Involving two or more academic, scientific, or artistic disciplines.

interest – The charge or cost of borrowed money.

interest rates – The price charged by lending institutions for use of money.

intrinsic value – The amount by which an option is in-the-money.

inventory – A physical count of all assets in a business with their estimated worth.

investment portfolio – A record of all investments with the goal of diversification to get greater returns.

investors – One who provides money to others, usually a business, for later income or profit.

irrigate – To furnish water to the soil for plant growth in place of, or in addition to, natural precipitation, by surface flooding or sprinkling and subirrigation methods, using surface water or that from underground sources.

journals – Books where accounting data are first entered.

Keynesian – Classification of economics that contends that economic systems are not always self-sufficient and sometimes need outside help.

kilns – An oven, furnace, or large heated room for the curing of lumber, tile, or bricks.

laissez faire – Noninterference by government, leaving coordination of an individual's wants to be controlled by the market.

lathes – Machines in which work is rotated about a horizontal axis and shaped by a fixed tool.

law of comparative advantage – Principle which states that producers tend to produce that product, or those products, for which they have the highest economic advantage or the least disadvantage.

law of diminishing returns – States that as you apply (input) additional units of a variable resource (fertilizer) to the production of a particular product (corn), there will come a point when the next unit of input will not increase total output as much as the previous one did.

law of increasing cost – Law that states that to produce equal extra amounts of one product or service, an increasing amount of another product or service must be given up.

ledger – An accounting system which includes a separate record for each item; It consists of three parts: a title (name of the particular asset, liability, or owner's equity), the debit (left) side, and the credit (right) side.

legal classification – Refer to *legal structure.*

legal entity – Whereby a corporation is separate from the people who own it or work for it.

legal structure – The way a business is set up for legal or tax purposes, such as a corporation, cooperative, or sole proprietorship.

liabilities – Amounts that are owed or debt. Examples include accounts payable, notes payable, and mortgages.

liaison – A close bond or connection.

lien – The lender's right to take possession of the asset in collateral if the borrower fails in repayment of the loan.

limited partnership – Special type of partnership in which some partners are not completely liable for their partners' debts. This partnership is for investment purposes only.

liquid income – Money that is available quickly such as cash.

locked into – When someone cannot get out of what he/she is doing because the debts exceed the assets.

long hedge (buying hedge) – Buying of a futures contract.

long-line companies – Produce and sell a wide variety of general farm equipment, including self-propelled combines, but no tractors.

long run – Period during which all resources are variable. There are no fixed costs.

long-term credit – The extension of credit by the decision of the lender. One source includes Farm Credit Services.

machine shop – Place containing the lathes, milling machines, cutters, planes, broaches, and automatically controlled machine tools.

macroeconomics – Concerned with the study of the economy on a large scale, or nationally.

maintenance margin – A customer's set minimum margin in his or her account.

marginal analysis – The examination of the consequences of adding to or subtracting from the current state of affairs.

marginal benefits – The value of the additional goods or services that could be produced.

marginal cost – The change in total cost associated with each additional unit of output.

marginal product – The change in total product associated with each additional unit of input.

marginal revenue – Change in total revenue associated with each additional unit of output.

marginal revenue product – The increase in income of the product created by adding an additional unit of something, such as fertilizer.

margining system – A system initiated to eliminate problems of buyers and sellers not fulfilling their contracts. It requires traders to deposit funds with the exchange or an exchange representative to guarantee contract performance.

market analysis – Collecting information to determine if a product will sell.

marketing – Starts with analyzing the market to see what is needed before beginning production, not just selling of the product.

marketing cooperatives – Assist production agriculturalists in marketing agricultural products by finding buyers who will pay the highest price.

mechanical power – Power originating from mechanized objects, such as the steam engine.

marginal input cost – Increase in the additional unit cost (fertilizer) with each additional unit of input.

margin call – The request for additional money.

market basket – About 400 goods, which are sold in about 21,000 outlets, including food, housing, transportation, clothing, entertainment, medical care, and personal care.

merchandising – Sales promotion as a comprehensive function including market research, development of new products, coordination of manufacturing and marketing, and effective advertising and selling.

metric ton – Equal to 1 million grams or 1.102 short (U.S.) tons.

microeconomics – Concerned with the study of the economy on a small scale.

millinery – Women's apparel for the head.

milling machines – A machine tool on which work usually of metal secured to a carriage is shaped by rotating milling cutters.

minimum price fluctuations – The smallest price unit a futures contract trades at. See also *tick.*

monetary policy – Management of the money supply and interest rates; government economic policy designed to influence economic activity which involves controlling the supply of money and credit to influence the economy.

monetary systems – Refer to *monetary policy.*

mutual funds – Investments made solely by a company on behalf of others. By law, a company must invest in a least thirty different stocks or other investments; most invest in many more.

needs – Things that are really crucial to daily living.

net pay – Amount taken home in the paycheck after deductions.

net profit – Revenue minus the cost of sales; also called profit margin.

net worth – The difference between total assets and total liabilities.

niche – A place or position in the community suitable for a person or thing.

nitrogen fixation – The metabolic assimilation of atmospheric nitrogen into ammonia by soil microorganisms.

noncommercial food establishments – Category of food service establishment in which meals and snacks are served as a supportive service rather than as the primary service. These include education, extended care, vending, plants, and office buildings.

nondeductible expenses – Expenses not considered business expenses, such as family living expenses—housing, personal transportation, food, clothing, recreation, and education.

nonrecourse loans – Loans designed to prevent farmers from having to sell their commodities immediately upon harvest when prices are usually at their lowest seasonal level.

nonruminant – An animal, such as a pig, without a functionable rumen.

note payable – A formal written promise to pay a certain amount of money, plus interest, at a definite future time.

offset – Most common method of closing out an option position by purchasing a put or call identical to the put or call you originally sold or by selling a put or call identical to the one you originally bought.

open outcry – Method of public auction for making verbal bids and offers in the trading pits or rings of futures exchanges.

operating budget – This budget summarizes the expected sales or production activities and related cost for the year; it is an estimate of sales and income plus the fixed and variable expenses that the agribusiness should experience during the year.

operating expenses – Any regular expenses associated with the operation of the business, including rent, salaries, utilities, insurance, and depreciation.

opportunity cost – The value of a service or product that must be given up in order to obtain another good or service.

organic fertilizer – First developed by putting the dead fish into the ground along with corn seed; usually include only natural, organic, proteinaceous materials of plant and animal origin and exclude synthetic, organic, nonproteinaceous materials such as urea.

output – A marketable product of a farming operation, such as cash crops, livestock, etc.; the result of the production process.

outstanding loans – Loans that have not yet been paid.

owner's equity – Money the owner invests in the agribusiness.

parent company – The franchisor; prepackages all the business planning, management training, and assistance with advertising, selling, and day-to-day operations.

partnership – An organization whereby two or more people legally agree to become co-owners of a business.

part-time (avocational) enterprises – Enterprises that provide a supplemental income to a full-time occupation.

pathogens – Disease-causing organisms.

Payment In Kind (PIK) – A government program established to help farmers by giving them products as payment for reducing acres planted of certain crops, for example, giving soybeans instead of cash.

peak – Highest level of economic activity in a business cycle; it indicates prosperity and means the economy is rapidly expanding.

performance bond margin – A financial guarantee required by both buyers and sellers to ensure they fulfill the obligation of the futures contract.

perishable – Liable to spoil or decay, such as fruit, vegetables, butter, and eggs.

perseverance – Steady persistence in a course of action.

pessary – A sponge or other material to which hormones have been added and that is implanted under the skin or inserted in the female reproductive tract.

pesticides – Chemicals used to control weeds, insects, and diseases that affect crops, livestock, or people. Pesticides include herbicides, insecticides, fungicides, nematocides, and rodenticides.

pheromone – A chemical substance that is used by animals and insects to attract mates through their sense of smell.

physiologically – Characteristic of, or appropriate to, an organism's healthy or normal functioning.

physiologist – Person who is concerned with the functions and activities of life or of living matter.

"plastic prosperity disease" – The impression that "no cash" passed through the hand so no money was spent; as with credit cards.

pneumatic – Adapted for holding or inflated with compressed air.

policy analysis – Analysis from which forecasts are made which serve as a basis for economic decision making by production agriculturalists, consumers, and others.

policy statements – Deal with issues that affect modern agriculture both directly and indirectly including such areas as transportation, national welfare, conservation, foreign affairs, taxation, education, health and safety, and numerous other issues.

political power – Power relating to the government procedures and administration of government policies.

political science – A social science concerned chiefly with the description and analysis of political and governmental institutions and processes.

porcine somatotropin (PST) – Type of growth-enhancing hormone used on pigs.

posting – The process of transferring the debit and credits from the general journal to the ledger account.

power take-off (PTO) – A supplementary mechanism allowing the operator to control mounted and drawn equipment with the tractor's engine.

precision farming – A crop management system that adjusts applications of fertilizers and other crop inputs on the basis of production differences that exist within a field.

preferred stock – Gives a person the opportunity to invest in the business, and, hopefully, to receive a reasonable return on investment.

premium – The cost of providing insurance for the stored crop in the futures market. It is also the cost of a futures option.

premix – Feed containing only vitamins and minerals and is used at the rate of less than 100 pounds per ton; these are mixed with feed grains and a protein source such as cottonseed or soybean meal to provide a complete ration; these account for about 2 percent of feed produced.

prepackages – This generally includes management training and assistance with advertising, selling, and day-to-day operations, which is done by the parent company or franchisor.

prerequisite – Something that is necessary to an end or to the carrying out of a function.

price discovery – The generation of information about future cash market prices through the futures markets.

price support – The price for a unit of a farm commodity which the government will support. These are determined by law and set by the secretary of agriculture.

pricing efficiency – How close an economic system comes to achieving the point when prices reflect the full value of resources and resources are allocated to their highest and best use.

principal – A capital placed at interest, due as a debt, or used as a fund.

private agriservices – Any services not included in federal, state, or local government; for example, the Farm Credit System.

producer price index – An index that shows the cost of resources needed to produce manufactured goods during the previous month.

product advertising – A way of advertising which focuses on the product itself, such as its usefulness, durability, price, value, and a customer's need for the item.

production agriculturalists – The farmers; they produce food and fiber (cotton, wool, etc.), and the output is taken by agribusiness companies that process, market, and distribute agricultural products.

production controls – Government policies that control production of agricultural commodities.

production efficiency – Receiving optimum output from a reasonable input.

production function – The relationship between inputs and outputs.

production possibilities – A type of economic analysis in which the possibilities of the producer are determined by the type and amount of resources available in combination with a specific level of technology.

product-product decisions – Decision of what enterprise or enterprises to produce.

profit – A valuable return.

profitable – Yielding advantageous returns or results.

prominent – Standing out or readily noticeable.

prospectus – Document explaining how mutual funds work and where the money is invested.

prosperity – The condition of being successful.

protectionism – The idea of establishing government economic protection for domestic producers through restrictions on foreign competitors.

psychology – The science of mind and behavior.

public agriservices – Group that provides special services at the federal, state, and local levels. The major areas of emphasis include research, education, communication, and regulation.

put – An option that gives the option buyer the right (without obligation) to sell a futures contract at a certain price on or before the expiration date of the option.

rapport – A harmonious or sympathetic relationship.

rationale – An explanation of controlling principles of opinion, belief, practice, or phenomena; an underlying reason.

raw material – Crude or processed material that can be converted by manufacture, processing, or combination into a new and useful product.

reaper – Used to cut or harvest, as a crop of grain; Cyrus McCormick invented the mechanical reaper to reduce hand labor in the harvest of grain.

receivables – This is an example of an asset, which is owned by a business and is expected to benefit future operations.

recession – A period of reduced economic activity.

reconcile – Compare, as in balancing a checkbook to see if it's the same as the bank records.

reconciling – Checking a financial account against another for accuracy.

regulatory medicine – Covers medicines that are regulated by governmental laws.

resources – The available means for production, such as land, labor, capital, fertilizer, chemicals, machinery, transportation, marketing, etc.

resource specialization – The economic theory of producing one agricultural commodity or resource rather than diversifying into two or more agricultural commodities.

restriction enzyme – A specific enzyme capable of removing a particular gene from its location on a chromosome.

retailers – Person or store that sells directly to the consumer.

revenue – The value of what is received from goods sold, services rendered, and other financial sources; money coming into the agribusiness.

risk management – Making investments to get the highest returns with less risk.

risk transfer – A concept used in futures trading to transfer the risk of the crop (hedger) to the risk of the speculator.

roughages – A feed high in fiber and low in digestible nutrients such as straw, hay, haylage, and silage.

ruminant – Any one class of animals, including sheep, goats, and cows, that have multiple stomachs.

rural sociology – Area of study dealing with the social relationship of country life and its people.

salvage value – The value of a piece of equipment once it has been fully depreciated.

sanctions – Embargoes that prohibit shipments of commodities to certain other countries.

scarce – Deficient in quantity or number compared with demand.

scarcity – Economic term that describes a situation where there are not enough resources available to satisfy people's needs or wants.

scenario – A synopsis of a possible course of events or actions.

secured – Relieved from duty.

securities markets – Investments backed by some tangible asset that is pledged to the inceptor if the principal is not paid back, example, secured bonds.

selective breeding – The breeding of selected plants or animals chosen because of certain desirable qualities or fitness, as contrasted to random or chance breeding.

selling – Taking the farm product to the market and getting whatever the price is for that day.

semen – A fluid substance produced by the male reproductive system containing spermatozoa suspended in secretions of the accessory glands.

service cooperatives – Cooperatives that provide their farmer members with a specific service, rather than a product that members cannot afford to provide individually.

services – Nontangible products that cannot be held, stored, or touched. They can include benefits, activities, or satisfactions that are offered for sale.

short hedge (selling hedge) – Selling of a futures contract.

short-line companies – Produce highly specialized equipment such as planters and cultivators, forage equipment and milking equipment.

short run – Period during which certain inputs or resources are fixed and others are variable.

sickle – A sharp, curved metal blade fitted with a short handle; used for cutting weeds or grasses. It is one of the earliest hand implements used for harvesting small grain.

simple interest – Interest that applies to loans with a single payment. Since no borrowing fees or down payment are required, then the APR would be the rate charged.

single entry bookkeeping system – The simplest bookkeeping system, where the entry is made only in the checkbook or the checkbook and the journal.

single (sole) proprietorship – An organization that is owned, and usually managed, by one person.

small business – A business that is independently operated, is not dominant in its field, and meets certain size standards in terms of number of employees and annual receipts.

Small Business Administration (SBA) – A division of the federal government that oversees small businesses.

socialism – Economic system in which you have public ownership of all productive resources. The government, or the state, directs all decision making.

sociology – The science dealing with society, social institutions, and social relationships.

sodium citrate – A crystalline salt used chiefly as a buffering agent, emulsifier, or alkalizer; used as a semen extender.

Soil Conservation Service (SCS) – A bureau of the U.S. Department of Agriculture established by the Soil Conservation Act of 1935. Its basic purpose is to aid in bringing about physical adjustments in land use and treatment that will conserve natural resources, establish a permanent and balanced agriculture, and reduce the hazards of floods and sedimentation.

solarization – A cultural method that inhibits the accumulation of starch in leaves in the presence of intense illumination.

sole proprietor – A professional in your own practice; one person is the owner and manager.

solvency – If management converts all assets to cash, there is more than enough cash to cover all liabilities.

specialization – Specializing in only one or two crops from which most of the income is derived.

specialized food stores – These stores make up about 6 percent of retail grocery sales. They usually sell a single food category. Examples include retail bakeries; meat and fish markets; candy and nut stores; natural health and food stores; coffee, tea, and spice stores; and ice cream stores.

speculative investments – Assumption of unusual business risk in hopes of obtaining commensurate gain.

speculators – Those who are willing to accept price risk. They just buy and sell futures contracts and hope to make a profit on their expectations of future price movements.

spot price – Another name for the prevailing cash price.

standardization – The establishment of standards for matters such as quality, size, weight and color.

standard of living – How well people are living in a particular nation, dependent upon the amount of goods and services that people of a nation enjoy.

start-up expenses – Expenses which occur before the business begins, such as attorney's fees, incorporation expenses, and cost for development site.

statement of cash flows – Reports cash receipts and disbursements, related to the agribusiness's major activities: operations, investments and financing.

stock – Shares in the ownership of the corporation.

stockholders – The owners of the stock who pay a set price for their shares; each stockholder has one vote in the major decisions made by the corporation for each share of purchased stock.

stock prices – Price of a share of ownership in the corporation issuing the stock.

straight fertilizer – Fertilizers containing only one of the three primary materials—nitrogen, phosphoric acid, and potash.

Subchapter C – Regular corporation which sells stocks to investors.

Subchapter S – Small business or family corporation.

Subchapter T – Type of corporation involving cooperatives.

subsectors – Parts of the economy: household sector (all consumers), business sector (all firms), and the government sector (all government agencies).

subsidies – Government grant of money to aid or encourage a private enterprise that serves to benefit the public.

substitution – Replacing one input with another in an attempt to increase profit.

superettes – A type of retail grocery store which is a smaller grocery store, which offers a variety of food and nonfood grocery products. These are usually located in areas not adequately served by supermarkets.

supermarket – Type of retail grocery store which offers the customer a full line of 10,000 to 15,000 items, plus many nonfood items such as household cleaners, health and beauty aids, and so on.

superstores – Offer a greater variety of products than conventional supermarkets; account for 25 percent of all supermarkets resulting in 33 percent of supermarket sales.

supplementary enterprises – Enterprises which use different resources or the same resources at different times.

supplements – Formula feeds requiring the addition of grain to a complete ration; these account for 18 percent of feed produced.

supply – Amount of a product available at a specific time and price.

supply (purchasing) cooperatives – Act mostly as purchasing associations for such things as feed, seed, fertilizer, and fuel.

surveying – The practice of measuring a tract of land for size, shape, or the position of boundaries.

synthesis – The combination of parts or ideas into a whole.

synthetic chemicals – Chemical produced artificially in order to control weeds, insects, and diseases.

target customers – The full-time and part-time production agriculturalists whom farmer supply chain stores target.

tariffs – Government-imposed taxes on goods incurred when they cross national boundaries.

taxable income – The difference between revenue and expenses; loss of an agribusiness.

tax-deferred – Postponing taxes until retirement.

tax-deferred savings – Savings that are tax deferred until retirement.

Tax Shelter Account (TSA 403b) – An investment allowed by the government for teachers and hospital employees allowing 20 percent investment of income yearly and taxes are deferred until retirement.

teleconferencing – Communication technology involving the use of telephone in order for several people to talk to each other as in a conference.

terminal – A station for delivery or receipt of produce; a depot or warehouse needed to store items for later distribution in smaller quantities to retail food stores or to restaurants.

term life insurance – Insurance in which a yearly premium is paid, but the premium (money) never accumulates a cash value.

therapeutic activity – Something done as a stress reliever.

threshing machine – Developed in order to separate the grain from the waste of the plant.

ticks – The smallest price unit at which a futures contract trades. See also *minimum price fluctuations.*

time value – Also called extrinsic value. It reflects the amount of money that buyers are willing to pay in hope that an option will be worth exercising at or before expiration.

tissue culture – The development of roots, stems, and leaves from callus tissue using a solution containing nutrients and hormones.

torque amplification – A feature that expands or increases the force that tends to produce rotation.

total cost – Sum of the fixed cost and the variable cost.

total fixed cost – Those costs that do not change as level of production changes.

total product – Represents the total amount of a product (corn, cotton, soybeans, etc.) produced in a given period of time with a given group of resources.

total revenue – Represents the total amount of money coming into a business in a given period of time. It equals price multiplied by quantity.

total variable cost – Costs that change as production levels or the amount of use change.

trade barriers – Ways to protect or restrict trade.

trading pit – The heart of the futures exchange, where buyers and sellers meet each day to conduct business.

transformation – The act of changing one configuration or expression into another.

transgenic – An animal that has developed from a genetically modified cell to which a gene has been transferred from another living organism.

trial balance – The proof of the equality of debit and credit balances.

trichinae-safe pork – Pork that is safeguarded from trichinosis by means of irradiation and testing.

tricycle-type – A tractor introduced by International Harvester in 1924, which was very popular for cultivating as well as plowing.

trough – The lowest level of business activity in a particular business cycle.

turbocharger – A centrifugal blower driven by exhaust gas turbines and used to supercharge an engine.

turpentine – A substance used as a fuel in the early internal combustion engines.

undercapitalization – Not enough capital (assets) to get a company started, thus leading to possible business failure.

universal life insurance – Insurance in which a yearly premium is paid and, besides providing money for one's family at death, a cash value accumulates.

unlimited liability – The claim that creditors can make on the owner's personal assets, as well as the business assets, in payment of business debts.

utilization of existing resources – The use of resources already available to you in whatever business venture you plan.

utilization of resources – Using whatever resources you already have to limit expenses or purchases for greater economic efficiency.

vaccines – A substance that contains live, modified, or dead organisms or their products that is injected into an animal in an attempt to protect the host from a disease caused by that particular organism; discovered by Edward Jenner.

value-add – Increasing the value of a product through processing, packaging, or other improvement after the product leaves the site of production.

value judgments – Used in normative economics to assess the performance of the economy and economic policies.

variable expenses – Expenses that change regularly, such as recreation and entertainment, in which you have control over the amount spent.

variables – Quantities that may assume any one of a set of values.

variable substitution – When one resource can be substituted for part of another at different rates for each additional unit of input.

vending – Selling of foods or snacks, as from a coin-operated machine or on the sidewalk.

venture capital – Money invested into one's business by family, friends, or investors.

vertical integration – Where several steps in production, marketing, and processing of animals are joined together.

vertically integrated – Controlled by a single firm of two or more stages in the chain of production, processing, and distribution.

viable – Capable of working, functioning, or developing adequately.

wants – Things which are not crucial to daily living.

warehouse/limited-assortment supermarket – Offer larger sizes of fewer items at lower prices than the typical supermarket.

wholegoods – Machinery or equipment in its completely assembled state.

wholesalers – Middle person between producer and retailer; sells in large quantities to volume buyers and retail stores.

write off – To depreciate or reduce the estimated book value of.

xenotransplantation – Transplantation of organs and tissues between species.

yardage fee – The fee a terminal market charges the seller for caring for animals until they are sold.

Index

A